Environmental Engineering for Pathogen Control

Charles N. Haas

LD Betz Professor of Environmental Engineering and
Distinguished Professor in the Department of Civil,
Architectural & Environmental Engineering
Drexel University

WILEY

Published by John Wiley & Sons, Inc., Hoboken, New Jersey.

Library of Congress Cataloging-in-Publication Data has been applied for:

Paperback: 9781394253388
ePDF: 9781394253401
epub: 9781394253395
oBook: 9781394253418

Cover Design: Wiley
Cover Image: STEVE GSCHMEISSNER/SCIENCE PHOTO LIBRARY/Getty Images

Contents

List of Figures

List of Tables

Preface

Environmental engineering, with some of its roots in sanitation and public health, has long been concerned with intervention to protect from disease. However, especially since the SARS-CoV-2 pandemic, it has been clear that this locus of the field is not often understood, even by environmental engineers as well as others.

Having long been involved in protecting populations from infectious organisms, I realized that compiling the subjects that could constitute a subfield of environmental engineering for pathogen control would be helpful. This book aims to inform engineers about how many of their skills can be used to reduce harm to populations from pathogens and to inform those outside the engineering community (public health specialists, physicians) about how engineering may help address problems often relegated solely to the purview of the health professions.

In writing this book, I recognize I draw on many years of beneficial interactions with colleagues, students, and collaborators. At the risk of failing to mention all those who have influenced me, I first acknowledge my long-time collaborators in risk assessment and environmental health microbiology, Joan Rose and Chuck Gerba. My Drexel University colleagues, especially Patrick Gurian, Chris Sales, Michael Waring, and James Lo, have provided me with much insight into their respective domains, and over the years, I have benefited from two incredibly supportive former deans, Joe Hughes and Sharon Walker. I have had decades of great students who have walked the path of this field with me, including Chris Crockett, Mukul Gupta, Tom Armstrong, Tim Bartrand, Mark Weir, Yin Huang, Kerry Hamilton, David Kahn, and many others. Two professional mentors have been very important to my development: my late doctoral advisor, Richard Engelbrecht, who first introduced me to the field of health microbiology-related environmental engineering, and Cecil Lue-Hing, who spurred me on in many ways to consider microbial risk-related problems.

Last, but certainly not least, my wife, Victoria, has been incredibly patient with my devotion to this project for over two years and has been so supportive of the effort.

January 2026

Charles N. Haas
Philadelphia, PA

Disclosure for AI usage

Drafts of abstracts for chapters were prepared using Claude and ChatGPT, and final edits were made by the author. Certain figures were generated using Google Gemini and ChatGPT.

1

Introduction

The origins of modern environmental engineering go back to the fields of sanitary engineering and even earlier public health engineering. Arguably, engineering was central to the coalescence of a distinct public health field, distinct from medicine [1]. The focus was on protecting people from the adverse effects of infectious microorganisms and gross environmental pollution. As early as the 1920s, it was recognized that water purification (filtration and disinfection) and early sewage collection and treatment were instrumental in the dramatic reduction of typhoid fever and other major infectious diseases [2].

Sanitary engineering was more firmly codified in the 1950s with the formal certification of credentials, and several subspecialties were recognized [3]:

- Water Supply and Waste Water Disposal
- Public Health
- Industrial Hygiene
- Air Pollution Control
- Radiation Hygiene and Hazard Control

With the environmental awakening of the late 1950s and 1960s, the field of sanitary engineering broadened and transformed into environmental engineering. Currently, the American Academy of Environmental Engineers and Scientists [4] certifies the following subspecialties (which are broader than the list mentioned earlier):

- Air Pollution Control
- Environmental Sustainability
- Industrial Hygiene Engineering
- Radiation Protection Engineering
- Site Remediation (Hazardous Waste Management and Site Remediation)
- Solid Waste Management
- Water Supply/Wastewater Engineering

The current field of environmental engineering has embraced a broad area of concern as defined by a 2018 report by the US National Academies of Science, Engineering, and Medicine:

> "Environmental engineers support the well-being of people and the planet in areas where the two intersect. Over the decades, the field has improved countless lives through innovative systems for delivering water, treating waste, and preventing and remediating pollution in air, water, and soil." [5]

The need for continued vigilance in controlling pathogens has been downplayed, as environmental engineering concerns broaden. However, with the emergence of antibiotic resistance, the possible role of the environment in exacerbating its prevalence [6], and the significance of climate change in broadening the range of pathogenic prevalence [7], a re-evaluation of this aspect of practice is essential.

This book aims to enable engineers, especially environmental engineers, to regain a vital role at the table, along with health professionals, in protecting the public from infectious diseases. A secondary objective is to acquaint health professionals with some of the skills and attitudes that engineering may bring to a collaborative approach to solving such problems.

1.1 Scope of Coverage

The focus of this book will be on diseases in which the environment serves as an intermediate in their exposure to populations. Typically, modern environmental engineering focuses on exposures that can occur via water, air, and land. Many environmental engineering situations also involve indirect exposures that can occur via the ingestion of foods (animal or plant) contaminated by environmental conditions.

While foodborne exposure to infectious agents can be substantial [8], this book will only cover aspects of this pathway in passing, because substantial literature in food microbiology and food safety already exists [9].

Similarly, although there are significant infectious diseases in wildlife and domesticated animal populations [10], these are not the focus of the book, and further applications of the methods discussed herein to those problems would undoubtedly be appropriate. In some cases, however, exposure of humans to animals or to animal wastes may result in adverse effects, and those pathways are included in coverage.

1.2 Instructor Notes

This book is divided into four parts. The first part summarizes the essential basics of environmental engineering and the properties of critical pathogens of concern, including how they are measured. In the second part, sources of pathogens and their fate and transport in the environment are considered. The third part considers technologies and processes that can be used as interventions to reduce exposures, as well as the formal procedures for conducting an exposure assessment. The final part covers the conduct of the overall risk characterization, including the potential of transmission of contagious pathogens, and how decisions can be made regarding what interventions, if necessary, should be implemented.

The overall book is suitable for a dedicated semester-long course. If students have a good understanding of microbiological issues, Chapters 4–6 may be skipped or briefly treated — and the remainder of the material covered in either a semester or a term.

References

1 Richard J Gelting, Steven C Chapra, Paul E Nevin, David E Harvey, and David M Gute. Back to the future: Time for a renaissance of public health engineering. *International Journal of Environmental Research and Public Health*, 16(3):387, January 2019.

2 George W Fuller. The influence of sanitary engineering on public health. *American Journal of Public Health*, 12(1):16–23, January 1922.

3 Earnest Boyce. The American Sanitary Engineering Intersociety Board. *American Journal of Public Health and the Nations Health*, 53(7):1127–1131, July 1963.

4 American Academy of Environmental Engineers and Scientists. Become board certified in environmental engineering, September 2023.

5 Committee on the Grand Challenges and Opportunities in Environmental Engineering for the Twenty-First Century. *Environmental Engineering for the 21st Century: Addressing Grand Challenges*. National Academies Press, Washington, D.C., 2018.

6 D G Joakim Larsson and Carl-Fredrik Flach. Antibiotic resistance in the environment. *Nature Reviews Microbiology*, 20(5):257–269, May 2022.

7 Xiaoxu Wu, Yongmei Lu, Sen Zhou, Lifan Chen, and Bing Xu. Impact of climate change on human infectious diseases: Empirical evidence and human adaptation. *Environment International*, 86(1):14–23, 2016.

8 Daniel Dewey-Mattia, Virginia A Roberts, Antonio Vieira, and Kathleen E Fullerton. Foodborne (1973–2013) and waterborne (1971–2013) disease outbreaks — United States. *MMWR. Morbidity and Mortality Weekly Report*, 63(55):79–84, October 2016.

9 Michael P Doyle, Francisco Diez-Gonzalez, and Colin Hill. *Food Microbiology: Fundamentals and Frontiers*. John Wiley & Sons, 2020.

10 Anke K Wiethoelter, Daniel Beltrán-Alcrudo, Richard Kock, and Siobhan M Mor. Global trends in infectious diseases at the wildlife–livestock interface. *Proceedings of the National Academy of Sciences*, 112(31):9662–9667, 2015.

Part I

Environmental Engineering and Pathogen Basics

2

Who, How, Where

As a starting point, three basic questions about the interaction between microorganisms and hosts can be asked as a frame for further discussion:

- What are the potential infectious microorganisms that could be of concern?
- How can people become exposed to (receive a dose of) infectious microorganisms?
- Where can such exposures occur — under what circumstances?

Classic consideration of infectious disease transmission uses the concept of an epidemiologic triad [1]. This is frequently represented as intersecting circles or as the vertices of a triangle. In Figure 2.1, this is represented by a surface (disease) that, to be supported (occur), requires the adequacy of a microorganism, the sensitivity of a host, and the propensity of the environment to allow the organism to persist (and perhaps multiply). Without all these three factors ("failure" of one of the columns), disease cannot occur. Of course, each of these attributes is variable. To reduce the occurrence of disease, factors that weaken one (or more) of these supports can be adopted:

- The environment may be made unfavorable for the pathogen to persist and spread (perhaps by ventilation, disinfection, or control of temperature and humidity).
- Host susceptibility may be modified by medical interventions to provide immunization or pre-exposure prophylaxis or the use of personal protective equipment.

In more modern conceptions, variability in the environment, the population of microorganisms, and the population of hosts is recognized. Therefore, this framework should be regarded as an idealization of a much more complex and probabilistic set of processes.

For example, there is a fundamental question of who is a host capable of becoming infected. As noted by Casadevall and Pirofski [2], numerous intrinsic (e.g., genetic) and extrinsic (e.g., other

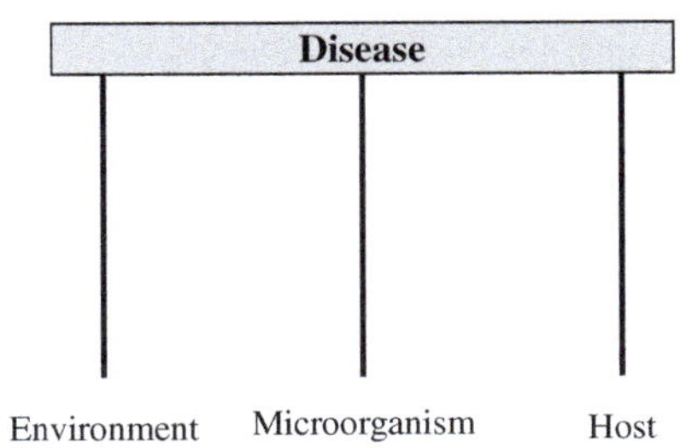

Figure 2.1 The disease triad.

stressors, age, nutrition) factors can mitigate or potentiate a host's ability to become infected. Furthermore, the likelihood of becoming infected is a function of the dose to which an individual is exposed — this will be discussed further in Chapter 14.

2.1 Pathogens of Concern

Classically, pathogens were considered to be organisms capable of causing disease. In contrast, opportunistic pathogens were considered capable of causing disease only in hosts that were somewhat compromised, and non-pathogens were considered incapable of causing disease. But it has become clear that the competency of a microbial population to cause disease may evolve via the acquisition of or selection for virulence factors [3], and that the spectrum of host competency may erase firm distinctions between pathogenic and non-pathogenic microorganisms [4]. Furthermore, in some environments, microbial growth and proliferation of organisms not pathogenic at lower doses may occur such that harmful (to some portion of hosts) doses can be produced [5]. Thus, absolute distinctions between pathogens solely based on taxonomy may no longer be possible.

Nonetheless, we can categorize microorganisms into various types, recognizing that pathogenic potential (at appropriate doses) can exist in all kinds. The common feature is that within (or upon) a host, these organisms can proliferate and cause an adverse physiological effect.

It is noted parenthetically that some organisms can proliferate externally, producing a toxin. The toxin, at a high enough dose, may have adverse effects when a host is exposed; these are more akin to chemical intoxications than infections and will not be considered further in this book. In addition, antibiotic-resistant bacteria and antibiotic-resistance genes are a growing concern for environmental protection [6]. However, since these are not infectious agents *per se*, they will not be considered further in this book.

The purpose of classifying microorganisms is to make inferences and generalizations across similar situations. The mode of classification should be viewed in this light. There are several approaches to classifying organisms that may cause disease upon multiplication in a host. One approach is through high-level taxonomic or structural differences.

2.1.1 Subviral Agents

The simplest infectious agents[1] are the organisms called "viroids," which are small circular RNA elements. Initially, the only viroids known to cause disease were plant pathogens [8]. However, it has now been found that hepatitis D results from the coinfection of an individual with hepatitis B and an enveloped viroid — hepatitis delta [7].

Infectious agents that can multiply in a host are at a greater level of complexity. These are viruses, bacteria, fungi, and protozoa, in order of complexity.

2.1.2 Viruses

Viruses are obligate parasites that can only grow with a host and contain nucleic acid (DNA or RNA) surrounded by a protein coat (called the capsid) and perhaps other materials [9]. Some viruses may have additional proteins associated with their nucleic acid within the capsid. Viruses were

1 Even simpler agents of disease, the prions, are not considered here since they are not infections in a classical sense, but the result of a form of transmissible aberrant protein folding [7].

originally (in the late 19th century) identified from diseases that could be caused by agents capable of passing through filters [10].

It is now recognized that viruses may have a broad range of sizes and shapes, as shown in Figure 2.2.

Shape and size are not the only descriptors by which viruses may be classified. The approach used to classify viruses depends on the objective of the classification.

In reproducing within a host, viruses must use the host machinery to produce proteins for new virions. Proteins are made from templates provided by messenger RNA. An early classification of viruses by Baltimore [12] was based on the type of nucleic acid carried by the virus. In his original scheme, there were six groups, which have been extended to a seventh group [13].

Positive- and Negative-sense Nucleic Acids

The concept of "sense" when referring to nucleic acids relates to how they participate in protein synthesis. In this process, a positive-sense strand of DNA is *transcribed* into a positive-sense strand of RNA, which is then *translated* into a protein. The DNA and RNA strands complementary to the positive-sense strand are termed negative-sense.

The seven Baltimore groups are depicted schematically in Figure 2.3 and described as follows:

Group I: Contains double-stranded DNA transcribed and translated using ordinary host machinery.

Group II: Contains single-stranded DNA, which is then replicated within the host to double-stranded DNA, and subsequently transcribed and translated.

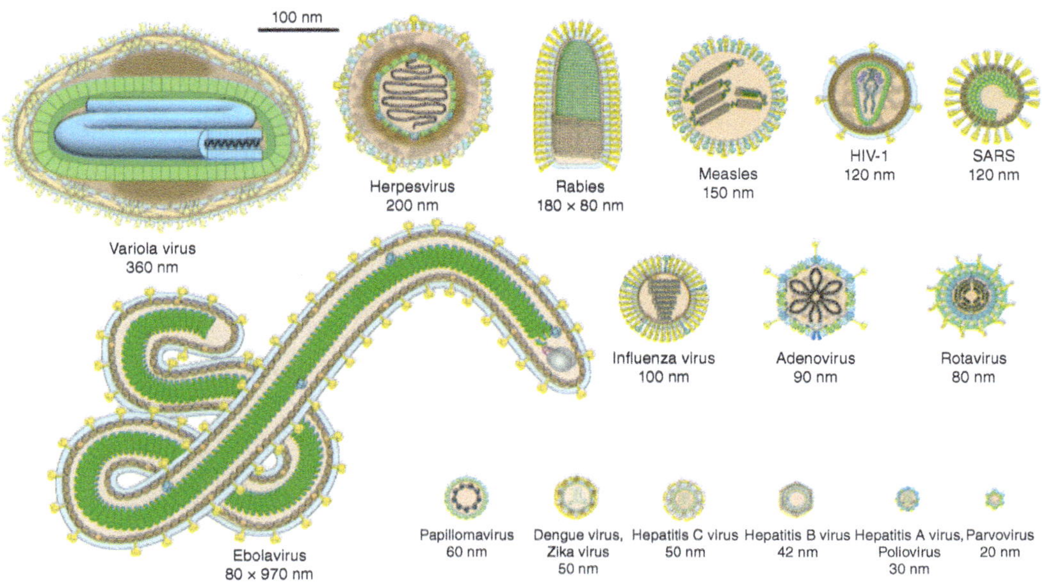

Figure 2.2 Various shapes and sizes of pathogenic viruses (nucleic acids not to scale). *Source*: [11]/ViralZone/SIB Swiss Institute of Bioinformatics/CC BY 4.0.

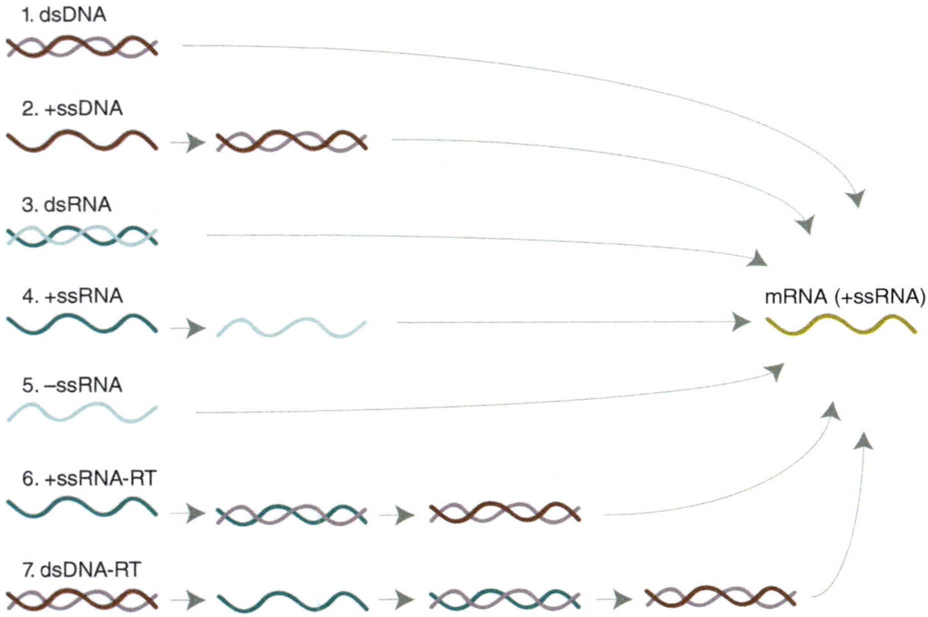

Figure 2.3 Schematic definition of Baltimore virus groups. *Source*: Modified from [14] under Creative Commons license.

Group III: Contains double-stranded RNA, which is then "unzipped" and the positive-sense strand is translated.

Group IV: Contains a single positive-sense strand of RNA, which is then translated.

Group V: Contains a single negative-sense strand of RNA, which is first transcribed (to a positive-sense strand) and then translated.

Group VI: Often termed "retroviruses." Contains a single strand of negative-sense RNA, as well as a reverse transcriptase enzyme. The enzyme is used within the host to synthesize a positive-sense DNA strand that then proceeds through transcription and translation.

Group VII: Contains double-stranded DNA of which one strand is incomplete. Reverse transcription is used to produce a complete double-stranded DNA template prior to transcription and translation.

As noted by Koonin et al. [13], other viruses are not completely defined in these seven groups. As shown in Table 2.1, there are viral pathogens of importance in all of the seven designated Baltimore groups of viruses.

More formal and systematic taxonomic approaches focus on classifying organisms based on evolutionary commonality — especially using molecular biological (e.g., sequencing) techniques. For viruses, the central international taxonomic authority is the International Committee on Taxonomy of Viruses (ICTV).

Formal taxonomy is one approach, and for viruses (as well as other organisms), the central focus has been on describing the evolutionary relationships among a large diversity of organisms. There has been a consensus established [15] that taxonomic assignment of viruses should be based on four principles:

Table 2.1 Example pathogens in each Baltimore group.

Group	Example pathogenic viruses
Group I	Adenovirus, herpesvirus
Group II	Parvovirus
Group III	Rotavirus
Group IV	Coronavirus, norovirus
Group V	Ebola virus, influenza virus
Group VI	HIV
Group VII	Hepatitis B

Table 2.2 Viral realms and example human pathogens.

Realm	Major properties	Example pathogens
Adnaviria	Viruses of Archaea	None
Duplodnaviria	dsDNA virus with HK97 protein	Herpesvirus, Varicella (chickenpox, shingles)
Monodnaviria	ssDNA viruses with an HUH endonuclease	Parvovirus, human cyclovirus
Riboviria	RNA viruses with RNA-dependent RNA polymerase or reverse transcriptase	Coronavirus, HIV, rabies
Ribozyviria	Viroids	Hepatitis D
Varidnaviria	dsDNA viruses encoding a "jelly roll" capsid protein	Adenovirus, pox viruses

Source: Adapted from [16].

- Taxonomic assignment should be based on evolutionary history.
- Viral properties may guide assignment to ranks.
- Taxonomy is only one means to classify viruses.
- When genetic sequencing is used, data should be subject to quality control.

The third principle allows for the continued use of alternative classifications, such as the Baltimore group. This also implies that such alternative classifications, albeit useful for some purposes, may not be congruent to formal taxonomic assignments.

Analogous to the taxonomy of other forms of life, ICTV has used eight taxonomic ranks for categorization:

$$\text{Realm} > \text{Kingdom} > \text{Phylum} > \text{Class order} > \text{Family} > \text{Genus} > \text{Species}$$

There are also "sub" levels for all these ranks except for species. The concept is that all members within a higher-level rank share specific traits or genetic sequences that have been conserved over a longer time.

At the highest rank, there are seven defined realms of viruses. Brief descriptions and examples of human pathogens are shown in Table 2.2. In addition, there are several viruses not yet assigned to a particular realm but only to a subsidiary taxonomic rank (e.g. family).

Viral classification based on evolutionary taxonomy has limitations, as does the classification based on the mode of genetic replication (per Baltimore). If the goal is to make inferences based on similarities in fate, transport, potency, and persistence in the environment, other approaches or supplemental information are needed.

Other viral attributes to consider when evaluating pathogen characteristics are discussed subsequently.

As noted, the viral nucleic acid is packaged within a protein container, the capsid. The capsid protects the genetic material from some damage. In addition, the proteins may function as antigens to which the host immune system can respond. As shown in Figure 2.4, the viral capsid provides a "shell" around the genetic material. For poliovirus, as in the illustration, the capsid is relatively smooth. Some viruses have "spikes" extending from the relatively smooth surface to facilitate interaction with and entry into the host cells.

The viral capsid proteins can also serve as antigens to induce an antibody response by the host.

Some viruses, termed "enveloped," have glycoprotein coatings surrounding the protein capsid, often with spikes extending beyond the envelope. There are many examples of enveloped viruses of human health and environmental significance, including SARS-CoV-2, influenza, and measles [17]. Note that neither the Baltimore group classification nor the ICTV taxonomy (at least at the level of realm) use the presence of an envelope as a distinguishing feature.

The structure of the enveloped influenza A virus is shown in Figure 2.5. Note the "spikes" extending beyond the envelope. Also, since this virus is a single-stranded RNA virus (with eight segments), there is a viral RNA polymerase packaged within the virion.

The presence of a viral envelope may alter (and perhaps enhance) the persistence of the virus in the air [19], on surfaces, or in waters [20, 21].

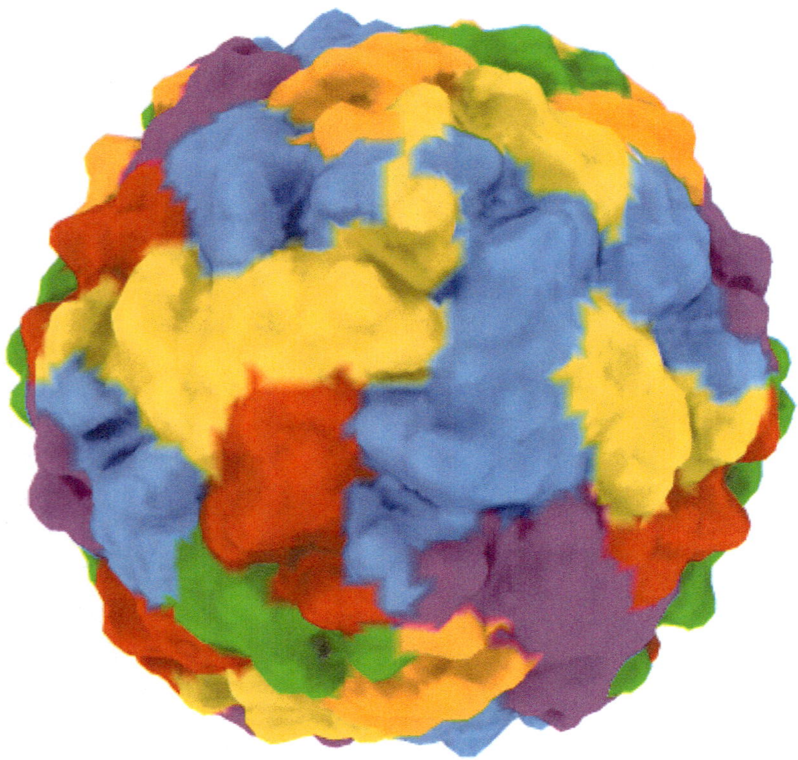

Figure 2.4 Depiction of poliovirus type 3. Colors represent different viral capsid proteins. *Source*: National Institute of Health (NIH)/CC0 1.0/https://3d.nih.gov/entries/3DPX-000710/last accessed Feb 9, 2026.

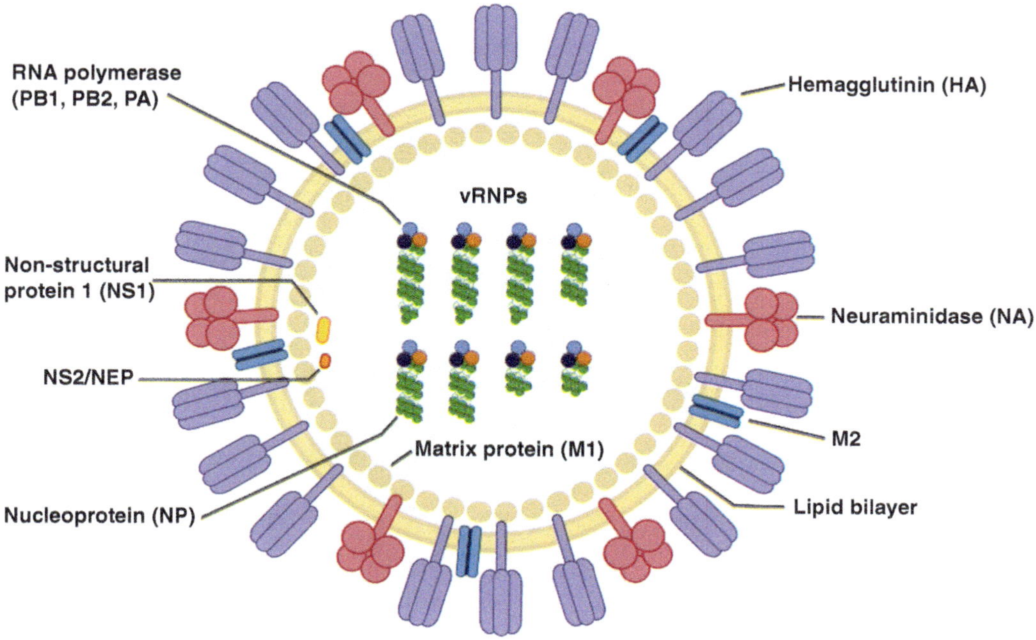

Figure 2.5 Structure of the encapsulated influenza A virus. *Source*: [18]/MDPI/CC BY 4.0.

2.1.3 Bacteria

Bacteria form the simplest form of cellular life capable of existing outside a host and are regarded as a taxonomic domain.[2]

Bacteria are simple unicellular organisms that lack a membrane around their nucleus and lack true cell walls. Historically, the word "prokaryotic" was used to characterize both bacteria and *Archaea*, but that word has proven problematic [22].

The domain of bacteria is highly diverse, and different types have many properties that can affect their behavior in the environment and their role as pathogens. Historically, some of these characteristics were used to classify bacteria, but modern approaches to "bin" bacteria into taxonomic ranks rely heavily on genetic similarity.

Bacteria, in general, are microscopic organisms. However, there are examples of much larger bacteria, such as the*Thiomargarita* [23], whose cells can grow as large as 750 μm, although such large bacteria do not appear to have pathogenic significance. The first differentiator among bacterial types is their shape (as observed under a microscope). These include spherical (frequently but not always aggregated into clusters of various types) — termed cocci; rod-shaped — termed generically as bacilli; spiral-shaped; and others. Examples are shown in Figure 2.6. They may or may not have one or two sets of flagella and/or smaller hair-like projections termed cilia.

2 Domain is the highest taxonomic rank used for organisms other than viruses, and corresponds to the rank of Realm in viral taxonomy. It is noted in passing that the domain of *Archaea*, previously called archaebacteria, are also simple organisms — however, to date, there are no known human pathogens in this domain, so they will not be considered further in this book. However, the *Archaea* are ecologically important.

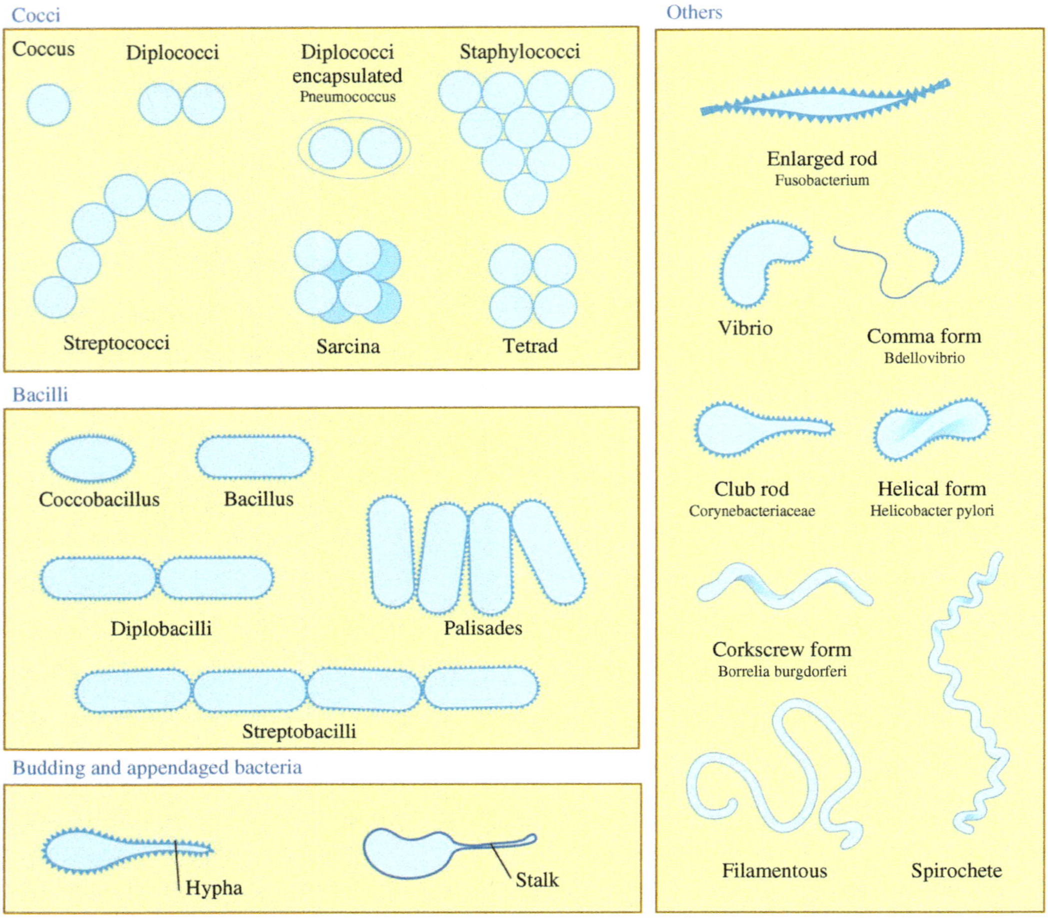

Figure 2.6 Different bacterial shapes. *Source*: Michal Maňas/Public domain/https://commons.wikimedia.org/wiki/File:Bacterial_morphology_diagram-cs.svg/last accessed Feb 09, 2026.

Another significant property that differentiates between bacteria is the type of outer cell envelope present. In the 19th century, Hans Christian Gram developed a microscopic procedure for visualizing bacteria, which is referred to as the "Gram stain." In this procedure, a dried suspension of bacteria on a microscope slide is first treated with the crystal violet dye, rinsed, then treated with iodine, followed by a decolorization [24]. Some types of microorganisms retain the violet-iodine complex and appear purple under a microscope (these are termed Gram-positive). In contrast, other types of microorganisms do not retain the dye complex and only retain some of the uncomplexed iodine, appearing red under the microscope (termed Gram-negative). Some cells, either by their intrinsic properties or physiological conditions, may be weakly stained or variable in appearance. The Gram stain property reflects differences in the outer layers of the cell envelope, with Gram-positive organisms generally having thicker, less permeable outer layers. There are differences in the sensitivity of Gram-positive and Gram-negative organisms to many antibiotics and antimicrobial agents.

There are examples of both Gram-negative and Gram-positive pathogens, as shown in Table 2.3.

Table 2.3 Examples of Gram-negative and Gram-positive genera with pathogenic bacteria.

Gram-negative	Gram-positive
Shigella	*Streptococcus*
Escherichia	*Bacillus*
Campylobacteria	*Listeria*
Legionella	*Clostridium*

2.1.3.1 Classification by Metabolic Strategy

Bacteria and organisms in general can be categorized by their metabolic strategy. Metabolism can be classified by how organisms obtain energy and by what raw materials they use to synthesize new biomass. The purpose of this discussion is not to consider the details of metabolism, which can be found elsewhere [25], but rather to provide a framework for understanding the factors involved in enumerating microorganisms from environmental samples and in influencing microbial growth and decay.

The energy sources for bacteria may be either from the decomposition of materials (and abstraction of their embodied energy), termed "chemotrophy," or from the energy in light, termed "phototrophy." There are no known pathogenic phototrophic bacteria.

Among the chemotrophs, they can be classified based on whether they can persist only on inorganic carbon or whether they require preformed organic carbon for biosynthesis. The former organisms are termed "chemoautotrophs," while the latter are termed "chemoheterotrophs." Nearly all bacterial pathogens are chemoheterotrophic, although the organism *Clostridium difficile,*[3] which is important in healthcare-related infections and may have an environmental component, is a chemoautroph [27]. There may be other such organisms emerging in the future.

The chemoheterotrophs are the largest category of pathogenic bacteria. They obtain their energy by the transformation of chemicals and form their cells by biosynthesis of chemical intermediates from preformed organic compounds. The process of obtaining energy and decomposing exogenous organic materials is termed "catabolism," while the process of synthesizing cellular material is termed "anabolism."

The chemoheterotrophs can further be categorized by how they engage in catabolism. The general strategy is shown in Equation 2.1

$$\text{Electron donor} + \text{electron acceptor} \rightarrow \text{decomposition products} \tag{2.1}$$

In the catabolic process, this reaction is exergonic, resulting in the release of chemical energy, which, in an organism (i.e., a bacterial cell), can be captured for cellular maintenance and biosynthesis. The electron donors are generally organic compounds, while the electron acceptor can be one of several possibilities.

There are different subsets of chemoheterotrophs depending on the nature of the electron acceptor they use. They can be categorized into aerobes, microaerophiles, facultative organisms, and anaerobes. Organisms that use a distinct electron acceptor, such as oxygen, are called "respiring." Other organisms that do not use a distinct electron acceptor are termed "fermentors." One type of fermentation metabolism would use an organic compound such as glucose and transform it both

3 This organism has been reclassified as *Clostridioides difficile* [26].

to a more oxidized and more reduced compound, as in the following reaction in which ethanol is produced as a product:

$$C_6H_{12}O_6 \rightarrow 2\,C_2H_5OH + 2\,CO_2 \tag{2.2}$$

Aerobes require oxygen as an electron acceptor and, from catabolism, produce carbon dioxide. By definition, they can also tolerate and counteract the effects of atmospheric concentrations of oxygen (which intrinsically could react with various biological materials). There is a broad spectrum of such organisms, including many pathogens.

Microaerophillic organisms are those that also engage in catabolism using oxygen as an electron acceptor. However, they are more sensitive to the toxicity from oxygen and therefore prefer concentrations less than atmospheric [28]. Examples of microaerophilic pathogens include *Campylobacter*, *Vibrio*, and *Legionella*.

Anaerobic organisms do not grow using oxygen as an electron acceptor. Those that are readily inactivated by the presence of oxygen are termed fastidious anaerobes. Anaerobes may grow using other electron acceptors — sulfate (SO_4^{-2}) and nitrate (NO_3^-) being common. Note that such organisms that use alternative electron acceptors also engage in respiration. The products of such respirations may include, for example, hydrogen sulfide (H_2S) or nitrite (NO_2^-), respectively. The most important genus of anaerobic bacteria is *Clostridium*.

Facultative organisms, alternatively called facultative anaerobes, are capable of persisting and growing either in the presence or the absence of oxygen [29]. This may give them a selective advantage. The common gut organism *Escherichia coli*, of which some strains are pathogenic, is an example of a facultative pathogen. Other important pathogens in this group are [29]:

- *Pseudomonas aeruginosa*
- *Staphylococcus aureus*
- *Salmonella*
- *Shigella*
- *Klebsiella*

2.1.3.2 Other Important Characteristics

Other important characteristics of bacteria affect their environmental behavior and pathogenicity. The first such characteristic is hydrophobicity — the tendency, based on the outer layers of a cell, to repel water. Measurement of this property in films of bacteria of diverse strains by van Loosdrecht et al. [30] showed a broad range of behavior over the hydrophobic–hydrophilic spectrum (measured by contact angle). As hydrophobicity increases, there is a greater tendency for bacteria to partition to liquid–solid or air–liquid interfaces and to adhere with each other to form aggregates. This may aid in environmental persistence, e.g., the occurrence in and formation of biofilms. There also appears to be a relationship between hydrophobicity and some types of pathogenicity [31].

Certain bacteria, in response to stressors, can form resistant stages called spores. This ability is primarily restricted to particular species of *Bacillus* and *Clostridium* within the phylum *Bacillota*.[4]

The formation of a spore under stress conditions imposes an energetic cost on a bacterial cell, which is offset by enhanced survival (albeit at a much reduced metabolic state). During spore formation, there is partial dehydration resulting in a "core" with a gel-like or semisolid consistency and the formation of a mechanically rigid and more impermeable coat and crust layer [32]. These structures also impart high degrees of resistance to heat (wet or dry), desiccation, chemical agents, ultraviolet light, and ionizing radiation.

---

4 Until 2021, the bulk of organisms within this phyla were classed within the former Phylum of *Firmicutes*.

Despite this relatively inert state, under favorable conditions, spores can rapidly revive (termed germination) to produce vegetative, metabolically active bacteria. This may be triggered by exposure to specific amino acids, sugars, or peptides [33].

Nonsporulating bacteria may also enter a state of dormancy, termed "viable but non-culturable" (VBNC) [34]. In such a state, metabolic activity is low, and these organisms may not show growth under standard incubation conditions and culture media. Bacteria capable of entering VBNC are many, including (for example) *Legionella* and *Escherichia coli* [35, 36].

2.1.3.3 Formal Taxonomy and Classification

Formal taxonomy generally has three objectives. The first is to define characteristics that most directly separate individual species from one another. The second is to determine which factors groups of species share within larger taxonomic units. A third objective is to arrange these larger taxonomic units in a way that might reflect their evolutionary emergence over time, although this generally results in unprovable hypotheses.

To some degree, the taxonomy of bacteria is much more complex than that of viruses. A classic definition of species [37] is an actual or potential population reproductively isolated from other groups. However, as noted, e.g., by Riley and Lizotte-Waniewski [38], for asexual populations such as bacteria, which have a strong ability to exchange genetic material even between putative species, a better approach is to focus on core genes that define a species.

Indeed, prokaryotes such as bacteria are distinct in that they can carry extrachromosomal DNA that is replicated and can be transmitted by exchange with other cells (including to other bacteria). This may result in molecular evolution and the transfer of innovative traits, such as antibiotic resistance or the ability to metabolize particular substrates [39].

Genetic material may be exchanged between cells via horizontal gene transfer in at least three ways, which may or may not involve plasmid intermediates [40]:

Transformation is the uptake of environmental DNA (released by donor cells) by recipient cells and integration into the recipient cell genome.

Transduction is the process whereby a bacteriophage infecting a donor cell integrates a portion of the donor DNA into its genome. Then, progeny infect recipient cells and transfer the donor DNA into the recipient.

Conjugation is the process where "male" (donor) and "female" (recipient) cells exchange DNA via specific sex pili and receptors.

An Example of Genetic Exchange

One example of the importance of genetic exchange in pathogens (of which there are many) is the recognition of the strain *Escherichia coli* O157:H7 through acquisition of a virulence gene from *Shigella* [41].

Microscopic examination of bacteria is first attributed to van Leeuwenhoek in the 17th century. However, only with the acceptance of the germ theory of disease, propounded by Pasteur in the 19th century, was there recognition of the importance of bacteria in disease. Numerous pathogens were identified. In 1923, with the first publication of *Bergey's Manual of Determinative Bacteriology*, tools were available to systematically enable differentiation of the different bacterial species and genera [42]. These early differentiation methods were based on physiological properties such as:

- Shape of cells under a microscope
- Ability to grow under particular conditions (temperature, presence or absence of oxygen, requirements for particular carbon sources or other nutrients, etc.)
- Morphology of colonies grown on a solid medium
- Appearance after use of various stains (including Gram stain) under a microscope
- Formation of particular end products from metabolic activity

With the development of molecular biological approaches starting in the 1950s, it became possible to use genetic information alongside physiological traits to classify bacteria. This approach, used in modern taxonomy, is termed "polyphasic taxonomy" [43].

Perhaps the earliest and most straightforward use of genetic information is the calculation of the proportion of DNA base pairs that are guanine (G) or cytosine (C) as opposed to adenine (A) or thymine (T).[5] Since the GC pair is triply bonded, whereas the AT pair is doubly bonded, the GC pair requires more energy to disassociate than the AT pair. This can be determined by isolating DNA from a pure culture of bacteria and measuring its "melting point."[6]

As shown by a systematic review by Barcelo-Antemate et al. [44] in Figure 2.7, there is some separation of values of GC ratio by taxon, although certain taxa exhibit high variability. These variations manifest in different amino acid ratios in translated proteins. However, clearly, due to high variability among certain taxa — notably the Proteobacteria, which are a large and diverse group, GC cannot always be used alone as a discriminator.[7]

Several more sophisticated approaches to the use of nucleic acids for bacterial classification have been devised. One of the first was the sequencing of genes for 16S ribosomal RNA (16S rRNA). The 16S[8] unit of RNA contains sequences that are strongly conserved, at least down to genus level [45], and was proven useful in developing taxonomies based on genetic information. However, some genera may exhibit greater variability in 16S RNA gene sequences, and thus universal cutoffs for differentiating species based on this method may not be possible without additional information [46].

Over the past 20 years, technological improvements have made it less expensive to sequence entire bacterial genomes (and those of higher organisms). This has enabled the whole genome, using appropriate bioinformatics software [47], to be used to identify unknown microorganisms and to classify organisms based on whole-genome sequencing, along with other information in the polyphasic approach [43]. As with any other analytical technique, attention to quality control and use of high-resolution sequencing methods are needed to provide good results [48].

---

5 In DNA, a guanine nucleotide on one strand is bonded to a cytosine on the complementary strand, and vice versa. Similarly, an adenine on one strand is bonded to a thymine on the complementary strand.

6 Double-stranded DNA has UV absorbance at 260 nm. When the temperature is increased, thermal energy causes the unraveling into single strands, a process termed melting. The temperature at the midpoint of the change in optical absorbance is termed the melting point and can be related to the fraction of G+C in the DNA.

7 This graph, and the paper use taxa under an older system, rather than the currently accepted phyla described below.

8 "S" is the abbreviation for Svedberg, which is 10^{-13} s. In ultracentrifugation, the relationship between sedimentation velocity, υ_s, acceleration, $\omega \cdot r$, where ω is the rotation rate and r is the radius of the centrifuge tube, and time, t, to completely sediment is $t = \frac{\upsilon_s}{\omega \cdot r}$. This time is conventionally expressed in Svedbergs. Early in the study of ribosomal RNA, ultracentrifuges were used, and this portion of RNA was the slower to sediment.

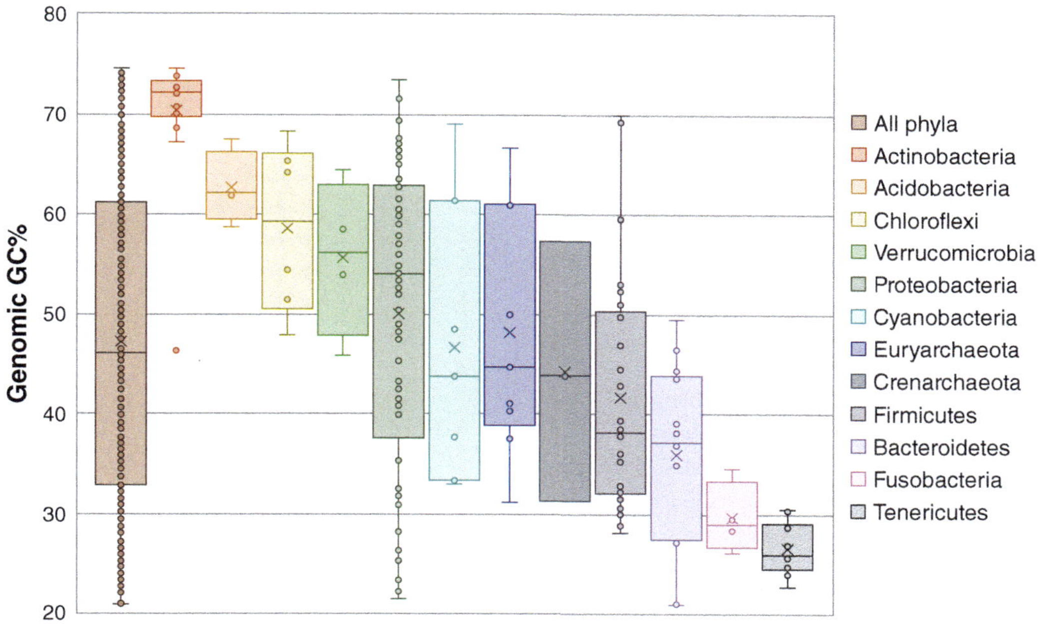

Figure 2.7 Distribution of GC ratio by bacterial group. *Source*: [44]/Public Library of Science (PLOS)/ CC BY 4.0.

Using such polyphasic approaches, the International Committee on Systematics of Prokaryotes (ICSP) has reformulated bacterial taxa and organized bacteria into 42 phyla [49].[9] There are 33 phyla without currently known pathogens (although noting the discussion early in this chapter, absolute differentiation between pathogens and non-pathogens is not possible), which are enumerated in Table 2.4.

There are nine phyla that contain known pathogens. These are enumerated in Table 2.5 with example genera containing pathogens (though not all members of the identified genus may be pathogenic).

2.1.4 Protozoa

Protozoa are single-celled Eukaryotic organisms that have a defined nucleus and nuclear membrane and do not have either a cellulosic cell wall or photosynthetic functionality. The taxonomy of Eukaryotes in general [50] and protozoa in particular [51] is somewhat unsettled with respect to assignment to higher taxonomic ranks. Protozoan illnesses involving environmental transmission are significant in the developed world [52] as well as less developed countries (LDCs).

9 As with other taxa, such as the viruses, there are a series of subsidiary taxonomic groups, e.g., class>order>family>genus>species.

Table 2.4 Bacterial phyla with no currently known human pathogens.

Acidobacteriota	Fibrobacterota
Aquificota	Fusobacteriota
Armatinonadota	Gemmatimonadota
Atribacterota	Ignavitbacteriota
Balneolota	Kiritimatiellota
Bdellovibrionota	Lentisphaerota
Caldisericota	Myxococcota
Calditrichota	Nitrososphaerota
Chlorobiota	Nitrospinota
Chloroflexota	Nitrospirota
Chrysiogenota	Rhodothermota
Coprothermobacterota	Synergistota
Deferribacterota	Thermodesulfobacteriota
Deinococcota	Thermomicrobiota
Dictyoglomota	Thermoproteota
Elusimicrobiota	Thermotogota
	Verrucomicrobiota

Table 2.5 Bacterial phyla with known pathogens.

Phylum	Example genera containing pathogens
Actinomycetota	*Bifidobacterium, Mycobacterium*
Bacillota[a]	*Bacillus, Clostridium*
Bacteroidota	*Bacteroides*
Campylobacterota	*Campylobacter, Helicobacter*
Chlamydiota	*Chlamydia*
Mycoplasmatota	*Mycoplasma*
Planctomycetota[b]	
Pseudomonadota	*Pseudomonas, Salmonella, Legionella*
Spirochaetota	*Leptospira, Borrelia*

[a] This phylum was previously named Firmicutes.
[b] Suspected to contain pathogens, though not confirmed.

Rather than discussing formal assignment into higher taxonomic groups, it is thus more appropriate to distinguish protozoan pathogens operationally — CDC has suggested the mode of motility is one useful categorization [53]. This is described as follows:

Sarcodina are "amoeboid protozoa which trap food by some form of pseudopodial action, pseudopodia ...[and] have no cytostomes and have bodies of a less organized shape than most protozoa" [54]. *Entamoeba histolytica* is a classical pathogen and the cause of amoebic dysentery. A micrograph of the negative state of this organism is shown in Figure 2.8.

Mastigophora are protozoa that move by means of one or more flagella. These have also been referred to as flagellated protozoa. An example is the protozoan *Giardia muris*, which can be transmitted by food and water, an image of which is in Figure 2.9.

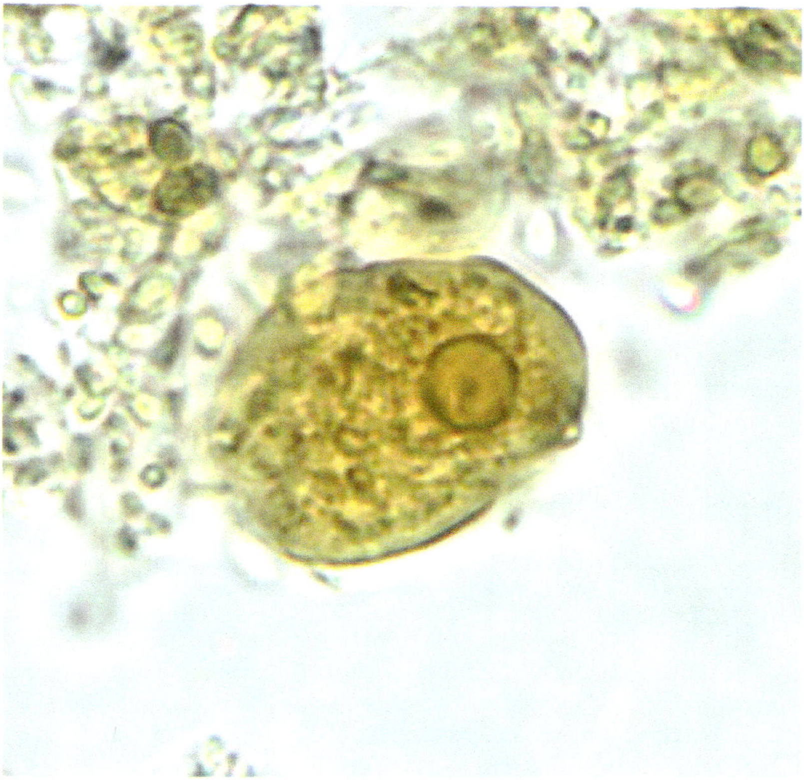

Figure 2.8 Image of *Entamoeba histolytica. Source*: Centers for Disease Control and Prevention/Public domain/https://www.cdc.gov/dpdx/amebiasis/index.html/last accessed Feb 12, 2026.

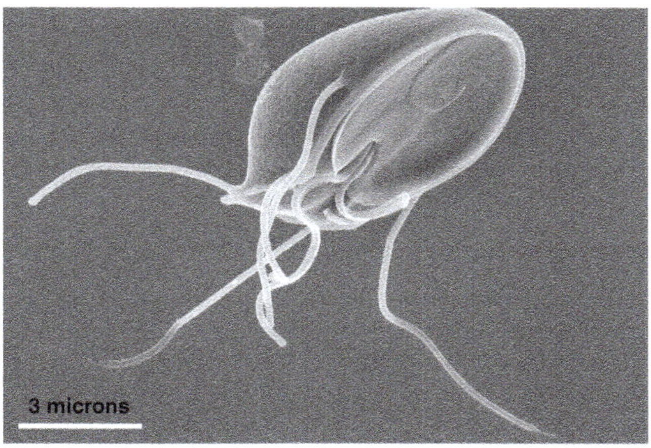

Figure 2.9 Microscopic image of *Giardia muris. Source*: Centers for Disease Control and Prevention/Dr. Stan Erlandsen/Public domain/https://wwwn.cdc.gov/phil/details.aspx?pid=11643/last accessed Feb 12, 2026.

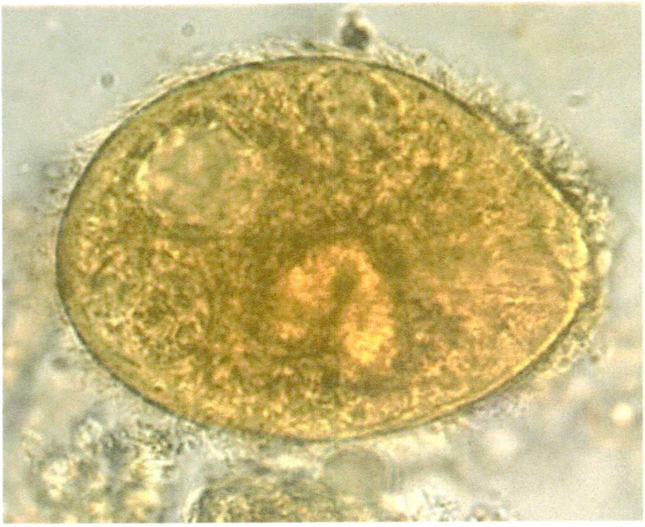

Figure 2.10 Microscopic image of *Balantidium coli*. *Source*: Centers for Disease Control and Prevention/Public domain/https://www.cdc.gov/dpdx/balantidiasis/index.html /last accessed Feb 12, 2026.

Ciliophora are protozoa that move by many cilia on their surface. An example is the pathogen *Balantidium coli* that can be transmitted by food and water, an image of which is in Figure 2.10.

Sporozoa are non-motile protozoa, without pseudopods, flagella, or cilia. They often have resistant stages and a more complex life cycle (but these are traits not confined to this group).

There are several important pathogenic protozoans. These are summarized in Table 2.6.

As noted in this table, many protozoans have resistant stages, which, depending on their detailed structure, are referred to as cysts, oocysts, or spores.

Many protozoa have complex life cycles and can exhibit relatively resistant forms. This is illustrated in the case of *Giardia* in Figure 2.11. The organism infects the host as a cyst, where it emerges as a vegetative form (trophozoite) and multiplies. It is excreted into the environment as cysts, which have enhanced survival (vs. trophozoites) and can then infect other hosts. Other protozoa may form resistant forms denoted as oocysts or spores — the different nomenclature reflecting differences in the detailed morphology of their resistant stages.

2.1.5 Fungi

Fungi are eukaryotic organisms distinguished by cell walls that contain chitin (which is a polysaccharide containing nitrogen, formally a polymer of N-acetylglucosamine). They range from unicellular to multicellular, with examples of pathogens at both levels. The single-celled fungi include yeasts, some of which may be dimorphic — they can exist in two morphological forms. Historically, many fungal diseases have been regarded as localized in particular geographic areas; however, recent retrospective analyses of health records show expansion of such ranges [56].

As with protozoa, the taxonomy of fungi, especially at higher taxonomic levels, has undergone changes. One recent review places fungi within the Opisthokonta group, within the Amorphea supergroup that also includes animals and protozoans [50].

Table 2.6 Some important pathogenic protozoa genera.

Genus	Group	Comments
Acanthamoeba	Sarcodina	Eye and disseminated infections from environmental (especially water) exposure. Forms resistant cysts.
Balantidium	Ciliata	Gastrointestinal illness, primarily in LDCs, from water contaminated by swine waste. Forms resistant cysts.
Blastocystis	Sarcodina	Prevalent even in healthy individuals. Thought to cause gastrointestinal illnesses, some with chronic *sequelae*. From water and environmental contamination. Forms resistant cysts.
Cyclospora	Sporozoa	Transmitted by contaminated food and water. Gastrointestinal. Forms a resistant spore stage.
Cryptosporidium	Sporozoa	Transmitted by contaminated food and water. Gastrointestinal. Forms a resistant oocyst stage.
Cystoisospora	Sporozoa	From contaminated food and water. More common in tropical and semitropical settings. Forms a resistant oocyst stage.
Dientamoeba	Sarcodina	Mild to moderate gastrointestinal illness. Mode of transmission uncertain. No evidence for resistant stages.
Endamoeba	Sarcodina	Includes the classic agent of amoebic dysentery — water- and foodborne. Forms resistant cysts.
Giardia	Mastigophora	Water- and foodborne gastrointestinal pathogen. Forms resistant cysts.
Leishmania	Mastigophora	Insect-borne. Disseminated multiple organ sites of attack. Mainly in tropical and semitropical countries.
Various Microsporidia[a]	Unclear	A broad spectrum of organ sites can be affected. Forms resistant spores.
Naegleria	Various	Eye and nervous system infection from recreational and other uses of contaminated water. Complex life cycle.
Plasmodium	Various	Insect-borne agent of malaria with a complex life cycle.
Toxoplasma	Sporozoa	Transmitted from contaminated food and water and infected animals. Complex life cycle with a resistant oocyst stage.

[a] This is a group name for multiple protozoan species. *Source*: Data from [55].

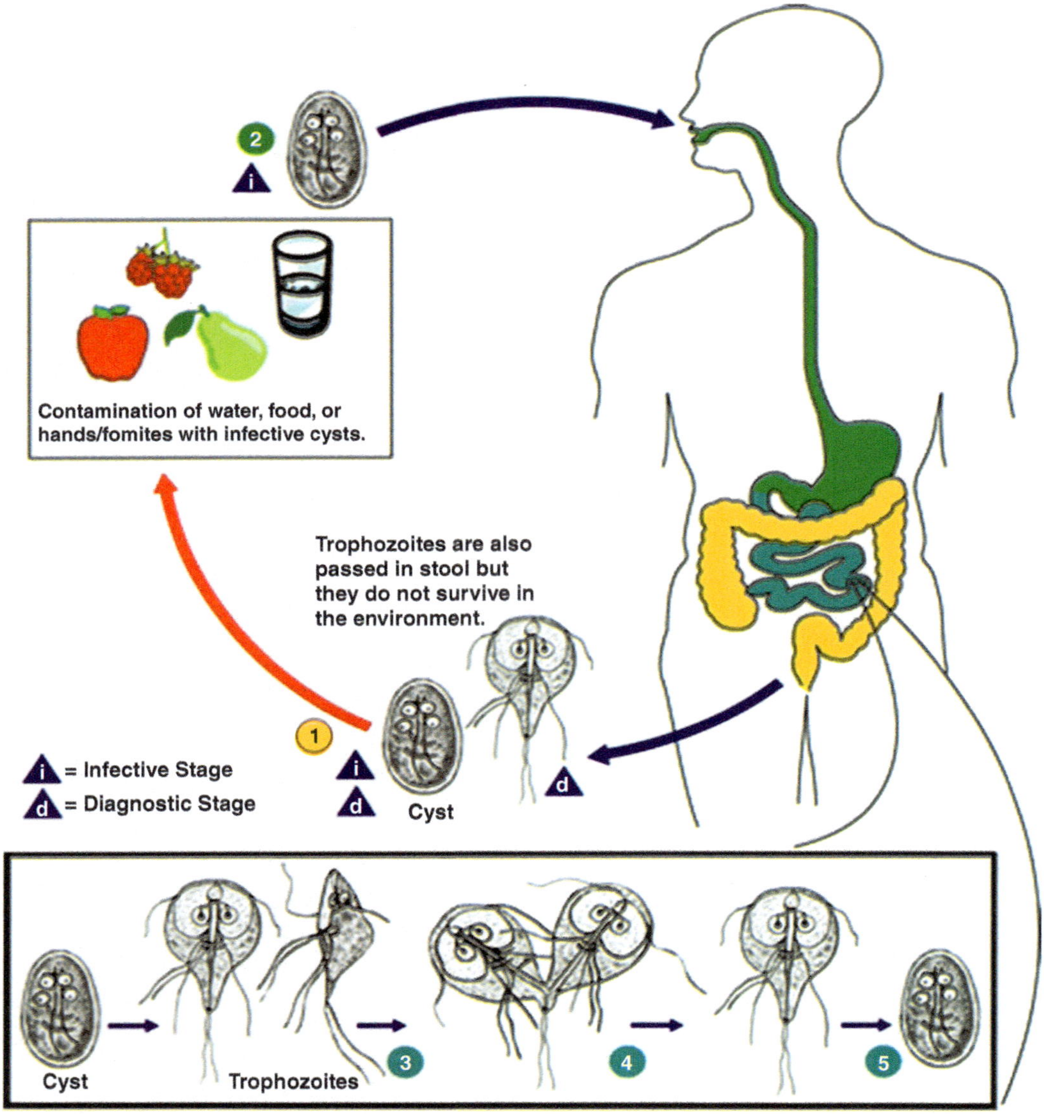

Figure 2.11 Life cycle of *Giardia*. *Source*: Centers for Disease Control and Prevention/Public domain/https://www.cdc.gov/dpdx/giardiasis/index.html/last accessed Feb 12, 2026.

Naranjo-Ortiz and Gabaldon [57] subdivide fungi into nine groups at a level possibly equivalent to a phylum separated into three broad categories, and this represents a useful division. These phyla and examples of pathogens in each are expanded upon as follows.

2.1.5.1 Zoosporic Fungi

These are fungi that at least in one of their life stages have one or more flagella — indicating some resemblance to protozoa.[10] Phyla in this group are:

---

10 Some would classify Microsporidia within this broad grouping, rather than protozoa, indicating the degree of taxonomic uncertainty for some organisms.

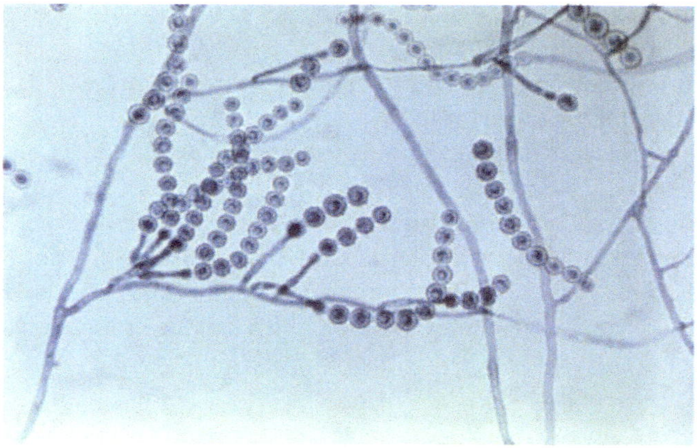

Figure 2.12 Photomicrograph (at 475X) of *Scopulariopsis* sp. showing septate, filamentous hyphae, and numerous chains of rough-walled conidial spores. *Source*: Centers for Disease Control and Prevention/Public domain/https://wwwn.cdc.gov/phil/Details.aspx?pid=23076/last accessed Feb 12, 2026.

Opisthosporidia. Microsporidia are included in this category. No other well-recognized human pathogens are included, though a number of genera are animal pathogens.

Chytridiomycota. This phylum comprises several genera that are isolatable from soil and water, although no recognized pathogens are known.

Neocallimastigomycota. Primarily cellulose degraders and no recognized pathogens.

Blastocladiomycota. Saprophytic, some animal but no known human pathogens.

2.1.5.2 Zygomycetous Fungi

These are primarily multicellular, which (except for a few members of the group) produce spores, often by a hyphae — an example of which is shown in Figure 2.12.

Phyla classed in this group are listed as follows:

Zoopagomycota contains many saprophytes and pathogens of insects and aquatic animals. However, genera of *Basidiobolus* and *Condidobolus* have been identified as human pathogens [58].

Glomeromycota are primarily plant pathogens and saprophytes.

Mucormycota contains a number of general features that have been identified as pathogenic [58]. In particular, organisms in the group Mucorales were responsible for severe illnesses in patients hospitalized from COVID-19 who were diabetic [59].

2.1.5.3 Dikarya

This is a complex and very diverse phylum, which also includes mushrooms and many yeasts. It can be subdivided into the following:

Basidiomycota are primarily plant pathogens and saprophytes.

Agaricomycotina contains many yeasts, including those of economic value. The pathogenic genus *Cryptococcus*, which is an emerging pathogen with environmental and animal reservoirs, is in this group [60].

Ascomycota contains numerous genera of pathogenic fungi. Important examples include the following:

- *Pneumocystis* is an important cause of pneumonia in AIDS patients [61].
- *Candida* contains many species that can be frank or opportunistic pathogens [62].

- *Aspergillus* can be commonly found in soil or dust, but also can cause respiratory allergies or infections [63].
- *Coccidiodes* is the infectious agent of valley fever, historically prevalent in the Southwest US [64], but apparently increasing in geographic range [56].
- *Paracoccidiodes* is a rare pathogen in the United States, although somewhat prevalent in Central and South America [65].
- *Histoplasma* is prevalent in nitrogen-rich soils often contaminated by bird guano. It is the agent for histoplasmosis, which is widespread in the Americas, Asia, and Africa.
- *Stachybotrys chartarum* is a black mold seen in buildings after flooding or dampness, and either the organism itself or the products or fragments of the organism can result in various lung symptoms [66].

2.1.6 Other Agents of Concern

Helminthic diseases can be transmitted via contaminated soils, sludges, etc., and are frequently transmitted by exposure to or ingestion of the eggs of these worms [67]. They are much more common in less-developed settings. However, since they are not microorganisms per se, they will not be considered in more detail.

Algae of a diversity of species can be found (and proliferate) in the environment. However, human adverse effects are almost fully confined to toxins that algae produce, or allergic reactions to the algae themselves, rather than infection [68]. Therefore, they will not be considered further.

2.2 Portals of Entry

The "how" of exposure to an infectious agent refers to the means by which a host gets exposed to a pathogen. This is often (though a bit misleadingly[11]) referred to as the portal of entry. This is one of the key links in a chain of infection (Figure 2.13).

Exposure occurs when an individual experiences a contaminated environment (soil, water, air, food, or surfaces) or contact with, or in the vicinity of, an infected individual or animal [69]. This may result in lodging within or on the host via several mechanisms. In Chapter 13, these mechanisms will be quantified.

- Inhalation occurs when pathogens in the air are breathed and enter the respiratory system.
- Direct deposition on mucus surfaces (e.g., eyes, nostrils) in which pathogen-laden particles "land" on a suitable body surface.
- From water or food contaminated with pathogens, exposure may occur by ingestion. Due to the connection of the gastrointestinal tract and respiratory tract at the mouth, this may also result in inadvertent exposure to the respiratory tract by aspiration.
- The deposition of organisms (or particles containing organisms) onto the skin, either directly or indirectly, may result in those organisms finding a suitable site for colonization and multiplication. This may result in direct adverse skin effects, or the colonized organisms may subsequently be transported through the skin if there are wounds or abrasions.

11 Misleading since deposition on and colonization of skin can be a means of exposure and ultimately illness, without necessarily entering into the body of a host at that initial site.

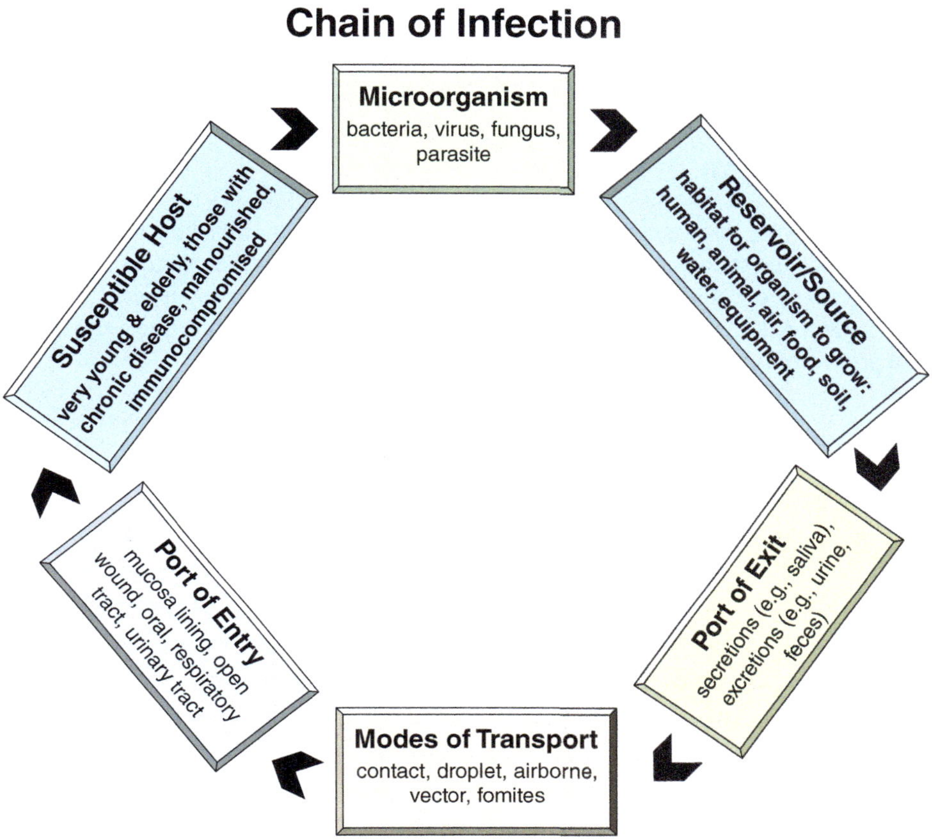

Figure 2.13 Chain of infection. *Source*: Centers for Disease Control and Prevention/Public domain/ https://www.cdc.gov/niosh/learning/safetyculturehc/module-2/3.html/last accessed Feb 12, 2026.

- Some organisms, notably *Pseudomonas aeruginosa*, can cause ear infections (otitis externa). This can occur in recreational water activities such as swimming [70], where it is called "swimmers' ear."
- Fomites are inanimate surfaces on which infectious organisms can survive. They might be deposited from the settlement of air particles, dust, or animal activity. The hands of a potential host may then become colonized after touching a surface. The organisms may merely be deposited on the hands from touch. Once on the hands, subsequent touching of the hands to the eyes, nose, or mouth may result in internal delivery of the infectious agent.

There are other potential portals of entry or exposure to pathogens that, while significant [1], do not involve an intermediate environmental step and are therefore out of scope for this book. For example, pathogens can enter the bloodstream via the bites of insects (or animals) harboring them, which they may have obtained via prior contact or bites with other infected humans.[12] Blood-borne

12 An environmentally mediated blood-borne route is also possible in which the insect or animal vector gets exposed by feeding or landing on or in a contaminated environment.

exposure can also occur via contaminated needles, transfusions, infections on wounds, or accidental cuts with contaminated medical instruments. Another route excluded from further discussion is direct sexual contact with or without the exchange of body fluids.

2.3 Venues of Concern

Having discussed the types of organisms of concern and the portals of entry, the actual venues in which exposure occurs — this is the "where" of the chapter. Knowing the venues of interest will allow one to frame the scenario and mechanisms through which exposure can occur, and further quantify potential risks and the potential benefits of intervention. In each venue, exposure to pathogens can occur through the portals of entry discussed in Section 2.2.

2.3.1 Indoor Environments

It is logical to first consider the indoor environment as a venue of exposure, since it is clear, across multiple countries, that people spend the most time indoors [71], whether at home, at school, at work, or elsewhere. This can be depicted using average values and average lifespans in Figure 2.14.

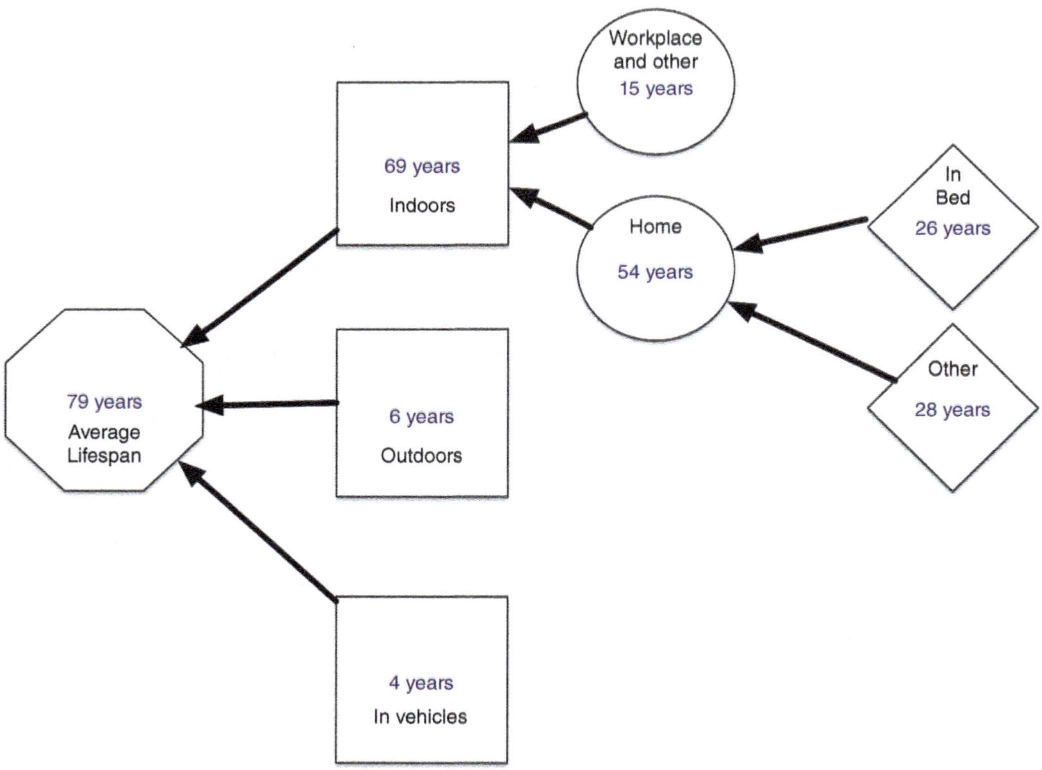

Figure 2.14 Time indoors vs. outdoors based on average US lifespan. *Source*: From Unpublished Presentations by Richard Corsi.

As also implied by this figure, exposure in the indoors itself can be subdivided into different categories, e.g., home, work, vehicles, indoor recreation (arenas, concert halls). Within each of these venues, exposures can occur by many routes and pathways:

- If an infected person enters the room, they may (depending on the organism) emit airborne particles that can be inhaled by others in the room, can land on mucus membranes, or can deposit on surfaces and serve as fomites.
- Infected pets or pests may serve as sources similar to the aforementioned point.
- The water supply to the facility may contain pathogens. This can be ingested, contaminate surfaces that would then serve as fomites, or be used in a way that generates airborne particles. Such latter uses could include sink hoses, showers, toilets, or humidifiers [72]. In various nondomestic residential buildings, exposure to airborne particles from other water uses may occur, such as from ornamental fountains or water features, grocery store produce misters, and architectural features.
- Wetted surfaces within buildings may serve as locales for the growth of pathogens, which can then be released (either by sloughing or by active dispersal of spores or organisms) into the air for exposure. Fungi can particularly be suited to multiplication in such niches. Such locales can include plant media, on air conditioning condenser coils, or damp indoor surfaces (from flooding or poor humidity control).

The indoor work environment may have distinctive venues of exposure. In addition to being potentially more densely occupied, activities may result in the generation of dust or particles that could be transported via air or settle on surfaces for fomite exposure.

In the school environment, vocalization by instructors or students may constitute an additional source of exposure. The production of airborne particles by talking has been documented and will be considered in more detail in Chapter 7.

2.3.2 Outdoor Environments

Outdoor venues are diverse and may encompass multiple types of exposures and portals of entry.

Aquatic environments can include natural (beaches, lakes, seashores) or man-made (pools, spas). The nature and extent of exposure, as well as the portal of entry, will depend on the activities undertaken and their duration. Primary-contact exposure features intense full immersion into the water and may involve entry via accidental ingestion or aspiration or exposure by contact with mucus membranes or skin. Examples include swimming, diving, waterskiing, kayaking, and personal powered watercraft (e.g., Jet Ski's™). Secondary-contact exposure involves activities such as boating, wading (without head immersion), and fishing.

Outdoors on land (including on beaches) and direct contact with soils and ground surfaces may occur. This can result in exposure to skin, via ingestion or inhalation (regarding the soil or sand as a fomite). Note also that, especially in warmer environments, some organisms (particularly some bacteria and fungi) can multiply on or within sands and soils [73]. Direct proximity to infected individuals outdoors may result in exposure to exhaled or shed (from skin) particles from infected individuals.

On land, there are many additional sources of exposure. Some of these include:

- Cooling towers, whether used at commercial power plants or, more importantly, at the building scale, can create conditions that promote the proliferation of a rich biota [74]. Water particles may be emitted and carried to other venues where (depending on their size) they may be inhaled or deposited on skin or mucus membranes.

- Irrigation of agricultural land, parkland, lawns, golf courses, etc., could result in the "drift" of aqueous particles to other locales and be deposited on inhaled. The water deposited on the irrigated land may result in a fomite that can be responsible for subsequent exposure if the land is used while any pathogens are still viable.
- Wastewater treatment plants may emit aerosols from a variety of processes, including turbulence at pump stations and aeration of activated sludge. These aerosols may contain pathogens [75] that can be conveyed both to individuals at the treatment plant and to populations outside.
- Particularly in warm arid regions, water mists may be used as a local form of outdoor cooling. If there are pathogens in the feedwater (either from the domestic water supply itself or from growth in conveyances), this may result in potential inhaled exposures, as well as deposition on skin and mucus membrane surfaces.
- The outdoor environment in proximity to water bodies (seashore areas, areas near rivers or lakes) may receive airborne particles resulting from wind action, wave action, and turbulence on the water surface [76, 77].
- Agricultural activities such as plowing and animal feeding operations [78] can result in dust that can be spread by wind to expose off-site individuals.
- Wind-blown dust from natural soils, or exacerbated from construction activities, may present an exposure source to off-site individuals.

There is evidence, at least in certain cases, that wind can carry viable microorganisms long distances. Fungi and bacteria have been measured in the Caribbean from dust originating in Africa [79]. Dillon and Dillon have done a systematic review noting many cases of windblown diseases at all scales [80].

2.3.3 In Vehicles

Vehicles represent a particular type of indoor environment in which different exposures can occur. This can encompass a wide variety of venues, including personal automobiles, ride-shares, taxis, local and long-distance buses, local and long-distance trains, trolleys and trams, and airplanes. Often, especially in multi-passenger and transport vehicles, the effective occupant density is greater than in other indoor environments. In addition, cleaning, operation, or maintenance of vehicles can introduce other hazards to the occupants; for example, it has been found that *Legionella pneumophila* [81] can survive and grow in windshield washer fluid, and presumably some portion can be transported into the passenger areas by air circulation and present an opportunity for exposure.

2.4 Discussion Questions and Problems

1 Find a recent outbreak of environmentally transmitted disease. A useful source is *Morbidity and Mortality Weekly Reports*, published by CDC. For the virus that is the causal agent, report which Baltimore group it belongs to and which taxonomic realm.

2 Keep a diary for a week, recording the time spent in different venues noted in Figure 2.14. How do your results compare with the average?

References

1 Jean Maguire Van Seventer and Natasha S Hochberg. Principles of infectious diseases: transmission, diagnosis, prevention, and control. In Stella R Quah, editor, *International Encyclopedia of Public Health*, pages 22–39. Academic Press, 2nd edition, 2017.

2 Arturo Casadevall and Liise-Anne Pirofski. What is a host? Attributes of individual susceptibility. *Infection and Immunity*, 86:1–17, 2018.

3 Sam P Brown, Daniel M Cornforth, and Nicole Mideo. Evolution of virulence in opportunistic pathogens: Generalism, plasticity, and control. *Trends in Microbiology*, 20(7):336–342, July 2012.

4 Arturo Casadevall and Liise-Anne Pirofski. What is a pathogen? *Annals of Medicine*, 34(1):2–4, January 2002.

5 L M Feazel, L K Baumgartner, K L Peterson, D N Frank, J K Harris, and N R Pace. Opportunistic pathogens enriched in showerhead biofilms. *Proceedings of the National Academy of Sciences of the United States of America*, 106(9):16393–16399, September 2009.

6 Emily Garner, Marisa Organiscak, Lucien Dieter, Carley Shingleton, Madison Haddix, Sayalee Joshi, Amy Pruden, Nicholas J Ashbolt, Gertjan Medema, and Kerry A Hamilton. Towards risk assessment for antibiotic resistant pathogens in recycled water: A systematic review and summary of research needs. *Environmental Microbiology*, 23(12):7355–7372, October 2021.

7 Alice U Lee and Caroline Lee. Hepatitis D review: Challenges for the resource poor setting. *Viruses*, 13(10):1912, September 2021.

8 Ahmed Hadidi, Liying Sun, and John W Randles. Modes of viroid transmission. *Cells*, 11(4):719, February 2022.

9 National Human Genome Research Institute. Talking glossary of genomic and genetic terms virus talking glossary of genomic and genetic terms, October 2023.

10 Charles E Simon. The filterable viruses. *Physiological Reviews*, 3(4):483–508, October 1923.

11 ViralZone, SIB Swiss Institute of Bioinformatics. Virions of some of the most common human viruses with their relative size, February 2021.

12 David Baltimore. Expression of animal virus genomes. *Bacteriological Reviews*, 35:235–241, 1971.

13 Eugene V Koonin, Mart Krupovic, and Vadim I Agol. The Baltimore classification of viruses 50 years later: How does it stand in the light of virus evolution? *Microbiology and Molecular Biology Reviews*, 85(3):e00053-21, August 2021.

14 Gita Mahmoudabadi and Rob Phillips. A comprehensive and quantitative exploration of thousands of viral genomes. *eLife*, 7:e31955, April 2018.

15 Peter Simmonds, Evelien M Adriaenssens, F Murilo Zerbini, Nicola G A Abrescia, Pakorn Aiewsakun, Poliane Alfenas-Zerbini, Yiming Bao, Jakub Barylski, Christian Drosten, Siobain Duffy, et al. Four principles to establish a universal virus taxonomy. *PLoS Biology*, 21(2):e3001922, 2023.

16 Wikipedia Contributors. Realm (virology) — Wikipedia, the free encyclopedia. https://en.wikipedia.org/w/index.php?title=Realm_(virology)&oldid=1176018836, 2023. [Online; accessed 17-October-2023].

17 Krista R Wigginton and Alexandria B Boehm. Environmental engineers and scientists have important roles to play in stemming outbreaks and pandemics caused by enveloped viruses. *Environmental Science & Technology*, 54(7):3736–3739, 04 2020.

18 Hi Eun Jung and Heung Kyu Lee. Host protective immune responses against influenza a virus infection. *Viruses*, 12(5):504, 2020.

19 Dylan H Morris, Kwe Claude Yinda, Amandine Gamble, Fernando W Rossine, Qishen Huang, Trenton Bushmaker, Robert J Fischer, M Jeremiah Matson, Neeltje van Doremalen, Peter J Vikesland, et al. Mechanistic theory predicts the effects of temperature and humidity on inactivation of SARS-CoV-2 and other enveloped viruses. *eLife*, 10:e65902, 2021.

20 Lisa Marie Casanova and Scott R Weaver. Inactivation of an enveloped surrogate virus in human sewage. *Environmental Science & Technology Letters*, 2(3):15–20, 2015.

21 Andrea I Silverman and Alexandria B Boehm. Systematic review and meta-analysis of the persistence of enveloped viruses in environmental waters and wastewater in the absence of disinfectants. *Environmental Science & Technology*, 55(21):14480–14493, 2021.

22 W Ford Doolittle and Olga Zhaxybayeva. What is a prokaryote? In Eugene Rosenberg, Edward F DeLong, Stephen Lory, Erko Stackebrandt, and Fabiano Thompson, editors, *The Prokaryotes*, pages 21–37. Springer, Berlin, Heidelberg, 2013.

23 Jake V Bailey, Samantha B Joye, Karen M Kalanetra, Beverly E Flood, and Frank A Corsetti. Evidence of giant sulphur bacteria in neoproterozoic phosphorites. *Nature*, 445(7124):198–201, 2007.

24 Manfred Rohde. The gram-positive bacterial cell wall. *Microbiology Spectrum*, 7(3):GPP3–0044–2018, May 2019.

25 E C S Chan. Microbial nutrition and basic metabolism. In Duncan Mara and Nigel J Horan, editors, *Handbook of Water and Wastewater Microbiology*, pages 3–33. Academic Press, San Diego, CA, 2003.

26 Paul A Lawson, Diane M Citron, Kerin L Tyrrell, and Sydney M Finegold. Reclassification of clostridium difficile as clostridioides difficile (hall and o'toole 1935) prévot 1938. *Anaerobe*, 40:95–99, 2016.

27 Michael Köpke, Melanie Straub, and Peter Dürre. Clostridium difficile is an autotrophic bacterial pathogen. *PLoS ONE*, 8(4):e62157, April 2013.

28 N R Krieg and S Hoffman. Microaerophilly and osygen toxicity. *Annual Review of Microbiology*, 40:107–130, 1986.

29 Antonin C André, Lorine Debande, and Benoit S Marteyn. The selective advantage of facultative anaerobes relies on their unique ability to cope with changing oxygen levels during infection. *Cellular Microbiology*, 23(8):1–8, August 2021.

30 M C Van Loosdrecht, J Lyklema, W Norde, G Schraa, and A J Zehnder. The role of bacterial cell wall hydrophobicity in adhesion. *Applied and Environmental Microbiology*, 53(8):1893–1897, August 1987.

31 Ron J Doyle. Contribution of the hydrophobic effect to microbial infection. *Microbes and Infection*, 2(4):391–400, April 2000.

32 Peter Setlow and Graham Christie. New thoughts on an old topic: Secrets of bacterial spore resistance slowly being revealed. *Microbiology and Molecular Biology Reviews*, 87(2):e00080-22, June 2023.

33 Lior Sinai, Alex Rosenberg, Yoav Smith, Einat Segev, and Sigal Ben-Yehuda. The molecular timeline of a reviving bacterial spore. *Molecular Cell*, 57(4):695–707, February 2015.

34 James D Oliver. Recent findings on the viable but nonculturable state in pathogenic bacteria. *FEMS Microbiology Reviews*, 34(4):415–425, July 2010.

35 Sheng Chen, Xi Li, Yahong Wang, Jie Zeng, Chengsong Ye, Xianping Li, Lizheng Guo, Shenghua Zhang, and Xin Yu. Induction of Escherichia coli into a VBNC state through chlorination/chloramination and differences in characteristics of the bacterium between states. *Water Research*, 142:279–288, October 2018.

36 Elisabeth Dietersdorfer, Alexander Kirschner, Barbara Schrammel, Anna Ohradanova-Repic, Hannes Stockinger, Regina Sommer, Julia Walochnik, and Sílvia Cervero-Aragó. Starved viable but non-culturable (VBNC) Legionella strains can infect and replicate in amoebae and human macrophages. *Water Research*, 141:428–438, September 2018.

37 Ernst Mayr. *Systematics and the Origin of Species from the Viewpoint of a Zoologist*. Columbia University Press, New York, 1942.

38 Margaret A Riley and Michelle Lizotte-Waniewski. Population genomics and the bacterial species concept. In John M Walker, Maria Boekels Gogarten, Johann Peter Gogarten, and Lorraine C Olendzenski, editors, *Horizontal Gene Transfer*, Volume 532. Humana Press, 2009.

39 Jerónimo Rodríguez-Beltrán, Javier DelaFuente, Ricardo León-Sampedro, R Craig MacLean, and Álvaro San Millán. Beyond horizontal gene transfer: The role of plasmids in bacterial evolution. *Nature Reviews Microbiology*, 19(6):347–359, June 2021.

40 J T Trevors. Evolution of gene transfer in bacteria. *World Journal of Microbiology and Biotechnology*, 15:1–6, 1999.

41 P S Mead and P M Griffin. Escherichia coli O157:H7. *Lancet*, 352(9135):1207–1212, October 1998.

42 R G E Murray and John G Holt. The history of Bergey's manual. In William B. Whitman, editor, *Bergey's Manual of Systematics of Archaea and Bacteria*, pages 1–20. Wiley, 2015.

43 P Vandamme, B Pot, M Gillis, P De Vos, K Kersters, and J Swings. Polyphasic taxonomy, a consensus approach to bacterial systematics. *Microbiology Review*, 60(2):407–38, 1996.

44 Diana Barceló-Antemate, Fernando Fontove-Herrera, Walter Santos, and Enrique Merino. The effect of the genomic GC content bias of prokaryotic organisms on the secondary structures of their proteins. *PLOS ONE*, 18(5):e0285201, May 2023.

45 Jill E Clarridge. Impact of 16S rRNA gene sequence analysis for identification of bacteria on clinical microbiology and infectious diseases. *Clinical Microbiology Reviews*, 17(4):840–862, October 2004.

46 Morgane Rossi-Tamisier, Samia Benamar, Didier Raoult, and Pierre-Edouard Fournier. Cautionary tale of using 16S rRNA gene sequence similarity values in identification of human-associated bacterial species. *International Journal of Systematic and Evolutionary Microbiology*, 65(6):1929–1934, 2015.

47 Laura Uelze, Josephine Grützke, Maria Borowiak, Jens Andre Hammerl, Katharina Juraschek, Carlus Deneke, Simon H Tausch, and Burkhard Malorny. Typing methods based on whole genome sequencing data. *One Health Outlook*, 2(3):s42522, December 2020.

48 Cristiane C Thompson, Gilda R Amaral, Mariana Campeão, Robert A Edwards, Martin F Polz, Bas E Dutilh, David W Ussery, Tomoo Sawabe, Jean Swings, and Fabiano L Thompson. Microbial taxonomy in the post-genomic era: Rebuilding from scratch? *Archives of Microbiology*, 197(3):359–370, April 2015.

49 Aharon Oren and George M Garrity. Valid publication of the names of forty-two phyla of prokaryotes. *International Journal of Systematic and Evolutionary Microbiology*, 71(10):005056, October 2021.

50 Fabien Burki, Andrew J Roger, Matthew W Brown, and Alastair G B Simpson. The new tree of eukaryotes. *Trends in Ecology & Evolution*, 35(1):43–55, January 2020.

51 Sina M Adl, Alastair G B Simpson, Mark A Farmer, Robert A Andersen, O Roger Anderson, John R Barta, Samuel S Bowser, G U Y Brugerolle, Robert A Fensome, Suzanne Fredericq, et al. The new higher level classification of eukaryotes with emphasis on the taxonomy of protists. *Journal of Eukaryotic Microbiology*, 52(5):399–451, 2005.

52 Stephanie M Fletcher, Damien Stark, John Harkness, and John Ellis. Enteric protozoa in the developed world: A public health perspective. *Clinical Microbiology Reviews*, 25(3):420–449, July 2012.

53 Centers for Disease Control and Prevention. About parasites. Centers for Disease Control and Prevention, March 2022.

54 Brenda Nisbet. Sarcodina. In Brenda Nisbet, editor, *Nutrition and Feeding Strategies in Protozoa*, chapter 10, pages 185–201. Springer Netherlands, Dordrecht, 1984.

55 Centers for Disease Control and Prevention. Microsporidiosis. https://www.cdc.gov/dpdx/microsporidiosis/, May 2019.

56 Patrick B Mazi, John M Sahrmann, Margaret A Olsen, Ariella Coler-Reilly, Adriana M Rauseo, Matthew Pullen, Julio C Zuniga-Moya, William G Powderly, and Andrej Spec. The geographic distribution of dimorphic mycoses in the United States for the modern era. *Clinical Infectious Diseases*, 76(7):1295–1301, April 2023.

57 Miguel A NaranjoOrtiz and Toni Gabaldón. Fungal evolution: Diversity, taxonomy and phylogeny of the fungi. *Biological Reviews*, 94(6):2101–2137, December 2019.

58 Corina E Gonzalez, Michael G Rinaldi, and Alan M Sugar. Zygomycosis. *Infectious Disease Clinics*, 16(4):895–914, 2002.

59 Asma Al Balushi, Ahmed Al Ajmi, Qais Al Sinani, Varna Menon, Zainab Al Berieki, Asma Al Shezawi, Saleh Al Azri, Azza Al Rashdi, Amina Al Jardani, Thuwaiba Al Baluki, et al. COVID-19-associated mucormycosis: An opportunistic fungal infection. a case series and review. *International Journal of Infectious Disease*, 121(8):203–210, 2022.

60 Maurizio Del Poeta and Arturo Casadevall. Ten challenges on cryptococcus and cryptococcosis. *Mycopathologia*, 173(5-6):303–310, June 2012.

61 Charles F Thomas and Andrew H Limper. Pneumocystis pneumonia. *New England Journal of Medicine*, 350(24):2487–2498, 2004.

62 Cristina Nicoleta Ciurea, Irina-Bianca Kosovski, Anca Delia Mare, Felicia Toma, Ionela Anca Pintea-Simon, and Adrian Man. Candida and candidiasis—opportunism versus pathogenicity: A review of the virulence traits. *Microorganisms*, 8(6):857, 2020.

63 Jean-Paul Latgé and Georgios Chamilos. Aspergillus fumigatus and aspergillosis in 2019. *Clinical Microbiology Reviews*, 33(1):e00140-18, 2019.

64 Samantha L. Williams and Tom Chiller. Update on the epidemiology, diagnosis, and treatment of coccidioidomycosis. *Journal of Fungi*, 8(7):666, June 2022.

65 Anderson Messias Rodrigues, Ferry Hagen, Rosana Puccia, Rosane Christine Hahn, and Zoilo Pires de Camargo. Paracoccidioides and paracoccidioidomycosis in the 21st century. *Mycopathologia*, 188(1):129–133, 2023.

66 Mariusz Dyląg, Klaudyna Spychała, Jessica Zielinski, Dominik Łagowski, and Sebastian Gnat. Update on stachybotrys chartarum—Black mold perceived as toxigenic and potentially pathogenic to humans. *Biology*, 11(3):352, February 2022.

67 Alex Loukas, Rick M Maizels, and Peter J Hotez. The yin and yang of human soil-transmitted helminth infections. *International Journal for Parasitology*, 51(13):1243–1253, 2021.

68 Wolfgang Karl Hofbauer. Toxic or otherwise harmful algae and the built environment. *Toxins*, 13(7):465, 2021.

69 D H Walker. Interactions of pathogens with the host. In Linda M McManus and Richard N Mitchell, editors, *Pathobiology of Human Disease*, pages 214–216. Elsevier, 2014.

70 S A Rice, B van den Akker, F Pomati, and D Roser. A risk assessment of *Pseudomonas aeruginosa* in swimming pools: A review. *Journal of Water and Health*, 10(2):181–196, June 2012.

71 Sabine Brasche and Wolfgang Bischof. Daily time spent indoors in German homes – Baseline data for the assessment of indoor exposure of German occupants. *International Journal of Hygiene and Environmental Health*, 208(4):247–253, July 2005.

72 J. O'Toole, M. Keywood, M. Sinclair, and K. Leder. Risk in the mist? Deriving data to quantify microbial health risks associated with aerosol generation by water-efficient devices during typical domestic water-using activities. *Water Science and Technology*, 60(11):2913–2920, 2009.

73 K M Yamahara, S P Walters, and A B Boehm. Growth of enterococci in unaltered, unseeded beach sands subjected to tidal wetting. *Applied and Environmental Microbiology*, 75:1517–1524, 2009.

74 Han-Fei Tsao, Ute Scheikl, Craig Herbold, Alexander Indra, Julia Walochnik, and Matthias Horn. The cooling tower water microbiota: Seasonal dynamics and co-occurrence of bacterial and protist phylotypes. *Water Research*, 159(Aug 1):464–479, August 2019.

75 Mengmeng Lou, Shuai Liu, Chunjie Gu, Huimin Hu, Zhengkun Tang, Yaopeng Zhang, Chenye Xu, and Fang Li. The bioaerosols emitted from toilet and wastewater treatment plant: A literature review. *Environmental Science and Pollution Research*, 28(3):2509–2521, 2021.

76 Renee B Pietsch, Hinrich Grothe, Regina Hanlon, Craig W Powers, Sunghwan Jung, Shane D Ross, and David G Schmale III. Wind-driven spume droplet production and the transport of *Pseudomonas syringae* from aquatic environments. *PeerJ*, 6:e5663, September 2018.

77 Matthew A Pendergraft, Derek J Grimes, Sarah N Giddings, Falk Feddersen, Charlotte M Beall, Christopher Lee, Mitchell V Santander, and Kimberly A Prather. Airborne transmission pathway for coastal water pollution. *PeerJ*, 9:e11358, June 2021.

78 Dick Heederik, Torben Sigsgaard, Peter S Thorne, Joel N Kline, Rachel Avery, Jakob H Bønløkke, Elizabeth A Chrischilles, James A Dosman, Caroline Duchaine, Steven R Kirkhorn, et al. Health effects of airborne exposures from concentrated animal feeding operations. *Environmental Health Perspectives*, 115(2): 298–302, 2007.

79 Joseph M Prospero, Edmund Blades, George Mathison, and Raana Naidu. Interhemispheric transport of viable fungi and bacteria from Africa to the Caribbean with soil dust. *Aerobiologia*, 21(1):1–19, 2005.

80 Charles F Dillon and Michael B Dillon. Multiscale airborne infectious disease transmission. *Applied and Environmental Microbiology*, 87(4):e02314-20, January 2021.

81 David Otto Schwake, Absar Alum, and Morteza Abbaszadegan. Automobile windshield washer fluid: A potential source of transmission for Legionella. *Science of The Total Environment*, 526(Sept.1):271–277, 2015.

3

Key Paradigms of Environmental Engineering

There are two major paradigms of environmental engineering that are important and adaptable to specific pathogen control cases. These are the idea of risk and the concept of a source→transport→fate→receptor process. The two concepts will be discussed in turn. The chapter will conclude by discussing the concepts of uncertainty and variability underlying these paradigms.

3.1 Risk Framework

To discuss risk, it is necessary to define terms. The first is defining risk. The Society for Risk Analysis [1] has developed a glossary, and even this foundational term has several definitions. The most useful in the context of this book is:

> "Risk is uncertainty about and severity of the consequences of an activity with respect to something that humans value."

Examples of activities that environmental engineering may be concerned with include the release (unintended or deliberate) of contaminants into the environment, the failure of a mitigation system, or the deliberate violation of a permit for an activity. Consequences could include ecological impacts (outside the scope of this book), illness (of varying severity and duration), and death. This can be made more systematic by considering that an activity in a given state of nature can result in a spectrum of possibilities that can be characterized by sets of characteristics [2] that (ideally exhaustively) enumerate the possible outcomes, the likelihood of that particular outcome, and the consequence of that outcome. The likelihood and the consequence may each have their own uncertainties.

As Kaplan and Garrick note [2], this formalism is embedded in a decision-theoretic context, and we can reconsider the spectrum of risk tuples[1] under alternative states of nature, for example, after the implementation of some particular interventions aimed at reducing consequences. Then, by

1 In the original formulation, this was termed a risk "triplet" embodying a scenario, a consequence, and a likelihood. But as examples in this chapter and later will show, the result of a scenario may require a vector attribute, and the outcomes and consequences may themselves be vector-valued. Therefore, the word "tuple" is used for greater generalizability.

comparing the consequence reduction with the cost of interventions, the desirability of a specific decision can be ascertained (along with associated uncertainties).[2]

A simple illustration of this approach is given in Example 3.1.

Example 3.1 Example Risk Scenario Problem.

Exposure to a particular microorganism can occur by inhalation of contaminated air in an auditorium during a concert. The expected number of illnesses and their likelihood are given in the following table. Also shown is the projected impact if a particular engineering intervention (e.g., improving ventilation) occurs. If the cost of each illness is $1000, what would the justifiable cost of the intervention be?

Action	Illnesses	Likelihood	Value of consequence
Do nothing	0	0.2	$0
	50	0.7	$35,000
	100	0.1	$10,000
Intervene	0	0.8	$0
	50	0.15	$7500
	100	0.05	$5000

To assign a value to the consequence, the number of illnesses and their likelihoods are multiplied, and then multiplied by the cost per illness. If the costs are summed, the do-nothing case costs $45,000, and with the intervention, the cost of illnesses drops to $12,500. Therefore, any action that costs less than $30,000 is economically justifiable.

The assessment of the likelihood of particular numbers of illnesses and their uncertainties is the central aspect of risk characterization, which is the focus of Chapter 14. The process of assigning values to consequences and the various metrics (including but not only monetary) that might be used are discussed in Chapter 15.

There are two types of canonical problems in the risk analysis of environmental systems:

- Given a potential exposure scenario and a choice of management alternatives, what are the resultant consequences of each alternative? This can inform ultimate risk management decisions.
- Given current "facts on the ground" involving an exposure and existing mitigation or management measures, what is the residual risk that may remain and the consequences? Does this justify further action?

These might be termed a "prospective" and a "retrospective" analysis, although the methods involved in both would be very similar. The process of risk analysis is a broad activity involving risk assessment, risk communication, and risk management [1, 3]. A recognized framework is shown in Figure 3.1. The following are the key component parts, recognizing that at all stages there should be two-way public communication to facilitate acceptance of an ultimate resolution [4]:

2 Note the absence of discussion on metrics that might be used for costs and benefits. These might be in commensurable units or not, and in various forms such as dollars, health consequences, etc. This will be explicitly considered in Chapter 15.

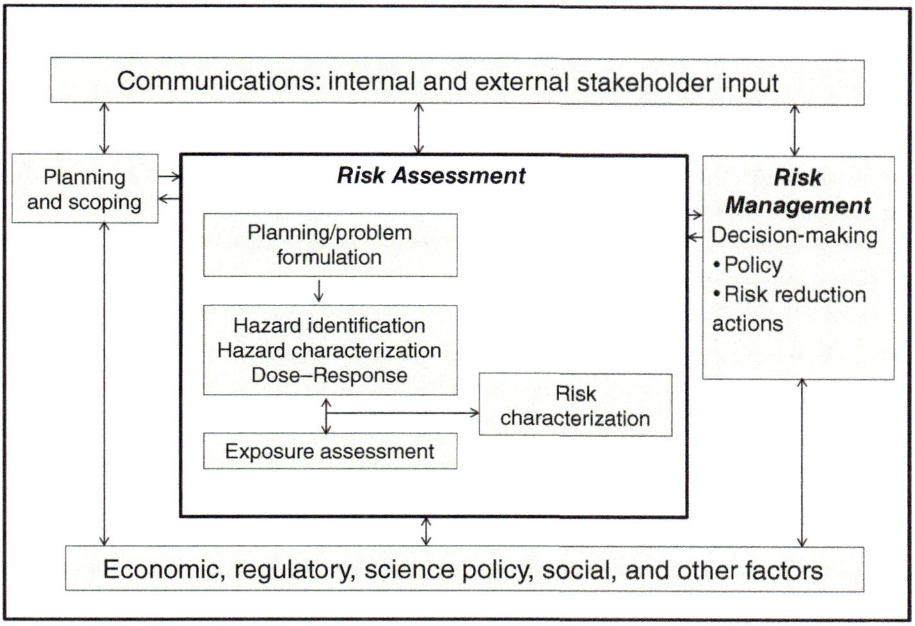

Figure 3.1 Risk analysis framework. *Source:* Adapted from [5].

3.1.1 Problem Definition

The specific problem under consideration is outlined. What is the nature of the decision to be made, and what constraints are present? Who are the decision-makers and the other relevant stakeholders? What are the desired metrics of consequence most important to decision-makers and stakeholders? What are conceptual alternatives? What are the constraints on alternatives, such as regulations, laws, specifics of the venue, and social factors? What pathogen(s), route(s) of exposure, venue(s), and outcome(s) are under consideration?

3.1.2 Risk Assessment

This is the centrality of the technical tasks to be undertaken and is divided into four activities. The original intent of delineating risk assessment as a formal task [3] was to separate what were thought to be purely technical tasks from those that required more subjective judgment. However, it is now recognized that within risk assessment, the choices of models, approaches, and parameters involve judgment, and hence the entire risk analysis enterprise needs transparency and stakeholder engagement at all steps. The formal risk assessment tasks are discussed as follows.

3.1.2.1 Hazard Characterization

What are the organisms[3] of concern, what are the sources, what are the outcomes (illness, loss of productivity, death), what are the routes of exposure, venues, and portals of entry?

3 For some organisms, as noted in Chapter 2, there has been a significant change in taxonomy over time, and so the risk assessor needs to be cognizant of these changes and alternative historical names in doing a literature review.

Are there sensitive subpopulations of concern? Perhaps the pathogen is one that has a differential impact on individuals based on their preexisting conditions, nutrition, or exposure to concomitant environmental stressors.

How are the organisms measured? Can the viable concentrations be measured? This will be discussed further in Chapter 6.

Does the exposure occur once or multiply (periodic or aperiodic)? For example, do emissions result from the occasional presence of a source (an infected individual or a malfunctioning control system)? Perhaps exposure arises from weather variability and needs to be considered.

Are there multiple pathogens that could be present — as in the case of swimming in waters contaminated by poorly treated sewage, which can result in exposure to multiple pathogens (as well as other hazardous substances)?

3.1.2.2 Dose–Response Assessment

What is the quantitative relationship between the exposure that individuals are exposed to and their likelihood of having particular adverse effects? The details of this task will be discussed in Section 14.1.

The main task is to establish a relationship between exposure level and the probability of an adverse outcome. This is conceptually illustrated in a very simple form for a single-point exposure in Figure 3.2. Information to inform the development of this relationship may come from experimental testing or from epidemiological information.

3.1.2.3 Exposure Assessment

What are the distributions of exposures to individuals, and what are their frequency and duration? This activity can be conducted using either or both experimental/measurement approaches or modeling approaches. If different management alternatives are under consideration, the exposure under each alternative needs to be assessed.

3.1.2.4 Risk Characterization

Combining information from dose–response and exposure assessment, what are the consequences of various outcomes, and what are the uncertainties in the magnitude of those consequences? This may require the assessor to characterize the uncertainties from the dose–response and exposure assessments and propagate them to produce uncertainty in the characterized risk outcomes. For

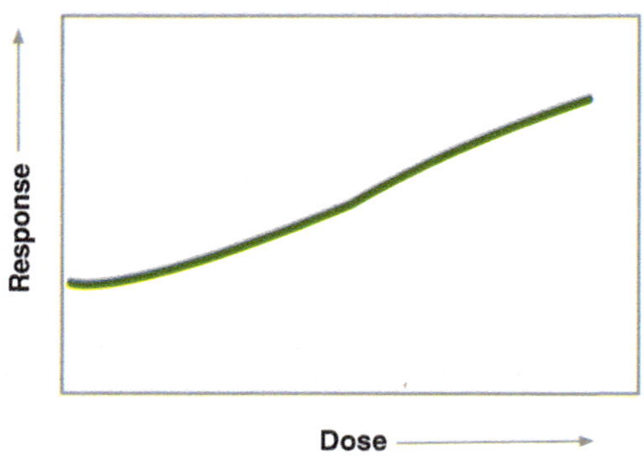

Figure 3.2 Conceptual dose–response for a single exposure.

different management alternatives under consideration, a separate formal characterization would be needed, as shown in the simplified case in Example 3.1.

The risk characterization step ends with an inventory of all possible outcomes and their frequencies and likelihoods. The valuation of these consequences (the rightmost column in Example 3.1) should be regarded as part of the risk management task, since it heavily involves value judgments entering the process.

The risk characterization step can be more complex than illustrated by the example in a number of ways:

- There may be a time series of consequences at regular or irregular intervals. For example, consider the potential exposure via inhalation of emissions from an infected individual. The impact of an exposure on the outcome from subsequent exposures needs to be known or assumed.
- There may be a spectrum of outcomes that may occur, from mild to severe illnesses to death. The likelihoods of the various outcomes need to be conveyed to the risk manager and be subject to evaluation.
- There may be countervailing risks incurred when a particular intervention is used to reduce a risk. This is often the case as seen by several examples:
  - When drinking water disinfection is used to reduce microbial risk, often disinfection byproducts are produced, which have their own adverse effects [6].
  - The use of "boil water" notifications to reduce exposure to pathogens in contaminated drinking water entails increased risks of fire and scalding that should be considered [7].
  - The shutdown of in-person education during the peak COVID-19 pandemic may have reduced the burden from that illness; it may also have reduced learning (and increased disparities) in the impacted students [8].

  Therefore, a fair calculation of the impacts of an intervention needs to consider both the increased and the reduced risks (saving their valuation for the risk management phase).
- There may be chronic, even delayed, outcomes from infection with pathogens. For example, infection with certain enteric bacteria is associated with the initiation of an arthritic response [9]; individuals infected with poliomyelitis may suffer from long-term muscular degenerative effects [10]; and there is growing evidence of the significance of "long COVID" in individuals who have been infected with SARS-CoV-2 [11].

3.1.3 Risk Management

Once a risk (with its uncertainty) has been quantified, and perhaps the remaining risks after possible policy and technology implementation have been determined, a decision is needed on which (if any) policies to pursue.

There may be a variety of factors beyond the degree of risk reduction and the difficulty of implementing interventions. These might include:

- Legal tools and mechanisms available for implementation
- Costs of enforcement and potential levels of compliance
- Differential costs and benefits to subpopulations of particular concern
- Potential for and impact of adverse impacts from the implementation of a particular intervention
- Time required for implementation (versus urgency of action)

A common approach is to view the problem solely as one of balancing costs and benefits on a single scale. However, as has been pointed out by many, including over 40 years ago by Lord Eric Ashby [12]:

> "There was a time — it's over now — when cost-benefit analysis was regarded as a reliable technique for making choices in such matters as the siting of airports, highways, power plants, and reservoirs. Imponderables, such as noise, disruption of the skyline, the hazards of smoke, and the amenities of wilderness — none of these dismayed the economist. He found ingenious ways to quantify them. How much would you pay to put in double glazing to cut out the racket of trucks, or, for that matter, how much would you pay to move away from the noisy neighborhood altogether? What is a reasonable compensation for having to repaint your home once a year, or for enduring the inconvenience of bronchial attacks brought on by smoke and sulfur dioxide? What price the risk of your death on a highway without crash barriers?"
>
> It isn't just that cost-benefit analysis is unreliable when it involves projection into the future (though this is true); it is also — and this is the decisive educational point — an inappropriate technique. It sets out to answer the question: What is efficient for society? But this is not what the public asks about environmental issues nowadays; they ask: What is good for society? And the most precious kinds of good cannot be quantified without distortion.
>
> The seminal question is this: "In making environmental impact assessments, is it better to restrict economic analysis to values that are unquestionably quantifiable, leaving the decision-maker to integrate the imponderables into the decision by subjective means? Or is it better to stretch economic analysis to cover values that have to be stripped of their human meaning in order to quantify them?"

Put briefly, aside from the numerous technical issues of conducting cost-benefit analysis (more subsequently), there is an ethical and philosophical question: While cost-benefit analysis might lead to delineation of the most *efficient* outcome, it might not necessarily lead to what leads to the most *good for society*.

Even if a risk-benefit approach is to be taken, there are a number of issues to be considered, including the ones that Ashby mentions:

1. How do we "weight" consequences of different severities and perhaps durations together?
2. How do we consider a series of consequences that may happen over time — is the approach of discounting appropriate and are there ethical issues?
3. There are potential equity issues (who is suffering the harm and who may have to pay for the intervention) that are irreducible to monetary terms, but nonetheless important to decision-making.

If cost-benefit analysis is to be used, there are many technical issues to consider, including those raised by Ashby. These points will be returned to in Chapter 15.

If risk management is not focused on the centrality of a cost-benefit approach, there are alternative frameworks that can be employed.

In the decision, the risk managers need to involve all stakeholders in the deliberation as to the decisions to be made. This process of detailed analysis (in the risk assessment step) accompanying deliberation with stakeholders is important in gaining acceptance for the ultimate decision and has been termed an analytic-deliberative process [13].

Who Is a Stakeholder?

Stakeholders are those who may be affected by a decision. Other than risk managers, these have been termed "risk losers, risk gainers, risk perpetrators" [14]. These should not be viewed as sets of entities with coherent or consistent interests. For example, there may be differential categories of risk gainers and risk losers who have different interests, concerns, and perceptions, and this needs to be considered during risk communication plan development and risk management.

Perhaps the broadest approach is to fully and openly outline the costs of alternatives and the risks (and reduction of risks) *in their actual terms* and engage in full deliberation with stakeholders — in this way, integrating risk communication (to be discussed below) directly into the risk management stage.

There are alternative approaches to reaching decisions at the risk management stage. Further discussion of these and the methods is covered in Chapter 15.

3.1.4 Risk Communication

Risk communication is a central activity throughout the risk analysis process, and it needs to be a two-way street between formal risk managers and the various publics and stakeholders. This has increasingly been recognized as important as it has become understood that the levels and severities of risk as perceived by various stakeholders/public(s) may be quite different depending upon the attributes of the risk itself [15].

As an example, Slovic and colleagues [16] in the 1980s asked experts and various "lay" populations to rank-order their ratings of the risk of various hazards. A portion of the results is shown in Figure 3.3. Several things are noteworthy: "lay" publics tend to rate greater risks as less severe and smaller risks as more severe than they are, i.e., the regression line has a slope less than 1.0. This has been seen in other studies. Also, certain risks are dramatically over-ranked or under-ranked compared to their "true" position.

Slovic and others have further characterized the reasons for certain risks to be amplified (or modulated) in perception, and there are two dominant clusters of factors that seem to explain amplification — broadly speaking, these are unobservable/unfamiliar/delayed, and uncontrollable/dread/catastrophic/severe/involuntary. The attributes of these factors are shown in Table 3.1.

Communication may help modulate the gap between perceived risk and quantified risk, although irreconcilable differences may remain. The risk communication process, as summarized by Fischoff [17], requires more than just doing the assessment itself correctly (Table 3.2). In any complex risk problem, those knowledgeable about best practices in risk communication should be engaged at the outset.

One structured approach to risk communication aimed at closing gaps between expert-determined risks and risks as perceived by publics is termed a "mental models" approach, which proceeds by the following steps [18]:

- Develop an expert and literature-informed decision model that is technically accurate.
- Elicit, from the publics and the decision-makers, their hazard beliefs and compare to the expert model.
- Survey the broader public for their key perceptions.
- Draft messaging.
- Evaluate message effectiveness and refine as needed.

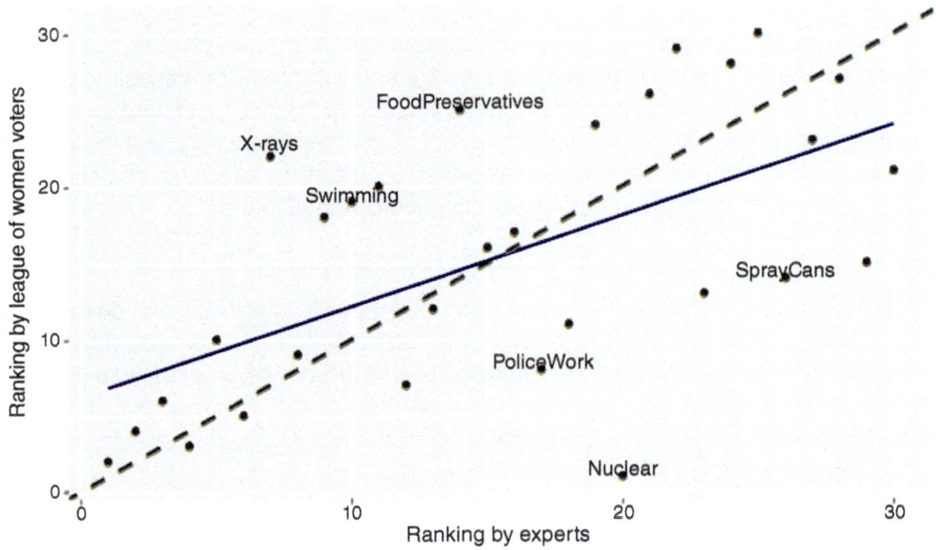

Figure 3.3 Ratings of different risks by experts and members of the League of Women Voters. 1 is the highest and 30 is the lowest risk. The solid line is the best-fit line, and the dashed line is the line of equality. *Source:* [16]/ with permission of Springer Nature.

Table 3.1 Attributes of two factors associated with risk amplification or attenuation.

Factor	Attenuation traits	Amplification traits
	Controllable	Uncontrollable
	Not dread	Dread
	Not global catastrophic	Catastrophic
	Not fatal	Fatal
"Dread"	Equitable	Inequitable
	Individual	Many impacted
	Low risk to future generations	High future risk
	Easily reduced	Difficult to reduce
	Risk decreasing	Risk increasing
	Voluntary	Involuntary
	Observable	Not observable
	Known	Unknown to those exposed
"Familiar"	Immediate	Effect delayed
	Old risk	New risk
	Known to science	Previously unknown to science

There is empirical evidence [19] that, when such a framework is followed, a useful result can be obtained.

In subsequent chapters, the application of the elements of risk analysis will be set forth with particular respect to pathogens.

Table 3.2 Stages in risk communication.

All we have to do is get the numbers right
All we have to do is tell them the numbers
All we have to do is explain what we mean by the numbers
All we have to do is show them that they've accepted similar risks in the past
All we have to do is to show them that it's a good deal for them
All we have to do is treat them nice
All we have to do is make them partners
All of the above

Source: [17]/with permission of John Wiley & Sons.

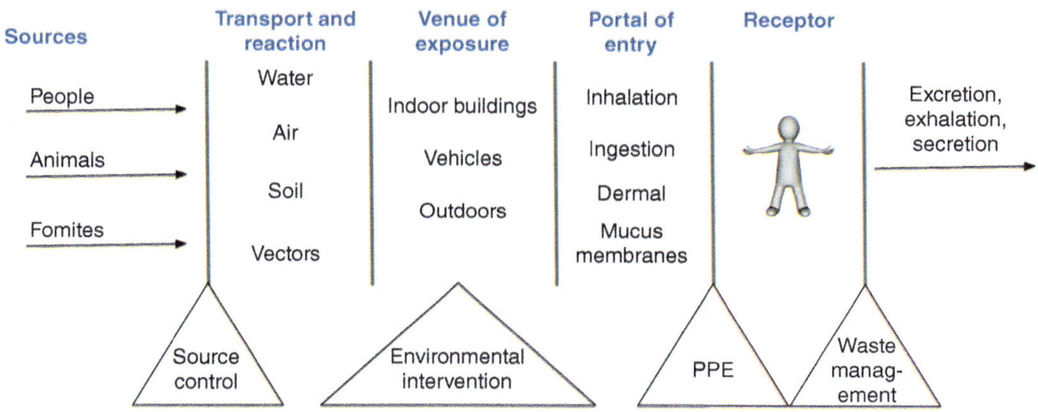

Figure 3.4 Source transport receptor framework.

3.2 Source, Fate, Transport, Receptor

From the risk paradigm, it is clear that a key task needs to be estimation of the levels of pathogens that a population experiences (exposure assessment), which is a central task to be undertaken. The paradigm of source→fate→transport→receptor is commonly used in environmental engineering and is applicable to the understanding and control of pathogens.

The operationalization of this framework in the context of pathogens is illustrated in Figure 3.4.

Sources of pathogens can be infected people, animals that have been colonized or infected, or fomites (soil, sand, surfaces). They can be transported through the environment on land, in water, in the air, or through the action of pests or vectors (insects, rodents) that may be actively colonized by pathogens or act as passive conveyors.[4]

After transport to a venue in which exposure occurs, an individual (receptor) may become colonized and infected. The individual infected with a contagious agent might secrete, excrete, or exhale pathogens that have grown on their body, and in that way propagate environmental contamination.

This figure highlights the need to understand the amounts and extents of the source terms (to be discussed in Chapter 7), the processes of transport and survival or decay in the environment (to be

4 It may not be evident to modern students and practitioners, but environmental engineering as a lineal descendant of sanitary engineering has always regarded vector and pest control within its purview [20].

addressed in Chapter 8), the dynamics of pathogens immediately proximate to the portal(s) of entry (to be discussed in Chapter 13), and the potential growth and emission from infected individuals.

At the bottom of Figure 3.4, several triangles indicate points of potential engineering intervention that can be used to reduce risk. Intervention processes will be discussed in Chapters 9–11. From left to right:

- Source control can be used to reduce pathogen levels from human and animal wastes, emissions, and secretions from being transported into the environment; for example, the disinfection of wastewater prior to discharge into receiving waters. Face masks could be used on infected individuals to reduce pathogen emissions to the air.
- Within individual venues, active decontamination, ventilation, or cleaning can be used to reduce concentrations of infectious agents. When a (e.g., hospital) space is occupied by an infectious individual, total room decontamination can be deployed prior to occupancy by a another person.
- Personal protective equipment could be used by potential receptors (masks, gloves, gowns, eyeshields, handwashing) to reduce exposure.
- Finally, the emissions from infected individuals can be reduced by the collection and treatment of their discharges. For highly contagious and severe pathogens, this could even include the use of dedicated, controlled quarantine orders and facilities, although this comes with substantial practical and legal issues [21].

The introduction of interventions will incur costs but also permit risk reduction. The quantification of the risk reduction will allow a judgment to be made as to the overall desirability of a particular course of action. This balancing concept will be covered in Chapter 15.

It should also be emphasized that, rarely, is any single barrier, including a pharmaceutical intervention, perfect. Therefore, the best practice is to employ multiple barriers to increase the likelihood and effectiveness of interrupting the transmission pathways of infectious diseases. This will be covered further in Chapter 9.

3.3 Uncertainty and Variability

An overarching concept in much of what environmental engineers do, particularly with pathogens, is that anything measured or predicted often has embedded levels of uncertainty and/or variability that prevent it from being known with exactitude.

To proceed further, a set of definitions is needed. Two broad phenomena lead to a lack of exactitude in a quantity. These can be termed "epistemic uncertainty" and "aleatory uncertainty" (often also called "variability").

3.3.1 Epistemic Uncertainty

Epistemic uncertainty is "a lack of precise knowledge as to what the truth is, whether qualitative or quantitative" [22]. Epistemic uncertainties can arise from multiple sources. A general taxonomy is shown in Figure 3.5. At the highest level, uncertainty can occur from either the measurement of parameters directly or from models used to assess the parameters.

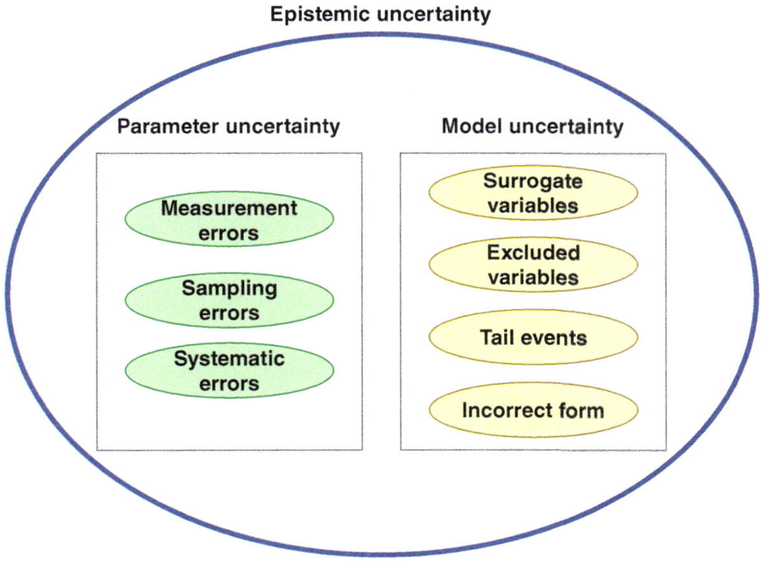

Figure 3.5 Taxonomy of forms of epistemic uncertainty.

For the first type of uncertainty (due to parameters), errors (either bias or random) may occur with the methods used to measure a parameter. There may be errors in obtaining representative samples from the environment on which measurements are being made.[5]

For the second type of uncertainty (due to models), there may be errors in the model form itself. This often leads to problems when the model(s) are used to provide predictions beyond the ranges of inputs on which they were developed.

There may be model errors associated either with the exclusion (failure to observe) or giving unusual weight to (perhaps by oversampling or over-including) rare extreme events. There may be important variables that would increase explanatory power (and reduce the uncertainty associated with a model) and are excluded, or there may be improper use of surrogate measures for other necessary (but difficult to measure) inputs.

3.3.2 Aleatory Uncertainty or Variability

Aleatory uncertainty, often called variability, results from the intrinsic random behavior of a quantity. Some sources may be purely intrinsic, while others could be reduced if the system could be characterized at a finer level of detail (perhaps incorporating currently unknown or "lurking" variables). As an example, consider that the distribution of some property has been measured and is described by the dashed curve in Figure 3.6 — which is a fairly broad distribution. But in reality, this might be the result of an equal mixture of two subpopulations (given by the solid curves in this figure), and if it were known which subpopulation was being measured, the "spread" of that subpopulation is substantially less than the undifferentiated total population.

5 Consider a situation in which an environment, such as a landfill, has "hot spots" or when a water supply has "hot moments" — spatial or temporal occurrences of very high concentrations, but when sampling is insufficiently intense or inadvertently fails to capture such occurrences.

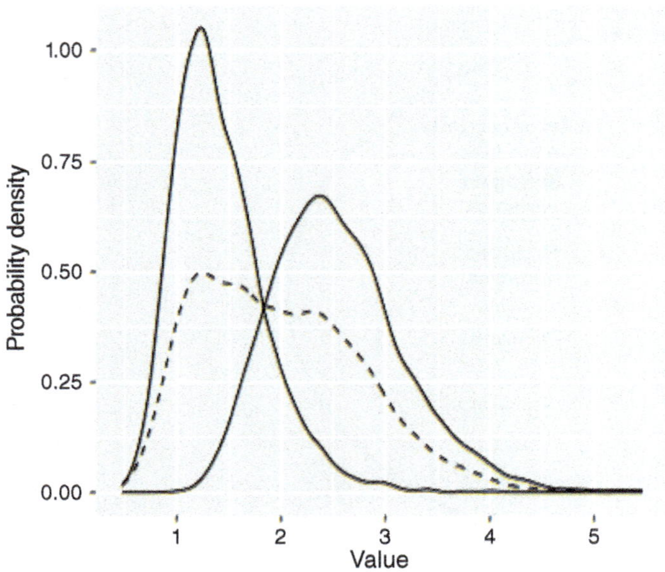

Figure 3.6 Effect of subdividing populations on distribution. Solid lines are two subpopulations. The dashed line is the overall distribution of an equal mixture.

Consider, for example, a distribution of drinking water consumption in a population — it would be broad. But if we subdivided the population in various ways, such as by gender, by age, etc., the individual distributions conditioned on those descriptors would be much narrower [23].

In other words, there is potential to reduce variability by increasing systematic knowledge of a system. However, there is always an irreducible level of variability in real systems.

Generally, sources of epistemic uncertainty are more reducible by additional investment of resources than are sources of aleatory uncertainty. In Chapter 14, these concepts will be used again in the context of risk characterization.

3.4 Discussion Questions and Problems

1 Consider Example 3.1. A vendor proposes installing a technology that would cost $40,000 per use. What efficacy (reducing illnesses) would be needed versus the do-nothing alternative to make that intervention economically justifiable?

2 In 2001, there were multiple facilities that were contaminated with spores of *Bacillus anthracis* from postal mail that contained the agents. About a dozen deaths occurred. This resulted in a cleanup of multiple federal and privately owned facilities. Write a brief report focusing on what stakeholders were involved in developing cleanup criteria and actions. Given what we know now as best practices, how might this have proceeded differently — particularly, which stakeholders would have been involved?

3 Given the factors associated with risk amplification in Table 3.1, give examples of risks, not among those highlighted in Figure 3.3, that would be expected to be amplified or attenuated, and why.

References

1 Committee on Foundations of Risk Analysis. Society for Risk Analysis Glossary, 2018.

2 Stanley Kaplan and B John Garrick. On the quantitative definition of risk. *Risk Analysis*, 1(1):11–27, March 1981.

3 National Academy of Sciences. *Risk Assessment in the Federal Government: Managing the Process*. National Academy Press, Washington, D.C., 1983.

4 Presidential/Congressional Commission on Risk Assessment and Risk Management. Framework for environmental health risk management. *Technical Report*, Presidential/Congressional Commission on Risk Assessment and Risk Management, Washington, D.C., 1997.

5 National Academies of Science, Engineering and Medicine. *Science and Decisions: Advancing Risk Assessment*. National Academies Press, Washington, D.C., 2009.

6 Xing-Fang Li and William A Mitch. Drinking water disinfection byproducts (DBPs) and human health effects: Multidisciplinary challenges and opportunities. *Environmental Science & Technology*, 52(4):1681–1689, February 2018.

7 Michael G Ryan, Patrick L Gurian, Charles N Haas, Joan B Rose, and Phillip J Duzinski. Acceptable microbial risk: Cost-benefit analysis of a boil-water order for *Cryptosporidium*. *Journal of the American Water Works Association*, 105(4):E189–E194, 2013.

8 Sébastien Goudeau, Camille Sanrey, Arnaud Stanczak, Antony Manstead, and Céline Darnon. Why lockdown and distance learning during the COVID-19 pandemic are likely to increase the social class achievement gap. *Nature Human Behaviour*, 5(10):1273–1281, 2021.

9 Anuli N Ajene, Christa L Fischer Walker, and Robert E Black. Enteric pathogens and reactive arthritis: A systematic review of Campylobacter, Salmonella and Shigella-associated reactive arthritis. *Journal of Health, Population, and Nutrition*, 31(3):299, 2013.

10 Daria A Trojan and Neil R Cashman. Post-poliomyelitis syndrome. *Muscle & Nerve*, 31(1):6–19, 2005.

11 Hannah E Davis, Lisa McCorkell, Julia Moore Vogel, and Eric J Topol. Long covid: Major findings, mechanisms and recommendations. *Nature Reviews Microbiology*, 21(3):133–146, 2023.

12 Eric Ashby. What price the Furbish lousewort? *Environmental Science & Technology*, 14(10):1176–1181, October 1980.

13 Ortwin Renn. A model for an analytic deliberative process in risk management. *Environmental Science & Technology*, 33(18):3049–3055, September 1999.

14 M R English. Who are the stakeholders in environmental risk decisions how should they be involved. *Risk: Health, Safety, Environment*, 11(3):243–54, 2000.

15 Roger E Kasperson, Ortwin Renn, Paul Slovic, Halina S Brown, Jacque Emel, Robert Goble, Jeanne X Kasperson, and Samuel Ratick. The social amplification of risk: A conceptual framework. *Risk Analysis*, 8(2):177–187, 1988.

16 Paul Slovic. Risk perception. In Curtis C Travis, editor, *Carcinogen Risk Assessment*, pages 171–181. Springer US, Boston, MA, 1988.

17 Baruch Fischhoff. Risk perception and communication unplugged: Twenty years of process. *Risk Analysis*, 15(2):137–145, November 1995.

18 M Granger Morgan. *Risk Communication: A Mental Models Approach*. Cambridge University Press, 2002.

19 Nick Boase, Mathew White, William Gaze, and Clare Redshaw. Evaluating the mental models approach to developing a risk communication: A scoping review of the evidence. *Risk Analysis*, 37(11): 2132–2149, 2017.

20 Mark D Hollis. The sanitary engineer in relation to public health. *Transactions of the American Society of Civil Engineers*, 118(2):624–633, January 1953.

21 Wendy E Parmet. Legal power and legal rights — isolation and quarantine in the case of drug-resistant tuberculosis. *New England Journal of Medicine*, 357(5):433–435, 2007.

22 Committee on Risk Assessment of Hazardous Air Pollutants Commission on Life Sciences. *Science and Judgment in Risk Assessment*. National Research Council, Washington, D.C., 1994.

23 Ann M Roseberry and David E Burmaster. Lognormal distributions for water intake by children and adults. *Risk Analysis*, 12(1):99–104, March 1992.

4

Unique Features of Pathogens

While the key paradigms discussed in Chapter 3 are applicable to pathogens, the consideration of these organisms in environmental engineering introduces new concepts that are not often seen with other contaminants:

- Exposure to small numbers of pathogens, even in a single dose, may cause adverse effects in some individuals.
- The "concentrations" that may be of interest in controlling are lower than those for other contaminants (when expressed on a molar basis).
- At such low concentrations, pure stochastic variability cannot be neglected and is superimposed on other sources of variability.
- Upon exposure to a host, the pathogen may multiply *in vivo,* and therefore the internal body burden is much greater than the amount from initial exposure.
- The amplified pathogens may be emitted, excreted, or secreted by an infectious host back into the environment and therefore pose a risk to other individuals.
- Analytical methods are different and require other types of laboratory skills and equipment than many other contaminants.

The first five of the aforementioned distinctions will be covered in this chapter. The particular analytical methods will be discussed in Chapter 6.

4.1 Stochastic Variability and Low Concentrations

As will be discussed in Section 14.1, in general, one viable organism has the potential to cause an effect (albeit perhaps with a small probability) in an individual host. Therefore, we are concerned with small concentrations. This has several significant ramifications.

The first ramification is that an exposure of a single organism can be expressed in "molar" terms by dividing by Avogadro's number[1] as 1.7×10^{-24} mole, or 1.7 yoctomole. There are no chemical environmental contaminants of significance at such low molar levels.

Such small doses can be experienced via short-term exposure to an environment even contaminated with low concentrations of pathogens. Unlike some chemical exposures, where chronic

1 Recall that this is 6.02×10^{23}, which is the number of particles of a material in a mole.

exposure may be necessary to accumulate sufficient damage, this is generally not the case with infectious microorganisms.

The importance of even low-level exposures, coupled with the fact that an individual microorganism is an indivisible entity, means that the statistical properties of small numbers become important. This can be seen in two ways.

If particles are distributed randomly in a system (either over space or time), then the distribution of *numbers* in a specific unit of a system (unit of time, mass, or volume) is given by a Poisson distribution, which has the form of Equation 4.1:

$$P(k) = \frac{\lambda^k}{k!} \exp(-\lambda) \tag{4.1}$$

where $P(k)$ is the probability of observing exactly "k" organisms in a unit of a system, and λ is the mean concentration of organisms per unit. Note that while "k" must be an integer, λ can be any positive real number.

Equation 4.1 can be plotted for different values of λ, as shown in Figure 4.1. At small values of this parameter, the distribution is quite asymmetric and concentrated toward the left. As λ increases, the distribution becomes more symmetric and can be approximated by a normal distribution.

The Poisson distribution has the property that the average number that would be observed in a unit of the system equals λ and the standard deviation of the number observed (e.g., among multiple units of the system) is $\frac{1}{\sqrt{\lambda}}$.

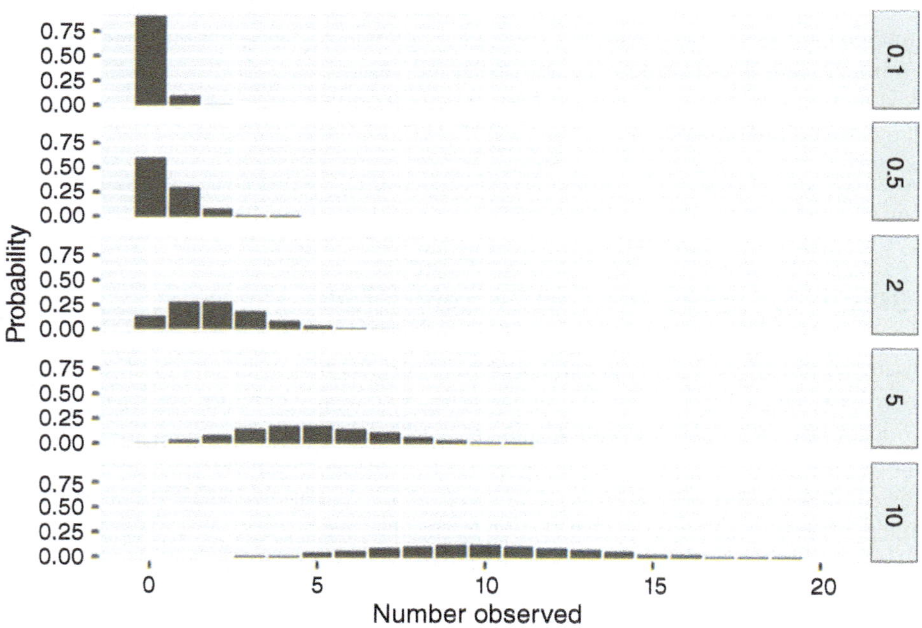

Figure 4.1 Poisson distribution for different values of λ (shown on right).

Example 4.1 Illustration of the Poisson Distribution.

Samples of 1 L of water are taken for microbial analysis. The average concentration is 1 organism per 100 L. What is the probability that a sample will contain 0, 1, 2, or 3 organisms?

Solution:

1. Since we have a 1 L sample, we expect the average in such a sample to be $\lambda = 0.01(100\ \mathrm{L/L}) \times 1 = 0.01$.
2. Equation 4.1 is used with $k = 0, 1, 2, 3$:

$$P(0) = \exp(-0.01)\frac{(0.01)^0}{0!} = 0.9900$$

$$P(1) = \exp(-0.01)\frac{(0.01)^1}{1!} = 9.9 \times 10^{-3}$$

$$P(2) = \exp(-0.01)\frac{(0.01)^2}{2!} = 4.95 \times 10^{-5}$$

$$P(2) = \exp(-0.01)\frac{(0.01)^3}{3!} = 1.65 \times 10^{-7}$$

Note that, although almost all samples would be expected to contain zero organisms, there is still a small probability of obtaining some organisms in a sample (which would increase as the average concentration increases).

As an inverse example, if a number of samples are taken and assayed for organisms, the mean concentration can be inferred. Note that, from Equation 4.1, the probability of observing zero in a set of samples (each of constant amount) is given by:

$$P(0) = \exp(-\lambda) \tag{4.2}$$

$$\therefore \lambda = -\ln[P(0)] \tag{4.3}$$

If we set the observed fraction of zeros in a set of samples to $P(0)$ in Equation 4.3, by assuming the Poisson distribution, we get a naive estimate of the mean (λ). This can be extended to more complex sample designs, e.g., a set of samples with subsets of different amounts, which will be discussed in Chapter 6.

The property of small numbers implies that there may be high variability between sample replicates even when the organisms are randomly distributed according to the Poisson distribution (Figure 4.2).

This same phenomenon can be depicted by plotting the relative standard deviation of the number of organisms versus the number of organisms expected in a sample (Figure 4.3). This stochastic

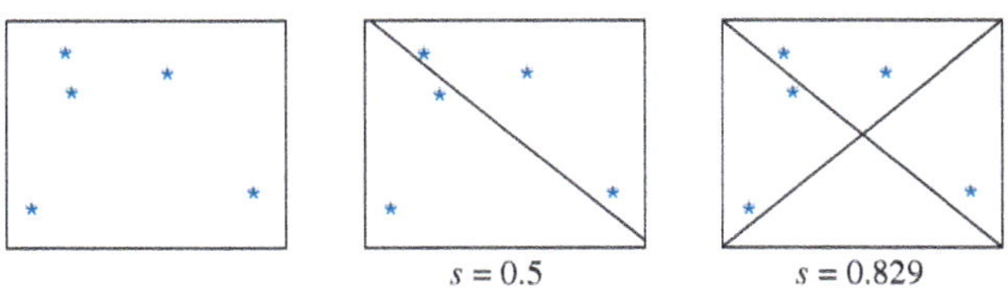

Figure 4.2 Impact of subsampling from environment with few organisms. "s" is the population standard deviation between individual samples.

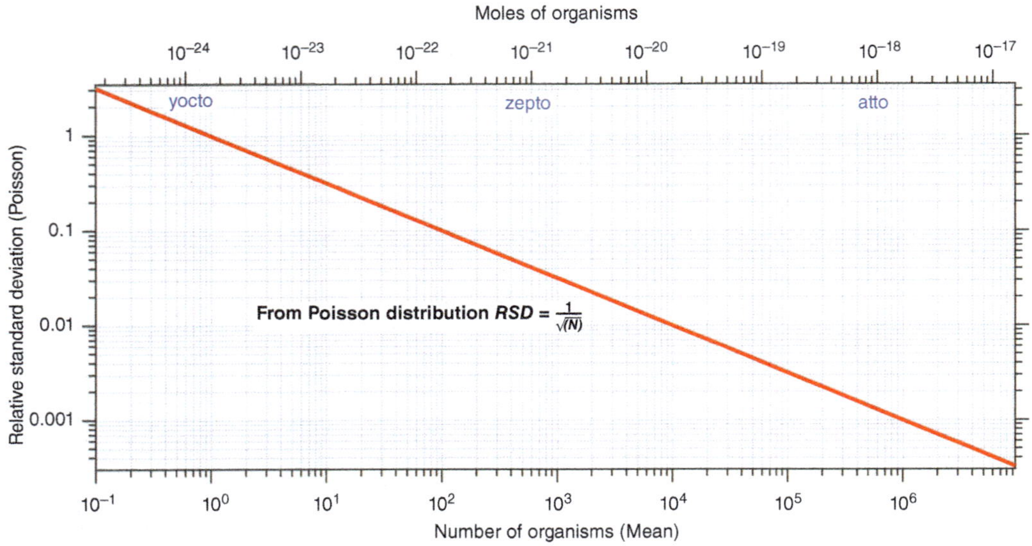

Figure 4.3 Relative standard deviation based on Poisson distribution.

variability is superimposed upon other sources of variability, such as those associated with the analytic measurements, source terms, and environmental properties, as well as sources of uncertainty. At low concentrations of organisms, as would be expected for pathogens, this can represent a substantial contribution to variability—quite unlike chemical contaminants in the environment.

4.2 Growth and Decay

Within a host, all microorganisms must be capable of growth and have the potential to colonize, infect, and initiate disease. This is in contrast to chemical pollutants, which do not multiply but rather degrade *in vivo*.[2]

It should also be noted that the Poisson distribution describes the "random" distribution of particles (pathogens) among portions. There are, however, other processes that may lead to underdispersed or overdispersed distributions. Underdispersion occurs when the variability in actual numbers between replicates is smaller than the Poisson distribution; this is a relatively rare occurrence. Overdispersion occurs when the variability is greater than the Poisson distribution and is a frequent finding in environmental sampling [2]. A schematic illustration of this effect is shown in Figure 4.4.

As an aside, a formal definition of some key terms is needed. The framework of Ref. [3] is adopted:

Colonization occurs when a pathogen to which a host is exposed adheres at a site near the portal of entry and begins multiplication.

Infection occurs when the colonized pathogen and progeny begin to invade and become established *within* host tissue.

Disease occurs when infection proceeds to the point of observable disruption of function with symptoms and signs of illness.

2 Parenthetically, some chemical contaminants, such as vinyl chloride [1], undergo partial metabolism *in vivo* to form more biologically potent products. But they do not "multiply" as pathogens do.

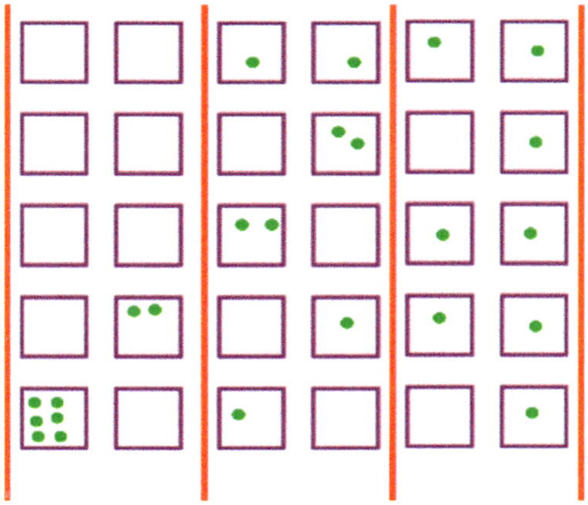

Figure 4.4 Effect of overdispersion (left panel) and underdispersion (right panel) relative to Poisson (center) at constant mean. Each square represents a subsample.

Also, it should be noted that the aforementioned three processes occur sequentially, and so it is entirely possible to observe the process of infection, perhaps by monitoring excretions or secretions for the signs of multiplied pathogens before individuals develop signs and symptoms, or indeed even in individuals who never become symptomatic.

Within the host, there is also a competitive dynamic between pathogen reproduction and immune induction to combat colonization and infection. This can result in the pathogen overcoming the immune responses, or vice versa, and the relative dynamics will influence the time course of the process [4]. Ultimately, this is related to the incubation time for any illness, which will be discussed in Section 14.1.

In addition to growth within a host, some pathogens can grow in the environment. Hubálek [5] uses the descriptor "sapronoses" to define

> "human diseases transmissible from abiotic environment (soil, water, decaying plants, or animal corpses, excreta, and other substrata). The ability of the agent to grow saprophytically and replicate in these substrata (i.e., not only to survive or contaminate them secondarily) are the most important characteristics of a sapronotic microbe."

Ashbolt [6] uses a related term "saprozoic" to describe pathogens that can grow in the environment in biological material, particularly within protozoa that subsist in biofilms. The key point is that, for pathogens, the growth processes in the environment and the host, as well as the decay processes, need to be understood. The level of exposure from the environment may thus exceed the level that was directly emitted by sources.

4.3 Humans as Sources and Receptors

As a result of the multiplication of pathogens *in vivo* within a host, it is entirely possible for humans to serve as sources of pathogens, as well as receptors.

It has long been established that human pathogens emanating from excreta can be isolated from sewage [7]. A wide variety of pathogens may be excreted in the urine or feces, or in respiratory secretions or skin sloughing that may eventually find their way into wastewater streams [8]. This concept has been successfully deployed in response to the COVID-19 pandemic, where monitoring wastewater for this virus's nucleic acid has been shown to provide an effective means of monitoring community case prevalence [9].

Frequently, however, detection is based on measuring nucleic acids rather than viable organisms themselves. In such cases, the mere isolation of nucleic acids cannot be a signal of infectiousness or risk. More on this topic will be discussed in Chapter 6.

In exhaled breath, pathogens (or their biochemical signatures) can be detected. The amount of liquid in expired breath of healthy volunteers was measured in a small study [10] summarized in Figure 4.5. Even normal breathing can result in the production of liquid. Therefore, if there is significant colonization and infection in the respiratory tract, some infectious agents can be emitted. This has been confirmed by experimental studies on individuals who contracted symptomatic influenza. Yan et al. [11] measured such subjects by collecting and assaying their expired breath. A summary of some results is shown in Table 4.1, where both infectious virus and viral RNA (at higher concentrations) could be measured in fine aerosols. This will be expanded upon in Section 7.3.

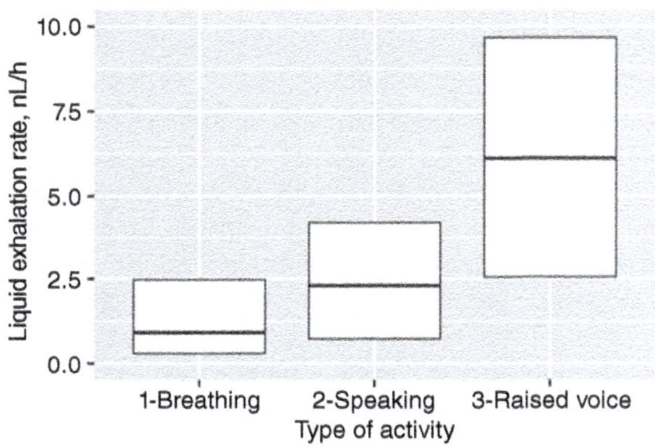

Figure 4.5 Exhaled liquid volume produced during certain activities. Mean and ranges of observed values. *Source:* [10]. MDPI (CC BY 4.0).

Table 4.1 Measurement of influenza virus—infectious particles and RNA—in symptomatic individuals.

	Coarse aerosol	Fine aerosol
Positive on culture	N/A	39%
FF Assay		
Positive	N/A	30%
Geometric mean of positives	N/A	37
RNA Assay		
Positive	40%	76%
Geometric mean of positives	12,000	38,000

The data are for 30 minutes of sampling. FF = "fluorescent focus" assay on cell culture.
Source: Data from [11].

Similar findings have been made with respect to patients with symptomatic COVID-19 illness [12].

Investigators have developed methods for sampling and microbial assays of filters used in central heating, ventilation, and air conditioning systems [13]. It is therefore not surprising that nucleic acids from SARS-CoV-2 can be isolated from such filters, the bulk of which presumably originates from exhaled breath [14].

Another source of pathogens from colonized or infected humans arises from the shedding of skin, hair, and nasal secretions—all of which may contain pathogens. Hospodsky et al. [15] analyzed indoor air and floor dust of occupied rooms and found that a substantial portion of the bacteria measured emanated from skin, hair and scalp, and nasal cavities. They further called attention to the possibility of dust resuspension in the room, which could serve as another route for occupant exposure.

It is well established that pathogenic bacteria resident on the skin and nostrils (specifically *Staphylococcus aureus*) can be isolated from indoor air [16]. It would seem likely that other pathogens colonizing the skin could also end up in the room air and dust from shedding.

4.4 Dose–Response

As depicted in Figure 3.1, dose–response is a central task of risk assessment. The nature of organisms as individual particles, each of which has the potential to multiply and initiate colonization, poses particular challenges for formulating a dose–response function.

A dose–response function, depicted conceptually in Figure 3.2, relates the average dose experienced by a population from environmental exposure to risk. This can be viewed in two ways. It can be viewed as the probability that *an individual* experiences an exposure from the average dose and can have the risk, or alternatively, it can be viewed as the *proportion of a population* that is so exposed that it will succumb to the risk.

As shown in Example 4.1, if a population is exposed to an average dose of 0.01 organisms, there is still a small fraction of individuals in that population who are exposed to 1, 2, etc., microorganisms. Hence, a dose–response relationship for pathogens must also consider the stochastic variability intrinsic to the dose scale at low doses.

Another key difference in dose–response for pathogens versus other environmental contaminants is that the infection (and disease) process requires the *in vivo* proliferation of the infectious agent and the attainment of a body burden or the accumulation of damage. However, this may result from the initial exposure to a small number of organisms from the environment.

The detailed development of appropriate microbial dose–response functions is covered in Section 14.1.

4.5 Contagiousness and Population Spread

The fact that individuals colonized and infected by pathogens themselves may function as sources of exposure to additional individuals means that the impacts on individuals not initially exposed to the pathogen should be considered in a complete risk characterization.

There are numerous examples where such secondary cases occur. For example, early in the COVID-19 outbreak, an investigation identified a cluster of cases among individuals who had eaten at a restaurant, including an individual who was, unbeknownst to them, infected [17]. In

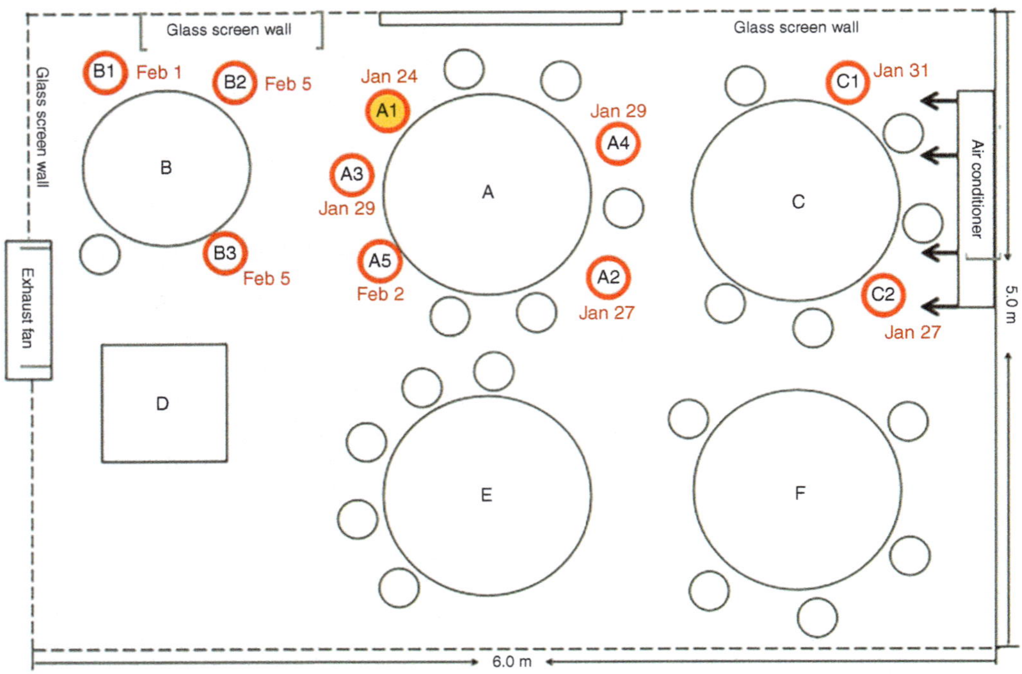

Figure 4.6 January/February 2020 COVID-19 cluster in Guangzhou Restaurant associated with a lunch on January 24. The initial (index) case is noted at Table A, and other patrons subsequently diagnosed are noted with the date of onset of symptoms. *Source:* [17]. Centers for Disease Control and Prevention. Public Domain.

Figure 4.6, the cases occurring subsequent to a luncheon at a restaurant in the same room as the presumed index case (initial source of infection) are noted.

There have also been well-documented secondary cases resulting from norovirus in households [18] and potentially from individuals infected with tuberculosis on aircraft [19]. However, not all pathogens are contagious; if they are not amplified and can exit the body, for example, there is no evidence of contagiousness for *Bacillus anthracis* or *Legionella pneumophila*.

An approach generally taken to describe transmission in a large population is to develop a disease transmission model describing proportions of the population who are susceptible, exposed, infected, and recovered, which may be embellished by further states and processes [20]. There are interchanges in persons between the states that may include the following:

- Susceptible individuals may become exposed either by contact (direct or indirect) with infected persons or from overall environmental exposure. This is linked to the concept of dose–response modeling.
- Exposed individuals, after suitable incubation, may become infected or may recover. If they recover, the recovered individuals may again become susceptible.
- Mortality may reduce the population.
- Due to migration or births, there may be an introduction or depletion of individuals in any state.
- The introduction of vaccination, public health policies, or therapies may alter the dynamics between states.

This can lead to rich dynamic models that can be informative. The *dynamics* with which illnesses occur can influence the total consequences of the exposure in a population. This will be expanded upon in Section 14.3.

4.6 Key Takeaways

To properly treat pathogens as pollutants in an environmental engineering context, the key elements described in this chapter must be taken into account. In later chapters, these will be integrated into a detailed consideration.

4.7 Discussion Questions and Problems

1 For each of the groups of 10 portions in Figure 4.4, compute the standard deviation between portions. How does that compare to what would be expected for the Poisson distribution with that mean?

2 Consider the exhaled liquid volume from Figure 4.5 for normal breathing.
   a Assuming an emission rate of pathogen RNA of 15,000 copies in 30 minutes (see Table 4.1), what would the expected concentration in the emitted liquid be?
   b If an aerosol particle were 2μm in diameter, how many RNA copies would be expected in one such particle (assuming no preferential selection)?

3 Water is sampled (1 mL samples) over time. If 100 samples are taken, and two are found to be positive for an organism, what is an estimate for the mean concentration if the Poisson distribution is assumed? What other assumptions need to be made to do this calculation?

References

1 Hermann M Bolt. Vinyl chloride—a classical industrial toxicant of new interest. *Critical Reviews in Toxicology*, 35(4):307–323, 2005.
2 Wesley O Pipes, Pamela Ward, and S H Ahn. Frequency distributions for coliform bacteria in water. *Journal (American Water Works Association)*, 69(12):664–668, 1977.
3 Jean Maguire Van Seventer and Natasha S Hochberg. Principles of infectious diseases: Transmission, diagnosis, prevention, and control. In Stella R Quah, editor, *International Encyclopedia of Public Health*, pages 22–39. Academic Press, 2nd edition, 2017.
4 Josep M Pujol, Joseph E Eisenberg, Charles N Haas, and James S Koopman. The effect of ongoing exposure dynamics in dose response relationships. *PLoS Computational Biology*, 5(6):e1000399, 2009.
5 Zdenek Hubálek. Emerging human infectious diseases: Anthroponoses, zoonoses, and sapronoses. *Emerging Infectious Diseases*, 9(3): 403–404, March 2003.
6 N J Ashbolt. Environmental (Saprozoic) pathogens of engineered water systems: Understanding their ecology for risk assessment and management. *Pathogens*, 4(2):390–405, June 2015.
7 W J Wilson. Isolation of enteric bacilli from sewage and water and its bearing on epidemiology. *British Medical Journal*, 2(3794):560–562, 1933.

8 Ryan G Sinclair, Christopher Y Choi, Mark R Riley, and Charles P Gerba. Pathogen surveillance through monitoring of sewer systems. *Advances in Applied Microbiology*, 65:249–269, 2008.

9 Colleen C Naughton, Fernando A Roman, Ana Grace F Alvarado, Arianna Q Tariqi, Matthew A Deeming, Krystin F Kadonsky, Kyle Bibby, Aaron Bivins, Gertjan Medema, Warish Ahmed, et al. Show us the data: Global COVID-19 wastewater monitoring efforts, equity, and gaps. *FEMS Microbes*, 4:xtad003, 2023.

10 Thomas Eiche and Martin Kuster. Aerosol release by healthy people during speaking: Possible contribution to the transmission of SARS-CoV-2. *International Journal of Environmental Research and Public Health*, 17(23):9088, December 2020.

11 Jing Yan, Michael Grantham, Jovan Pantelic, P Jacob Bueno de Mesquita, Barbara Albert, Fengjie Liu, Sheryl Ehrman, Donald K Milton, and EMIT Consortium. Infectious virus in exhaled breath of symptomatic seasonal influenza cases from a college community. *Proceedings of the National Academy of Sciences*, 115(5):1081–1086, January 2018.

12 Kristen K Coleman, Douglas Jie Wen Tay, Kai Sen Tan, Sean Wei Xiang Ong, The Son Than, Ming Hui Koh, Yi Qing Chin, Haziq Nasir, Tze Minn Mak, Justin Jang Hann Chu, et al. lViral load of severe acute respiratory syndrome coronavirus 2 (SARS-CoV-2) in respiratory aerosols emitted by patients with coronavirus disease 2019 (COVID-19) while breathing, talking, and singing. *Clinical Infectious Diseases*, 74(10):1722–1728, May 2022.

13 James E Farnsworth, Sagar M Goyal, Seung Won Kim, Thomas H Kuehn, Peter C Raynor, M A Ramakrishnan, Senthilvelan Anantharaman, and Weihua Tang. Development of a method for bacteria and virus recovery from heating, ventilation, and air conditioning (HVAC) filters. *Journal of Environmental Monitoring*, 8(10):1006–1013, 2006.

14 Jin Pan, Seth A Hawks, Aaron J Prussin, Nisha K Duggal, and Linsey C Marr. SARS-CoV-2 on surfaces and HVAC filters in dormitory rooms. *Environmental Science & Technology Letters*, 9(1):71–76, January 2022.

15 Denina Hospodsky, Jing Qian, William W. Nazaroff, Naomichi Yamamoto, Kyle Bibby, Hamid Rismani-Yazdi, and Jordan Peccia. Human occupancy as a source of indoor airborne bacteria. *PLoS ONE*, 7(4):e34867, April 2012.

16 Anna Kozajda, Karolina Jeżak, and Agnieszka Kapsa. Airborne Staphylococcus aureus in different environments—A review. *Environmental Science and Pollution Research*, 26(34): 34741–34753, December 2019.

17 Jianyun Lu, Jieni Gu, Kuibiao Li, Conghui Xu, Wenzhe Su, Zhisheng Lai, Deqian Zhou, Chao Yu, Bin Xu, and Zhicong Yang. COVID-19 outbreak associated with air conditioning in restaurant, Guangzhou, China, 2020. *Emerging Infectious Disease Journal*, 26(7):1628–1631, 2020.

18 Z A Marsh, S P Grytdal, J C Beggs, E Leshem, P A Gastanaduy, B Rha, M Nyaku, B A Lopman, and A J Hall. The unwelcome houseguest: Secondary household transmission of norovirus. *Epidemiology and Infection*, 146(2):159–167, 2018.

19 Cynthia R Driver, Sarah E Valway, W Meade Morgan, Ida M Onorato, and Kenneth G Castro. Transmission of mycobacterium tuberculosis associated with air travel. *JAMA*, 272(13):1031–1035, 1994.

20 Jonathan L Zelner, Aaron A King, Christine L Moe, and Joseph N S Eisenberg. How infections propagate after point-source outbreaks: An analysis of secondary norovirus transmission. *Epidemiology*, 21(5):711–718, September 2010.

5

Pathogen Sampling

Pathogens can be measured in a variety of settings, including:

- Wastewater or raw sewage
- Drinking water
- Indoor air
- Ambient air
- From fabrics or filter media
- From dusts
- From surfaces
- In solids or semisolids, such as sludges, soils, manures, or waste materials

While the details of handling each type of setting may differ, the common features of the workflow will first be discussed, followed by peculiarities for each setting — which occur primarily at the early stages of the workflow. This chapter presents details on how pathogens are sampled from various environments. In Chapter 6, the methods for quantifying those pathogens will be discussed. This quantification will enable the environmental engineer to use the information to develop risk estimates, validate, design, and implement control techniques.

5.1 Workflow

Regardless of the particular pathogen(s) and the sampled environment, the measurement workflow has many commonalities. This is illustrated in Figure 5.1. Sampling from the environment is the first step. Once taken, the sample(s) may need to be concentrated, extracted, and/or subject to an initial decontamination step. Subsequently, samples may be taken for culture (viability assessment) and/or for molecular analysis (details of which are the focus of Chapter 6); these two alternatives have somewhat different approaches and requirements. The details of these steps follow.

5.2 Taking the Sample

Most frequently, the objective of a sampling program is to get a representative measurement of the concentration of pathogens in a medium. Representativeness may be both with respect to space and time. To understand progress toward a goal, the nature of factors that may contribute to the diversity of pathogens across samples should be understood. This can be particularly challenging when there is strong spatial and temporal clustering [1]. The difficulty can also be seen by reference to

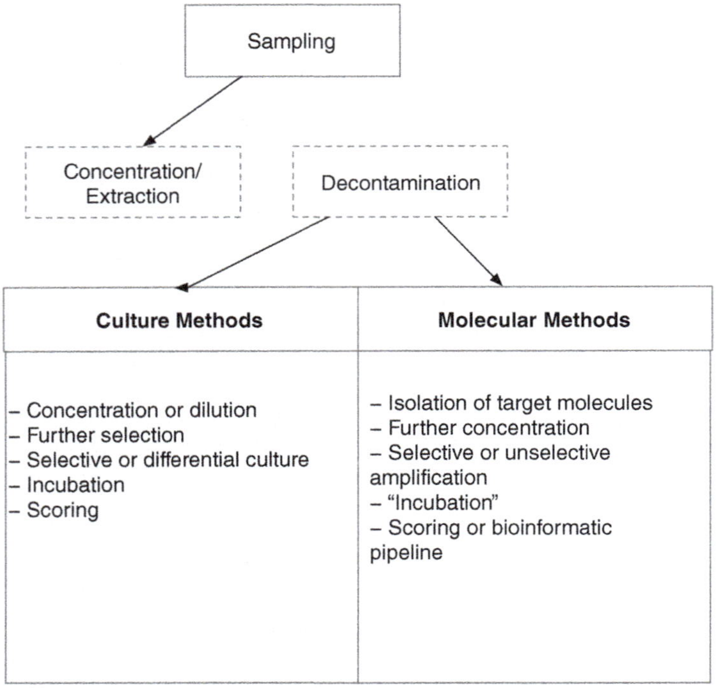

Figure 5.1 Conceptual workflow pipeline for methods.

Figure 4.2—compare the likelihood of finding pathogens in two aliquots in the case of overdispersion, random dispersion, and underdispersion. The issue of capturing variability will influence whether single "grab" samples or short-term samples can be used, whether an intensive set of grab samples is used, or whether long-term sampling or compositing is feasible. The particular strategy will be media-specific.

Another overarching concern in sampling strategy is whether the final analyses will be via culture, molecular methods, or both. Different sampling methods may influence the persistence of various analyte types during collection and handling, even before subsequent steps.

While most sampling occurs by examination of a known or measured collected amount of material from the environment, an alternative valuable approach in some circumstances is "passive sampling." In this approach, a material is placed in the environment, and intrinsic diffusive, gravitational, or convective fluxes bring it into contact with environmental pathogens, which are then collected by the material. The details of these systems are medium-specific and are discussed in that context.

5.2.1 Wastewater

The role of raw sewage in transmitting pathogens from excreta was known by the latter decades of the 19th century [2]. In the first decades of the 20th century, typhoid bacilli were definitively isolated from sewage and waters, and the relationship to cases was hypothesized [3]. By the early 1940s, in connection with outbreaks of poliomyelitis, the virus was found in sewage and connected to community prevalence [4]. More recently, with the greater use of "live" poliovirus vaccine,[1] the uptake of vaccine in the population could be tracked by monitoring for the vaccine virus strains in wastewater [5].

Frequent wastewater monitoring for non-polio enterovirus occurred in Milwaukee from 1994 to 2002 [6]. This program showed good correspondence between the frequency of isolation and the clinically reported illness frequencies, as well as between the types of viruses isolated from wastewater and those in clinical cases.

Given this experience, it was unsurprising that with the onset of the COVID-19 pandemic, wastewater sampling commenced early [7]. This has progressed to a vast international network of monitoring sites [8] and has been regularized to several formal governmental networks.

A wastewater system is a complex network of pipes connecting individual buildings, feeding into progressively larger pipes. Depending on the community's topography, intermediate pumping stations may convey wastewater to higher elevations, where it ultimately flows into a wastewater treatment plant. A simple network is shown in Figure 5.2.

Individual sewer lines may serve a variety of buildings. Then, the individual lines combine to meet or form larger pipes (often called interceptors). They may be further joined into larger and larger networks until they are finally conveyed to the wastewater treatment plant (shown at upper right). The entire network of buildings and other generators (offices, parks, etc.) contributing to the system is termed a "sewershed." Individual sections of this network may be called a "subsewershed" — for example, all of the buildings draining into the pipe and collecting at point "A."

Multiple collection points for a system such as this may be considered:

- Sampling at an individual building or facility.
- Sampling at a location within the sewershed (points "A" or "B").

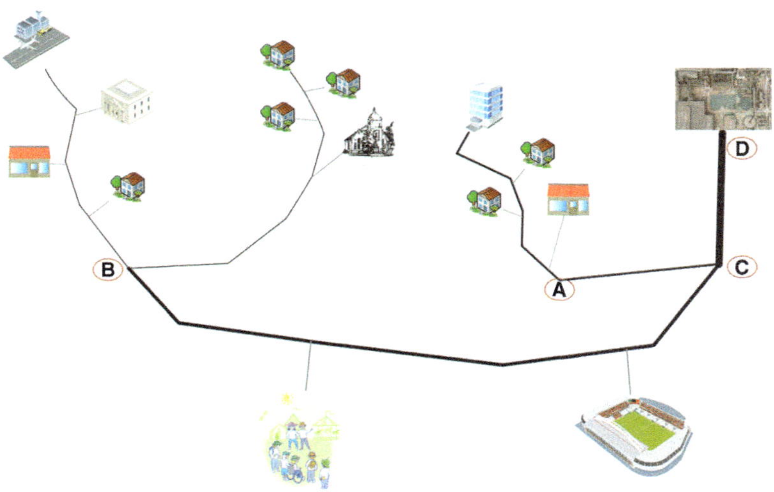

Figure 5.2 Schematic of an idealized sewer network. Labels suggest potential sampling locations as discussed in text. The thickness of lines suggests the diameter of pipes as flow goes to a central treatment plant – at the upper right. *Source*: (A) Clker-Free-Vector-Images / Pixabay. (B) Quique / Pixabay. (C) OpenClipart-Vectors / Pixabay. (D) OpenClipart-Vectors / Pixabay.

---

1 The initial polio vaccines used inactivated virus — the "Salk" vaccine. This was supplanted by the use of live, nonpathogenic polio strains for vaccination — the "Sabin" vaccine.

- Sampling within the main interceptor before conveyance to a wastewater treatment plant (point "C"). This might be at a pumping station where sewage is collected before final conveyance.
- At the influent of the wastewater treatment plant (point "D").

The logistics of sampling at each of these locations would be different.

Several other details of sewer networks should be understood. In many cases, sewers are designed to carry only wastewater generated by buildings, occupied spaces, and their residents. These are called "separate" sewer systems. The sewerage may contain contributions from industries, hospitals, and other facilities beyond human excreta and secretions. There is another piping network designed to carry storm drainage and runoff.

In some cases, especially in older cities, sewers are intentionally designed to carry drainage from roads and streets; these are called "combined" sewer systems. Even in separate sewer systems, leaks and precipitation infiltration can occur. In combined systems, road and street runoff may contain animal waste (pets, wild) that can harbor pathogens [9]. In addition, in areas with informal settlements of unhoused people, uncontrolled defecation may contribute human waste to sewers (as well as to adjacent watersheds) [10]. Therefore, any sewer sampling program must consider the waste sources and their variability that contribute to flow and microbial loads.

A second aspect of wastewater collection systems is the variability in flow rates over multiple time scales. With combined systems, the precipitation and drainage of stormwater can result in a significant increase in flows during wet weather compared to dry weather. However, even in separate or combined systems during dry periods, fluctuations can occur at multiple scales. These fluctuations can be due to ordinary variations in domestic water consumption and wastewater generation rates (showering, dishwashing, clothes washing, etc.) and to systematic variations among industrial and commercial users. An old but still commonly used equation [11] for the ratio of peak hourly flow to average daily flow (termed "peaking factor") is given by Equation 5.1:

$$\frac{\text{Peak} \quad \text{Hourly} \quad \text{Flow}}{\text{Average} \quad \text{Daily} \quad \text{Flow}} = \frac{18 + \sqrt{P}}{4 + \sqrt{P}} \tag{5.1}$$

where P is the population in thousands of persons. There is a similar peaking of pollutant concentrations relative to averages, but, particularly for pathogens, there is no comparable empirical relationship with the peaking factor.

The relationship between the peaking factor and population size indicates that in smaller systems and in small segments of an extensive system (e.g., point A in Figure 5.2), there will be greater variability than in large systems or the system as a whole.

Therefore, there has been discussion about the value of grab sampling (where single samples are collected on an infrequent — daily or less frequent — basis) versus composite sampling. In composite sampling, frequent samples of a system are collected over time, and the samples are pooled to form a mixture for analysis. It is easiest to perform time-weighted compositing, in which the same sample volume is taken at each time point (e.g., 15-minute intervals). When flow metering is available, such as at a wastewater treatment plant, flow-weighted compositing can also be performed, whereby more frequent or larger-volume samples are collected during higher-flow periods. While this may be correct for hydraulic variations, flow-weighted compositing may not be correct for concentration variations.

A downside of compositing is that samples must be stored for the time required to composite, which may allow some degradation of the material being analyzed [12]. While there have not been exhaustive tests for many microorganisms, with SARS-CoV-2 RNA as an endpoint, it has been found that the results from grab sampling and composite sampling were comparable [13]. Another downside of automatic composite samplers used in unstaffed locations is the potential vulnerability of the expensive instrument to theft or damage.

The smallest sampling scale is generally a single building. The details of where and how to sample in such settings are highly context-specific — it may be that a single point of access can be obtained in the basement of a building or at a maintenance hole.[2] By opening the hole cover, samples can be taken by threading a hose into the flow as in Figure 5.3, or by inserting a collection bottle or beaker via a long pole; grab samples may be taken. Parenthetically, it should be noted, especially when sampling is done at the building or comparable small scales, that there are substantial privacy and ethical considerations that arise [14, 15]. These issues should be addressed at the outset of any sampling campaign conducted at this scale.

During the early months of the SARS-CoV outbreak, investigators at the University of Arizona sampled wastewater from dormitories [16]. By focusing follow-up clinical testing on individuals in dormitories that had positive wastewater samples, they were able to conserve testing resources.

At a somewhat larger scale, sampling at a neighborhood level can be done if collection occurs at a point where the sewage from multiple buildings is combined. This can be at a maintenance hole, as in Figure 5.3

An example of this is provided by Oh et al. [17] who sampled at the neighborhood scale (<1 km) in multiple areas of Champaign County, Illinois, during periods of low incidence. They used autosamplers at sewers triggered to sample four-hour composites over a four-day period and found good relationships between detection probability and case incidence rate.

Sampling at the subsewershed (and building) scales raises practicalities that are not often discussed in the literature. Sewage contains all waste discharged by residents and occupants. Some materials may be coarse and/or abrasive. Therefore, to get a representative sample, it is necessary to use large-diameter tubing or bulk sampling with large containers. In addition, in the subsequent steps (decontamination and initial processing), the sample must either be settled or thoroughly homogenized. This becomes less of an issue with samples collected farther into the sewer network

Figure 5.3 Sampling from a sewer maintenance hole. *Source:* Centers for Disease Control and Prevention / Public domain / https://www.cdc.gov/nwss/wastewater-surveillance.html / last accessed Feb. 12, 2026.

2 This term is adopted instead of the historical and gendered term, "manhole," which should be regarded as deprecated.

(or at wastewater treatment plant influents) because the turbulence and agitation in sewage conveyance allow some mixing and homogenization.

A number of university-scale wastewater sampling campaigns for SARS-CoV-2 were initiated during the pandemic. Wartell et al. summarize some key lessons learned, which have generalizability to all subsewershed sampling efforts, including the following:

- Sewer maps may not be current and may receive wastewater from multiple buildings.[3]
- Access points may be in or near high-traffic (pedestrian or automotive) areas, which can be problematic for implementing collection.
- If battery-operated autosamplers are used (due to lack of power connections), recharging or replacement may need to be frequent, which is difficult under harsh weather conditions.
- If tubing is used for sample collection, strainers may be needed to avoid collecting abrasive debris, and the removed material may frequently clog the strainers.

Some universities (which are *de facto* small communities) can be sampled at a centralized facility. Sampling at the University of Notre Dame using such an approach tracked the reduction in positivity with the onset of the first waves of COVID-19 vaccination [19].

There are other locations where individuals reside for prolonged periods that can serve as sites for wastewater surveillance. These include prisons [20], nursing and retirement facilities, and hospitals [21].

Sampling directly at the influent of a facility is more straightforward in several respects. First, there is often dedicated power and protected spaces so that autosamplers can be used to composite samples. Second, the waste emanates from a large catchment area. Third, operational personnel are already experienced with collecting samples from all points, including raw wastewater.

However, particular details need to be understood in implementing such a program, in particular, where it is best to collect samples of the sewershed and any flow fluctuations that need to be considered. Many large cities responded to the challenges posed by SARS-CoV-2 by implementing centralized monitoring. In September 2020, the US Centers for Disease Control and Prevention established the National Wastewater Surveillance System (NWSS) as a response [22]. NWSS has maintained an updated website outlining best practices for this program (https://www.cdc.gov/nwss/sampling.html). Several of the critical points on wastewater sampling that have been identified in this program include the following:

- To detect trends, weekly sampling may suffice, although more frequent sampling may permit more rapid trend identification.
- Sampling may be from the raw wastewater or possibly preliminary treatment.[4] Alternatively, some have found it useful to sample from sludge from primary sedimentation [23, 24], which has been found to have higher concentrations of viral RNA, provides reasonable correlation to health status in the community, and in some sense provides a time-integrated sample that has been partially purified by the sedimentation process.[5]

---

3 In some locations, illicit connections to sewer systems have been known to occur [18].

4 Many wastewater treatment plants have bar screening, comminution, or fine screening of the raw sewage as an immediate step.

5 However, these findings should be verified with other pathogens. Also, sampling needs to account for the fact that at some wastewater treatment plants, primary sedimentation is the point at which certain other residuals are recycled back to treatment, such as supernatants from anaerobic digestion. In addition, wastewater treatment plants that receive septage frequently introduce this upstream of primary sedimentation.

- Sample volumes are typically 1 L of composite wastewater, 100 mL of wastewater grab samples, or about 2 g of primary solids.
- Samples should be stored under refrigerated conditions for up to 24 hours before processing — including time for shipping from sample collection to the laboratory. These conditions should also be verified with the particular pathogens of interest.

Another sampling approach is the use of a passive sampler (in the wastewater context, this is called a "Moore swab" [25]). Bivins et al. have done a systematic review in the particular context of SARS-CoV-2. They note the long history of such sampling:

> "Passive sampling is premised on the spontaneous exchange of an analyte between the bulk medium being sampled (i.e., wastewater) and a collecting medium (i.e., the passive sampler).... This exchange is mediated by immersing the sampler within the sampled medium, hence the term "passive" sampling (without active response or resistance) although it has also been referred to as "trap" sampling...
>
> Using passive samplers for disease surveillance via wastewater dates to the earliest days of bacteriology. Following the invention of selective media to isolate *Salmonella typhi*... Many of these early efforts were focused on linking downstream cases of typhoid fever with upstream sources of fecal contamination. In the first account of passive sampling for *S. typhi*."

Their review indicated that, across the four studies with side-by-side comparisons of passive wastewater sampling and composite sampling, the viral RNA signal was comparable. This strongly suggests that passive sampling may be well-suited to standardization, especially in resource-challenged settings. However, work remains to be done on optimizing collection media, whether a single medium would be near-optimal for a broad spectrum of pathogens, and the mechanical means by which the sampler contacts a flowing stream.

Wastewater can also be sampled in other venues than sewers or treatment plants. One promising avenue is lavatory waste from airplanes, especially those used on long-distance flights [26, 27]. This may be a particularly useful place to detect novel agents imported into a new venue. Sampling of these systems is similar to sampling from individual buildings, except that the waste is of higher strength and has often been treated with antimicrobial compounds for holding.

5.2.2 Drinking Water, Other Liquids

Other environmental liquids may be sampled for pathogens. Perhaps the most important of these are source waters used for drinking water. It is rarely informative to sample treated drinking water for pathogens because of the very low concentrations expected. However, sampling for microbiological indicators may be helpful.

The presence of bacterial pathogens in source waters used for drinking water was known in the early 20th century [28]. However, due to methodological difficulties, for most of the last century, microbiological quality of drinking water focused on the use of what are now called coliform organisms as indicators of acceptability [29]. When drinking water or source water (river, lake, stream, well) is to be sampled for coliform, generally, grab, samples of 100 mL to 1 L are taken for analysis.

In the 1950s, it became increasingly apparent that the focus on bacterial contamination did not adequately address the risks posed by the potential presence of viral pathogens. Melnick et al. [30] review the early literature on virus sampling. Unlike coliform and bacterial pathogen sampling in drinking water systems, viral concentrations are expected to be lower. Therefore, concentration at the sampling point is generally needed (except in highly polluted systems).

While concern for protozoan pathogens existed from an early date, with the recognition that the agent of amoebic dysentery (*Entamoeba histolytica*) could be waterborne, interventions (disinfection, filtration) generally achieved good control of these pathogens. Thus, further concern about waterborne pathogens did not arise until the late 1970s, with the recognition that *Giardia* is a waterborne pathogen that prior treatment approaches might not sufficiently remove [31]. Somewhat later on, *Cryptosporidium* emerged as a significant waterborne protozoan pathogen [32]. As with viruses, detecting protozoan in source waters, and even more so in drinking waters, requires concentration from large volumes at the sampling point so that tens to hundreds of liters of water can be processed for concentration.

In the drinking water context, because source waters are expected to have low concentrations of viruses, the concentration is expected to be relatively low after treatment. Thus, the practicality of using direct pathogen measurements for process monitoring is unrealistic. Instead, given the source water concentrations, the individual processes used to reduce pathogens are controlled by monitoring their characteristics (e.g., turbidity, disinfectant residual), and the degree of removal expected is projected to be that required to reduce the risk [33]. This has been embodied in US regulations on drinking water treatment.

After water treatment, water enters a network of pipes, where it is distributed to individual buildings and users. A simplified distribution system is shown in Figure 5.4. A real system may consist of multiple "legs" of these pipes, which could be connected in a branched or looped fashion, and there may also be tanks, reservoirs, or towers at various locations within the distribution system to maintain pressure and to supply emergency water in the event of firefighting needs or outages. In such a complex network, computer simulations are used to assess the flow and pressure distributions and chemical reactions (including disinfectant decay and potentially microbial growth) under different conditions. EPANET is a popular and open-source platform for executing these simulations [34]. Such simulations would be valuable in assessing where to sample.

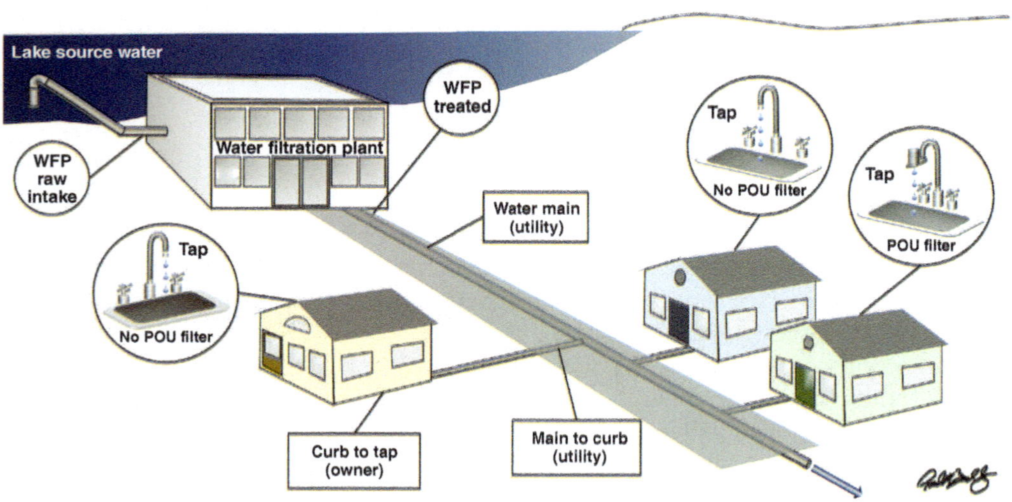

Figure 5.4 Simplified water distribution system. *Source:* USGS / Paul M Bradley / Public domain / https://www.usgs.gov/media/images/idealized-diagram-a-publi c-water-supplys-water-distribution-system / last accessed Feb. 10, 2026.

Water distribution system samples are generally obtained from individual buildings. By letting building taps run until the temperature stabilizes (and verifying the absence of any intermediate storage tanks between the distribution system and the tap), distribution system microbial quality samples can be obtained, as noted previously, for indicator organisms, such as coliforms. Smaller-volume samples can be taken. In contrast, for pathogen samples (particularly viruses and protozoa), it may be necessary to filter large volumes of water on-site.

Finished water, or water at building taps, generally contains detectable disinfection residuals (chlorine, chloramine). These may cause continued inactivation of microorganisms after the sample is collected and before analysis. Therefore, when sampling finished drinking water (including at user taps), the sample should be immediately mixed with a sterile dechlorinating agent to remove any residual chlorine. The most typical agent is sodium thiosulfate ($Na_2S_2O_3$).

As shown in Figure 5.4, individual buildings have (smaller diameter) pipes branching from the distribution system—these are called "service lines." While the distribution system itself is owned and operated by the utility, only a portion of the service line (shown in gray in the figure) is owned and operated by the utility, whereas the portion closer to the building is owned by the property owner. Corrosion of service lines can release lead or copper, as in Flint, Michigan [35], along with potential regrowth of *Legionella* bacteria [36]. Hence, there is increasing interest in sampling for water quality changes that may occur within the service lines themselves.

The water distribution itself should not be regarded as a sealed, biologically inert conveyance but instead can be the locus of many chemical and biological changes.

In some cities, water distribution systems can be old (75 years or longer) and deteriorate over time, leading to leakage and infiltration. This can occur due to internal or external corrosion, mechanical damage from construction, excessive road traffic, or seismic activity; damage from sudden freeze-thaw events; or simply the accumulation of routine deterioration over time. National studies of distribution systems estimate an average failure rate (defined as breaks in a main that require repair or replacement) of 0.14 breaks/mile/year [37]. Break rates vary with material, age, diameter, and region of the United States (or Canada) — and appear to be increasing with time.

Before a main break, there may be undetectable leaks that allow infiltration. While the goal of operating a distribution system is to maintain it pressurized to the surroundings (and to permit flow into individual user buildings), there may be short-term transients that occur, and these can lead to intrusion of materials from the surroundings, including pathogens. Maintaining a disinfection residual reduces the risks from these transients, at least for the more disinfectant-sensitive pathogens [38].

5.2.2.1 Backflow and Cross-connections

After entering a distribution system, water may become contaminated through cross-connections and backflow. "Cross-connections are actual or potential connections between a potable water supply and nonpotable water plumbing. Backflow is the unintended reversal of water flow through a cross-connection, which can result in a potentially severe public health hazard [39]." These situations can be reduced by installing backflow prevention devices that block reverse flow into distribution systems. While this would protect systems upstream of such devices, there remains the potential for contamination downstream (particularly within individual buildings).

Some canonical examples of cross-connections or backflows are the following:

- Within a building, there are inadvertent direct connections between potable water piping and either waste piping or a nonpotable supply (possibly rainwater or reclaimed wastewater being reused).

- At an individual tap, there is an unintended siphonage of contaminated water into the building service line during negative pressure events. This can occur, for example, if a user attaches a hose to the supply line and allows it to dangle into a pool of contaminated water.
- A water distribution system may encounter negative pressure (with respect to surroundings) under conditions where there is a main break or a very heavy use of water during a large fire, which may result in backflow conditions.

When a cross-connection or backflow is suspected, sampling should rely on detailed hydraulic knowledge to locate the point(s) at which contamination occurs and sample upstream and downstream of the source. Often, however, these phenomena are transient, while setting up for sampling requires time for deployment, and therefore microbiological sampling is of limited value in many cases.

5.2.2.2 Groundwater

Groundwater is the source of drinking water for nearly 40% of the US population [40]. It may be used for individual buildings, small clusters of residences, or an entire community. In the latter case, water may be drawn from multiple wells for treatment. There may be little treatment for individual buildings and very small communities.

A schematic of a simple residential well is shown in Figure 5.5.

Upon examination, there are numerous points at which, even if the aquifer were protected from contamination, degradation of quality could occur. Without intensive effort, sampling directly from

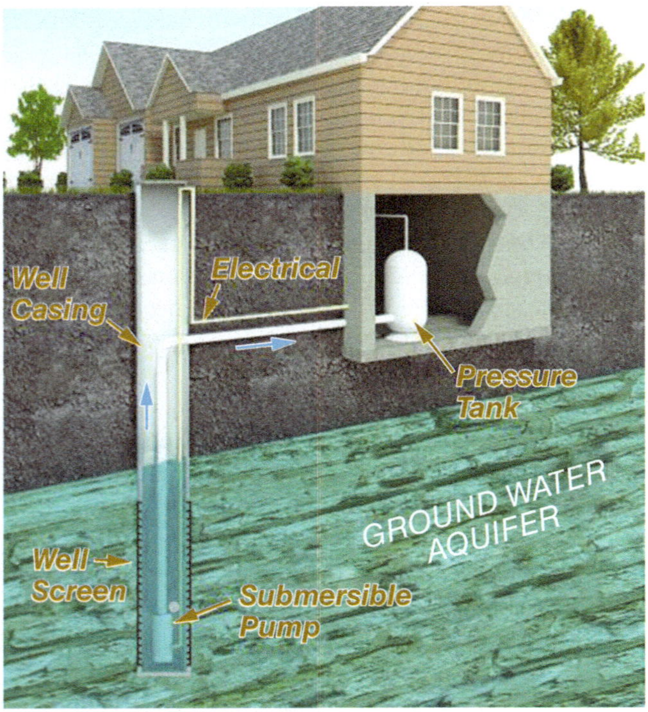

Figure 5.5 A simple residential well. *Source:* USGS / Public domain / https://www.usgs.gov/media/images/a-home-or-city-wells-generally-work-same-way / last accessed Feb. 10, 2026.

the aquifer is difficult. Furthermore, individual wells and wells serving fewer than 25 connections are not subject to federal or many state regulations, although relevant local sanitary codes may exist. Additionally, depending on the chemical water quality, microbial growth may occur on well casings, screens, and appurtenances, including pathogens [41].

Even community groundwater systems may not be required to treat other than by disinfection. As a result, outbreaks have occurred in such systems.

Hynds et al. performed a systematic review of groundwater systems in the United States and Canada contaminated by enteric pathogens [42]. Overall, septic tank and sewage system infiltration into groundwater was the most prevalent source of pathogen contamination, whereas agriculture/livestock and wild animal contamination were minor sources. The pathways of contamination were, in the majority of cases, geological (karst, fractured or unconsolidated formations), precipitation (causing infiltration into the aquifer), and source-specific (inadequacies in the design and construction of well casings and wellheads). Pathogens of all types were noted.

Aside from the complexity of determining sampling locations, the details of groundwater sampling are the same as those for drinking water in general. For pathogens, large volumes may need to be collected (in excess of 1 L) or filtered on-site using filters designed to remove pathogens and then transported to a laboratory.

5.2.2.3 Rainwater Collection

There is a long history of rainwater collection for water supply, especially in more water-scarce, under-resourced countries. However, in recent years, intense interest has also emerged in developed areas that are not water-stressed, where the benefits of avoiding urban stormwater runoff are sought [43].

A rainwater collection system, whether from an individual residence, as in Figure 5.6, or from a larger building, consists of the collection of rain impinging on roof surfaces and conveyance through initial devices into a storage tank. The devices are designed to screen out larger debris and keep out insects, wild animals, and other vermin.

While even rainwater itself may not be genuinely "pristine" (due to scrubbing of particles from the atmosphere, as well as solubilization of gaseous constituents), rain impinging on a surface may further degrade in quality due to contact with roofing materials and carriage of materials deposited from the atmosphere and from plant and animal contamination. Typically, the "first flush" of roof runoff is more heavily contaminated and is therefore diverted away from storage [44]. The collected rainwater may then be used for irrigation, for separate indoor piping for toilet flushing, or, with appropriate treatment, for potable or nonpotable use.

Further degradation may occur due to defects in storage, microbial growth in storage, and from holes or breakage of systems to control animal entrance [45–47]. Therefore, it may be of interest to sample from various storage locations and devices to assess the potential for exposure. Techniques similar to drinking water source sampling are appropriate, although the particulate loading may be greater for roof-harvested rainwater.

5.2.2.4 Premise (Building) Plumbing

The internal building or premise plumbing system can be complex, even in small buildings. As depicted in Figure 5.7, there may be both a hot water loop and cold water circulation within a building. Any disinfectant residual entering the distribution system often dissipates in one or both lines due to reactions with plumbing materials [48] and to holding time on the premise. Especially when water use is low (such as during absences or vacations), there may be an opportunity for pathogen growth due to stagnation. Maintaining a minimum temperature in the hot water line will

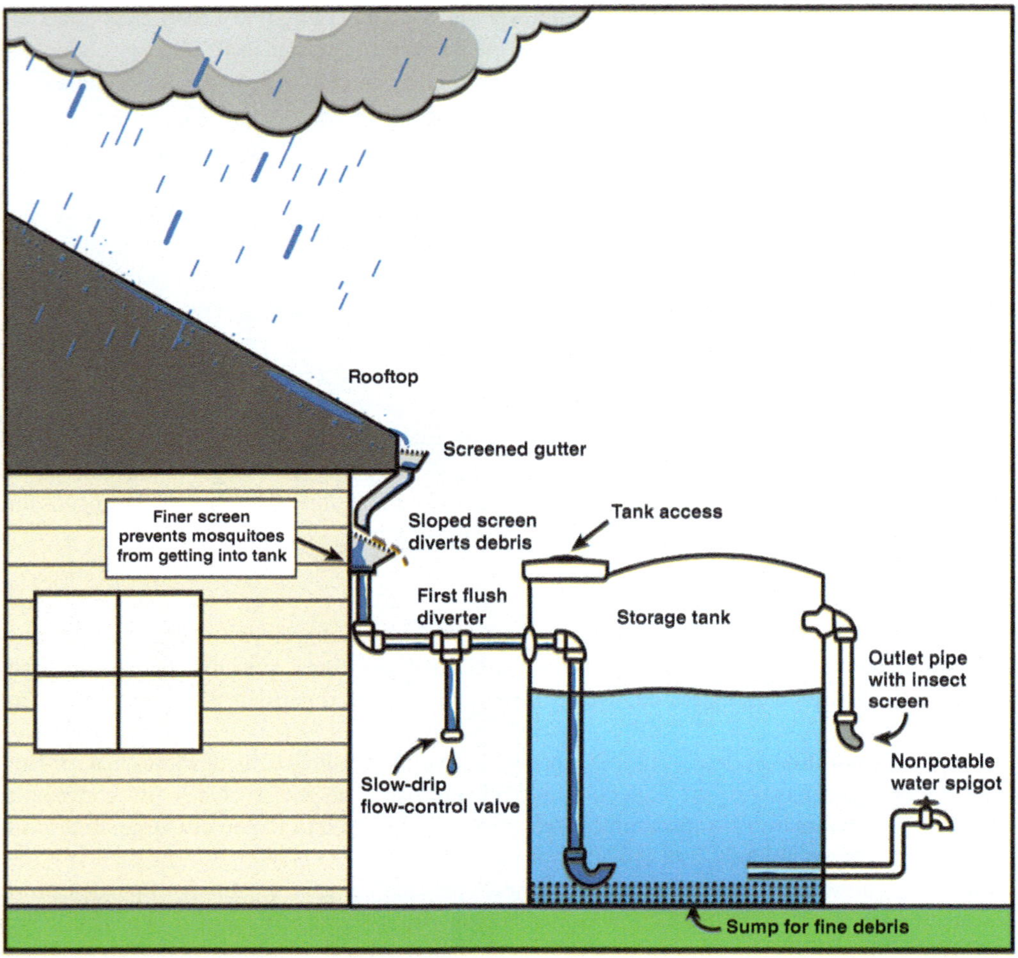

Figure 5.6 Residential rainwater harvesting system. *Source:* [47]/ North Carolina State University / https://content.ces.ncsu.edu/show_ep3_pdf/1702757536/25034/last accessed Feb. 12, 2026.

reduce the potential for such growth [49]. In addition, building plumbing is of a lesser diameter than the piping in water mains, therefore presenting a larger surface area to volume ratio, with more opportunity for the growth of biofilms [50].

Note that within a building, there are opportunities for aerosolization of water (with its included contaminants) through showers, faucet aerators, toilets, and other uses. This will represent a driver for indoor air sampling. Especially in commercial buildings, there may be other uses (including aerosol-generating uses), such as ornamental water fixtures and fountains, "green walls," misters (for produce or humidification), waterfalls, etc.

Hence, a sampling program for individual buildings must take account of the detailed building plumbing configuration as it exists, noting that this may differ due to renovation and modification from the initial building plans. Once sampling points are identified, samples can be taken, as in the case of drinking water sampling from a distribution system.

If water is drawn immediately from a cold water tap within a premise, especially after a period of little or no use, the sample will be from water that has been within the premise piping itself. However, if the tap is left running for a period of time, water will be drawn from the distribution system

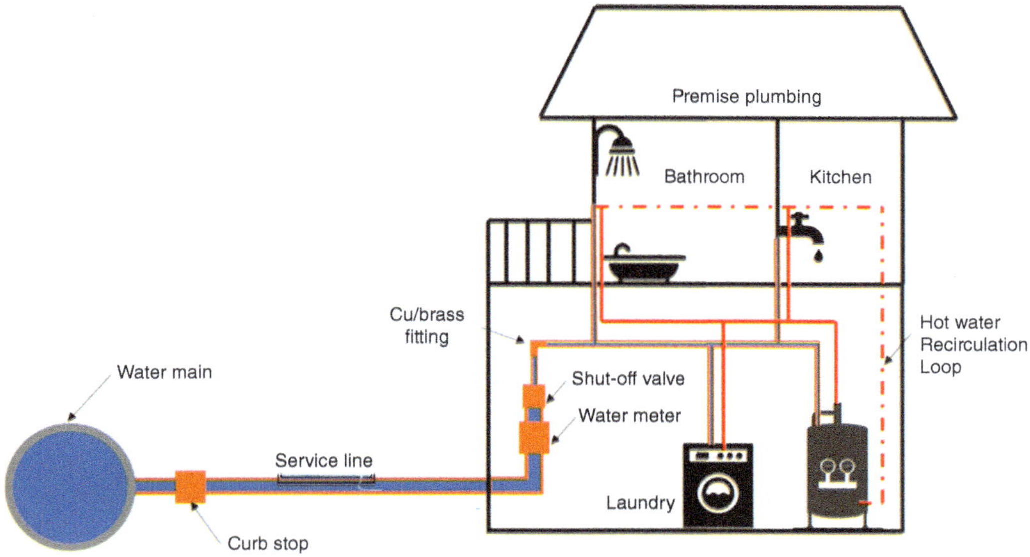

Figure 5.7 Schematic of a residential water system. *Source:* Modified from [50].

to the premise tap. The progression of water from the local piping to water drawn more directly from the distribution main can be determined by following a time progression of temperature or disinfectant residual. When either of these signals shows the attainment of a new steady state, the water can be inferred to have been drawn from a distribution system.

When sampling from individual premises, a critical detail is the exact configuration of the terminal fixture from which water is drawn [51]. Faucets (on sinks and showers) may have mixing valves that blend water from the cold and hot water lines to achieve a desired intermediate temperature—therefore, a mixed sample from both lines may exist. Further, such valves may have "dead volumes" of stagnant water that can foster pathogen growth and must be cleared to obtain a more accurate sample of the system water. Shower heads and hoses, often infrequently cleaned, can also foster biofilms and contain "dead volumes" permissive of pathogen growth. Therefore, persons sampling from individual building locations must maintain cognizance of the terminal fixtures and, if possible,[6] avoid such sites.

5.2.3 Recreational Waters

Recreational waters comprise a broad spectrum of venues, as shown in Figure 5.8. They can be natural waters, such as at beaches, piers, and boating venues, or engineered, in which water is abstracted either from nature or from treated supplies into a contained system, such as pools, water parks, and spray parks.

Within the categories of natural and engineered waters, the systems may be freshwater, saline, or brackish and may be heated either naturally (hot springs and thermal spas) or artificially (hot tubs and spas).

6 A critical exception being when the desire is to estimate the actual exposure to occupants who may be using those waters without deliberate flushing prior to use.

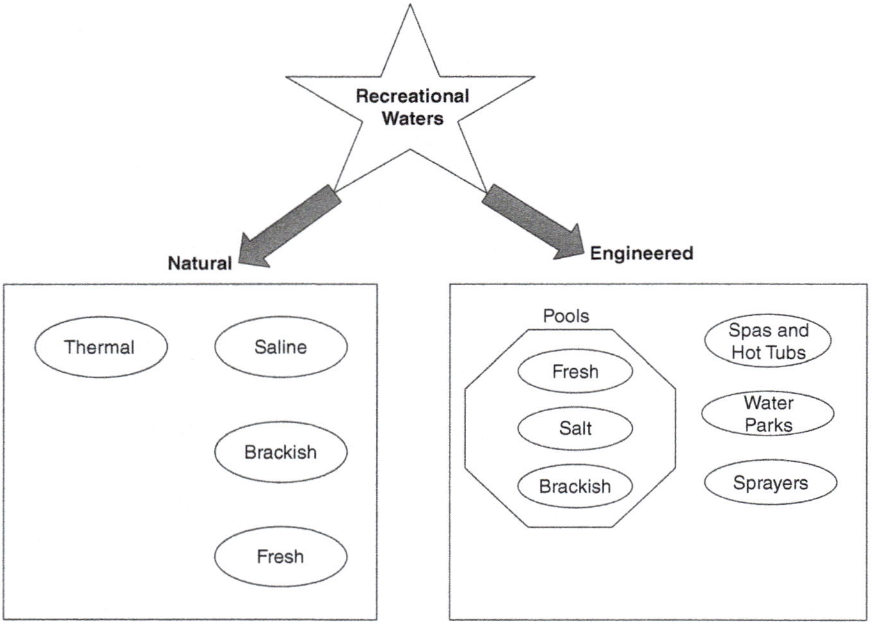

Figure 5.8 Classification of types of recreational waters.

Engineered recreational water systems may have various types of treatment, including filtration, disinfection, and pH adjustment. Often, engineered systems operate continuously, so there is an associated turnover time.

Recreational waters can also be classified by intended use relative to the potential for human exposure to contamination. A common framework used [52] is:

Non-contact: No intended direct contact with the water under ordinary use, e.g., walking. However, note that "spray" from the water could constitute an inhalable or deposition exposure.

Secondary contact: Only limbs are generally in contact with the water, and swallowing is unusual. Boating and fishing may often be considered in this category.

Primary contact: Full-body immersion where the head can be immersed with the potential for ingestion. Swimming, kayaking, waterskiing, and diving can be considered as examples.

While sampling from recreational waters is similar to sampling from drinking water, especially in natural waters, elevated turbidity may limit the volume that can be sampled. In addition, the specific sampling location (depth below the surface, distance from walls, shorelines, or inlets) will affect the concentration of microorganisms sampled. During sampling, additional metadata would be helpful to record (or obtain samples for measurement) to better interpret the results. These can include the following:

- For engineered systems, the types of water treatment in use (processes, chemical doses) and the turnover time of water in the enclosure
- Salinity (for saline and brackish systems) and other water chemistry parameters
- Solar insolation (for outdoor systems) and turbidity
- Density of recreational users, especially in primary and secondary contact areas

5.2.4 Air

As noted in Chapter 2, there are many different venues for air exposure that can occur — with a first dichotomy between indoor and outdoor air.

5.2.4.1 Ambient Air Sampling

Regardless of whether the air to be sampled is indoors or outdoors, it is necessary to process a large volume of air and collect the particles containing microorganisms. Following collection, they can be assayed via methods discussed later in this chapter. Important attributes of samplers include the following:

- Minimal to low degradation of the biological signal to be measured from the air sampled until taken and processed in the laboratory.
- Efficient capture of particles in the size ranges that bioaerosols are expected to be in.
- Ability to process sufficient air to detect the expected level of microorganisms present.
- Cost, size, and (especially for remote field deployment) portability.
- Where desired, the ability to collect size-specific particles.

Given these criteria, many sampler types have been deployed. First, active sampling technology will be discussed, and then passive sampling technologies will be reviewed.

Active air sampling relies on the physical or chemical characteristics of the aerosol particles to concentrate them from large volumes of flowing air.

The most basic of these is the use of impingers (Figure 5.9), in which air containing aerosols is fed into a collection fluid designed to retain the aerosols before the air is exhausted from the device. Details of performance for various designs are discussed in the literature [53], including problems of collection efficiency, the effect of detailed geometry on performance, and the potential for resuspension and loss. However, the advantage of impingers is that retention in a collection fluid can help maintain viability prior to analysis, making them suitable for both culture-based and molecular methods.

Filtration represents a second type of collection system for atmospheric particles, including bioaerosols. Either a single filter or a cascade of filters of different effective sizes can be used. This filtration system is routinely used in the sampling of the atmosphere for total particulate matter, particulate matter less than 10 μm (PM_{10}), or particulate matter less than 2.5 μm ($PM_{2.5}$). These samplers will cause loss of viability for many pathogens due to desiccation during collection, and loss of nucleic acid signal may also occur.

The effectiveness of filtration is a function of device design and the filter media used. Some filter media (e.g., membranes) have relatively "sharp" cutoffs for the size distribution captured. Other filter media have relatively less sharp cutoff size versus efficiency curves.

If filtration is used for bioaerosol sampling, then the filters need to be recovered, and the particulate material extracted, and the abstract taken for further analysis.

Electrostatic precipitation can be used to sample bioaerosols across an entire size range. This occurs when air flows between a channel with electrodes, since most particles either have or can be induced to have a charge, resulting in migration to an oppositely charged surface [54]. The recovered particles can then be taken for further analysis. As with filtration, there will likely be a loss of viability and perhaps of some molecular signal, which should be tested via appropriate challenge tests.

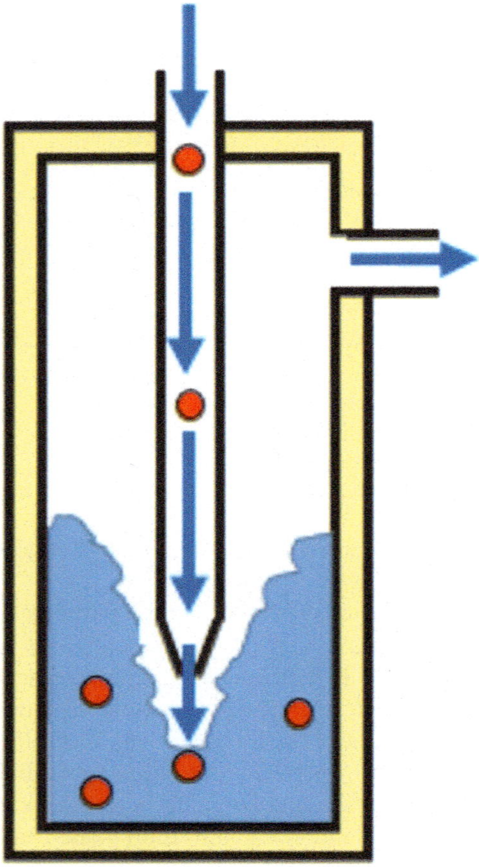

Figure 5.9 Schematic of impinger used for bioaerosol sampling. *Source:* [53]/ Centers for Disease Control and Prevention / Public domain / https://www.cdc.gov/niosh/nmam/pdf/chapter-ba.pdf / last accessed Feb. 10, 2026.

Impactors rely on the greater inertia of larger particles, which prevents them from completely following curved flow paths and thus causes them to collide with collection plates. Coating collection plates will reduce the potential for particle resuspension into the higher-velocity airflow. By arranging a cascade of nozzles (with different diameters) and collection plates, various "cuts" of particles of different sizes can be collected, as shown schematically in Figure 5.10. The collection surfaces may be coated plates or filter media and, after time, may be removed from the device for particle extraction and further workup. Due to the system having the potential for drying of the particles, loss of viability and some molecular response is possible in these systems. There are various designs for these cascades, including the widely used (both for bioaerosols and total particulates) Andersen impactor [55].

As with many of the collection methods, the potential for dehydration and accompanying loss in viability and molecular signature exists. However, the Anderson impactor may also use Petri plates with microbiological media as collection surfaces to permit viability to be (somewhat) maintained and thus allow viable counts and molecular counts (by washing from the plates). This also reduces the resuspension problem mentioned earlier.

Cyclone samplers also rely on the inertia of particles relative to the air in which they flow and on the centrifugal forces of rotation to provide an additional driving mechanism for removal. A schematic is shown in Figure 5.11. Particles above a particular effective size are deposited on the wall surface and can then be removed by washing the sides of the sampler. These have been widely used for bioaerosol sampling [53]. They can be coupled with an effluent filter to get two sizes of a sample.

Since there is no viability preservation step during collection, this (as well as possible deterioration of molecular signatures) needs to be assessed.

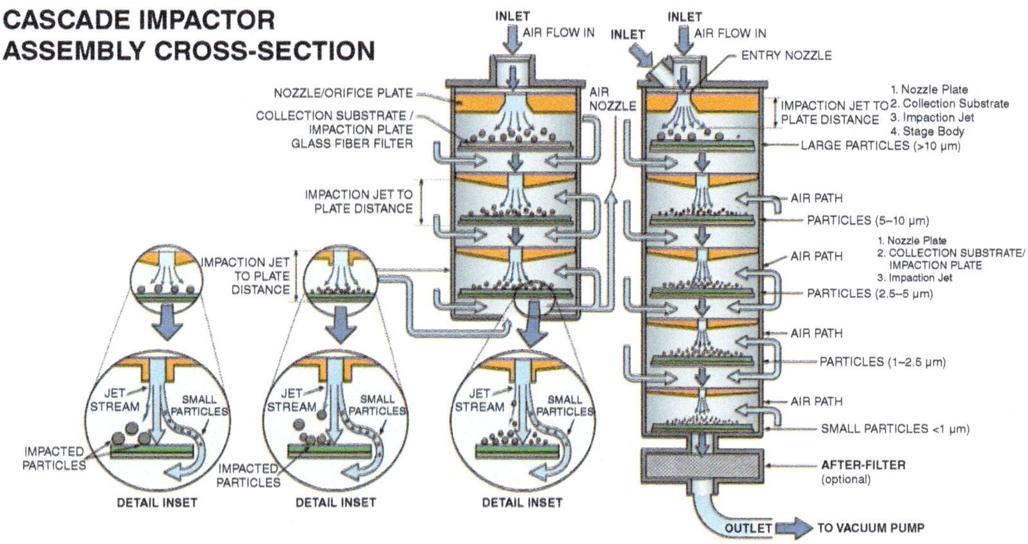

Figure 5.10 Schematic of a cascade impactor. *Source:* Benjamin Haywood. CC BY-SA 4.0. https://commons.wikimedia.org/wiki/File:Cascade_Impactor.png (last accessed on February 10, 2026).

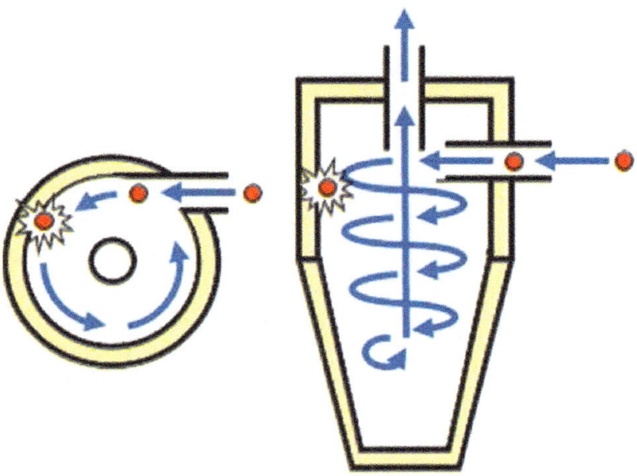

Figure 5.11 Cyclone sampler. *Source:* [53] / Centers for Disease Control and Prevention / Public domain / https://www.cdc.gov/niosh/nmam/pdf/chapter-ba.pdf / last accessed Feb. 10, 2026.

Variations of the cyclone and impaction samplers exist, in which there is a continual flow of collection fluid (perhaps with recirculation) over the surfaces onto which particle deposition occurs. This can help maintain viability from collection until processing [53].

The samples discussed earlier all have difficulty sampling ultrafine aerosols, i.e., those with diameters below 1 μm. Such particles may include exhalation of material from infected hosts at larger diameters, followed by evaporation to very small diameters shortly thereafter. As a result, condensation samplers have been designed [56], in which the air stream is humidified to supersaturation, then the temperature is reduced to promote condensation and particle growth via the retained water. The resulting larger particles are then collected into a liquid medium. This can substantially increase the collection efficiency of bioaerosols when the smallest-diameter particles dominate. Using such a system, viable viruses in the air of a student healthcare clinic serving SARS-CoV-2 patients were detected [57].

Many of the methods for active air sampling have been embodied in commercial devices. These have been reviewed by Grinshpun et al. [58]. Particular practical considerations include airflow rate, sampling time required to achieve the expected microorganism concentration in the air, and interference from other air contaminants (such as high dust concentration).

5.2.4.2 Passive Air Samplers

As with wastewater, the use of passive air samplers for detecting microorganisms has a long history. Since the late 19th century, investigators found that exposing plates containing solid growth media (originally gelatin and then agar) could collect microorganisms, some of which could grow into observable colonies [59]. Particles containing microorganisms can deposit on such plates either by gravity (for sufficiently large particles) or by random collisions—indeed, the processes of fate and transport that will be covered in Chapter 8 can be used to assess the deposition flux and thus, at least approximately, quantify the air that could have been sampled for such an exposure period. Upon incubation of the plates, deposited microorganisms that can grow on the media (bacteria and fungi) will form colonies. In Figure 5.12, such a plate used to sample outside air is shown. Individual colonies can be counted or sampled for molecular analysis to characterize them. This method, known as the use of settle plates, remains useful for semiquantitative assessment of air microbiology [60].

In addition to more deliberately designed and engineered settling systems, there are passive sampling devices using either electrostatic elements or adhesive surfaces to enhance collection. These have been reviewed by Manibusan and Mainellis [61].

5.2.5 Dusts, Surfaces, Fomites

Environments may contain pathogens on surfaces from prior touching by infected individuals, from the settling of emitted airborne particles, or by association with larger particles that subsequently settle out. There are a variety of approaches for collecting samples in such situations. First, the potential for opportunistic harvesting of dusts collected during routine vacuuming exists and has been found to contain a detectable signal [62]. Dust in buildings represents a time-integrated sample and can contain a broad diversity of microbial signals (as well as chemicals) [63]. Walking [64] or infant crawling [65] can induce resuspension of dust, including contained contaminants, into the breathing zone and be a source of subsequent exposures. Resuspension of outdoor soil dusts has also been observed as a vector for long-range dispersal of pathogens [66].

A deliberate sampling of dust can be done using vacuuming and collection, as per the protocol of Rintala et al. [63]. In this manner, the mass of dust collected per unit area can be quantified, and

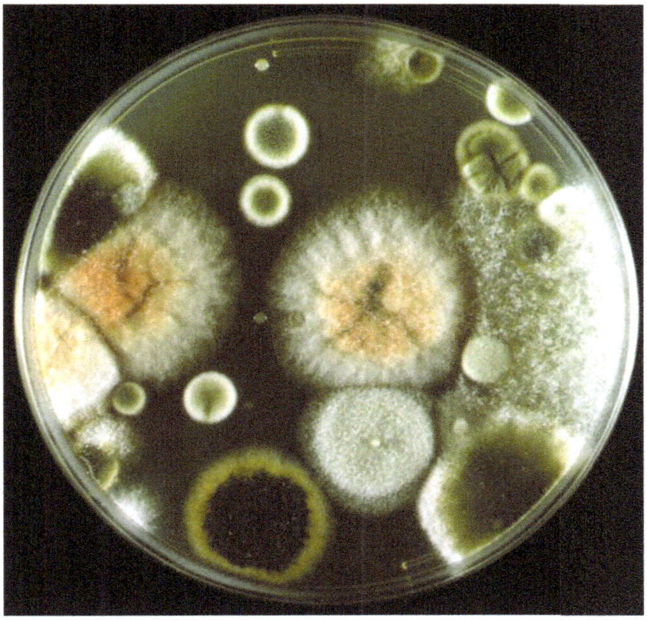

Figure 5.12 Petri dish culture plate left on a rooftop for a period of 30 minutes, then incubated at a temperature of 25 °C. *Source:* Centers for Disease Control and Prevention / Public domain / https://phil.cdc.gov/Details.aspx?pid=21387 / last accessed Feb. 10, 2026).

then the microbial concentration per unit area can be assessed. However, this will generally only sample the loosely adherent fractions.

In most cases, one of several other methods will be used for sampling of surfaces [53]:

- Swabs
- Wipes
- Adhesive tape
- Contact plates

An example of the use of swabs is shown in Figure 5.13, where a Q-tip™-like device is rubbed across an interior surface of the NASA International Space Station. By swabbing a defined area of surface, ultimately, the microorganism concentration per unit area can be developed. Wipes are similar to swabs but are used for larger surface area sampling. The wipe material (paper, cloth) may be chemically or electrostatically treated for better properties.

Adhesive tape sampling is used when the endpoint detection method is intended to be a microscopic examination, but the suitability for molecular and culture-based methods is uncertain.

Contact plates are agar plates in which the medium is filled to the rim and allowed to solidify. With gentle pressure, the surface microorganisms can be sampled by inverting the plate onto the surface. Alternatively, membrane filters (possibly moisturized) can be placed on a surface and then removed for incubation in a growth medium [53]. This approach is a descendant of the "replica plating" technique long used in microbiology [67].

With all these methods, standardization and challenge testing are important. Jones et al. have reviewed the literature on swabs and wipes and identified key areas of research that remain to be undertaken [68].

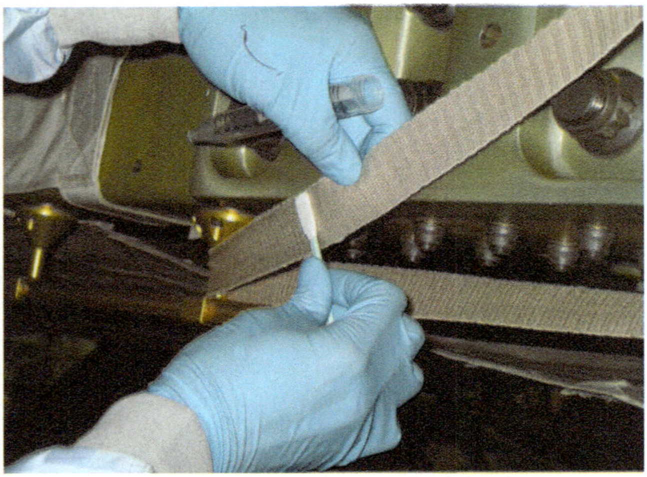

Figure 5.13 Technician swabbing mockup of indoor surface of NASA International Space Station. *Source:* NASA / Public domain / https://www.nist.gov/image/indoormicrobialsamplingjpg / last accessed Feb. 10, 2026.

The use of these other methods requires balancing two properties: an efficient collection, i.e., dislodgement, of microbial-containing particles from a surface, and the microorganisms removed from the collector must be capable of being efficiently (ideally completely) extracted from the collection system. Substantial development work is needed to continue optimizing swab and wipe systems and conditions across various circumstances. As one example of such testing, Weir et al. [69] studied the effect of wipe/swab material, surface, microorganism, and eluent on the recovery of a known number of microorganisms. Further work to extend and generalize such findings is needed.

Some examples of surface sampling in the context of this work are listed as follows:

- Lopez et al. [70] studied the transference of microorganisms from surfaces to fingers during touching. They then swabbed the fingertips to quantify the transferred microorganisms.
- In the investigation of a *Klebsiella* outbreak in a hospital, contact plates were used, as well as swabs, to determine whether organisms are consistent with the cases that were present in environmental samples [71].
- During the early part of the COVID-19 outbreak, Harvey et al. measured the presence of viral nucleic acid over time using a swabbing technique. They identified clear differences in occurrence by surface type and also a reduction in occurrence coinciding with the implementation of nonpharmaceutical (masking, distancing, closures) interventions to reduce disease spread.

5.2.6 Solids and Semisolids

Semisolids and solids comprise a highly diverse category. Operationally, these can be considered any material that cannot flow or be pumped easily, generally with solids over 5%. There are numerous potential materials that could be sampled that are in this category, including sewage sludges and biosolids, septate, manures, soils, and landfills.

These materials are likely to exhibit substantial spatial heterogeneity, as shown by the overdispersion in Figure 4.4; therefore, multiple samples representative of the venue of interest should be collected. Even within a sample taken, there should be some degree of homogenization and mixing during the initial steps of analysis to ensure that good representation *within* the sample occurs.

Various mechanical methods may be used to sample these materials, depending on the mixture's consistency. These include:

- Dip bottles or buckets are lowered into the material, which is allowed to enter by gravity, and then the sampler is removed—this can be done using a sampler on a pole or a rope.
- A shovel can be used for more solid material for direct collection near the surface. Alternatively, to collect samples at a greater depth, augers or split-spoon samplers can be used.

Examples of sampling and workflows can be illustrated with several case studies.

In 2007–2010, there was an outbreak of Q fever in the Netherlands. This disease is caused by the bacterium *Coxiella burnetii*. Uncertainty existed as to the importance of domestic animals as a reservoir for the amplification and transmission of this pathogen. Therefore, sampling of farm manures, specifically goats, was conducted [72]. The manures were directly collected manually in the farm area and immediately placed in sample vials. After extraction and processing, PCR was used to detect the signal from *C. burnettii*, which was positive.

During the early part of the COVID-19 outbreak, Peccia et al. monitored primary sewage sludge from a wastewater treatment plant for the molecular signal of SARS-CoV-2 [24]. They noted a good relationship between the viral signal and the community case rate (based on hospital admissions) when the lag times (for disease incubation and notification and lags in the sewer system and sewage treatment plant) were considered. The primary sludge was collected directly from the thickener underflow line.

Wan et al. sampled landfills in China at defined depths using a split-spoon device, which is capable of opening up to take a sample at a defined subsurface depth [73]. They were then able to use a metagenomic approach to identify bacterial taxa present in the samples and detected a number of putative pathogens.

5.3 Decontamination and Biosafety

This section is not intended to be a comprehensive exposition of biosafety. Nonetheless, those who seek to measure pathogens or pathogen signals in environmental samples should recognize the potential presence of viable infectious agents. The primary source of guidance on biosafety is the reference *Biosafety in Microbiological and Biomedical Laboratories* [74]. These programs are generally overseen by Institutional Biosafety Committees, which should be regarded as the definitive source of requirements for individual researchers.

In general, the biosafety level (BSL) required is determined by the microorganisms to be investigated, their severity, the availability of medical countermeasures in the event of accidental exposure of workers to the agent, and the nature of the laboratory activities undertaken. Procedures that result or may result in the generation of air particles containing pathogens are intrinsically of a greater level of concern. BSLs are given the numbers 1–4, with increasing stringency of control barriers—some of which are noted in Table 5.1.

For example, the processing of wastewater samples has been regarded by the CDC as requiring enhanced BSL2 precautions. However, some have found that when the signal to be used is nucleic acid, initial sample pasteurization can be used as a first step, conducted under appropriate precautions, with subsequent processing under BSL1 conditions, thereby reducing the stringency of facility requirements. However, it should be noted that this method of decontamination may not be generalizable to other microorganisms or sample types and would clearly not be applicable when it is desired to culture organisms to assess viability.

Table 5.1 Some significant requirements of different biosafety levels.

BSL1	
	Controlled access
	Hand-washing sink
	Sharp hazards warning policy
	Personal protective equipment
	Laboratory bench
	Autoclave
BSL2 (additional requirements)	
	Physical containment device (hood/glove box)
BSL3 (additional requirements)	
	Airtight when disinfecting
	Self-closing, double-door access
	Personal shower-out
	Sealed penetrations
	Powered air-purifying respirator
	Exhaust HEPA filter
	Effluent decontamination system
BSL4 (additional requirements)	
	Positive-pressure protective suit
	Chemical (decontamination) shower-out
	Supply and exhaust HEPA filter

Source: Modified from CDC (https://www.cdc.gov/orr/infographics/biosafety.htm).

5.4 Isolation, Concentration, and Separation

Following initial collection and possibly decontamination, the sample may need to undergo steps of concentration and separation—not necessarily in that order (Figure 5.1). Or, as noted previously, there may have been sufficient concentration in the field collection method that isolation and separation can occur without further concentration. Since each handling step introduces the potential for loss and greater variability, it is generally desirable to reduce the number of processing steps to the minimum needed to achieve a target sensitivity.

If the sample contains large amounts of solids, such as sewage, water filter backwash, air filters, sludges, or dusts, a first step is to dislodge organisms from the solids for further processing. As shown, for example, in the case of a water filter (Figure 5.14), significant material can be removed by filtration, requiring processing and cleanup. The dislodgement can be done using a variety of methods.

For the type of filter shown in Figure 5.14, a simple washing in a buffer or extraction medium may suffice to dislodge. In this case, using a small volume of liquid will facilitate further processing. This approach can also be used to dislodge organisms collected using swabs and cloths from surfaces.

Figure 5.14 Filter that has been used to sample water showing captured solid material. *Source:* USGS / Public domain / https://www.usgs.gov/media/images/particulate-organic-matter-ocean-water-samples-captured-filter / last accessed Feb. 10, 2026.

For processing larger volumes of materials, such as from sludge, filtered drinking water, or filtered wastewater, a stronger dislodgement may be required. One common approach is to take such materials, after a possible coarse "chopping," and process them using a stomacher [75].

What Is a Stomacher?

In microbiology, a stomacher is a specialized blender used for a crucial step in sample preparation: homogenizing solid samples to release and suspend microorganisms for further analysis. The sample is placed in a sterile plastic bag along with a specific buffer solution or diluent, which is secured in the instrument. Two reciprocating paddles move back and forth, mimicking a kneading motion. This agitation disrupts the sample and dislodges microorganisms into the overlying solution.

For viruses, a common sampling method, especially in cleaner waters, is adsorption after preconditioning onto filter media. Then the adsorbed material, leaving the microbial-containing supernatant, is transferred to a specific extraction medium [76].

Once this concentrate is obtained, with microorganisms dislodged from the solids, simple sedimentation or low-speed centrifugation can be used to remove the coarse solids, with the microbial-containing supernatant retained.

In the specific case of virus analysis of aqueous samples, a variety of more direct methods have been developed that are particularly suitable for samples with low solids content, such as settled wastewater, primary effluent, or natural waters. Early methods involved samples being processed directly through charged filters (with or without preconditioning), with viruses allowed to sorb to the filter medium [76]. The sorbed viruses are then eluted with a smaller volume of extraction fluid for further processing.

More recent approaches for direct processing (following removal of coarse solids) use precipitation flocculation (with agents such as polyethylene glycol or skim milk) followed by centrifugation to pellet the flocculant containing the virus. The concentrated pellet can then be further processed, generally using molecular methods [77]. The various approaches (prior to the concern with SARS-CoV-2) were systematically reviewed by Haramoto et al. [78].

After decontamination, isolation, and concentration, microorganisms from a large sample are concentrated into a smaller amount of material. It is important to track the concentrations so that they can ultimately be used to refer back to the original amount sampled from the environment. Two examples illustrate this.

Example 5.1 Sampling from Wastewater.

10 L of wastewater is obtained and filtered for processing. The filtered material is then resuspended into 1 mL of buffer. From the 1 mL of resuspended sample, 100 μL samples are assayed. What is the equivalent volume in each of those samples?

Solution:
There is a concentration of $10\,\text{L}/0.001\,\text{L} = 10^4$ between the collected wastewater and the concentrated sample. Therefore, each 100 μL of the concentrate is the equivalent to $10^{-4}\,\text{L} * 10^4 = 1\,\text{L}$.

Example 5.2 Sampling from Air.

100 m^3 (100,000 L) of air is sampled using an impinger (Figure 5.9). There is 100 mL of liquid in the impinger. If a 1 mL of liquid is taken for assay, what is the equivalent volume of air that it represents?

Solution:
The concentration factor is $100,000\,\text{L}/0.1\,\text{L} = 10^6$. Therefore, 1 mL sampled is equivalent to $0.001\,\text{L} \times 10^6 = 1000\,\text{L}$ or 1 m^3.

5.5 Discussion Questions and Problems

1 Materials and procedures for passive sampling of various environments are evolving. Do a literature review of novel methods for sampling wastewater.

2 Assume the concentration of a pathogen in air is one organism per cubic meter. A filtering air sampler with a flow rate of 0.2 m^3/ min is used. If an overall efficiency of collection and recovery from the device is 67%, and it is possible to quantify five pathogens in the collected sample, how long should the sampler be operated to collect that minimum number?

3 Assume a Poisson distribution. If the pathogen concentration in finished water is 0.001 organisms per liter, what volume of water must be sampled to have a 95% chance of detection if the sampler were perfect?

4 Repeat the calculation in the previous problem, if the efficiency of capturing the pathogen from the water is 80% and the efficiency of release of the pathogen from the collection device is 90%.

References

1 Molly K Leecaster and Stephen B Weisberg. Effect of sampling frequency on shoreline microbiology assessments. *Marine Pollution Bulletin*, 42(11):1150–1154, 2001.

2 Thomas F Harrington. An epidemic of typhoid fever from sewage pollution of city water supply. *Journal of the American Medical Association*, XLI(25):1511–1513, December 1903.

3 W J Wilson. Isolation of enteric bacilli from sewage and water and its bearing on epidemiology. *British Medical Journal*, 2(3794):560–562, 1933.

4 John R Paul, James D Trask, and Sven Gard. II. Poliomyelitc virus in urban sewage. *The Journal of Experimental Medicine*, 71(6):765–777, 1940.

5 Tomofumi Nakamura, Mitsuhiro Hamasaki, Hideaki Yoshitomi, Tetsuya Ishibashi, Chiharu Yoshiyama, Eriko Maeda, Nobuyuki Sera, and Hiromu Yoshida. Environmental surveillance of poliovirus in sewage water around the introduction period for inactivated polio vaccine in Japan. *Applied and Environmental Microbiology*, 81(5):1859–1864, 2015.

6 G Sedmak, D Bina, and J MacDonald. Assessment of an enterovirus sewage surveillance system by comparison of clinical isolates with sewage isolates from Milwaukee, Wisconsin, collected August 1994 to December 2002. *Applied and Environmental Microbiology*, 69(12):7181–7187, 2003.

7 Warish Ahmed, Nicola Angel, Janette Edson, Kyle Bibby, Aaron Bivins, Jake WO'Brien, Phil M Choi, Masaaki Kitajima, Stuart L Simpson, Jiaying Li, et al. First confirmed detection of SARS-CoV-2 in untreated wastewater in Australia: A proof of concept for the wastewater surveillance of COVID-19 in the community. *Science of The Total Environment*, 728:138764, 2020.

8 Colleen C Naughton, Fernando A Roman, Ana Grace F Alvarado, Arianna Q Tariqi, Matthew A Deeming, Krystin F Kadonsky, Kyle Bibby, Aaron Bivins, Gertjan Medema, Warish Ahmed, et al. Show us the data: global COVID-19 wastewater monitoring efforts, equity, and gaps. *FEMS Microbes*, 12(4):xtad003, 2023.

9 Edwin E Geldreich, L C Best, Bernard A Kenner, and D J Van Donsel. The bacteriological aspects of stormwater pollution. *Journal (Water Pollution Control Federation)*, 40(11):1861–1872, 1968.

10 Drew Capone, Anna Ferguson, Matthew O Gribble, and Joe Brown. Open defecation sites, unmet sanitation needs, and potential sanitary risks in Atlanta, Georgia, 2017–2018. *American Journal of Public Health*, 108(9):1238–1240, 2018.

11 Gordon M Fair and John C Geyer. *Water Supply and Wastewater Disposal*. John Wiley & Sons, New York, 1st edition, 1954.

12 Kristina M Babler, Mark E Sharkey, Samantha Abelson, Ayaaz Amirali, Aymara Benitez, Gabriella A Cosculluela, George S Grills, Naresh Kumar, Jennifer Laine, Walter Lamar, et al. Degradation rates influence the ability of composite samples to represent 24-hourly means of sars-cov-2 and other microbiological target measures in wastewater. *Science of The Total Environment*, 867:161423, 2023.

13 Brittany L Kmush, David Monk, Hyatt Green, Darcy A Sachs†, Teng Zeng, and David A Larsen. Comparability of 24-hour composite and grab samples for detection of SARS-2-CoV RNA in wastewater. *FEMS Microbes*, 3:xtac017, June 2022.

14 Danielle Jacobs, Troy McDaniel, Arvind Varsani, Rolf U Halden, Stephanie Forrest, and Heewook Lee. Wastewater monitoring raises privacy and ethical considerations. *IEEE Transactions on Technology and Society*, 2(3):116–121, 2021.

15 Teresa Scassa, Pamela J Robinson, and Ryan Mosoff. The datafication of wastewater. *Technology and Regulation*, pages 23–35, 2022.

16 Walter Q Betancourt, Bradley W Schmitz, Gabriel K Innes, Sarah M Prasek, Kristen M Pogreba Brown, Erika R Stark, Aidan R Foster, Ryan S Sprissler, David T Harris, Samendra P Sherchan, et al. COVID-19 containment on a college campus via wastewater-based epidemiology, targeted clinical testing and an intervention. *Science of The Total Environment*, 779:146408, 2021.

17 Chamteut Oh, Aijia Zhou, Kate O'Brien, Yusuf Jamal, Hayden Wennerdahl, Arthur R. Schmidt, Joanna L Shisler, Antarpreet Jutla, Arthur R Schmidt IV, Laura Keefer, et al. Application of neighborhood-scale wastewater-based epidemiology in low COVID-19 incidence situations. *Science of the Total Environment*, 852:10, 2022.

18 Sundra R Tatiparthi, Yashika G De Costa, Colin N Whittaker, Shihu Hu, Zhiguo Yuan, Ray Y Zhong, and Wei-Qin Zhuang. Development of radio-frequency identification (RFID) sensors suitable for smart-monitoring applications in sewer systems. *Water Research*, 198:117107, 2021.

19 Aaron Bivins and Kyle Bibby. Wastewater surveillance during mass COVID-19 vaccination on a college campus. *Environmental Science & Technology Letters*, 8(9): 792–798, 2021.

20 Ocean Thakali, Shalina Shahin, and Samendra P Sherchan. Wastewater surveillance of SARS-CoV-2 RNA in a prison facility. *Water*, 16(570):1–6, February 2024.

21 Dennis Schmiege, Timo Haselhoff, Alexander Thomas, Ivana Kraiselburd, Folker Meyer, and Susanne Moebus. Small-scale wastewater-based epidemiology (WBE) for infectious diseases and antibiotic resistance: A scoping review. *International Journal of Hygiene and Environmental Health*, 259:114379, June 2024.

22 Amy E Kirby, Maroya Spalding Walters, Wiley C Jennings, Rebecca Fugitt, Nathan LaCross, Mia Mattioli, Zachary A Marsh, Virginia A Roberts, Jeffrey W Mercante, Jonathan Yoder, et al. Using wastewater surveillance data to support the COVID-19 response - United States, 2020-2021. *Morbidity and Mortality Weekly Report*, 70(36):1242–1244, September 2021.

23 Sooyeol Kim, Lauren C Kennedy, Marlene K Wolfe, Craig S Criddle, Dorothea H Duong, Aaron Topol, Bradley J White, Rose S Kantor, Kara L Nelson, Joshua A Steele, et al. SARS-CoV-2 RNA is enriched by orders of magnitude in primary settled solids relative to liquid wastewater at publicly owned treatment works. *Environmental Science: Water Research & Technology*, 8:757–770, 2022.

24 Jordan Peccia, Alessandro Zulli, Doug E Brackney, Nathan D Grubaugh, Edward H Kaplan, Arnau Casanovas-Massana, Albert I Ko, Amyn A Malik, Dennis Wang, Mike Wang, et al. Measurement of SARS-CoV-2 RNA in wastewater tracks community infection dynamics. *Nature Biotechnology*, 38:1164–1167, 2020.

25 B Moore. The detection of paratyphoid carriers in towns by means of sewage examination. *Monthly Bulletin of the Ministry of Health and the Public Health Laboratory Service*, 7:241–248, 1948.

26 Warish Ahmed, Paul M Bertsch, Nicola Angel, Kyle Bibby, Aaron Bivins, Leanne Dierens, Janette Edson, John Ehret, Pradip Gyawali, Kerry Hamilton, et al. Detection of SARS-CoV-2 RNA in commercial passenger aircraft and cruise ship wastewater: a surveillance tool for assessing the presence of COVID-19 infected travelers. *Journal of Travel Medicine*, 27(5):taaa116, 2020.

27 Aaron Bivins, Robert Morfino, Andrew Franklin, Stuart Simpson, and Warish Ahmed. The lavatory lens: Tracking the global movement of pathogens via aircraft wastewater. *Critical Reviews in Environmental Science and Technology*, 54(4):321–341, 2024.

28 W T Sedgwick and J S Macnutt. On the Mills-Reincke phenomenon and Hazen's theorem concerning the decrease in mortality from diseases other than typhoid fever following the purification of public water-supplies. *Journal of Infectious Diseases*, 7(4):489–564, 1910.

29 William J Orchard, F W Green, Chester G Wigley, Edward Bartow, Joseph Race, and LI Birdsall. Water supply standards and their improvement [with discussion]. *Journal (American Water Works Association)*, 5(1):49–62, 1918.

30 Joseph L Melnick, Charles P Gerba, and Craig Wallis. Viruses in water. *Bulletin of the World Health Organization*, 56:499–508, 1978.

31 G F Craun. Waterborne giardiasis in the United States: A review. *American Journal of Public Health*, 69:817–819, 1979.

32 Joan B Rose. Occurrence and significance of *cryptosporidium* in water. *Journal of the American Water Works Association*, 80:53–58, 1988.

33 Martin J Allen, Jennifer L Clancy, and Eugene W Rice. The plain, hard truth about pathogen monitoring. *Journal of the American Water Works Association*, 92(9):64–76, 2000.

34 US Environmental Protection Agency. EPANET 2.2.0: An EPA and water community collaboration. https://www.epa.gov/sciencematters/epanet-220-epa-and-water-community-collaboration, July 2023.

35 Kelsey J Pieper, Min Tang, and Marc A Edwards. Flint water crisis caused by interrupted corrosion control: Investigating "ground zero" home. *Environmental Science & Technology*, 51(4):2007–2014, 2017.

36 William J Rhoads, Emily Garner, Pan Ji, Ni Zhu, Jeffrey Parks, David Otto Schwake, Amy Pruden, and Marc A Edwards. Distribution system operational deficiencies coincide with reported legionnaires' disease clusters in Flint, Michigan. *Environmental Science & Technology*, 51(20):11986–11995, 2017.

37 Steven Folkman. Water main break rates in the USA and Canada: A comprehensive study. *Technical Report 174*, Utah State University, Mechanical and Aerospace Engineering Faculty Publications, 2018.

38 Jian Yang, Mark W LeChevallier, Peter F M Teunis, and Minhua Xu. Managing risks from virus intrusion into water distribution systems due to pressure transients. *Journal of Water and Health*, 9(2):291–305, April 2011.

39 US Environmental Protection Agency. Distribution system water quality protecting water quality through cross-connection control and backflow prevention. *Technical Report EPA 815-F-21-004*, US Environmental Protection Agency, October 2021.

40 National Ground Water Association. Groundwater facts. https://www.ngwa.org/what-is-ground water/About-groundwater/groundwater-facts, December 2023.

41 Kelly R Bright, Francine Marciano-Cabral, and Charles P Gerba. Occurrence of *Naegleria fowleri* in Arizona drinking water supply wells. *Journal of the American Water Works Association*, 101:43–50, 2009.

42 Paul Dylan Hynds, M Kate Thomas, and Katarina Dorothy Milena Pintar. Contamination of groundwater systems in the US and Canada by enteric pathogens, 1990–2013: A review and pooled-analysis. *PLoS ONE*, 9(5):e93301, May 2014.

43 Jiayi Xu, Jiangyu Dai, Xiufeng Wu, Shiqiang Wu, Yu Zhang, Fangfang Wang, Ang Gao, and Yanping Tan. Urban rainwater utilization: A review of management modes and harvesting systems. *Frontiers in Environmental Science*, 11:1025665, February 2023.

44 Georgios D Gikas and Vassilios A Tsihrintzis. Assessment of water quality of first-flush roof runoff and harvested rainwater. *Journal of Hydrology*, 466-467: 115–126, 2012.

45 Kerry A Hamilton, Kerrianne Parrish, Warish Ahmed, and Charles Nathan Haas. Assessment of water quality in roof-harvested rainwater barrels in Greater Philadelphia. *Water*, 10(92):1–15, 2018.

46 Kerry A Hamilton, Warish Ahmed, Andrew Palmer, Kylie Smith, Simon Toze, and Charles N Haas. Seasonal assessment of opportunistic premise plumbing pathogens in roof-harvested rainwater tanks. *Environmental Science & Technology*, 51(3):1742–1753, 2017.

47 Kathy D Gee and William F Hunt III. Mosquito control for rainwater harvesting systems. *Technical Report*, North Carolina State University College of Agriculture and Life Sciences, 2021.

48 William J Rhoads, Annie Pearce, Amy Pruden, and Marc A Edwards. Anticipating the effects of green buildings on water quality and infrastructure. *Journal AWWA*, 107(4):50–61, 2015.

49 Md Rasheduzzaman, Rajveer Singh, Charles N Haas, and Patrick L Gurian. Required water temperature in hotel plumbing to control Legionella growth. *Water Research*, 182:115943, 2020.

50 Abraham C Cullom, Rebekah L Martin, Yang Song, Krista Williams, Amanda Williams, Amy Pruden, and Marc A Edwards. Critical review: Propensity of premise plumbing pipe materials to enhance or diminish growth of legionella and other opportunistic pathogens. *Pathogens*, 9(957):1–34, 2020.

51 Alshae'R Logan-Jackson, Marylia Duarte Batista, William Healy, Tania Ullah, Andrew J Whelton, Timothy A Bartrand, and Caitlin Proctor. A critical review on the factors that influence opportunistic premise plumbing pathogens: From building entry to fixtures in residences. *Environmental Science & Technology*, 57(16):6360–6372, April 2023.

52 World Health Organization. *Guidelines for Safe Recreational Water Environments: Coastal and Fresh Waters*, volume 1. World Health Organization, Geneva, Switzerland, 2003.

53 William G Lindsley, Brett J Green, Francoise M Blachere, Stephen B Martin, Paul A Jensen, and Millie P Schafer. *Sampling and Characterization of Bioaerosols*, page Section 5A. National Institute of Occupational Safety and Health, 5th edition, 2017.

54 J M Roux, R Sarda-Estève, G Delapierre, M H Nadal, C Bossuet, and L Olmedo. Development of a new portable air sampler based on electrostatic precipitation. *Environmental Science and Pollution Research*, 23(9):8175–8183, 2016.

55 Ariel A Andersen. New sampler for the collection, sizing, and enumeration of viable airborne particles. *Journal of Bacteriology*, 76(5):471–484, 1958.

56 John Lednicky, Maohua Pan, Julia Loeb, Hsin Hsieh, Arantzazu Eiguren-Fernandez, Susanne Hering, Z Hugh Fan, and Chang-Yu Wu. Highly efficient collection of infectious pandemic influenza H1N1 virus (2009) through laminar-flow water based condensation. *Aerosol Science and Technology*, 50(7):i–iv, 2016.

57 John A Lednicky, Sripriya N Shankar, Maha A Elbadry, Julia C Gibson, Md Mahbubul Alam, Caroline J Stephenson, Arantzazu Eiguren-Fernandez, J Glenn Morris, Carla N Mavian, Marco Salemi, et al. Collection of SARS-CoV-2 virus from the air of a clinic within a University Student Health Care Center and analyses of the viral genomic sequence. *Aerosol and Air Quality Research*, 20(6):1167–1171, 2020.

58 Sergey A Grinshpun, Mark P Buttner, Gediminas Mainelis, and Klaus Willeke. Sampling for airborne microorganisms. *Manual of Environmental Microbiology* John Wiley & Sons, Inc., pages 3–2, 2016.

59 Grace C Frankland, Percy Faraday Frankland, and Edwin Ray Lankester. XI. Studies on some new micro-organisms obtained from air. *Philosophical Transactions of the Royal Society of London. (B.)*, 178(December 31):257–287, 1887.

60 Christian Napoli, Vincenzo Marcotrigiano, and Maria Teresa Montagna. Air sampling procedures to evaluate microbial contamination: a comparison between active and passive methods in operating theatres. *BMC Public Health*, 12:594, 2012.

61 Sydonia Manibusan and Gediminas Mainelis. Passive bioaerosol samplers: A complementary tool for bioaerosol research. A review. *Journal of Aerosol Science*, 163:105992, 2022.

62 Nicole Renninger, Nicholas Nastasi, Ashleigh Bope, Samuel J Cochran, Sarah RHaines, Neeraja Balasubrahmaniam, Katelyn Stuart, Aaron Bivins, Kyle Bibby, Natalie M Hull, et al. Indoor dust as a matrix for surveillance of covid-19. *mSystems*, 6(2): e01350-20, 2021.

63 Helena Rintala, Miia Pitkäranta, and Martin Täubel. Chapter 4 - Microbial communities associated with house dust. In Allen I Laskin, Sima Sariaslani, and Geoffrey M. Gadd, editors, *Advances in Applied Microbiology*, volume 78, pages 75–120. Academic Press, 2012.

64 Jing Qian, Jordan Peccia, and Andrea R Ferro. Walking-induced particle resuspension in indoor environments. *Atmospheric Environment*, 89:464–481, 2014.

65 Heidi K Hyytiäinen, Balamuralikrishna Jayaprakash, Pirkka V Kirjavainen, Sampo E Saari, Rauno Holopainen, Jorma Keskinen, Kaarle Hämeri, Anne Hyvärinen, Brandon E Boor, and Martin Täubel. Crawling-induced floor dust resuspension affects the microbiota of the infant breathing zone. *Microbiome*, 6(25):1–12, 2018.

66 Joseph M Prospero, Edmund Blades, George Mathison, and Raana Naidu. Interhemispheric transport of viable fungi and bacteria from Africa to the Caribbean with soil dust. *Aerobiologia*, 21(1):1–19, 2005.

67 Joshua Lederberg and Esther M Lederberg. Replica plating and indirect selection of bacterial mutants. *Journal of bacteriology*, 63(3):399–406, 1952.

68 Sarah L Jones, Steven C Ricke, D Keith Roper, and Kristen E Gibson. Swabbing the surface: Critical factors in environmental monitoring and a path towards standardization and improvement. *Critical Reviews in Food Science and Nutrition*, 60(2):225–243, January 2020.

69 Mark H Weir, Tomoyuki Shibata, Yoshifumi Masago, Dena L Cologgi, and Joan B Rose. Effect of surface sampling and recovery of viruses and non-spore-forming bacteria on a quantitative microbial risk assessment model for fomites. *Environmental Science & Technology*, 50(11):5945–5952, June 2016.

70 G U Lopez, C P Gerba, A H Tamimi, M Kitajima, S L Maxwell, and J B Rose. Transfer efficiency of bacteria and viruses from porous and nonporous fomites to fingers under different relative humidity conditions. *Applied and Environmental Microbiology*, 79(18):5728–5734, 2013.

71 Ricarda M Schmithausen, Esther Sib, Martin Exner, Sylvia Hack, Claudia Rösing, Patrick Ciorba, Gabriele Bierbaum, Mykhailo Savin, Sally F Bloomfield, Martin Kaase, et al. The washing machine as a reservoir for transmission of extended-spectrum-beta-lactamase (CTX-M-15)-producing *Klebsiella oxytoca* ST201 to newborns. *Applied and Environmental Microbiology*, 85(22):e01435-19, 2019.

72 A de Bruin, R Q van der Plaats, L de Heer, R Paauwe, B Schimmer, P Vellema, B J van Rotterdam, and Y T van Duynhoven. Detection of *coxiella burnetii* DNA on small-ruminant farms during a Q fever outbreak in the Netherlands. *Applied and Environmental Microbiology*, 78(6):1652–1657, March 2012.

73 Shan Wan, Min Xia, Jie Tao, Yanjun Pang, Fugen Yu, Jun Wu, and Shanping Chen. Metagenomics analysis reveals the microbial communities, antimicrobial resistance gene diversity and potential pathogen transmission risk of two different landfills in china. *Diversity*, 13(6):230, 2021.

74 Paul J Meechan and Jeffrey Potts, editors. *Biosafety in Microbiological and Biomedical Laboratories*. U.S. Department of Health and Human Services, 6th edition, 2020.

75 Lia Schmidt Fernandes, Ana Galvão, Ricardo Santos, and Sílvia Monteiro. Impact of water reuse on agricultural practices and human health. *Environmental Research*, 216(Part 3):114762, 2023.

76 Luisa A Ikner, Charles P Gerba, and Kelly R Bright. Concentration and recovery of viruses from water: A comprehensive review. *Food and Environmental Virology*, 4(2):41–67, June 2012.

77 Devin North and Kyle Bibby. Comparison of viral concentration techniques for native fecal indicators and pathogens from wastewater. *Science of The Total Environment*, 905:167190, 2023.

78 Eiji Haramoto, Masaaki Kitajima, Akihiko Hata, Jason R Torrey, Yoshifumi Masago, Daisuke Sano, and Hiroyuki Katayama. A review on recent progress in the detection methods and prevalence of human enteric viruses in water. *Water Research*, 135:168–186, 2018.

6

How Pathogens Are Quantified

In the previous chapter, methods for sampling pathogens across various environments were discussed. Given such a sample, whether from air, water, surfaces, or solids, there are commonalities in the methods used to quantify pathogens. These methods are the focus of this chapter.

6.1 Selection of Targets

In environmental sampling for microorganisms, an initial decision should be made about which targets to use. The primary distinction is the use of indicator organisms, gene markers for source tracking, or direct measurement of pathogens.

6.1.1 Indicator Organisms

Historically, bacteria were used as indicators for pathogens (particularly in water and foods)—these include coliforms, enterococci, and total aerobic count [1]. They offered the advantage of often being more prevalent and at greater concentrations than pathogens, and also being simpler to assay, particularly when culture-based methods were used.

But numerous other indicators have been investigated for particular purposes. These include:

- *Clostridium perfringens*, as indicators of quality and human health risks, especially in tropical waters [2], and also as indicators of water treatment efficiency [3].
- Aerobic spores, particularly as indicators of treatment performance in drinking water treatment [4].
- Acid-fast bacteria, more recently known as nontuberculosis *Mycobacterium* as indicators of water and wastewater disinfection efficiency due to their relative resistance [5].

Historically, indicator organisms have been used for at least three purposes, and this leads to different attributes being desirable:

Indicator of absolute quality: To determine the suitability of a water source as a water supply (perhaps conditional on a particular level of treatment).
Indicator of treatment efficiency: To determine how well a particular treatment process (or set of treatment processes) is performing with respect to microbial removal.
Indicator of human health risk: To provide a direct correlate to risks of exposure

In the late 19th century, it was observed that during typhoid fever occurrences, there were frequently substantial levels of what is now called the total coliform group [6]. In 1908, the US

Department of the Treasury established the first bacteriological standards for water used on interstate railroads (using total bacterial count). By 1918, as summarized by Morse and Wolman, there was enough experience with and standardization of methodology to argue for the use of a total coliform standard for drinking water quality [7]. For an indicator of absolute quality, early studies showed quantitative relationships between total coliform organisms and typhoid bacteria in sewage and polluted waters [8]. But almost from the beginning, it was recognized that coliforms were a taxonomically heterogeneous group, which also included some organisms that did not originate from warm-blooded animals (the hosts of concern for typhoid and other gastroenteric pathogens then of primary concern). Therefore, it was proposed to use a modified method with incubation at elevated temperature to form a "fecal coliform" or "thermotolerant coliform" group that might have more direct sanitary significance [9].

As treatment processes became more stringent and overall public health improved, it was recognized that coliform organisms were insufficient as indicators of treatment efficiency. For example, it was found that many viruses could survive disinfection processes under conditions in which coliforms and fecal coliforms were substantially inactivated [10]. As a result, researchers investigated the use of more resistant indicator organisms, including yeasts and acid-fast bacteria [11]. More recently, based on concern for waterborne protozoa, there has been some interest in the use of the inactivation of aerobic spore-forming bacteria as an indicator for the inactivation of *Cryptosporidium* [12].

As a direct indicator of human health risk, indicator approaches have been used to develop recreational water quality standards. This approach, however, requires conducting planned epidemiological studies to establish a criterion for acceptability.

Early epidemiological studies were conducted at Chicago (Lake Michigan), Cincinnati (Ohio River), and New York (Long Island Sound) bathing beaches [13]. This led to the observation of discernible increases in swimmer illness rates when the total coliform levels exceeded 2300–2700/100 mL. However, the underlying studies had methodological shortcomings, including the absence of non-swimming control groups.

As a result, there were subsequent studies conducted using more stringent frameworks, including:

- Use of non-swimming controls and comparison of "excess" rates in swimmers to non-swimming controls[1]
- Precise definition of "swimming"—typically immersion of the head below water
- Precise definition of the adverse outcome, frequently "highly credible gastroenteritis" (HCGI)—defined as a mix of symptoms including fever, nausea, and diarrhea
- Conduct of the study on a prospective basis, in which investigators enrolled subjects at the beaches, and also measured water quality using diverse indicator methods

Russo *et al.* [14] did a systematic review and meta-analysis of these studies, some of which included recreational activities other than swimming.

Based on US epidemiological studies, *Escherichia coli* and enterococci were identified as the strongest correlates of swimmer HCGI in freshwater, and enterococci were identified as the strongest correlate in saline water. From these studies, current recreational water criteria were formulated [15].

---

1 Albeit, this introduced another potential bias in that there might have been confounding factors that led to an individual going to a beach but not swimming.

There have been dramatic improvements in methods for direct measurement of pathogens, particularly the molecular techniques discussed later in this chapter. Therefore, the use of indicators in many contexts may continue to be supplanted.

6.1.2 Microbial Source Tracking

Another application of indicators has emerged in the past 20 years, although it has historical origins. When water is found to be microbiologically polluted, contamination may have arisen from multiple sources, e.g., human waste, agricultural animals, wild animals, etc. In order to take action to reduce the degree of contamination, the identification of the relative importance of different sources is important. Therefore, some measurements of characteristics of diverse sources, using microbial measurements alone or in combination with other information, may prove useful.

Early work in this area, now termed "microbial source tracking" (MST) found that in human wastes and human-dominated wastes, the ratio of fecal coliform to fecal streptococcus bacteria was typically at or in excess of 4.0 [16], while the ratio was typically below 0.7 in other sources.

In more modern applications of MST, specific genetic signatures, typically from indicator bacteria, are used to "fingerprint" likely sources of water pollution. This has been reviewed by Paruch and Paruch [17]. These authors noted that a broad diversity of environmental DNA markers have been used across different applications, and that there is no universal consensus on the best approaches, which are likely to be specific to the environment. However, the fundamental methods are the same as the use of nucleic acids for the direct assay of pathogens, and thus the details of the analysis present no additional issues.

6.1.3 Direct Measurement of Pathogens

Specific pathogens can also be selected as target analyte(s). In principle, any of the pathogens discussed in Section 2.1 could be selected. A choice to measure pathogens directly poses difficulties when culture or viability measures are used. In general, media, incubation times, selection procedures, and possibly confirmation protocols need to be optimized on a pathogen-by-pathogen basis.

However, if specific pathogens are to be measured using molecular approaches, then (as will be discussed) it may be possible to use a common set of conditions to detect and quantify a broad diversity of target organisms simultaneously — albeit possibly with a loss of information on viability.

6.2 Count, Quantal, and Time-to-response Methods

Regardless of the experimental approach, there are three broad classes of methods, depending on how quantitative information is extracted. These are called count methods, quantal methods, and time-to-response methods.

Once a sample from the environment is taken for assay, the techniques in this chapter can be used to assess the level of organisms present in that sample. Depending on whether the assays use a count, quantal, or time-to-response approach, different statistical and practical considerations arise. These cross-cutting considerations are the focus of this section.

If multiple environmental samples are collected, perhaps over time or across space, one might be interested in the temporal or spatial distribution of the estimated microbial levels. This aspect will be covered in more detail below.

6.2.1 Count Methods

In count methods, the actual (integral) number of organisms in a sample is obtained or inferred. One canonical example is the use of agar plates on which a sample is deposited, such as from the deposition of air particles as shown in Figure 5.12. It is inferred that each of the discernible areas (mainly circular) of growth represents the progeny of a viable organism that deposited and multiplied. For example, there appear to be about 19 colonies, and it is therefore estimated that deposition from the atmosphere (in this case, for 30 minutes) resulted in the accumulation of 19 viable organisms on the surface. If the rate of deposition were known or could be estimated, then the concentration of settleable viable organisms (number per unit volume) in the air could be assessed.

Another counting method may involve assaying the virus inoculated on the surface of suitable host cells. In such a system, areas of host cells could be cleared of infection and of individual virus particles. Each such cleared area is presumed to originate from one viable virus. By counting these areas ("plaques"), the number of initial viable viruses in the sample can be inferred.

A third type of count method is the use of volume-calibrated microscope slides, discussed in Section 6.3 for the direct optical counting of microorganisms in a given sample volume (perhaps after concentration).

All these approaches share the commonality that the raw data from the analyses consist of a number of microorganisms (or their progeny) found and the volume or amount of sample processed. The counts themselves have intrinsic statistical variability as might be characterized by the Poisson distribution (Section 4.1), upon which other experimental sources of variability and uncertainty may be superimposed.

Count methods have several important assumptions that are key to their interpretation. If these hold, then more straightforward approaches can be used for data interpretation.

One overarching assumption is that often a larger sample (e.g., 1 L of water or 100 g of a solid) is taken from the environment, and only a small portion of this is taken for enumeration. To use the results from the small subsample to infer the properties of the larger sample (and ultimately the environment from which the larger sample was taken), it is necessary to assume that the initial sample was, or was made (during processing), homogeneous such that the subsample was representative and a random one. This can be formally tested by taking multiple subsamples and comparing results (to be discussed subsequently).

A second issue with count methods that use growth on a medium or a host cell to assess viability, such as bacterial colony or viral plaque count, is the assumption that all viable organisms (of the target class) are enumerated and that one colony or plaque represents the progeny of one initially viable organism.

There is no universal medium or host cell system and set of incubation conditions that can enumerate *all* viable bacteria, or viruses, or fungi, etc. Different organisms have different growth rates, metabolic requirements, preferences, or tolerances for oxygen, or propensity to infect particular host cells. Therefore, a selectivity is almost always involved when culture techniques are used, perhaps deliberately in some cases. For example, if low concentrations of slow-growing pathogens are present, adding inhibitors for more prevalent organisms that could grow more rapidly and crowd out the desired targets would be desirable.

In addition, as discussed in Section 2.1.3, some bacteria can enter a viable but nonculturable (VBNC) state. If some active recovery step is not incorporated into a culture method, these organisms will not be observable.

With viruses, environmental exposures, such as disinfectants, can partially injure individual virions. However, the sites of injury to different virions may differ. If multiple virions are allowed to co-infect a host cell, they can engage in recombination to regain viability in a phenomenon called "multiplicity reactivation" [18].

It is also known that many microorganisms can form clumps with each other or adsorb to or be trapped within solids [19, 20]. If the sample preparation prior to culturing, such as by intense mixing, blending, or perhaps sonication, does not completely break apart such associations, then a single plaque or colony may in fact represent multiple viable organisms. Hence, microbial concentrations may be underestimated. Therefore, it is conventional to refer to "colony forming units" (CFU) for bacterial and fungal assays, and "plaque forming units" (PFU), rather than absolute bacterial or viral numbers from these culture assays.

Another issue with culture assays is the presence of chemical or biological inhibitors. These may inhibit the growth of target organisms or may actually crowd out the growth of slow-growing colonies. This can be seen qualitatively in Figure 5.12, where the larger colonies are clearly outgrowing the smaller colonies.

Actual antagonism can occur if some of the organisms growing on plates excrete products that inhibit the growth of other microorganisms. Indeed, the occurrence of this and diligent observation were responsible for Fleming's original discovery of penicillin [21].

Chemical or physical agents present in the same sample as the target microorganisms may also cause inhibition and depression of counts on culture media.

One approach to detecting inhibition (whether chemical, physical, or biological) is to conduct culture-based counts at closely spaced dilutions and observe the effect. Typically, samples are processed using decimal (tenfold) dilutions to ensure that at least some dilutions yield culture-based counts within an acceptable range (generally regarded as less than 300 colonies on larger plates or less than 100 on smaller plates). However, it is possible to use more closely spaced dilutions to test whether dilution affects the response. This is shown with a hypothetical example in Figure 6.1.

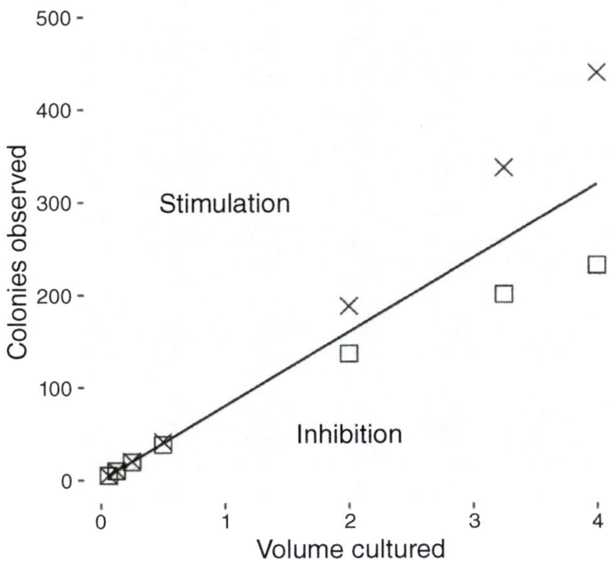

Figure 6.1 Effect of amount cultured on recovery. Line indicates ideal proportional recovery.

The straight line represents the result where the colonies (or plaques) observed would be perfectly proportional to the amount of sample examined. The squares (□) below the line indicate the effect of inhibition, with higher amounts examined showing a greater amount of inhibition, while the x's (×) show the converse effect of a stimulatory agent.

6.2.1.1 Statistics of Count Methods

Regardless of the particular technique, count assays can be analyzed by similar statistical approaches. There are some common issues and questions that arise. A primary concept that will be used for analysis of all types of analysis is the principle of maximum likelihood, which has often been attributed to Fisher, although it has a complex history both preceding and following his work [22]. Following the nomenclature of Stuart and Ord [23], if there is a set of n observations from the same distribution (characterized as $x_1, x_2, \ldots, x_n$) and the distribution parameters are given by (possibly a vector) θ, then the likelihood of those observations occurring from that distribution is:

$$L(x|\theta) = f(x_1|\theta)f(x_2|\theta) \ldots f(x_n|\theta) \tag{6.1}$$

The principle of maximum likelihood then states that the best estimate for the parameter(s) θ results from maximizing L in Equation 6.1. To employ this, the functional form of the distribution, f, needs to be specified. These estimators have many desirable statistical properties and can be used for inference [23].

Suppose an assay was done in which "n" replicate subsamples of each equivalent volume "V" were analyzed. A count assay found counts of $N_1, N_2, \ldots, N_n$. These are assumed to be independent samples from an identical Poisson distribution. Based on this, what is the best estimate for the mean volumetric concentration of organisms, $\hat{\upsilon}$?

To solve, the data are substituted in Equation 6.1

$$L(N|\theta) = f(N_1|\theta)f(N_2|\theta) \ldots f(N_n|\theta) \tag{6.2}$$

For each observation, in Equation 4.3, N_i is substituted for "k" and the product of mean concentration and volume (υV) is substituted for λ, and the likelihood function becomes:

$$\begin{aligned} L &= \left[\frac{(\upsilon V)^{x_1}}{x_1!}e^{-\upsilon V}\right]\left[\frac{(\upsilon V)^{x_2}}{x_2!}e^{-\upsilon V}\right] \ldots \left[\frac{(\upsilon V)^{x_n}}{x_n!}e^{-\upsilon V}\right] \\ &= \frac{(\upsilon V)^{\sum x_i}}{\prod x_i!}e^{-n\upsilon V} \end{aligned} \tag{6.3}$$

If this equation is differentiated with respect to υ, solving for the derivative equal to 0, we obtain the maximum likelihood estimator:

$$\hat{\upsilon} = \frac{\sum x_i}{nV} \tag{6.4}$$

This is simply the arithmetic mean of the counts divided by the total amount of material assayed (equivalent volume or mass) — which would be the naïve estimator. Note that some samples may have a zero count and still be included in this calculation.

Ideally, replicate analyses of the same sample using the same method should adhere to the Poisson assumption. This can be formally tested.

In the case of a single replicated sample, an index of dispersion (D^2) can be computed via:

$$D^2 = \frac{(n-1)s^2}{\overline{x}} \tag{6.5}$$

In this equation, s is the standard deviation of the counts, and $\overline{x}$ is the mean of the counts. A null hypothesis that the data are from an identical Poisson distribution is rejected if D^2 is in excess of an upper percentile of the χ^2 distribution with (n−1) degrees of freedom [24].

This is illustrated with an example.

Example 6.1 Replicates of single sample.

Kaper et al. [25] measured total aerobic viable counts in water samples from the Chesapeake Bay. In one experiment, at a 10^{-3} dilution, 19 enumerable plates were counted and the following results were obtained:

0	11	8
0	3	2
3	9	3
4	1	7
4	0	7
9	0	4
		12

Analysis of this data shows the mean was 4.57 organisms, with a standard deviation of 5.45. Applying this to Equation 6.5 gives an index of dispersion (D^2) of 55.99. For a χ^2 distribution, this has a p-value of 5.45×10^{-7}, which is highly significant, and therefore the null hypothesis of the Poisson is rejected. Based on this, it is likely that insufficient homogenization of the sample occurred before aliquots were taken for assay.

A second type of experimental design is when a series of samples (perhaps taken over time or space) is taken, from each of which replicate subsamples (aliquots) are analyzed. In aggregate, this data can be assessed to determine whether the within-sample error is consistent with a Poisson distribution.

In this case, the D^2 statistic is computed between aliquots for each sample. Then, there is a set of these dispersion indices, and the consistency of the *distribution of those dispersion indices* with a χ^2 distribution is assessed. This can get messy when replicate counts are unequal across samples. An illustration of handling such data is shown as follows.

Berg et al. [26] studied the enumeration of enteric viruses by a tissue culture plaque assay under controlled laboratory conditions. For the poliovirus LP Mahoney strain with a four-day incubation, they obtained the following data: note that they report means and standard deviations at successive dilutions directly.

This is analyzed for Poisson between replicate errors in the following box.

Example 6.2 Replicate Count Data Analysis – Multiple Experiments.

Using the means and standard deviations from Table 6.1, the dispersion index and the p-value for the χ^2 distribution at degrees of freedom equal to the number of replicates minus one are tabulated (rows correspond to rows in the original table):

Table 6.1 Plaque counts for Poliovirus Mahoney LP after four days of incubation as a function of sample dilution.

		Day 4 Counts	
Dilution	Replicates	Mean	Standard deviation
1.000	6	167	7
0.794	5	151	18
0.637	6	120	14
0.505	6	99	17
0.400	6	75	6
0.319	6	69	8
0.250	6	60	12
0.200	6	48	6
0.160	6	39	5
0.125	6	33	5

Source: Data from [26].

D^2	p
1.467	0.083
8.583	0.928
8.167	0.853
14.596	0.988
2.400	0.209
4.638	0.538
12.000	0.965
3.750	0.414
3.205	0.332
3.788	0.420

If there were a Poisson replication error between aliquots of the individual samples, then it would be expected that the "p" values would not differ from random samples from a uniform distribution over the range (0,1). A variety of statistics can be used to test this. One widely used test in this application is the Kolmogorov–Smirnov test [27].

This was applied to the D^2 values using R. The commands and output are shown as follows:

```
    > data=c(0.083161174,0.927582668,0.852713649,
+       0.987764282,0.208525879,0.538323909,
+       0.965212219,0.414058882, 0.331605183,
+       0.4196577)
> ks.test(data,"punif")

Exact one-sample Kolmogorov-Smirnov test

data:  data
D = 0.25271, p-value = 0.4706
alternative hypothesis: two-sided
```

Based on this, the uniform distribution cannot be rejected, and thus the Poisson distribution cannot be rejected for the within-sample replication in this data set.

As shown diagrammatically in Figure 4.4, it is entirely possible to have variability in excess of (more common) or less than (less common) Poisson. As a simple comparator, the Poisson distribution has a ratio of variance/mean (of counts) equal to 1.0. Overdisperse distributions (the statistical literature often uses the term "contagious distribution," which, in the context of this book, is confusing) have variance-to-mean ratios exceeding 1.0.

Re-examining the data in Example 5.1, the mean count was 4.57. The variance of the counts is 15.035, which produces a variance/mean ratio of 3.28 — clearly in excess of 1.0, and it was shown in that example that the index of dispersion shows a lack of coherence to the Poisson.

There are a number of alternative distributions that could be used to describe count (and other data). For a fuller description, see Haas et al. [28].

The most frequently used alternative overdispersed distribution is the negative binomial. This can be derived by assuming that the mean density in the Poisson distribution (Equation 4.1) is itself distributed by a gamma distribution [29]. This yields the alternative distribution for counts:

$$P_{NB}(x) = \frac{\Gamma(x+k)}{\Gamma(k)x!}\left(\frac{\bar{\upsilon}V}{k+\bar{\upsilon}V}\right)^{x}\left(\frac{k+\bar{\upsilon}V}{k}\right)^{-k} \tag{6.6}$$

As $k \to \infty$, the negative binomial distribution approaches the Poisson. This is shown graphically in Figure 6.2 with an example for a constant mean but varying k values. Note in particular that the negative binomial distributions have a greater proportion of both zeros and high counts compared to the Poisson — with these differences increasing as the k value decreases.

For the negative binomial, the relationship between the mean and the variance can be written as [30]:

$$\frac{s^2}{\upsilon V} = 1 + \frac{\upsilon V}{k} \tag{6.7}$$

Examination of this equation shows that as k decreases, the degree of overdispersion from the Poisson increases. Unfortunately, the maximum-likelihood fit of this distribution to even a simple data set cannot be expressed in closed form. However, it can be computed numerically. The R code for doing this is shown in Listing 6.1.

```
## ------Kaper et al Count Data----------------------------------------
counts=c(0, 0, 3, 4, 4, 9, 11, 3, 9, 1, 0, 0, 8, 2, 3, 7, 7, 4, 12)
## -------------------------------------------------------------------
PoissonMu=mean(counts)
VarEst=var(counts)
VarEst/PoissonMu;
PoissonP=(PoissonMu^counts)*exp(-PoissonMu)/factorial(counts)
Poisson_LogL=log(prod(PoissonP));    #Compute Poisson Log Likeihood
print(noquote(paste0('Poisson Log Likelihood = ',Poisson_LogL)))

## -----------------Compute Negative Binomial P------------------------
P_NB<-function(x,muV,k){
  A<-gamma(x+k)/(gamma(k)*factorial(x))
  B<-(muV/(k+muV))^x
  C<-((k+muV)/k)^(-k)
  return(A*B*C)
}

## ----Load Maximum Likelihood Package and Define Likelihood Function----------
loglikNB<-function(theta){
  mu<-theta[1]
  k<-theta[2]
  prob<-P_NB(counts,mu,k)
  return(sum(log(prob)))
}

## --------Calculate Maximum Likelihood for Neg. Binomial Fit-----------------
theta<-c(mu=10,k=1)
best<-maxLik(loglikNB,start=theta)
print(summary(best))
```

Listing 6.1 R program to fit Kaper et al. data to negative binomial.

The output of this code is shown as follows. The log-likelihood for the Poisson fit is computed as −60.98. The log-likelihood for the negative binomial (NB) fit is −49.663. As expected, the fit for the

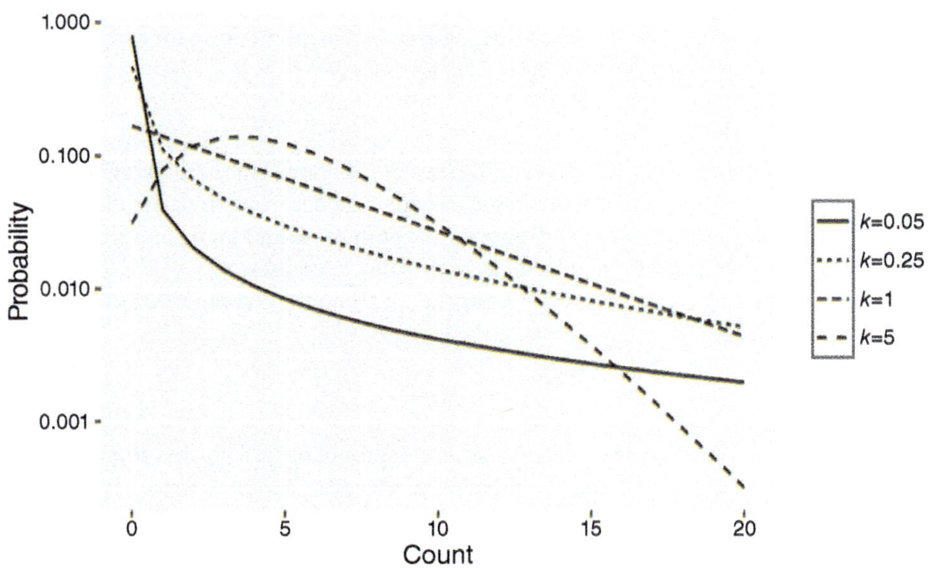

Figure 6.2 Effect of negative binomial k on probability for fixed mean = 5.

negative binomial, with two adjustable parameters and containing the Poisson as a special case, is better than the Poisson — i.e., the log-likelihood for the NB fit is greater than for the Poisson fit.

Output of Listing 6.1

```
[1] Poisson Log Likelihood = -60.9828748703943
-------------------------------------------
Maximum Likelihood estimation
Newton-Raphson maximisation, 7 iterations
Return code 8: successive function values within
relative tolerance limit (reltol)
Log-Likelihood: -49.6625
2  free parameters
Estimates:
   Estimate Std. error t value  Pr(> t)
mu    4.5789     1.0326   4.434 9.23e-06 ***
k     1.3384     0.6611   2.024   0.0429 *
---
Signif. codes:  0 *** 0.001 **
0.01 * 0.05 . 0.1   1
-------------------------------------------
```

A subsequent question is whether the improvement in fit is statistically significant. There are a number of ways to compare two models (fits) with each other, relying on the maximum likelihood fits. If one model can be regarded as a subset of another, then directly looking at the likelihoods for both of the fits can be used as a direct test of the significance of improvement. This, for example, is the case of the Poisson versus the negative binomial fit to replication data — where the former is a special case (when $k \rightarrow \infty$) of the latter.

Let the two models have parameter vectors θ_1 and θ_2 with dimensions N_1 and N_2 ($N_2 > N_1$), and model 1 is a special case of model 2. The fits have log-likelihoods for their best set of parameters equal to L_1 and L_2. Then the null hypothesis of equivalence between the fits can be rejected (i.e., the model with more parameters provides a statistically significant improvement in fit) if the deviance (D) as defined in Equation 6.8 is greater than the desired upper percentile of the χ^2 distribution with degrees of freedom equal to $N_2 - N_1$:

$$D = 2(L_2 - L_1) \tag{6.8}$$

Alternative approaches to selecting appropriate models, which are valid even when the models are not nested, are the use of what are termed "information criteria," which are derived from the likelihood function. Two of the most widely used are the Akaike information criterion (AIC) and the Bayesian information criterion (BIC) .

Both BIC and AIC methods start from the optimized log-likelihood function but add a penalty term based on model complexity, and the model that should be selected is the one yielding the *minimum* value of the criterion [31]. The BIC is computed from Equation 6.9.

$$\text{BIC} = -2\ln(L) + k\ln(N) \tag{6.9}$$

where L is the log-likelihood of the model, k is the number of parameters in the model, and N is the number of observations.

The AIC corrected for small sample numbers [32] (termed AICc) is computed from Equation 6.10

$$\text{AICc} = -2\log(L) + 2k\left(\frac{N}{N-k-1}\right) \tag{6.10}$$

Alternative techniques for model selection include cross-validation, a computationally intensive data-splitting method [33]. This is particularly suitable for larger and more complex data.

Example 6.3 Testing Improvement of Fit for Negative Binomial with Kaper et al. Data.

Given the fit of the Kaper et al. data in Listing 6.1, use the deviance criteria, AICc and BIC, to assess whether the negative binomial provides an improved fit vis-à-vis the Poisson distribution.

Solution

The Poisson and negative binomial log-likelihoods are −60.983 and −49.663, respectively. Applying Equation 6.8, the deviance is calculated to be:

$$D = 2[-49.663 - (-60.983)] = 22.64$$

For one degree of freedom, the upper 5th percentile of the χ^2 distribution is 3.84, so we reject the null hypothesis (of statistical equivalence of the Poisson and negative binomial).

The BIC is computed using Equation 6.9 for the Poisson and negative binomial as follows:

$$\text{BIC}_{\text{Poisson}} = -2(-60.983) + (1)\ln(19) = 124.910$$

$$\text{BIC}_{\text{NB}} = -2(-49.663) + (2)\ln(19) = 105.215$$

Hence, since the negative binomial BIC is less than the Poisson, the negative binomial is also preferred by this criterion.

The AICc is computed using Equation 6.10 for both distributions as follows:

$$\text{AICc}_{\text{Poisson}} = -2(60.983) + 2(1)\left(\frac{19}{19-1-1}\right) = 124.201$$

$$\text{AICc}_{\text{NB}} = -2(49.663) + 2(2)\left(\frac{19}{19-2-1}\right) = 104.076$$

Using the AICc criterion, the same conclusion of the superiority of the negative binomial distribution vis-à-vis the Poisson for this data is reached.

Previously, it was noted that the Poisson distribution could not provide a good fit to this data, based on the index of dispersion. The question arises as to whether the negative binomial provides a fit that cannot be rejected as being unacceptable. Another version of a χ^2 statistic is used to test this. The steps in this process are:

- Divide the count data into bins so that all bins contain at least five observations (hence, they may be of unequal width).
- Tabulate the observed number of observations with counts in each bin. This is denoted as O_i for the ith bin.
- From the fitted distribution, compute the number of predicted times (given the total number of observations) that counts would fall into the range of each bin. This is denoted as E_i for the ith bin.

- Compute the following quantity (where j is the number of bins):

$$X^2 = \sum_{i=1}^{j} \frac{(O_i - E_i)^2}{E_i} \tag{6.11}$$

- if X^2 is greater than the upper percentile of the χ^2 distribution at (j-k) degrees of freedom, where k is the number of distribution parameters estimated from the data, then the null hypothesis that the fit is indistinguishable from the data itself can be rejected.

Example 6.4 Goodness-of-Fit Test of Kaper et al. Data to Negative Binomial.

The observed and fitted frequency of occurrence of different counts are tabulated as follows, using three bins.

Range	Observed frequency	Expected frequency
0–3	9	9.822
4–7	5	5.367
≥ 8	5	3.811

From this table, applying Equation 6.11, an X^2 of 0.465 is computed. There is one degree of freedom (three rows minus two parameters for the negative binomial fit), so this is much less than the upper 95th percentile of the χ^2 distribution at one degree of freedom. Hence, the null hypothesis of adequacy of fit cannot be rejected.

As noted earlier, the negative binomial distribution can be regarded as a mixture of Poisson distributions where the Poisson mean is itself distributed according to the gamma distribution. This is a special case of a more general situation in which the Poisson distribution can be convolved with any other continuous distribution with support on the positive real line. A general relationship, for any such mixture distribution, "h" is shown in Equation 6.12.

$$P_M(x; V, \beta) = \int_0^{\infty} P_p(x; vV) h(v; \beta) dv \tag{6.12}$$

There are many such mixture distributions [34]. Three that have been covered in detail in the context of this section are the Poisson-inverse Gaussian (PIG), the Poisson-log normal(PLN), and the Poisson-generalized inverse Gaussian (PGIG). These are covered in more detail in the context of microbial quantification elsewhere [28]. Except for larger data sets, there will likely be no major differences between the negative binomial and the PIG and PLN distributions, all of which are two-parameter distributions. The PGIG distribution is a three-parameter distribution and so might provide some improvement in fit. These distributions are as follows:

- **Poisson-log normal**

  Parameters ζ, s

  equation $P(x) = \int_0^{\infty} \frac{(vV)^2 \exp(-vV)}{x!} \frac{}{vs\sqrt{2\pi}} \exp\left[-\frac{(\ln(v) - \zeta)^2}{2s^2}\right] dv$

  comments No analytical solution

- **Poisson-inverse Gaussian**

  Parameters υ, ϕ

  equation $P(x) = \sqrt{\frac{2}{\pi}}[\phi(\phi + 2\upsilon V)]^{-(x-1/2)}\left[\frac{e^{\phi}(\upsilon V\phi)^x}{x!}\right]K_{x-1/2}\left(\sqrt{\phi(\phi + 2\upsilon V)}\right)$

  comments K is the modified Bessel function of the third kind [35]

- **Poisson-generalized inverse Gaussian**

  Parameters υ ω, λ

  equation $P(x) = \frac{(\upsilon V)^x}{x!}\left(\frac{\omega}{\omega + 2\upsilon V}\right)^{(x+\lambda)/2}\left[\frac{K_{x+\lambda}\left[\sqrt{\omega(\omega + 2\upsilon V)}\right]}{K_\lambda(\omega)}\right]$

  comments Has, as a special case, the negative binomial and the PIG.

Several complications or elaborations exist on the basic scheme of count data outlined earlier. These are the use of multiple volumes/dilutions, the presence of too-numerous-to-count data , and the combination of these effects.

If there are multiple volumes, and it is presumed that there is a common density or concentration υ that characterizes all of the samples taken, even though they are at different volumes, V_i, then Equation 6.1 can be generalized to:

$$L\left(\overline{N}|, \overline{V}, \theta\right) = f(N_1|V_1, \theta)f(N_2|V_2, \theta) \ldots f(N_n|V_n\theta) \tag{6.13}$$

where the set of parameters (θ) includes a common mean concentration.

By extension, in this circumstance, the Poisson distribution (Equation 6.3) becomes

$$\begin{aligned} L &= \left[\frac{(\upsilon V_1)^{x_1}}{x_1!}e^{-\upsilon V_1}\right]\left[\frac{(\upsilon V_2)^{x_2}}{x_2!}e^{-\upsilon V_2}\right] \ldots \left[\frac{(\upsilon V_n)^{x_n}}{x_n!}e^{-\upsilon V_n}\right] \\ &= \frac{\prod (\upsilon V_i)^{x_i}}{\prod x_i!}e^{-\upsilon \sum V_i} \end{aligned} \tag{6.14}$$

The maximum likelihood estimate for the sample concentration can be obtained by differentiating Equation 6.14 with respect to υ and setting the derivative equal to zero. Upon doing this, it is found that:

$$\hat{\upsilon} = \frac{\sum x_i}{\sum V_i} \tag{6.15}$$

A second complication is that, in sample replicates with so many organisms that they cannot be counted, or with culture methods, adjacent colonies or plaques start to merge with each other. At best, the laboratorian can ascertain some (uncountable) number of organisms greater than or equal to a particular threshold. For nominal 50 mm diameter Petri plates used for culture, 100 is often regarded as a threshold, and for 100 mm nominal diameter plates, 300 is often regarded as a threshold—but this will depend very much on the diameter to which individual colonies of organisms grow under the incubation conditions.

Adequate consideration of such censored/TNTC counts requires a further generalization of Equation 6.13. Two additional quantities need to be defined:

- δ_i is an indicator variable, equal to 1 if observation i is known, and equal to 0 if it is only known to be $> N_i$ with "i" being the maximum countable number in the assay.

Table 6.2 Data on secondary effluent coliform measured by membrane filter.

Volume analyzed (mL)	Colonies
0.01	TNTC, TNTC, TNTC, TNTC
0.001	TNTC, TNTC, TNTC, TNTC
0.0001	72, 77, 66, 48

Source: Data from [36].

Note: The cutoff for TNTC was not given in the reference. For this analysis, an assumption that TNTC corresponds to >100 colonies is used.

- $\bar{F}(N_i)$ is the complementary cumulative distribution, i.e., the expected proportion of observations that would be *in excess* of N_i. For discrete distributions (such as Poisson, negative binomial, etc.), this can be written as:

$$\bar{F}(i) = \sum_{k=i+1}^{\infty} f(k) = 1 - \sum_{k=0}^{i} f(k) \tag{6.16}$$

Then the revised likelihood equation becomes

$$L\left(\overline{N}|, \overline{V}, \theta\right) = f(N_1|V_1, \theta)^{\delta_1} f(N_2|V_2, \theta)^{\delta_2} \ldots f(N_n|V_n, \theta)^{\delta_n} \times \bar{F}(N_1|V_1, \theta)^{1-\delta_1} \bar{F}(N_2|V_2, \theta)^{1-\delta_2} \ldots \bar{F}(N_n|V_n, \theta)^{1-\delta_n} \tag{6.17}$$

In general, even for Poisson-distributed counts, the maximum likelihood solution to this equation must be evaluated numerically. This can be done using standard non-linear optimization software. In R, this can be done by elaborating on Listing 6.1.

The use of this approach is illustrated with a specific data set. Clark [36] reported coliform tests at a wastewater treatment plant. For one sample of an undisinfected secondary effluent, the data in Table 6.2 was recorded.

The results of this analysis are shown in Listing 6.2 to the Poisson. With a reprogramming to include alternative distributions, the function `maxLik` can be used to maximize the log-likelihood function.

```
library(maxLik)

volumes=c(0.01,0.01,0.01,0.01,0.001,0.001,0.001,0.001,0.0001,0.0001,0.0001,0.0001)
counts=c(100,100,100,100,100,100,100,100,72,77,66,58)
deltaindex=c(0,0,0,0,0,0,0,0,1,1,1,1)
nobs=12
#-- Function for Complementary Cumulative Distribution
CompCumDistF_Poisson<-function(x,mu){ # calculate complementary cumulative
  y<-sum(dpois(0:x,mu)) # use built in function for Poisson
  return(1-y)}

#-- Define Likelihood Function
loglikPoiss<-function(mu){
  summation=0
  for (i in 1:nobs) {
    if(deltaindex[i]==1){
      ith_term=(dpois(counts[i],mu*volumes[i],log=TRUE))}
    else{ith_term=log(CompCumDistF_Poisson(counts[i],mu*volumes[i]))}
```

```
    print(ith_term)
    summation=summation+ith_term}
  return(summation)}

# -- Call Optimization and Display Results
mu_guess=50000
m <- maxLik(loglikPoiss, start=c(mu=mu_guess),method='NM')
summary(m)
```

Listing 6.2 R program to fit Clark data with TNTC's to Poisson distribution.

The output of this fit is shown as follows, with the maximum likelihood found as –13.60 and the value of $\hat{\mu}$ of 682,500 organisms per mL.

Output of Listing 6.2

```
Maximum Likelihood estimation
Nelder-Mead maximization, 36 iterations
Return code 0: successful convergence
Log-Likelihood: -13.60442
1  free parameters
Estimates:
    Estimate Std. error t value Pr(> t)
mu 682500.00      16.78    40680   <2e-16 ***
---
Signif. codes:  0 *** 0.001 ** 0.01 * 0.05 . 0.1   1
```

The improvement of fit with distributions other than Poisson, even in this censored case, can be determined using the AICc, BIC, or chi-squared approaches noted earlier. Goodness of fit can be ascertained using an approach as in Example 6.4 — recognizing that with censored data, the top range bin must be the censored data, e.g., > 100.

6.2.2 Quantal Methods

Quantal methods or assays are those in which the response of a measurement is only the presence or absence of the microorganism or signal of the microorganism of interest. With suitable statistical assumptions, results of such experiments can be used to estimate the *amount* of organisms or signals present per unit amount of sample.

Consider the hypothetical in the center panel of Figure 4.4. Among 10 replicate subsamples, 6 were positive. If a count assay was performed, it was clear that the average number of organisms per subsample was 0.8. In fact, this same 0.8 per subsample occurs in both the left (overdispersed) and right (underdispersed) panel of this figure, the proportion of positive subsamples is 0.2 in the overdispersed and 0.8 in the underdispersed case.

There are several specific examples of this type of assay. The experimental details of the individual methods will be discussed in the next section, but a common approach to data analysis from quantal methods exists, which is set forth.

There are numerous examples of quantitative assays for different microbial groups using various technologies.

Perhaps the earliest approach in this protocol is the use of fermentation tubes to assess gas formation during bacterial growth (possibly including a pH indicator to assess acid production), which

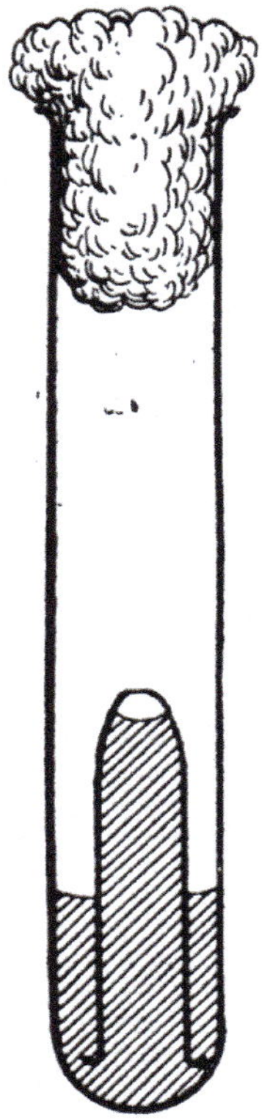

Figure 6.3 Fermentation tube design of Durham [37].

was an early approach for assessing coliform. Durham [37] designed an inverted small tube contained in a larger tube into which samples were inoculated (Figure 6.3). Various modifications to this design were made over subsequent years [38]. Growth results in discernible gas accumulation at the top of the small tube, indicating the presence of a particular organism. These tubes were elaborated upon in the early 20th century. By incorporating multiple tubes and using multiple dilutions, it was shown, e.g., by McCrady [39] that the results could be used to assess concentrations.

Quantal assays have been used for virus culture. In this embodiment, "lawns" of host cells are in replicate containers. Then, after incubation, the presence of any viral infectivity is indicated by the destruction of some or all of the host cells—in a process termed "cytopathic effect" [40] (CPE). A similar tissue culture-based assay for infectious oocysts of *Cryptosporidium parvum* was devised by Slifko et al. [41].

Quantal assays can also use colorimetric or other response measures to ascertain the "positivity" of the response of each subsample.

In recent years, small-volume 96-well plates have become available, which are especially suitable for polymerase chain reaction (PCR) methods (to be discussed in Section 6.6.2). A quantal analysis technique can be applied to this assay. As an example, Banting et al. [42] devised a quantal PCR assay in such a system for the quantification of *Campylobacter* in environmental waters.

More recently, droplet digital PCR (ddPCR) has been developed. In this system, a small sample is mixed with reagents for the PCR assay and then subdivided into a large number of nanodroplets (each with a volume much lower than 1 υL). Then the PCR reaction(s) are conducted, and the number of positive nanodroplets is counted, yielding a quantal measurement.

A final type of quantal assay involves using animals to bioassay microbial populations. The endpoint of such an assay is when an individual animal dosed with a subsample is positive, typically defined by observable symptomatology or death. An example of such an assay is the quantification of viable cysts of *Giardia* by inoculation into mice [43]. This use of animal assays is quite analogous to the experiments that are used to develop dose response relationships — which will be discussed in greater detail in Section 14.1.

6.2.2.1 Statistics of Quantal Methods

Having shown that many experimental techniques share the characteristics of quantal assays, a common framework for analyzing such results can be presented. There is a long history of analysis of quantal experiments in various contexts, perhaps the earliest being the interpretation of coliform MPN results by McCrady [39].

The layout of such experiments is that there are a number of "volumes" used per subsample.[2] There are "i" volumes per subsample, with the volumes denoted as $V_1, V_2 \ldots, V_i$. For each volume level, a total of T_i individual assays are conducted in which that volume, V_i is used per assay. At the end of the experiment, of the T_i individual assays, P_i are determined to have a positive response, i.e., to have contained one or more target organisms. By difference, there are also $T_i - P_i$ negatives at the ith level.

If the Poisson distribution can be applied to describe the distribution of infectious units among replicate subsamples, i.e. there is a random distribution, then from Equation 4.3, the probability of finding no infectious units in a subsample of volume V_i given a mean concentration of υ is:

$$p(0) = \exp(-\upsilon V_i) \tag{6.18}$$

And by application of the complement rule of probability, the probability of any non-zero number of organisms being present is:

$$p(\bar{0}) = 1 - \exp(-\upsilon V_i) \tag{6.19}$$

If each of the replicates in the subsample behaves independently and identically from one another, then the binomial equation can be used to construct the likelihood (in a single volume case) of observing the number of positive tubes:

$$L = [1 - \exp(-\upsilon V)]^P [\exp(-\upsilon V)]^{(T-P)} \left[\frac{T!}{P!(T-P)!} \right] \tag{6.20}$$

Using the maximum likelihood principle, Equation 6.20 can be differentiated with respect to υ, and setting the derivative equal to zero, the best estimate can be found to be:

$$\hat{\upsilon} = -\frac{1}{V} \log\left(\frac{T-P}{T}\right) \tag{6.21}$$

Note that the factorial terms in Equation 6.20 drop out since they are not a function of υ. Also note that if all replicate subsamples are positive, then the value of $\hat{\upsilon}$ is indeterminate.

---

2 The term volume here is generic and could include analysis of solid samples, with mass replacing volume as the index of amount assayed per subsample.

If there are multiple sets of subsamples at different volumes, i.e., $i > 1$, then the likelihood in Equation 6.20 and the log-likelihood can be generalized to (ignoring the factorial terms):

$$L = \prod_{k=1}^{i} \{[1 - \exp(-\upsilon V_k)]_k^P [\exp(-\upsilon V_k)]^{(T_k - P_k)}\} \tag{6.22}$$

$$\log(L) = \sum_{k=1}^{i} \{P_k \log([1 - \exp(-\upsilon V_k)]) - (T_k - P_k)(\upsilon V_k)\} \tag{6.23}$$

In general, this cannot be solved analytically and must be solved numerically. Historically, in the case of frequently used combinations of volumes and numbers of subsamples, tables of best estimates (MPN values) for υ as a function of positive numbers at each volume were prepared and published, as in one example [44]. However, it is now more straightforward to compute the estimates for any combination by a simple program.

In Listing 6.3, the maximum likelihood solution, using the aforementioned log-likelihood function for a sample problem, is computed using R. In this case, there are three volumes (10, 1, and 0.2) with 10, 9, and 11 subsamples at each of those volumes, of which 9, 4, and 1 are positive.

```
library(maxLik)

volumes<-c(10,1,0.2) # Volume of each subsample
aliquots<-c(10,9,11) # Number of subsamples
positives<-c(9,4,1) # Number of positives

logL<-function(mu){
  A=positives*log(1-exp(-mu*volumes))
  B=(aliquots-positives)*mu*volumes
  return(sum(A-B))
}
guessmu=0.3
best<-maxLik(logL,start=guessmu)
summary(best)
```

Listing 6.3 R program to fit multiple-volume quantal assay.

Output of Listing 6.3

```
Maximum Likelihood estimation
Newton-Raphson maximisation, 3 iterations
Return code 1: gradient close to zero (gradtol)
Log-Likelihood: -13.77473
1  free parameters
Estimates:
     Estimate Std. error t value Pr(> t)
[1,]   0.3253     0.1087   2.991 0.00278 **
---
Signif. codes:  0 '***' 0.001 '**' 0.01 '*' 0.05 '.' 0.1 ' ' 1
```

Note that this solution estimates confidence limits. Although there are other approaches, a common way to do this is to compute the inverse of the Hessian matrix (the matrix of second derivatives) or the parameter estimates, which is asymptotically the variance/covariance matrix. A fuller discussion of confidence limits of quantal assays is beyond the scope of this presentation, but is presented in more detail in [28].

The maximum likelihood approach for estimating the most probable concentration, v, from quantal experiments, relies on the assumptions of a random (Poisson) distribution of infectious units and the independence of subsamples within each dilution. As shown previously, these assumptions could be tested using count assays. There are several approaches to testing these assumptions in quantal assays. Further discussion is presented elsewhere [28], and these approaches are briefly outlined as follows.

In one approach, rather than the Poisson zero-probability term (Equation 6.18), the probability of zero organisms from an alternative distribution, such as the negative binomial, could be used to construct the likelihood function to be maximized. This would necessitate estimating other "nuisance" parameters, such as the negative binomial "*k*." The fit could then be compared to the fit using the Poisson assumption, such as by the use of the AICc or BIC tests, to ascertain whether there was evidence for rejecting a Poisson assumption.

A second approach was outlined by Stevens [45], wherein, for a series of samples, the range of transition over the dilution series is examined and compared with what a theoretical distribution would be under the Poisson assumption. The use of this approach in testing a long series of coliform analyses of drinking water intake in Chicago showed some differences from the assumptions [46]. Whether this resulted from methodological issues or intrinsic issues was unclear. However, in the presence of such deviations, more robust estimation methods might be used.

6.2.3 Time-to-response Methods

A third type of microbial measurement can be used to quantify microbial populations. The time course of development of some response in a system related to microbial characteristics can be associated with the initial microbial level in the system. At a basic level, when viable microorganisms are inoculated into a suitable growth medium, they undergo an initial adaptation (lag phase) followed by rapid growth, a stationary phase, and decline as noted in Figure 6.4.

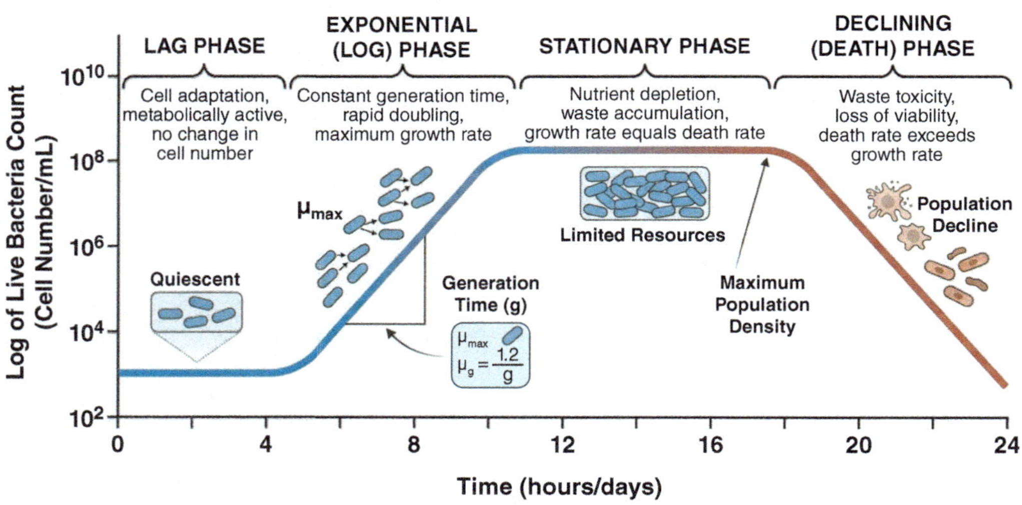

Figure 6.4 Schematic microbiological growth curve. *Source:* From wikipediacommons, Creative Commons 3.0. Attribution-ShareAlike (CC BY-SA 3.0).

It has long been known that turbidity can be used to measure bacterial numbers if a standard curve is developed [47]. Gavin [48] further noted that a quantitative method can be developed "based on the measurement of the rate of growth of the test organism [during] short, predetermined incubation periods or [when] incubation is terminated when a definite turbidity value is reached in a control tube."

There are several examples of this type of assay. The most basic, as implied earlier, is the time to reach a given turbidity in a suitable growth medium. This can be modified by using selective colorimetric indicators for specific groups of organisms. As an example, Warren et al. [49] developed an assay in which a specific compound metabolized by fecal coliform organisms generated a colored product. A standard curve could be created for the intensity of color produced after a specific, short incubation as a function of the initial coliform concentration and then used as an assay to estimate the initial coliform concentration in unknown samples.

More recently, as described further, the PCR for the specific amplification of targeted microbiological groups can be used in a format called quantitative PCR (qPCR). Essentially, this approach uses the time for the buildup of amplified nucleic acid as an indicator of the original concentration of target organisms. By calibrating against a standard curve, the initial concentration in the sample can be estimated.

Since time-to-response methods are highly assay-specific, details of the analytical approach for quantification will be presented in the context of the method—primarily for qPCR.

6.3 Direct Microscopy

The Dutch scientist Antonie van Leeuwenhoek is credited with the invention of the first microscope and observing what are now recognized as bacteria and protozoa by microscopy [50] about 350 years ago. From such an early invention, optical microscopy has evolved considerably, and the use of electron microscopy has enabled the study of smaller organisms and smaller features within microorganisms.

Microscopy can be used for qualitative analysis to detect what organisms are present and for quantitation. Due to the wavelength of visible light, optical microscopy is most useful for large viruses, bacteria, and other larger organisms, while electron microscopy is required for most viruses (Figure 6.5).

For ordinary light microscopy, a practical lower limit to resolution is about 230 nm in the horizontal direction and 1000 nm (1 υm) in the vertical direction [51].

6.3.1 Optical Microscopy

A schematic of a basic compound optical microscope is shown in Figure 6.6. Because this allows all light to pass through the sample, this is often called a bright-field microscope. Illumination from below passes through a microscope slide on a stage, fitted with devices to move the slide horizontally. The light passing through the slide first passes through the lower objective lens and then is further magnified by the eyepiece lens. Typically, the overall magnification ranges from 50-fold to 1000-fold, with higher magnifications requiring oil immersion, in which a special oil is used in contact with both the slide cover slip and the microscope objective to enable closer focus. One fundamental limitation of any microscopic technique is that the wavelength of the light used limits the size of material in a sample that can be differentiated. This is expressed in terms of the resolving power (RP) [52]. This is related to the numerical aperture of the microscope objective lens, a measure of how much light passing through the sample is collected. For optical microscopy

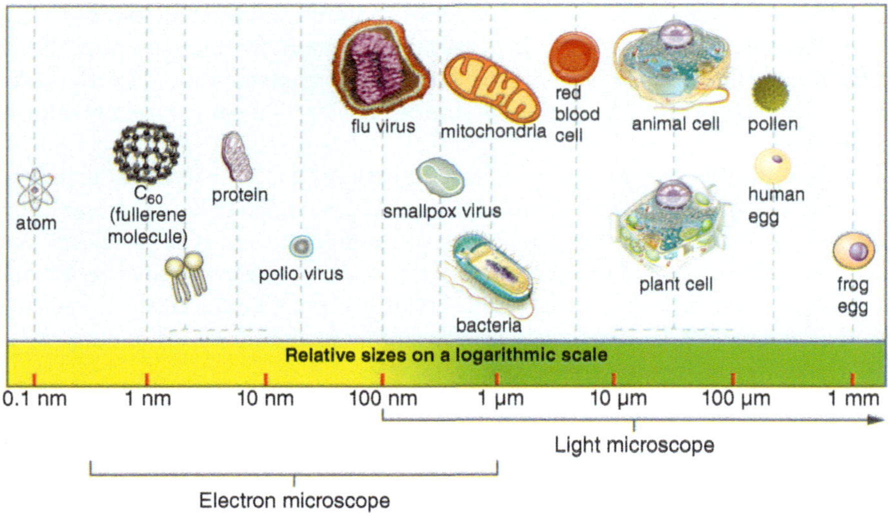

Figure 6.5 Relationship between sizes of microorganisms and use of types of microscopy. *Source:* OpenStax/Rice University / CC BY 4.0 / https://openstax.org/apps/archive/20240130.204924/resources/ccb2d41136898806a7dd14ca9743309fe9f1bc71 / last accessed Feb. 14, 2026.

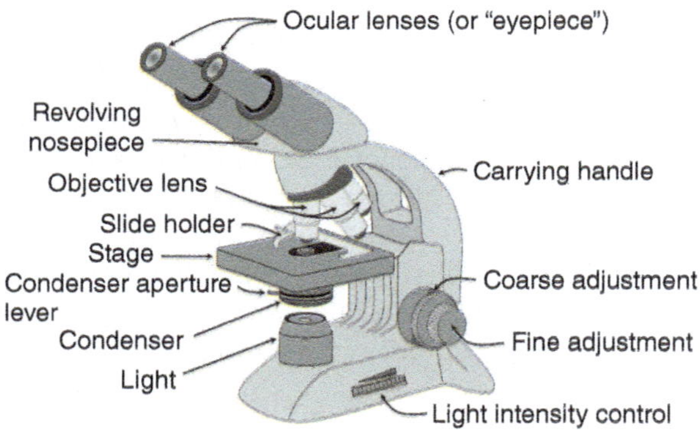

Figure 6.6 Parts of a basic optical microscope. *Source:* Mikael Häggström / CC0 1.0 / https://commons.wikimedia.org/wiki/File:Parts_of_a_light_microscope_(english)_-_larger_text.png / last accessed Feb. 10, 2026.

lens, the numerical aperture (NA) is a function of the lens design, focal length, magnification and the use of oil immersion, and generally ranges from 0.1 to 1.4. In turn, the resolving power of a microscope can be given by Equation 6.24.

$$RP = \frac{\lambda}{2\,NA} \tag{6.24}$$

As an example, for 600 nm wavelength light (orange-red on the spectrum), and an objective with numerical aperture of 0.5, $RP = \frac{600\,\text{nm}}{2 \times 0.5} = 0.6\ \mu\text{m}$. So details within individual bacterial cells, for example, would not be well differentiated under these conditions.

Another challenge in the microscopy of many samples to examine microbiological contents is the lack of contrast between the suspending fluid and the microorganisms. As a result, one technique that can be employed is chemical pretreatment of the sample with a stain prior to microscopy. This can also be used to assist in identifying specific organisms or characteristics. As noted in Section 2.1.3, the Gram stain for bacteria is used for taxonomic purposes and can also assist microscopy. Examples of staining, though not exhaustive, are given below. In addition, some individual stains may share multiple attributes of the list.

Contrast staining: One approach to increasing contrast and visual detection is to treat the sample with a dye that may be either selectively taken up or selectively rejected by the microorganism(s) under investigation. With negative staining, a dye that is typically negatively charged to the microorganism is added to color the suspending fluid, and is rejected by the microorganism [53].

General-purpose positive stains that are selectively taken up by microbial components will increase the contrast of the microorganism. One example is iodine, which can accumulate in starches and carbohydrate structures.

Staining of particular components: Dyes can be used that are selectively accumulated in different organelles or by different biomolecules. As a common example, there are a number of dyes that selectively associate with nucleic acids, and can be used to visualize nuclei and nuclear regions in cells—most of these require the use of fluorescent microscopes, described as follows [54]. In particular circumstances, stains also exist to visualize bacterial capsules and flagella, as well as eukaryotic mitochondria.

Differential staining: This technique typically uses two or more stains and often provides a visual difference between two or more taxonomic groups. One example is the Gram stain, which can differentiate bacteria into two groups based on their cell envelopes (Section 2.1.3). The acid-fast stain can be used to differentiate bacteria in the genera *Mycobacterium* and *Nocardia*, which both have extremely hydrophobic cell envelopes from other bacteria [55].

A more recent approach to differential staining is fluorescence in situ hybridization (FISH) staining. In this method, a DNA segment specific to the microbial group of interest is bound to a fluorescent chemical moiety, and this construct is contacted with the sample under conditions that allow it to chemically bind to the nucleic acid of the target organism (which may be DNA or ribosomal RNA). Then the preparation is examined under a fluorescence microscope to detect organisms in which that tagged probe has accumulated [56].

Viability or activity staining: The microorganisms to be visualized can be treated with chemicals that have reactions or binding properties indicative of specific metabolic activity of properties associated with viability. One commonly used class of dyes are tetrazolium salts [57]. These are compounds generally with at least four aromatic rings, which can be taken up by cells, and can serve as an alternative electron receptor *in vivo*. When reduced in this way during active metabolism, they become colored and can be visualized by microscopy. Hence, the presence of active metabolic activity can be assessed microscopically.

A second widely used approach is to treat cells with dyes that, when the membranes are intact, are ordinarily excluded— essentially a negative viability stain. Williams et al. showed that propidium iodide could be used as a negative stain, with exclusion being indicative of viability [58]. For easier visualization, this can be combined with a second "universal" stain that binds to all nucleic acids—such an approach has been commercialized, including in the kit trademarked "Bac Light" [59].

A limitation of viability staining is that the mechanisms of inactivation, such as by a disinfectant, must occur in a way that produces the appropriate response with the stain. For example, with UV inactivation, not all viability staining methods correlate well with loss of viability [60].

An alternative to, or a supplement for, staining for visualizing microorganisms in a sample is the use of equipment that provides different means of illumination. There are a number of examples of such alternatives, which generally require more complex equipment than a simple direct illumination optical microscope.

The simplest and perhaps earliest illumination modification was the development of dark field microscopy. In this approach, light from the central portion of the condenser (Figure 6.6) is blocked, and only an annular light is allowed to pass through the stage. Then the only visible imaging results from light scattered through the sample from the annular illumination. Therefore, material may be visualized against a dark background.

Phase-contrast microscopy can be used by modest adaptation of the configuration of the standard optical microscope [61]. It relies on the difference in refractive index between the particles or microorganisms being examined and the suspending fluid. This causes a phase shift in light passing through a sample, and when viewed with appropriate optics, it can provide a higher-contrast image in unstained samples compared to standard microscopy. The switch from bright field microscopy to phase-contrast microscopy requires the installation of two specialized components:

- A phase-contrast objective, which has a phase ring near the back lens that interacts with a matching phase-contrast condenser.
- A phase-contrast condenser, which has an annulus (a ring-shaped diaphragm) that matches the numerical aperture of the phase-contrast objective. It focuses light onto the phase ring in the objective.

The improvement in contrast of unstained microbial samples is shown in Figure 6.7.

Fluorescence microscopy relies on illumination of the sample with light of one wavelength (the excitation wavelength) and then the visualization of emitted light from the sample in response to that illumination [62]. Specialized optical paths are used to filter out the excitation illumination from the emitted illumination. There are multiple modes and types of instruments. The excitation can come from below the sample, and the emitted light comes directly from the sample. Alternatively, the excitation and emitted light can come from different directions. The fluorescence may arise from the emission of autofluorescing molecules within the sample or from dyes that selectively bind to cellular components (some of which are discussed earlier) and emit light upon excitation, which can assist visualization.

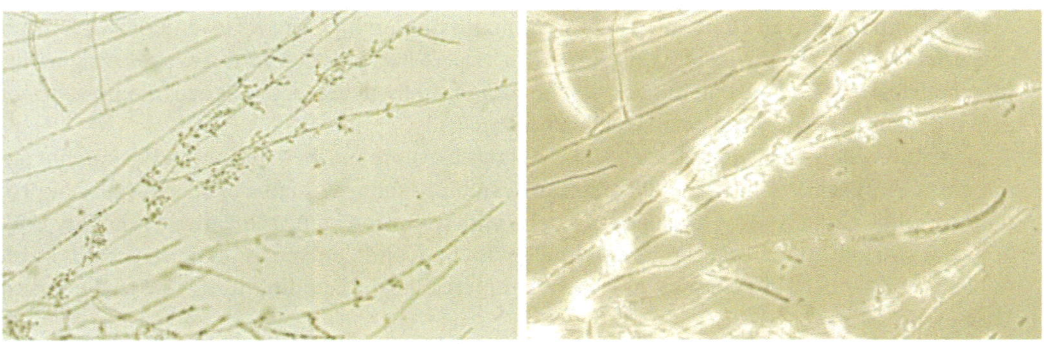

Figure 6.7 Micrograph of *Candida albicans* with bright field (left) and phase-contrast (right) illumination. *Source:* (Left) Centers for Disease Control and Prevention. Public domain. https://phil.cdc.gov/Details.aspx?pid=22947 (last accessed February 10, 2026). (Right) Centers for Disease Control and Prevention. Dr. Hardin. Public domain. https://phil.cdc.gov/Details.aspx?pid=22947 (last accessed February 11, 2026).

Differential interference microscopy (DIC) is a related technique to phase contrast. In DIC, a polarized light beam is split, with one portion of the light rotated. Then, one portion is used to illuminate the specimen and is recombined with the second portion. Interference patterns between the two beams depend on the optical path differences in the specimen, which are related to the product of path length and refractive index. This often provides greater contrast than in phase-contrast microscopy and produces an appearance of depth in the sample.

6.3.1.1 Quantification by Optical Microscopy

Most of the optical techniques described earlier can be used to quantify the number of microorganisms present in a sample. To do so, it is typical to use a special microscope slide constructed to contain known volumes of fluid in which the microorganisms are suspended.

One type of slide is termed a hemocytometer (since these were originally designed for quantifying blood cells). In Figure 6.8, a top view and a slide view of one type of such slide are shown. The depth between the bottom of the cover slip and the slide is 0.1 mm, and the slide is ruled with precise grinding.

A typical set of grid rules is shown in Figure 6.9. There are grid squares of various sizes. Each of the large nine squares is 1 mm on a side. Each of these is further divided into 0.25×0.25 mm (0.0625 mm^2), 0.25×0.20 mm (0.05 mm^2), and 0.20×0.20 mm (0.04 mm^2). The middle square is subdivided into 0.05×0.05 mm (0.0025 mm^2) squares. The depth of liquid on the slide is 0.1 mm. Hence, if organisms present in a single one of the largest squares are counted, then a total volume of 0.1 mm^3 (or 100 nL) is quantified.

The small volume that can be contained on the slide necessitates either a high concentration of organisms in the sample prior to examination or pre-concentration by one of the techniques described previously. An example illustrates this—if one organism is detected in a volume of 100 nL, this corresponds to 10,000 organisms per mL or 10^7 /L, which is very large.

6.3.2 Electron Microscopy

Two major types of electron microscopy have been used to analyze microorganisms. These are transmission electron microscopy (TEM) and scanning electron microscopy (SEM).

TEM operates by passing a high-energy electron beam through a thin sample. The interactions between electrons and atoms within the sample reveal information about the sample's structure

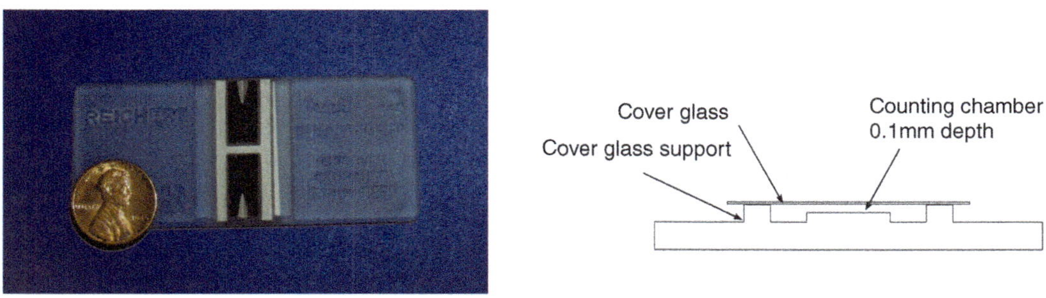

Figure 6.8 Hemocytometer top view and side view. *Source:* (a) Jeffrey M. Vinocur / CC BY 2.5 / https://commons.wikimedia.org/wiki/File:Hemocytometer_with_pennyJPG#/media/File:Hemocytometer_with_pennyJPG / last accessed Feb. 11, 2026. (b) Ewen. Public domain. https://commons.wikimedia.org/wiki/File:Haemocytometer_side.svg (last accessed February 10, 2026).

Figure 6.9 Example of a Neubauer hemocytometer. The corner squares (W1–W4) each contain 0.1 μL. The center square is divided into 25 smaller squares. *Source:* From https://en.wikipedia.org/wiki/File:Haemocytometer_Grid.png under Creative Commons Attribution-Share Alike 3.0 Unported license.

and composition and enable imaging at finer resolutions than optical microscopy. Some differences between TEM and optical microscopy are:

- TEMs use electrons instead of light. Because electrons have a much shorter wavelength than visible light, they can be used to achieve much higher resolution images.
- The sample must be very thin (less than 100 nm) for electrons to penetrate effectively.
- There is absorption and scattering depending on the composition and density of the sample. Typically, TEM is conducted on dried materials.
- The unscattered and minimally scattered electrons that pass through the sample create an image based on the variations in intensity. Denser or thicker regions of the sample block more electrons, appearing darker in the picture.
- Electromagnetic lenses, rather than optical lenses, are used to focus and direct the incident electron beam.
- Transmitted electrons are detected using a fluorescent screen or a specialized camera, creating a visible image that can be analyzed.

The effective wavelength (λ) in TEM is given by application of Planck's law (Equation 6.25) [63].

$$\lambda = \sqrt{\frac{h^2}{2m\, e\, V}} \tag{6.25}$$

where h is Planck's constant, m is the mass and e the charge of the electron, and V is the accelerating voltage of the TEM beam through the sample. This can be used in Equation 6.24 to compute the resolving power. In an example by Penczek [63]:

> "For an accelerating voltage 100 keV, λ = 0.0037 nm. The theoretical resolution of electron microscopes is 0.23 nm for an accelerating voltage 100 keV (λ = 0.003701 nm) and 0.12 nm for 300 keV (λ = 0.001969 nm), and is mainly limited by the quality of electron optics. "

This theoretical resolution enables electron microscopy to see finer details of viruses (as well as bacteria and protozoa). However, operational aspects of TEM and sample preparation are critical and typically require highly specialized knowledge. In addition, care must be taken to ensure that the higher voltages required to achieve fine resolution do not introduce artifacts due to sample alteration.

An example of TEM visualization is shown in Figure 6.10, which depicts an adenovirus micrograph. This virus is typically 90–100 nm in size, and some substructural features are visible.

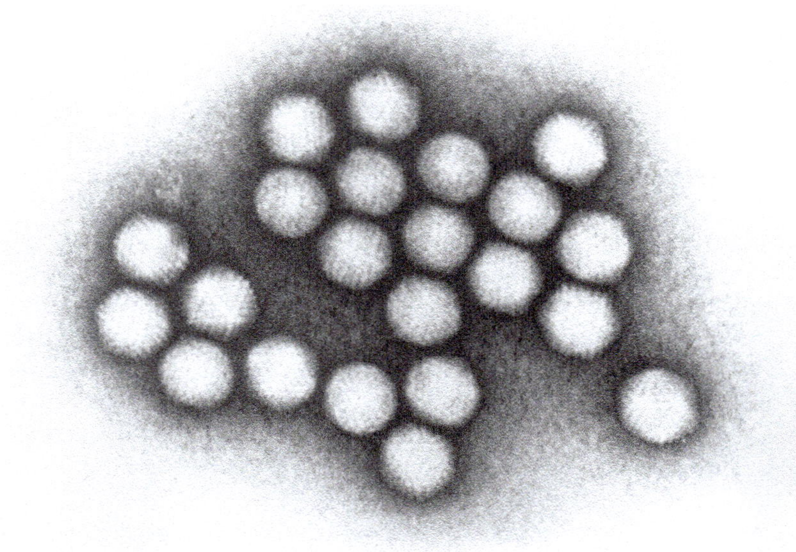

Figure 6.10 Transmission electron micrograph of adenovirus. *Source:* Centers for Disease Control and Prevention. Public domain. https://phil.cdc.gov/Details.aspx?pid=237 (last accessed February 10, 2026).

TEMs operate under high vacuum, and therefore, samples must be dried. Samples must be thin to facilitate penetration of the electron beam. As with optical microscopy, staining is needed to enhance contrast.

With viruses, the size of individual virions is small enough so that a concentrated sample can be placed directly on a prepared EM grid. There are different negative stains with various characteristics, including uranyl acetate, phosphotungstic acid, and ammonium molybdate. Following negative staining and drying, the grids can be examined by TEM [64].

With higher organisms, their size generally requires more elaborate preparation prior to TEM. Silva [65] has discussed workflows for bacteria, and Lambrecht [66] for protozoan cysts and oocysts. There are a number of options, but briefly, the flow goes through the following steps:

Fixation. The sample is treated with reactive chemicals, such as organic aldehydes or osmium tetroxide, to stabilize (and cross-link) key organic molecules. Osmium tetroxide will also act as a positive stain.

Dehydration and Embedding. The sample is dehydrated. A variety of techniques can be used, including exposure to a graded series of alcohols, replacing water with organic solvents. The sample is then embedded in a polymeric matrix.

Ultrathin sectioning, staining and mounting. The embedded sample is then sliced thinly using an ultramicrotome, mounted on a grid, and, if desired, stained further (with a positive or negative stain).

Visualization. The mounted sample is then inserted into a TEM for visualization.

This process can enable the investigation of structural features within microorganisms and the differentiation of taxa among organisms; it has been found particularly useful for viruses [67].

SEM uses a beam of electrons that impinge upon a sample, but rather than (as in TEM) examining the electrons that pass through the sample, it examines the electrons that scatter from the sample. By rastering the incident beam (similar to that in a cathode-ray television tube), the image of scattered electrons can be used to reconstruct a picture of the surface. Since there is no longer a need to pass electrons through the sample, preparation of thin sections or films is not required; instead, a different set of sample preparation workflows is used.

In conventional SEM (CSEM), as with TEM, the entire apparatus must be kept dry and under high vacuum to minimize electron beam dissipation and attenuation. More recent modifications, termed environmental SEM (ESEM), allow the sample to be contained with a pressure of 0.015–0.03 atm, which might include water vapor. Therefore, there are fewer constraints to sample preparation [68].

In conventional SEM, due to the high vacuum, sample preparation is required to remove water and fix the sample.

Suspensions of bacteria and higher organisms can be subjected to a successive process of fixation with aldehydes, staining, fixation with osmium tetroxide and uranyl acetate, dehydration with an ethanol series, and then with a series of organic nonvolatile solvents, such as hexamethyldisilazane. Between successive treatments, the bacteria are pelleted by centrifugation. Then, the final pellet is placed on electron microscope grids and sputter-coated with a thin metal layer, such as iridium or gold [69]. Similar preparations can be used for microorganisms on solid substrates, such as soils or particles [70].

An alternative approach can be used for viruses and bacteria, and these treatments are done after the microorganisms have been collected on a suitable membrane filter. After treatment, the filter is sputter-coated, and a portion is directly mounted on a support grid for examination [71].

An example of the detail that an SEM can provide is shown in Figure 6.11, which is an image of a cell of *Vibrio vulnificus*. This is a flagellated organism, and the flagella are clearly seen in the image.

While electron microscopy (TEM and SEM) is a powerful tool for visualizing microorganisms and understanding the environments and niches in which they exist, it has limited value for absolute quantification. The effective volume of material that can be examined in SEM and TEM is even smaller than in optical microscopy, so microorganism concentrations in a sample need to be much higher to be visualized.

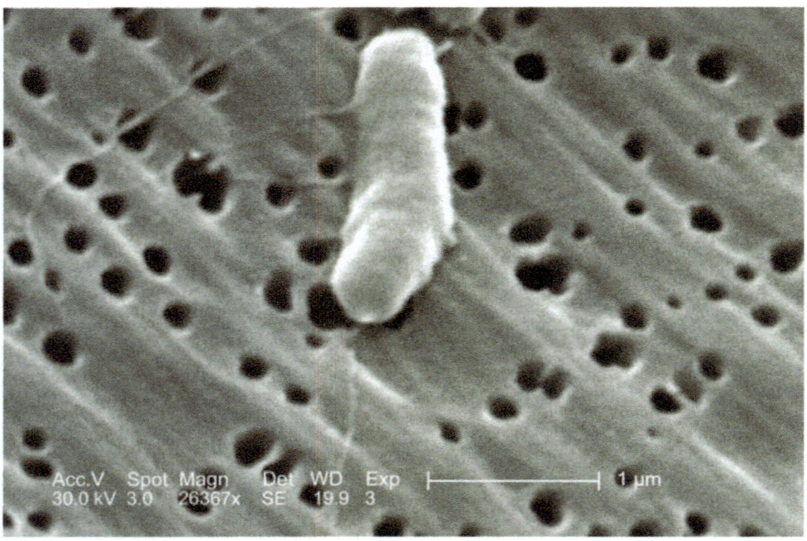

Figure 6.11 SEM of *Vibrio vulnificus*. *Source:* Centers for Disease Control and Prevention. Janice Haney Carr. Public domain. https://wwwn.cdc.gov/phil/Details.aspx?pid=6935 (last accessed February 10, 2026).

6.4 Flow Cytometry

Flow cytometry is a technique in which individual organisms are allowed to flow through an engineered channel in which focused light beams, typically from lasers, are impinged. The scattered light and the attenuated light can be measured. Fluorescent dyes can be used, or natural fluorescence can be examined. Multiple wavelengths of light can be used simultaneously. Instruments are available with multiple lasers and multiple channels for light measurement, enabling the simultaneous sensing of dozens of signals [72]. For each organism passing through the sensing channel, these optical responses can be determined, and thus, the distribution of each signal in the collection of organisms is assessed.

Since flow cytometry relies on the passage of the sample through small fluid channels, removal of other particles to the maximum extent possible would facilitate analysis. The residual particles that remain would generally make total light scattering not useful, and targeting of specific responses is required. While certain microorganisms (e.g., algae with chlorophyll and intrinsic nucleic acid fluorescence), generally, the use of dyes (analogous to stains in microscopy) to assess particular characteristics of the microorganisms can be effective. Such dyes include nucleic acid stains or DAPI/PI (to assess viability). If multiple dyes or detection channels are used, the response *of each particle* passing through a channel can be assessed, and this concomitant response can permit a more detailed characterization.

Flow cytometry has advantages over microscopy, PCR, or culture methods in that it can be conducted in minutes and may have a detection limit (in the fluid supplied to the instrument) as low as 10 organisms/mL [73].

By coupling fluorescent dyes to biomolecules targeting specific features of the organism, it is possible to achieve some level of taxonomic specificity. One approach is using fluorescently labeled antibodies [72, 73], often to cell membrane, wall or (for virus) capsid materials.

A more recent approach has been the adoption of FISH, a microscopy technique, for flow cytometry [74]. In this method, a dye is coupled to a nucleic acid probe. Then, the microbial sample is contacted with this probe under suitable conditions to allow the probe to penetrate and bond (hybridize) to the target nucleic acid. After washing the unbound probe, the sample can be analyzed by flow cytometry to detect the target microorganisms (by fluorescence) to which the probe has bound.

Safford and Bischel [75] did a systematic review of applications of flow cytometry in the water and wastewater sectors. There have been many applications with respect to the general microbial ecology of individual systems and processes. With particular focus on the subject of this book, they noted the following applications:

- Detection of protozoa (*Cryptosporidium parvum, Giardia, Toxoplasma gondii*), bacteria (*Legionella pneumophila, Escherichia coli O157:H7*), and virus (Adenovirus) in various water environments.
- Source water and receiving water pathogen measurements, including assessment of bacteria in VBNC states, have used this technique.
- Vital staining to assess responses of pathogens as well as heterotrophic degraders to conditions during treatment, including disinfection processes.
- Removal of pathogens during processes (including membrane treatment) employed for wastewater reuse.
- Regrowth of pathogens and heterotrophic organisms during drinking water distribution

In general, there is a lower limit to the size of particles that can be detected by flow cytometry. This is typically about 0.5 μm, though with particle treatment to achieve a higher refractive index, the resolution may decrease to 200 nm [76].

However, at small particle sizes, there is a potential that colloidal, virus-like particles may be produced by certain stains used and could generate artifacts. Therefore, viral analysis of environmental samples, especially for smaller-diameter viruses, may require specialized nanoscale flow cytometers that are under development [77].

Flow cytometry can be coupled with technology that separates individual organisms with specific signal profiles, a process called cell sorting. When signals are generated by fluorescence, this is termed fluorescence-activated cell sorting (FACS). The separated cells, now enriched from the original sample, can then be taken for further examination, possibly genetic characterization, or, when the organisms remain viable, for culturing. For example, in situ hybridization with a dye-tagged genetic element specific to a target organism can be used to pretreat cells passed through FACS, and the selected cells further examined [78].

6.5 Culture Methods

The importance of growing bacteria was appreciated as early as the mid-19th century by Koch, who also developed solid media using gelatin and then agar for the formation of bacterial colonies [79]. By definition, culture methods require the growth of microorganisms under suitable conditions for their examination, characterization, and quantification. This means that organisms that can exist in a VBNC, such as some bacteria (Section 2.1.3), may be difficult or impossible to culture. Furthermore, there is no universal culture medium or set of incubation conditions that allows *all* organisms to grow; therefore, culture responses are operationally defined by the conditions used.

6.5.1 Pre-enrichment or Pre-selection

Prior to placement of a sample on or into a culture medium, a pre-enrichment or pre-selection step may be useful. Pre-enrichment is the brief exposure of the microbial sample to relatively rich, non-selective conditions to enable resuscitation (e.g., of VBNC organisms), repair, or emergence from the stationary phase. An example of this is in the culturing of *Salmonella* or *E. coli* prior to plating or incubation under more selective conditions [80].

It is also possible to pretreat a sample prior to culture using a more selective approach (pre-selection). The objective is to reduce the presence of competing organisms that may interfere with the culturing of the target. An example of this is the use of an acid pretreatment step prior to the enumeration of nontuberculosis *Mycobacterium* species in water [81] . An alternative could be the use of a more highly selective incubation medium.

6.5.2 Culture Media Types, Selection, Incubation

In culture-based methods, after any pre-enrichment or pre-selection, there are several types of methods by which organisms can grow and be detected, and these may further incorporate additives that enable differentiation among types of organisms.

The media design is in part related to the nutritional types of organisms (carbon source, electron acceptor, etc.) as discussed in Section 2.1.3. Other key variables associated with culture media and conditions include the following.

Liquid vs. solid medium: Is the medium a liquid or a solid? For a liquid medium, a quantal approach to analysis is most likely (Section 6.2.2), whereas for a solid medium, a counting approach is most often used.

Oxygen, redox: Incubation may be performed in equilibrium with atmospheric oxygen, in an oxygen-depleted environment, such as under a CO_2 or N_2 gas blanket, or with the presence of strong reducing agents in the medium. Such conditions are useful for microorganisms that preferentially grow under microaerophilic or anaerobic conditions.

Temperature, time: Microorganisms have different temperature optima for growth and different ranges at which they can grow. Temperature can be used to select for or enrich microorganisms with such preferences. Some microorganisms grow rapidly, so visible colonies (on solid media) or turbidity (in liquid media) may be observed in 24 hours or less, whereas other organisms grow more slowly. As an example, one study of culture methods for nontuberculosis *Mycobacterium* strains isolated from water used an incubation time of six weeks [81]. With prolonged incubation, it is necessary to take precautions to prevent the culture medium from drying out.

Solid medium—spread, pour, membranes: For solid media cultures, there are multiple methods for inoculating a sample prior to incubation. In the spread plate method, a small amount of liquid sample (typically 1 mL or less) is placed on the media surface, and a sterile applicator (glass rod, wire loop) is used to evenly coat the medium with a sample. In the pour plate method, a small amount of agar is cooled to just above its melting point, the sample is added and mixed, and the mixture is evenly poured on and dispersed over the plate. In a membrane method, which is particularly useful for larger volume samples, the sample is filtered through a sterile membrane filter (generally 0.45 μm or 0.22 μm pore size). Then the membrane with retained microorganisms is placed on a media plate. In some cases, it can be a sterile pad saturated with liquid media; in other cases, an agar medium is used.

Addition of selective or differential agents. Selective media for culture select for the desired organism(s) or against undesired microorganisms. In the former category, if particular nutrients, growth factors, carbon sources, or electron acceptors are required by certain organisms and perhaps rarely by others, they can be included in the media. Antibiotics or other inhibitors can be used to select against undesired microorganisms.

Differential media contain substances that cause characteristic, distinguishable changes during the growth of the target organism. This frequently results from the partial metabolism of a compound by target organisms that change color.

Attributes of selective and differential media can be combined. Even with molecular biological techniques, these remain useful screening tools [82], and can more readily assess viability.

6.5.3 Whole Organism, Organ, or Cell Culture

The use of other biological materials as culture media has a long history. Prior to the use of gelatin or agar medium, early microbiologists used dried blood or boiled potatoes as surfaces for growth of colonies [83].

Inoculation of animals with protozoan cysts and oocysts has been used to assess viability [43]. For some viruses, including influenza, growth on embryonated chicken eggs has been used [84].

In the case of animal systems, responses are generally quantal. The analysis of data from such assays is closely related to the analysis of dose-response curves for risk assessment, and thus the methods discussed in Section 14.1 can be applied to these data.

Rather than using whole organisms or organs to quantify and grow target microorganisms, cells from suitable hosts can be prepared and used as a substrate to support pathogen growth.

One possibility is to take cells directly from animals, disaggregate them, and use them directly—these are termed "primary" cells [85]. These can be used directly or propagated for longer-term use. Some primary cell lines have a limited number of passages before they are no longer suitable hosts. Cells derived from embryonic material or from tumors can be passaged[3] longer times, often indefinitely. A number of cell lines are readily available commercially, such as [86]:

- Primary rhesus monkey kidney cells (RhMK)
- Primary rabbit kidney cells
- Human embryonic lung fibroblasts (MRC-5)
- Human foreskin fibroblasts
- Human epithelial carcinoma (A549)

Even some non-primary cell lines can be passaged only a limited number of times before they must be discarded and replaced with a new initial stock. The process of splitting and passaging cell lines for cell culture requires fastidious attention to aseptic technique and specialized training to avoid contamination by unwanted bacterial and fungal agents and cross-contamination between different cell lines [87].

Non-primary cell lines that are commonly used in environmental virology include the following [88]:

- Buffalo green monkey kidney (BGM)
- Fetal rhesus monkey kidney (FrhK-4)
- Human epithelial type 2 (HEp-2)
- Human lung carcinoma (A549)
- Rhabdomyosarcoma (RD)
- Human colon carcinoma (CaCo-2)
- Simian kidney epithelial (MA104)

In general, no single cell line can enumerate or propagate all viable viruses in a sample, just as no single agar formulation can enumerate all bacteria.

There are a number of physical systems available for the growth of host cells and the quantification of the virus. Commercially, host cells can be purchased already grown on test tube surfaces and used directly. Laboratory-prepared and propagated cells may be grown as a monolayer on petri dishes or flat flasks that can be sealed. Or, in some cases, suspended growths of host cells can be used.

For cells grown on a surface, the virus may grow to form discrete, countable plaques. These are discrete zones in the cell monolayers that resemble bacterial colonies, and each plaque is presumed to descend from one infectious virus. Often, cells are stained to aid enumeration, with viable host cells retaining the stain and killed (infected) areas remaining unstained.

Some viruses or some assay systems do not readily form plaques. Instead, some infectious viruses rapidly and completely infect a large proportion of the host cells in a system to produce a CPE. In such cases, replicate inoculations of multiple portions of host cells can be used, with multiple dilutions, and the results analyzed using a quantal approach (Section 6.2.2).

Cell culture of viruses can be coupled with other methods to increase assay sensitivity and specificity and/or decrease the time required for results [88, 89]. Some of these methods include:

---

3 In a passage, growth from a prior aliquot of cells is disaggregated, apportioned to multiple new containers with fresh growth media, and allowed to grow out before use.

- Engineering of host cells to include a (typically fluorescent) tagged genetic element, which, upon viral infection and multiplication, releases the signal for detection prior to visibility of plaques or CPE.
- After a short time for viral multiplication, disruption of the cell line and assay for production of viral nucleic acid by PCR.
- Use of fluorescent-labeled viral antibodies or nucleic acid probes to detect viral infection microscopically more rapidly than visible plaques or CPE.

Cell culture can be used for quantifying organisms other than viruses, provided that they can infect the particular cell line used. A notable example is the measurement of *Cryptosporidium* in water. Cell culture followed by PCR can be used for quantification, with the additional advantage that the products from nucleic acid amplification can be used for sequencing to identify species and strains [90].

6.6 Molecular Methods

Molecular methods focus on detecting the nucleic acid, or perhaps proteins or metabolites, uniquely produced by target microorganisms of interest. The use of these methods, as with others, presumes that target microorganisms have been sufficiently concentrated from the environmental sample of interest.

6.6.1 Microbial Nucleic Acids

A first step in the assessment of microbial nucleic acids is the extraction of nucleic acids from organisms, and possible separation and decontamination. This will be discussed after a review of the relevant molecules of interest.

6.6.1.1 Molecules of Interest

Prior to considering methods for nucleic acid detection to assess pathogen presence, a review of the basics of molecular genetics is needed. Genetic information in organisms is encoded in nucleic acids. For organisms other than some viruses, this information is encoded within DNA. For some viruses, as described in Section 2.1.2, the information may be encoded in RNA (specifically viral messenger RNA). The "Central Dogma" of molecular biology, first articulated by Crick, although now recognized to have exceptions, asserts that information flows from DNA to RNA via transcription and from RNA to proteins via translation, in one-way directions [91].

DNA is a polymer consisting of two "backbone" strands of phosphorylated deoxyribose (Figure 6.12). These are in opposite directions, with the strand ending with a phosphate group termed the 5' end, while the strand ending with a hydroxyl group is termed the 3' end. Covalently bound to each rung of a DNA strand is one of the four DNA bases: adenine, thymine, cytosine, or guanine. Thymine and adenine can form two hydrogen bonds, while cytosine and guanine can form three. Hence, a cytosine base on one strand of DNA is always paired with a complementary guanine base and vice versa, while an adenine is always paired with a thymine and vice versa. Note that with this complementarity, each strand of DNA has the complete genetic information *per se*. In natural replication, the two DNA strands unravel, and a complementary DNA strand is synthesized on each of the original strands via the action of the enzyme DNA polymerase.

As noted in Section 2.1.2, not all viruses use DNA to carry genetic information, and some DNA viruses may only possess a single strand, and so their mechanisms of processing information may

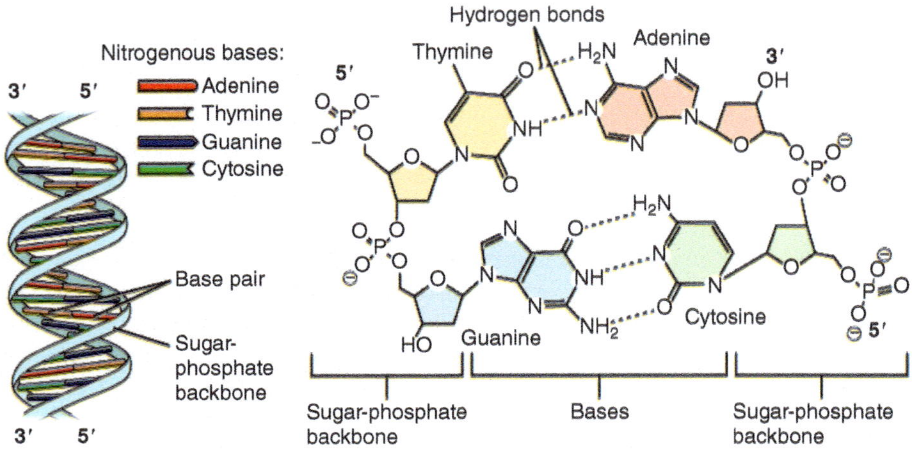

Figure 6.12 Structure of DNA and nucleotides. *Source:* From https://upload.wikimedia.o rg/wikipedia/commons/6/62/DNA_Nucleotides.jpg, CC BY 3.0 Deed.

differ from those in higher organisms. DNA is transcribed into messenger RNA (mRNA) via the enzyme DNA-directed RNA polymerase (often just called RNA polymerase). Like DNA, mRNA may be a long sequence of bases. Rather than deoxyribose, RNA has ribose backbones. Also, DNA does not have thymine, but instead it has uracil (pairing with adenine).

In addition to mRNA, organisms contain other forms of RNA. The two most salient to this discussion are transfer RNA (tRNA) and ribosomal RNA (rRNA). In the process of translation (of RNA to proteins), the sequence of RNA bases is "read" by the ribosome. Each triplet of RNA bases is read sequentially, and corresponds to either one amino acid or has another function.

Codons

Since there are four bases in RNA, there are $4^3 = 64$ possible triplets. There are 20 amino acids. Each amino acid is encoded by 2-4 triplets, introducing redundancy. There is one triplet that indicates the start of a protein to be synthesized (start codon), and three triplets that indicate the stop of a protein to be synthesized (stop codons). Since amino acids can be specified by multiple (synonymous)codons, it has been suggested that the relative proportion of synonymous codon uses could have value in the classification of taxonomic groups of organisms [92].

Protein synthesis from mRNA occurs in cellular organelles called ribosomes. Ribosomes consist of three or four RNA molecules (termed ribosomal RNA, or rRNA) of different sizes, and a large number of proteins [93]. One set of rRNA molecules are termed the small subunit rRNA molecules. Their sedimentation velocity in ultracentrifugation characterizes these in units of Svedbergs.[4] Prokaryote small subunit rRNA has a 16S sedimentation velocity while eukaryote small subunit rRNA has an 18S sedimentation velocity [94].

4 The Svedberg has units of time, formally 10^{-13} s and is a corrected form of sedimentation rate. The larger value of S corresponds to larger molecules.

In the ribosome, the organelle "reads" triplets from mRNA, which are paired with a corresponding transfer RNA (tRNA) molecule carrying its amino acid, and then the ribosome catalyzes the bonding of adjacent amino acids to form polypeptides and, thence, proteins.

Transfer RNA molecules are single-stranded with a complex secondary structure. They contain a triplet complementary to the mRNA codon, and they have a selective discriminator region that binds to their specific amino acid. In addition, tRNA molecules may have catalytic and other cellular functions beyond their role in translation [95].

6.6.1.2 Extraction, Separation, and Concentration of Nucleic Acids

Once a sample is obtained and potentially decontaminated, to employ molecular detection, it is necessary to isolate and potentially concentrate nucleic acids from the microorganisms (Figure 5.1). The flow pipeline typically involves a combination of four steps (not all of which may be present in a particular protocol [96]:

- Lysis of organisms or cells
- Separation of nucleic acids from other constituents, including washing if necessary
- Concentration of the separated and cleaned nucleic acids

In the past, these processes were often conducted manually using protocols developed by individual laboratories. Currently, however, a number of manufacturers supply kits for performing all of these steps for various types of samples and organisms. Often, when setting up an assay, multiple kits can be compared to assess their performance [97, 98]. Such kits can be used in conjunction with automated instrumentation for extraction, which clearly improves the possibility for consistent performance and enhances throughput. These systems may incorporate chemical or mechanical lysing, separation of nucleic acids by adsorption, and potential concentration by centrifugation.

6.6.2 Polymerase Chain Reaction-Based Methods

The discovery and practical implementation of the PCR underlie many of the tools for applying molecular methods to the analysis and quantification of microorganisms. Mullis realized that sequencing could be improved by denaturing double-stranded DNA and using a reverse primer to synthesize complementary DNA at each step. By reproducing this many times, the amount of DNA could be dramatically multiplied. This was further improved by the discovery of thermotolerant DNA polymerases that would remain stable over multiple cycles (undertaken by increasing temperature) of denaturation and synthesis [99].

The conduct of PCR requires multiple reagents. To initiate the reaction, short segments of double-stranded DNA, used as primers, are added. These are designed to ensure specificity for the genes or gene segments of interest. In addition, heat-resistant DNA polymerase, deoxynucleotide triphosphates, a buffer solution, and a suitable mix of monovalent and divalent cations are required. These latter sets of reagents are often supplied as a prepared mixture by commercial companies.

The PCR process in its various embodiments takes the mixture of sample DNA and the reagents through a cycle of temperatures such as:

1. Elevation of temperature to 95 °C to denature dsDNA (denaturation).
2. Lowering of temperature to 50–65 °C to allow the primers to bind to the sample ssDNA (annealing).
3. Elevation of the temperature to 75–80 °C to allow the DNA polymerase to extend the dsDNA from the partial form annealed with the primer (elongation). At the end of this stage, the amount of dsDNA is approximately doubled, and the next cycle can be commenced.

Due to temperature cycling, the entire process is generally conducted using instruments called thermocyclers—incubation systems that can rapidly change the temperature of the samples being analyzed.

By performing a large number of cycles, the DNA concentration of the target region can be increased manifold; e.g., 30 cycles can result in a theoretical $2^{30} = 1.07$ billion-fold increase, which is why PCR can be a highly sensitive technique. It should also be noted that RNA, typically either mRNA or rRNA, can be analyzed by PCR after an initial step of synthesizing DNA from the target RNA using a reverse transcriptase.

The optimization of a PCR assay involves designing primers and selecting temperatures (and holding times) for each step. Depending on how the resulting DNA is assayed, additional design considerations arise.

With PCR, several methods are available to assess the target DNA in the initial sample. In the initial implementation of PCR, the products of the reaction, possibly after treatment with specific DNA restriction enzymes (which cleave DNA containing specific short sequences), would be analyzed by gel electrophoresis, allowing determination of whether DNA with an appropriate fingerprint could be detected. By repeating this at several dilutions, a semiquantitative assessment of occurrence (in the original sample) could be made.

In more current implementations, used in qPCR, fluorescent chromophores are added to the PCR mix. There are two common approaches. One approach entails employing a fluorescent dye that binds indiscriminately to double-stranded DNA during its formation. SYBR™ Green stands out as the prevalent intercalating dye, emitting fluorescence upon binding to newly synthesized DNA. The fluorescence intensity increases with the amount of DNA produced during the qPCR reaction. Alternatively, another detection method involves proprietary hydrolysis probes bound to probe DNA sequences. The intact probe has no fluorescence signal on its own. However, upon incorporating the probe in synthesized DNA, the active chromophore is released, and a signal can be measured. With either approach, increasing DNA synthesis increases fluorescence.

In qPCR,[5] the thermocycling is conducted with a measurement of fluorescence intensity at each cycle. In conventional analysis, the number of cycles required for the intensity to rise above a threshold is recorded as in the left-hand panel of Figure 6.13, and by use of a standard curve based on response of the assay to known amounts of nucleic acid (right-hand panel), the unknown quantity of nucleic acid in a sample can be estimated. More complex data analysis of these experiments can be conducted, which arguably are able to extract more information from the entire intensity "growth curve" [100].

Like any analytical method, the appropriate use of qPCR requires adherence to concepts of quality assurance and quality control. In a general sense, though mainly from a clinical orientation, these have been outlined as "Minimum Information for Publication of Quantitative Real-Time PCR Experiments," abbreviated as MIQE [102]. A later group particularized these as "Environmental Microbiology Minimum Information" (EMMI) guidelines [103]. General concepts are that the use of this approach needs to incorporate consideration of the following attributes:

Sensitivity: The limit of detection (in terms of gene copies) that consistently produce a positive signal in the assay.

Specificity: The degree to which other (non-target) sequences do not result in a positive signal in the assay.

5 In some of the early literature, this was called real-time PCR, and given an acronym of rtPCR, but this was confusing since rtPCR has come to be used for reverse transcriptase PCR where the aim is to apply the PCR reaction with initial reverse transcriptase to measure RNA.

Accuracy: The presence of systematic differences between the quantitatively measured amount in a sample and the actual amount present.

Repeatability: The degree to which replicates of the same sample yield consistent results in a single laboratory using the same method.

Reproducibility: The stability of the signal when the assay is repeated in the same laboratory but in different "runs" (intra-laboratory reproducibility) or when repeated by different laboratories (inter-laboratory reproducibility).

A qPCR assay is the endpoint of a series of operations, and the particular conditions of each of the operations need to be documented. These include the elements in Table 6.3.

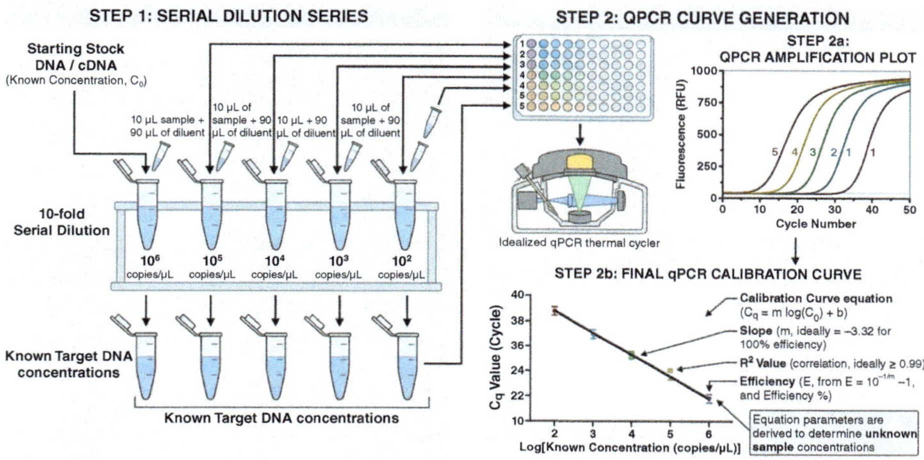

Figure 6.13 Schematic of workflow for development of a qPCR standard curve. *Source:* Modified from [101], CC BY-NC-ND.

Table 6.3 Some QA/QC considerations for each step in flow of a qPCR assay.

Step[a]	Some considerations for documentation
Environmental Sampling	Location, dates, and procedure. Storage and shipping times and conditions.
Sample pretreatment (e.g., filtration, precipitation, pasteurization)	Detailed protocol.
Sample reduction (e.g. coagulation, settling, centrifugation)	Detailed protocol. Concentration factor.
Nucleic acid extraction	Detailed protocol. Concentration factor. Total nucleic acid recovered.
Reverse transcription[b]	Detailed protocol, including time, temperature, enzymes. Assessment of inhibition.
Endpoint qPCR	Detailed protocol. Target genes, amplicons, and primer(s) used. PCR mix used. Instrumentation. Time/temperature profile. Calibration curve and limit of detection. Method for determination of cycle threshold.

Source: Modified from [103].

[a] Each step should have appropriate positive and negative controls.

[b] Where RNA is the target.

Controls at each step of the process are important. Positive controls assess any loss of material or signal at each stage, or the existence of inhibitors or the lack of efficacy of reagents. Negative controls assess the presence of contamination, and also permit the determination of the background that defines a positive response. An alternative approach to assessing microbial concentrations to qPCR is digital PCR (dPCR). The history of this approach has been concisely reviewed [104]. In the late 1990s, researchers found that in sufficiently diluted samples, single gene copies could be detected by PCR. It was then extended by taking a sample and subdividing it into a large number of very small samples, and noting the proportion of those small samples in which a positive PCR reaction occurred.

In modern commercial dPCR implementations, a sample with the appropriate PCR reagents is subdivided into a very large number of nL (or sub-nL) aliquots, carried through the PCR reaction conditions, and then "scored" as positive or negative. There are multiple types of technologies used to conduct this, including [105]:

- Forming an aqueous in oil emulsion of the nanodrops (this is called digital droplet PCR, abbreviated as ddPCR).
- Depositing the nanodrops on individual microwells of a specially manufactured plate.
- Using microfluidic technology for the dispersal of nanoscale drops and resulting scoring.

Regardless of the technology, a single assay may involve examining thousands or tens of thousands of nanodrops, which are scored as positive or negative after multiple PCR cycles. The data are thus quantal and can be analyzed using the approach discussed in Section 6.2.2. Use of dPCR requires specialized instrumentation and experienced personnel; however, a number of advantages are cited versus qPCR, including [105]:

- Quantification does not require a standard curve
- Ability to achieve higher precision
- Resistance to PCR inhibitors
- Greater sensitivity for rare agents or variants

As with qPCR, there have been guidelines developed for dPCR analyses [106] and, in particular, for environmental analytical applications [103]. Considerations on sample collection, pretreatment, concentration, and optimization of the PCR reagents and protocols are as with qPCR. Some other unique considerations are:

- Threshold settings for determination of a positive need to be documented.
- Number of nano-aliquots and the target level should be documented.
- For the use of Poisson statistics to apply, it is necessary that there be uniformity of the volume in each nano-aliquot, and that there be even (random) mixing of target and reagents between aliquots.

There have been several studies that have done comparisons between qPCR and dPCR in environmental contexts, which reinforce the general observations:

- Yang et al. [107] compared dPCR and qPCR in the enumeration of *Cryptosporidium* oocysts from animal fecal waste. They concluded that (in 2014) the cost per sample for dPCR was greater than qPCR, and that both techniques achieved similar detection limits. There was greater precision with dPCR, which was less sensitive to inhibitors but took somewhat more time to complete.

- Ahmed et al. [108] compared the performance of reverse transcriptase qPCR and dPCR for the enumeration of SARS-CoV-2 virus from wastewater. They concluded that the performance of the two methods was similar for gamma-irradiated wastewater, whereas dPCR was less sensitive to matrix interferences.

With either qPCR or dPCR, it is possible to simultaneously amplify and detect nucleic acid from multiple targets at once. This is termed "multiplexing." This requires the design of primers and probes specific to each target, amplification conditions that work well for each target system, and signals that are separable—for typical fluorescence detection, this means that the emission spectra of the different chromophores are separable. It also requires that there is no cross-interference between the targets being amplified. If these conditions are met, a single qPCR or dPCR run can generate data on multiple targets simultaneously.

A similar approach is the commercial availability of pre-designed systems for multiplex qPCR using individual wells of multi-well plates [109]. Alternatively, an investigator can use individually designed and optimized multiplex systems, in which multiple assays are conducted in the same volume, or in different volumes in a multi-well system. Multiplex dPCR is a more recent development and will likely gain importance over time [110].

As an example, Rao et al. report the detection of multiple pathogens (spanning a range of bacteria, protozoa, virus and helminths) in wastewater using both qPCR and dPCR multiplex systems [111]. This suggests the power of this approach for broad screening, although, as the authors note, the use of an extensive multiplex system results in a loss of sensitivity for each target due to a reduction in the equivalent volume that can be assayed.

6.6.3 Sequencing Approaches

The use of PCR methods for quantitation relies on prior knowledge of which organisms (genes) of interest might be present. A complementary approach is to sequence environmental (possibly amplified) nucleic acid directly and then to analyze the correspondence of sequences to know the genera/species/strains of organisms. This has been made possible by advances in technologies that have dramatically reduced the cost of sequencing large numbers of bases of DNA (Figure 6.14). Not shown in this figure is the dramatic reduction in sequencing time for large numbers of base pairs and the availability of commercial instruments at a reasonable cost to accomplish this task.

The use of sequencing methods is a rapidly developing area of technical advancement. This includes improvements in instrumentation, protocols, algorithms, and libraries. Therefore, those who work in this area, in particular (as with many other fields of knowledge and practice), should review the latest knowledge frequently.

The ideal goal of sequencing is to obtain data on the abundance of individual types of organisms without any prior bias (e.g., specific probes or primers used in PCR). Two broad sequencing strategies have been used—amplicon-based sequencing and shotgun (or whole genome) sequencing.

6.6.3.1 Amplicon-based Sequencing

Amplicon-based sequencing relies on PCR to amplify specific genes or portions of genes of interest and then applies sequencing to those products, which are now present at higher concentrations.

Woese pioneered the use of 16S rRNA sequences for taxonomic discrimination of prokaryotes [112]. It is highly conserved within prokaryotic taxa, and has nine variable regions (v1 through v9) which differ at the species and higher taxonomic levels. Specific primers can be used to amplify the entire rRNA genome (about 1500 base pairs) or only one or more of the variable regions. These results can be used to differentiate a broad spectrum of taxa [113].

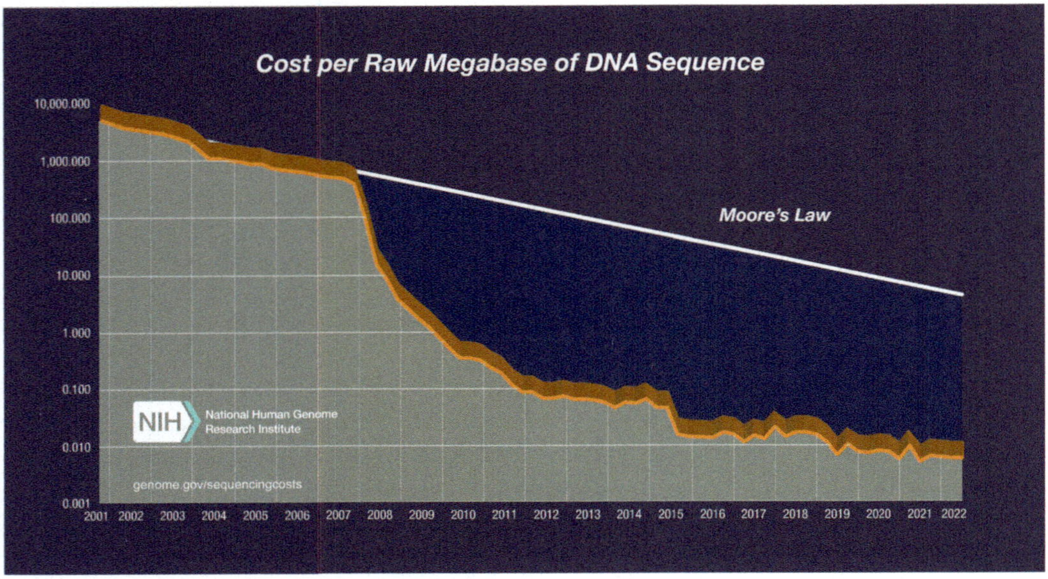

Figure 6.14 Cost of DNA sequencing. *Source:* National Human Genome Research Institute / Public domain / https://www.genome.gov/about-genomics/fact-sheets/DNA-Sequencing-Costs-Data / last accessed Feb. 10, 2026. Note that data are not inflation-adjusted.

For fungi, there have been several approaches developed. The analogy to 16S rRNA in eukaryotes is 18S rRNA. Another approach with fungi is the internal transcribed spacer (ITS) of rRNA. It has also been possible to use a larger rRNA component, the 26S large subunit. A comparison of these approaches shows that the ITS approach gives lower fidelity results than the 16S (or 26S) approach [114]. Amplicon-based sequencing of the 18S rRNA is also suitable for application to protozoa [115], as well as higher eukaryotes.

While viruses lack ribosomes, broad-spectrum methods for amplicon-based sequencing are available for specific virus groups [116].

The workflow of amplicon-based sequencing consists of a number of steps (subsequent to obtaining the concentrated and cleaned DNA samples as described earlier). First, PCR is used to amplify the target region(s) of interest (possibly the entire target subunit rRNA genome, or some of the variable regions). This is done using specified primers that target those sequences. Then the sample may optionally be "barcoded" to identify it.[6]

The samples may then be further cleaned up to remove DNA fragments that are too small or too big for the sequencing platform used. There are a variety of cleanup methods used, including sorptive magnetic (for separation) microbeads. The sample(s) (if multiple barcoded samples are used, they are mixed) are then a sequencer for processing.

Tools for sequencing have evolved from the early work of Sanger. Conventionally, approaches have been divided into three generations [118] as follows:

---

6 Barcoding is the addition of short, known segments of DNA to each DNA strand of an amplified sample to be able to identify the source of a sample [117]. This is beneficial because modern platforms can simultaneously and rapidly sequence a mixture of DNA from various sources. After sequencing, software can be used to identify sequences corresponding to individual samples.

- First-generation approaches were labor-intensive, requiring DNA fragmentation using enzymes, followed by electrophoresis, inference of the composition of small segments by radiolabeled complementary oligonucleotides, and assembly into larger sequences. Single runs could generate sequences in the range of 1000 base pairs (bp) and led to the first automated instrumentation.
- Second-generation technology used a sequencing-by-synthesis approach. The first embodiment used luminescent complementary bases and allowed reads of 400-500 bp per microwell plate, but by conducting sequencing of multiple DNA fragments, the results could be compiled by software into an entire sequence of interest. This technology has been termed pyrosequencing or 454 sequencing.
- Third-generation technology uses a variety of approaches to achieve longer reads more rapidly.

Third-generation sequencing can generate reads of up to 30 kbp, although some implementations produce higher error rates and generally have higher costs [119]. It is entirely possible that faster, more accurate, and more portable technologies may develop in the future and open up additional applications.

Amplicon-based sequencing results in "reads" of several hundred bp, with tens of thousands of reads per sample [120]. The reads are cleaned up to remove spurious DNA signals and spliced to yield the complete gene sequence of the target. If barcoding is used, the barcodes are removed by software. There are both proprietary and open-source software platforms to facilitate this process, and the resulting sequences are compared with libraries of known sequences to obtain taxonomic correspondence. With rRNA sequencing, the genus can often be identified with high confidence. Both instrumentation error and individual single-nucleotide variations (from point mutations) can result in some diversity from library sequences, so one criterion needed is the degree of similarity to ascertain that the unknown sequence is attributed to the genus in a library [121]. The following box illustrates an application of amplicon sequencing.

Example of Amplicon Sequencing

Free-living amoeba have been known to harbor a number of bacterial species, including pathogens, in various water environments [122]. Moreno-Mesonero et al. [123] sampled free-living amoeba (FLA) from 16 wastewater treatment plants in Spain and used 16S rRNA gene amplicon-based sequencing to characterize bacterial composition within the amoeba. Their overall data set comprised over half a million sequences, which were categorized at the genus level. Pathogenic genera (not all of which were in every sample) included *Aeromonas, Arcobacter, Campylobacter, Helicobacter, Legionella, Mycobacteriunm, Pseudomonas,* and *Salmonella.*

As was discussed previously, ordinary molecular methods, including sequencing (without further modification) do not necessarily provide quantitative concentration information, nor information about viability.

6.6.3.2 Shotgun Sequencing Methods

In contrast to amplicon-based sequencing, where the desire is to focus on a target gene or gene segment, shotgun sequencing is designed for the characterization of all DNA in a sample. When applied to complex environmental samples rather than single strains or pure cultures, the objective is to determine the entire metagenome of the sample. In shotgun-based sequencing, second or mostly third-generation sequencers are used; however, the nature of sample preparation, which determines the set of samples to be sequenced, differs from amplicon-based methods.

Principal differences between shotgun sequencing and amplicon-based sequencing lie in the methods of preparing the extracted sample nucleic acids and in the software used to interpret the data. For low amounts of DNA, an initial whole-genome amplification may be conducted. Subsequent to that, various fragmentation processes may treat the DNA to produce short[7] segment fragments [124]. There are instruments to automate this process and enhance reproducibility [125]. High-throughput third (and beyond) generation sequencers can generate tens of gigabytes to terabytes of data from a single run (which might be for multiple samples), perhaps each sample in a mixture having fragments that are barcoded.

The data produced will consist of a large number of DNA sequences from the (possibly mixed) sample. At this point, bioinformatic analysis software is needed. Liu et al. [120] has summarized some of the tools available. Briefly, there are multiple steps involved:

- If multiple samples are pooled in a run using barcoding, the fragments must be associated with a sample, and the barcode sequence stripped out.
- From the sequence of fragments, the longer genome is assembled. A simple example is illustrated in Figure 6.15. There are four short fragments, and common sequences are used to align them, yielding a longer sequence. Various control issues may arise. As noted earlier, there may be a small number of errors during sequencing. In addition, if (as likely in environmental samples) there are small variations in a sequence in different members of the same population, the fragments may not have totally identical matching; some degree of discordance is allowable during the process—this will also depend on how many overlapping bases must exist to declare a match. These are software tuning parameters. Extending this approach to a large number of fragments allows the assembly of the total genome.
- Using the resolved sequences, comparing similarity to libraries of known taxa, the taxonomic association can be made. In addition, an investigator may be able to discern the presence of particular genes of interest, for example, for resistance to certain antibiotics, or for the ability to metabolize particular compounds. Software tools to facilitate this matching are available, and various sequence libraries exist, perhaps the most significant being GenBank https://www.ncbi.nlm.nih.gov/genbank/.

Shotgun sequencing has been used to measure pathogens in a variety of environmental settings. Ko et al. [126] have summarized a number of these:

- Investigation of food and waterborne outbreaks
- Investigation of outbreaks originating from animals

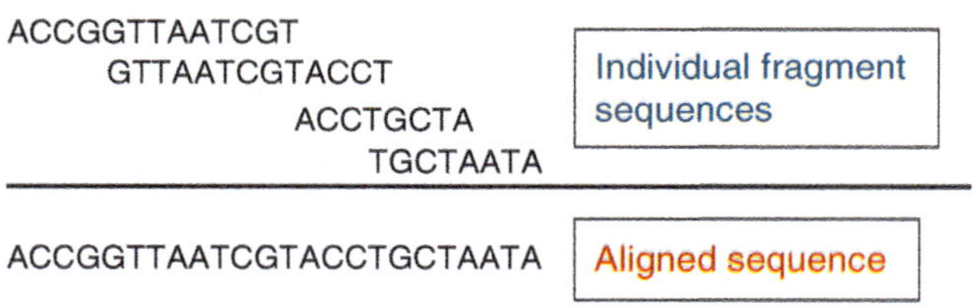

Figure 6.15 Simple example illustrating sequence assembly from fragments.

7 "Short" is a relative term and is a function of the sequencer used—it may be hundreds, or several thousand base pairs— nonetheless, shorter than a full chromosome.

- Occurrence of pathogens and antibiotic resistance genes in mass transport systems, typically from surface samples
- Monitoring for pathogens and antimicrobial resistance in environmental samples from healthcare environments
- Detection of human pathogens in animals and animal excreta that have the potential to be sources of human exposure

In the water context, Ko et al. have reviewed applications and potentials for metagenomic sequencing to monitor the quality of reclaimed water. Garner et al. [127] report a systematic review of sequencing in water and wastewater applications and identified a number of critical areas in need of advancement in knowledge and practice, including:

- Standardization of methods, including sample collection and preparation
- Standardization of bioinformatic tools
- Improvement of databases for environmental organisms, especially for non-bacterial taxa
- Development of methods for quantitative absolute abundance

There has been considerable advancement in the use of sequencing methods to detect and characterize viral pathogens in environmental samples [128]. For viruses and other pathogens, sequencing can be used to detect *strains* of individual species in environmental samples [129]. With sufficiently deep sequencing (i.e., use of very long or near-total genome lengths), it is possible to detect the emergence of mutations of differential potency, as was done during the progression of SARS-CoV-2 via wastewater monitoring [130].

6.6.4 Use for Quantitative Exposure Estimation

Regardless of whether amplicon-based or shotgun sequencing approaches are used, the objective (in the context of this book) is to use the information to generate quantitative measures of individual pathogen concentrations in environmental samples. However, there remain limitations to the direct use of sequencing for these purposes.

Earlier findings have been summarized [131]. Briefly, mock microbial community experiments have been conducted with different sequencing approaches. In these studies, a known mixture of taxa is added to a synthetic sample, and the quantitative relative abundances after sequencing are compared with the known relative abundances added. With 16S rRNA gene sequencing, substantial bias in results has been observed [132]. Long-read shotgun sequencing, however, appears to have less bias [133]. It is becoming clear that, especially for rarer taxa such as pathogens in environmental samples, shotgun sequencing, especially when large numbers of sequences are analyzed, is more effective at detecting pathogens [134].

Selective enrichment and depletion of nucleic acids may occur as a result of nucleic acid isolation. For example, the use of a common procedure for cell lysis and removal of putative contaminants may not be neutral with respect to different taxa that may be present [121]. There are a number of other sources of bias that may arise, including:

- Contamination during sample handling and processing
- Failure to match a sequence with library data due to mutations and misalignment due to overly stringent matching criteria
- Use of inadequately rich sequence databases for matching

Some of these issues may be detected or corrected using controls and, perhaps, even internal standards [135].

Despite qualitative discrepancies, sequencing can still be used to identify (possibly unexpected) pathogens and then follow up with qPCR or dPCR for quantification of specific agents.

Finally, there is the issue of assessing absolute abundance across all sequencing methods. In other words, the interest is not only in the relative proportions of individual taxa in a sample of interest but also in their absolute concentration in a sample, such as gene copies per mL or per gram. If there was the fidelity of recovering relative proportions, then it would be possible to multiply by total nucleic acid concentration (or total nucleic acid in a supertaxon, such as bacteria, virus, fungi, or protozoa) in a sample to get the taxon-specific gene copy concentration. Validation of such an approach remains a research need.

6.6.4.1 Relationship to Viability and Infectiousness

Even in the absence of the aforementioned bias in assessing relative abundance, there is a further issue: sequencing methods themselves do not provide information about whether individual organisms are viable and hence infectious. Several approaches have been used to overcome these barriers.

One approach is to couple a culture-based technique with qPCR, dPCR, or sequencing. With such methods, only a short-term culture incubation is needed to permit several cycles of pathogen replication sufficient for detection by molecular methods, rather than a more prolonged incubation that might be required to visualize a colony, a plaque, or a color change in a diagnostic medium. There have been several examples of this approach:

- Orlofsky et al. [136] devised a system in which pathogens in air, water, and soil could be pre-enriched using an MPN scheme (at different dilutions) for short-term (a day or less) incubation and then the presence or absence could be assessed via qPCR. Clearly, this could also be done using dPCR.
- For protozoa, a brief MPN protocol using competent host cells can be used with a qPCR end-point determination. This was found to be useful for *Cryptosporidium* in water using animal host cells [137].
- Shah et al. developed a procedure for rapid viability quantification of SARS-CoV-2 using a combination of cell culture and rt-PCR [138]. Host cells in a microwell plate were inoculated with a test suspension, and a 100-fold (2 $\log_1 0$) increase in qPCR measured RNA after c. 17 hours of incubation was used as a criterion for the presence of viable virus.

Another approach to viability assessment has been to add materials that block nucleic acid amplification. The most frequently used such modification is the addition of propidium monoazide (PMA). This aromatic compound can intercalate into nucleic acids and block replication; however, it is excluded from cells by the cell membrane [139], and therefore selectively inhibits DNA quantification in cells with damaged membranes. This results in a "viability" modification of qPCR or dPCR assays [121]. This has also been found to be effective as a "viable" modification for viral assays, at least for enveloped viruses [140]. A compound related to PMA, ethidium monoazide (EMA) can also be used for a viability modified assay [141]. It should be noted that if an organism is inactivated by a method that does not adversely affect membrane (or capsid, for a virus) impermeability, this method may give false positive results of viability. There has been extensive discussion of the limitations of this approach in the literature.

For example, a systematic review noted several limitations, often requiring system-by-system optimization of protocols, and highlighted a viability-PCR (or sequencing) approach using PMA or EMA [127]. However, a recent study on the occurrence of microorganisms in recreational waters suggests that a viability approach, when coupled with shotgun sequencing and robust controls (including known spike additions), may yield reproducible results [142]. More work and experience will help resolve the utility of this method.

6.7 Implications for Exposure

This chapter has described a variety of methods for quantifying pathogens in the environment. There is no single "perfect" method for any organism. The engineer and analyst often need to combine data obtained using multiple methods to develop reasonable estimates for concentrations of (viable) pathogens (and their distributions) in the environment. With such information, the implications for estimation of occurrence, transport, removal, and exposure can be judged. These further activities will be discussed in subsequent chapters.

6.8 Discussion Questions and Problems

1 Consider the data on replicate counts of poliovirus using the plaque assay shown in Table 6.1. Plot the mean values versus dilution. Using the concept of Figure 6.1, is there evidence of stimulation or inhibition? For advanced students, assess the statistical significance of any effect you posit.

2 Using the data from Berg on viral plaque counts, total aerobic counts in Chesapeake Bay water samples (Table 6.1), write code for the fit to a Poisson-inverse Gaussian (PIG) distribution. Compare the fit to the negative binomial, and discuss which fit gives a better description of the data.

3 A microscopic count using a hemocytometer is conducted. Each of the nine largest squares in Figure 6.9 is counted. The results in terms of number of organisms are 5,2,8,3,0,2,4,6,2.
   - a What is the mean concentration per mL of bacteria?
   - b What is the standard deviation of the estimate?
   - c Is this data consistent with a random distribution of organisms? Why or why not?

4 qPCR was used to quantify the number of target bacteria in a sample. The left-hand figure is the result of a qPCR run, where the solid horizontal line is the response above which the signal crosses a threshold. The right-hand figure is a standard curve with known numbers of bacteria plotted against the cycle threshold. Using this information, what is the best estimate for the bacterial level in the sample?

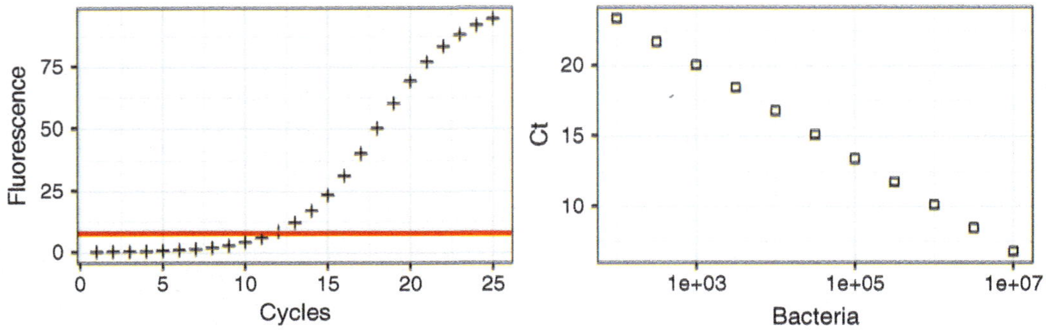

5 A dPCR assay was run with 10,000 drops each containing 1 μL equivalent volume of the original sample. If 2300 drops are positive, what is the estimate for the concentration in the original sample?

6 A duplex dPCR assay was run to quantify two viruses. In each drop, the positivity or negativity for each virus was recorded. If each drop contained 1 μg equivalent of an original air particulate sample, with the data given in this table, answer the following questions:

		Virus A	
		Negative	**Positive**
Virus B	Negative	3000	2000
	Positive	1500	1000

a What is the mean concentration (virus gene copies per gram of original sample) for each virus?

b Is the presence of each virus statistically independent of the presence of the other virus?[8]

7 For a current pathogen of concern where there is evidence or suspicion of the importance of an environmental route of transmission, do a short review of the literature to identify what methods are being used to analyze the pathogen in the environment, and what are their detection limits?

References

1 E E Geldreich. Bacterial populations and indicator concepts in feces, sewage, stormwater and solid wastes. In G Berg, editor, *Indicators of Viruses in Water and Food*, pages 51–97. Ann Arbor Science, Ann Arbor, MI, 1978.

2 Roger S Fujioka and Lyle K Shizumura. Clostridium perfringens, a reliable indicator of stream water quality. *Journal (Water Pollution Control Federation)*, 57(10):986–992, 1985.

8 This will require some elementary concepts from statistics not discussed in this book. A lack of statistical independence could be an indicator of problems with the methods being used.

3 Pierre Payment and Eduardo Franco. *Clostridum perfringens* and somatic coliphages as indicators of the efficiency of drinking water treatment for viruses and protozoan cysts. *Applied and Environmental Microbiology*, 59(8):2418–2424, 1993.

4 E W Rice, K R Fox, R J Miltner, D A Lytle, and C H Johnson. Evaluating plant performance with endospores. *Journal of the American Water Works Association*, 88(9):122–130, 1996.

5 R S Engelbrecht, C N Haas, J A Shular, D L Dunn, D Roy, A Lalchandani, B F Severin, and S Farooq. Acid-fast bacteria and yeasts as indicators of disinfection efficiency. *Technical Report* EPA-600/2-79-091, US Environmental Protection Agency, 1979.

6 E J McWeeney. Report on the bacteriological examination of the water supplied to the dublin military barracks: With observations on the biology of coliform organisms. *British Medical Journal*, 2(1863):660–663, September 1896.

7 Robert B Morse and Abel Wolman. The practicability of adopting standards of quality for water supplies. *Journal (American Water Works Association)*, 5(3):198–228, 1918.

8 Robert W Kehr and Chester T Butterfield. Notes on the relation between coliforms and enteric pathogens. *Public Health Reports (1896-1970)*, 58(15):589, 1943.

9 P W Kabler, H F Clark, and E E Geldreich. Sanitary significance of coliform and fecal coliform organisms in surface water. *Public Health Reports*, 79(1): 58–60, 1964.

10 G Berg, D R Dahling, G A Brown, and D Berman. Validity of fecal coliforms, total coliforms and fecal streptococci as indicators in chlorinated primary sewage effluent. *Applied and Environmental Microbiology*, 36:880–884, 1978.

11 R S Engelbrecht, B F Severin, M T Masarik, S Farooq, S H Lee, C N Haas, and A Lalchandani. New microbial indicators of disinfection efficiency. *Technical Report* EPA-600/2-77-052, US Environmental Protection Agency, 1977.

12 B Headd and S A Bradford. Use of aerobic spores as a surrogate for *Cryptosporidium* oocysts in drinking water supplies. *Water Research*, 90: 185–202, 2016.

13 A H Stevenson. Studies of bathing water quality and health. *American Journal of Public Health and the Nations Health*, 43:529–538, 1953.

14 Gary S Russo, Sorina E Eftim, Alexandra E Goldstone, Alfred P Dufour, Sharon P Nappier, and Timothy J Wade. Evaluating health risks associated with exposure to ambient surface waters during recreational activities: A systematic review and meta-analysis. *Water Research*, 176:115729, June 2020.

15 US Environmental Protection Agency. *Recreational Water Quality Criteria*. Number 20-F-12-058. US Environmental Protection Agency Washington, DC, 2012.

16 E E Geldreich and B A Kenner. Concepts of fecal streptococci in stream pollution. *Journal (Water Pollution Control Federation)*, 41(8):R336–R352, 1969.

17 Lisa Paruch and Adam M Paruch. An overview of microbial source tracking using host-specific genetic markers to identify origins of fecal contamination in different water environments. *Water*, 14(11):1809, June 2022.

18 D C Young and D G Sharp. Partial reactivation of chlorine-treated echovirus. *Applied and Environmental Microbiology*, 37(4):766–773, 1979.

19 M R Templeton, R C Andrews, and R Hofmann. Particle-associated viruses in water: Impacts on disinfection processes. *Critical Reviews in Environmental Science and Technology*, 38:137–164, 2008.

20 Rob Jamieson, Doug M Joy, Hung Lee, Ray Kostaschuk, and Robert Gordon. Transport and deposition of sediment-associated *Escherichia coli* in natural streams. *Water Research*, 39(12):2665–2675, 2005.

21 Robert Gaynes. The discovery of penicillin—new insights after more than 75 years of clinical use. *Emerging Infectious Diseases*, 23(5):849–853, May 2017.

22 Stephen M Stigler. The epic story of maximum likelihood. *Statistical Science*, 22(4):598–620, November 2007.

23 A Stuart and J K Ord. *Kendall's Advanced Theory of Statistics*, volume 2. Oxford University Press, Inc., New York, 5th edition, 1987.

24 Charles N Haas and B A Heller. Statistics of enumerating total coliforms in water samples by membrane filter procedures. *Water Research*, 20: 525–30, 1986.

25 James B Kaper, Aaron L Mills, and Rita R Colwell. Evaluation of the accuracy and precision of enumerating aerobic heterotrophs in water samples by the spread plate method. *Applied and Environmental Microbiology*, 35(4):756–761, 1978.

26 Gerald Berg, Eugene K Harris, Shih L Chang, and Kenneth A Busch. Quantitation of viruses by the plaque technique. *Journal of Bacteriology*, 85(3):691–700, March 1963.

27 M A Stephens. Use of the Kolmogorov–Smirnov, Cramér–Von Mises and related statistics without extensive tables. *Journal of the Royal Statistical Society: Series B (Methodological)*, 32(1):115–122, January 1970.

28 Charles N Haas, Joan B Rose, and Charles P Gerba. *Quantitative Microbial Risk Assessment*. John Wiley & Sons, Hoboken, NJ, 2nd edition, 2014.

29 Major Greenwood and G Udny Yule. An inquiry into the nature of frequency distributions representative of multiple happenings with particular reference to the occurrence of multiple attacks of disease or of repeated accidents. *Journal of the Royal Statistical Society*, 83(2):255–279, 1920.

30 C I Bliss and A R G Owen. Negative binomial distributions with a common k. *Biometrika*, 45(1/2):37–58, 1958.

31 Hirotugu Akaike. A Bayesian analysis of the minimum AIC procedure. *Annals of the Institute of Statistical Mathematics*, 30(Part A):9–14, 1978.

32 Clifford M Hurvich and Chih-Ling Tsai. Regression and time series model selection in small samples. *Biometrika*, 76(2):297–307, 1989.

33 Luke A Yates, Zach Aandahl, Shane A Richards, and Barry W Brook. Cross validation for model selection: A review with examples from ecology. *Ecological Monographs*, 93(1):e1557, February 2023.

34 Norman L Johnson, Samuel Kotz, and N Balakrishnan. *Discrete Univariate Distributions*, Volume 2. Wiley Interscience, New York, 2nd edition, 1994.

35 Wagner Barreto-Souza and Alexandre B Simas. General mixed Poisson regression models with varying dispersion. *Statistics and Computing*, 26(6):1263–1280, 2016.

36 Steven Paul Clark. An evaluation of a modified membrane filter procedure for enumerating stressed fecal coliforms in chlorinated sewage effluents. *PhD Thesis*, Virginia Polytechnic Institute and State University, 1977.

37 H E Durham. A simple method for demonstrating the production of gas by bacteria. *British Medical Journal*, 1(1952):1387–1387, May 1898.

38 Wm Mansfield Clark. A note on penfold's modification of *Bacillus coli communis*. *Science*, 38(984):669–672, 1913.

39 M H McCrady. The numerical interpretation of fermentation-tube results. *The Journal of Infectious Diseases*, 17(1): 183–212, 1915.

40 H G Pereira. The cytopathic effect of animal viruses. *Advances in Virus Research*, 8:245–285, 1962.

41 Theresa R Slifko, Debra E Huffman, and Joan B Rose. A most-probable-number assay for enumeration of infectious *Cryptosporidium parvum* oocysts. *Applied and Environmental Microbiology*, 65(9):3936–3941, 1999.

42 Graham S Banting, Shannon Braithwaite, Candis Scott, Jinyong Kim, Byeonghwa Jeon, Nicholas Ashbolt, Norma Ruecker, Lisa Tymensen, Jollin Charest, Katarina Pintar, et al. Evaluation of various *Campylobacter*-specific quantitative PCR (qPCR) assays for detection and enumeration of *Campylobacteraceae* in irrigation water and wastewater via a miniaturized most-probable-number–qPCR assay. *Applied and Environmental Microbiology*, 82(15):4743–4756, 2016.

43 C W Labatiuk, F W Schaefer, G R Finch, and M Belosevic. Comparison of animal infectivity, excystation, and fluorogenic dye as measures of *Giardia muris* cyst inactivation by ozone. *Applied and Environmental Microbiology*, 57(11):3187–3192, 1991.

44 J C de Man. MPN tables for more than one test. *European Journal of Applied Microbiology and Biotechnology*, 4(4):307–316, 1977.

45 W L Stevens. Dilution series: A statistical test of technique. *Journal of the Royal Statistical Society. Part B - Methodological*, 20:205–214, 1958.

46 Charles N Haas and B A Heller. Test of the validity of the Poisson assumption for analysis of MPN results. *Applied and Environmental Microbiology*, 54:2996–3002, 1988.

47 Mary Wotherspoon Colley. Notes on the technique of measuring the growth of bacteria with a nephelometer. *American Journal of Botany*, 18(3):205–210, 1931.

48 J J Gavin. Microbiological process report: Analytical microbiology. III. Turbidimetric methods. *Applied Microbiology*, 5(4):235–243, July 1957.

49 L S Warren, R E Benoit, and J A Jessee. Rapid enumeration of fecal coliforms in water by a colorimetric beta-galactosidase assay. *Applied and Environmental Microbiology*, 35(1):136–141, 1978.

50 J R Porter. Antony van Leeuwenhoek: Tercentenary of his discovery of bacteria. *Bacteriological Reviews*, 40(2):260–269, 1976.

51 Rainer Heintzmann and Gabriella Ficz. Breaking the resolution limit in light microscopy. *Briefings in Functional Genomics*, 5(4):289–301, 2006.

52 Mortimer Abramowitz, Kenneth R Spring, H Ernst Keller, and Michael W Davidson. Basic principles of microscope objectives. *BioTechniques*, 33(4):772–781, 2002.

53 Rita B Moyes, Jackie Reynolds, and Donald P Breakwell. Preliminary staining of bacteria: Negative stain. *Current Protocols in Microbiology*, 15(1):A–3F, 2009.

54 Martens-Habbena Willm and Sass Henrik. Sensitive determination of microbial growth by nucleic acid staining in aqueous suspension. *Applied and Environmental Microbiology*, 72(1):87–95, 2006.

55 Jackie Reynolds, Rita B Moyes, and Donald P Breakwell. Differential staining of bacteria: Acid fast stain. *Current Protocols in Microbiology*, 15(1):A–3H, 2009.

56 Hagen Frickmann, Andreas Erich Zautner, Annette Moter, Judith Kikhney, Ralf Matthias Hagen, Henrik Stender, and Sven Poppert. Fluorescence in situ hybridization (fish) in the microbiological diagnostic routine laboratory: A review. *Critical Reviews in Microbiology*, 43(3):263–293, 2017.

57 Michael V Berridge, Patries M Herst, and An S Tan. Tetrazolium dyes as tools in cell biology: New insights into their cellular reduction. In *Biotechnology Annual Review*, volume 11, pages 127–152. Elsevier, 2005.

58 S C Williams, Y Hong, D C A Danavall, M H Howard-Jones, D Gibson, M E Frischer, and P G Verity. Distinguishing between living and nonliving bacteria: Evaluation of the vital stain propidium iodide and its combined use with molecular probes in aquatic samples. *Journal of Microbiological Methods*, 32(3):225–236, 1998.

59 Lina Boulos, Michèle Prévost, Benoit Barbeau, Josée Coallier, and Raymond Desjardins. LIVE/DEAD® BacLight™: application of a new rapid staining method for direct enumeration of viable and total bacteria in drinking water. *Journal of Microbiological Methods*, 37(1):77–86, 1999.

60 Kris M Weigel, Felicia K Nguyen, Moira R Kearney, John S Meschke, and Gerard A Cangelosi. Molecular viability testing of UV-inactivated bacteria. *Applied and Environmental Microbiology*, 83(10):e00331-17, 2017.

61 C R Burch and J P P Stock. Phase-contrast microscopy. *Journal of Scientific Instruments*, 19(5):71, 1942.

62 Olli H Tuovinen, Ying Li, and Warren A Dick. Fluorescence microscopy for visualization of soil microorganisms? A review. *Biology and Fertility of Soils*, 39(5):301–311, April 2004.

63 Pawel A Penczek. Resolution measures in molecular electron microscopy. *Methods in Enzymology*, 482:73–100, 2010.

64 Cynthia S Goldsmith and Sara E Miller. Modern uses of electron microscopy for detection of viruses. *Clinical Microbiology Reviews*, 22(4):552–563, 2009.

65 Thiago P Silva, Natália P Noyma, Thabata L A Duque, Juliana P Gamalier, Luciana O Vidal, Lúcia M Lobao, Hélio Chiarini-Garcia, Fábio Roland, and Rossana C N Melo. Visualizing aquatic bacteria by light and transmission electron microscopy. *Antonie Van Leeuwenhoek*, 105:1–14, 2014.

66 Ellen Lambrecht, Julie Baré, Myriam Claeys, Natascha Chavatte, Wim Bert, Koen Sabbe, and Kurt Houf. Transmission electron microscopy sample preparation protocols for the ultrastructural study of cysts of free-living protozoa. *Biotechniques*, 58(4):181–188, 2015.

67 Alan Curry, Hazel Appleton, and Barry Dowsett. Application of transmission electron microscopy to the clinical study of viral and bacterial infections: Present and future. *Micron*, 37(2):91–106, 2006.

68 Lars Bergmans, P Moisiadis, Bart Van Meerbeek, Marc Quirynen, and Paul Lambrechts. Microscopic observation of bacteria: Review highlighting the use of environmental SEM. *International Endodontic Journal*, 38(11):775–788, 2005.

69 Monika Kąkol, Ezher Tagliasacchi, and Mirosław Słowakiewicz. Influence of different sample preparation techniques on imaging viruses and virus-like particles by scanning electron and scanning transmission electron microscopes. *Frontiers in Microbiology*, 14:1279720, 2023.

70 Sarah E Fratesi, F Leo Lynch, Brenda L Kirkland, and Lewis R Brown. Effects of SEM preparation techniques on the appearance of bacteria and biofilms in the Carter Sandstone. *Journal of Sedimentary Research*, 74(6):858–867, 2004.

71 Christine G Golding, Lindsey L Lamboo, Daniel R Beniac, and Timothy F Booth. The scanning electron microscope in microbiology and diagnosis of infectious disease. *Scientific Reports*, 6(1): 1–8, 2016.

72 Katherine M McKinnon. Flow cytometry: An overview. *Current Protocols in Immunology*, 120(1):5.1.1–5.1.11, 2018.

73 Yingying Wang, Frederik Hammes, Karen De Roy, Willy Verstraete, and Nico Boon. Past, present and future applications of flow cytometry in aquatic microbiology. *Trends in Biotechnology*, 28(8):416–424, 2010.

74 André C Pereira, Ana Tenreiro, and Mónica V Cunha. When FLOW-FISH met FACS: Combining multiparametric, dynamic approaches for microbial single-cell research in the total environment. *Science of the Total Environment*, 806:150682, 2022.

75 Hannah R Safford and Heather N Bischel. Flow cytometry applications in water treatment, distribution, and reuse: A review. *Water Research*, 151:110–133, 2019.

76 Natalya Sankova, Pavel Shalaev, Viktoriya Semeykina, Sergey Dolgushin, Elena Odintsova, and Ekaterina Parkhomchuk. Spectrally encoded microspheres for immunofluorescence analysis. *Journal of Applied Polymer Science*, 138(8):49890, 2021.

77 Elena Dlusskaya, Rafik Dey, Peter C Pollard, and Nicholas J Ashbolt. Outer limits of flow cytometry to quantify viruses in water. *ACS ES&T Water*, 1(5):1127–1135, 2021.

78 Giampiero Batani, Kristina Bayer, Julia Böge, Ute Hentschel, and Torsten Thomas. Fluorescence in situ hybridization (fish) and cell sorting of living bacteria. *Scientific Reports*, 9(1):18618, 2019.

79 Albert Johne, Robert Koch, the conductor of the German Cholera Commission, and his method of bacteria cultivation. *The Journal of Comparative Medicine and Surgery*, 6(2):113–128, April 1885.

80 Wallace H Andrews. Resuscitation of injured *Salmonella* spp. and coliforms from foods. *Journal of Food Protection*, 49(1):62–75, 1986.

81 Kimberly J Alexander, Jennifer L Furlong, Julianne L Baron, John D Rihs, Dominic Stephenson, John D Perry, and Janet E Stout. Evaluation of a new culture medium for isolation of nontuberculous mycobacteria from environmental water samples. *PLOS ONE*, 16(3):1–14, 2021.

82 Andrea Prinzi and Rodney Rohde. Why differential & selective media remain invaluable tools. https://asm.org/articles/2020/september/why-differential-selective-media-are-invaluable-to, September 2020.

83 George W Jr Lewis. The bacillus of typhoid fever: Some observations on its life history-a series of experiments, with their results. *Buffalo Medical and Surgical Journal*, 26(1):1–7, August 1886.

84 Erica Spackman and Mary Lea Killian. Avian influenza virus isolation, propagation, and titration in embryonated chicken eggs. *Animal Influenza Virus: Methods and Protocols*, pages 149–164. Springer, 2020.

85 Susan Payne. Chapter 4 - Methods to study viruses. In *Viruses*, pages 37–52. Elsevier, 2017.

86 Ali Hematian, Nourkhoda Sadeghifard, Reza Mohebi, Morovat Taherikalani, Abbas Nasrolahi, Mansour Amraei, and Sobhan Ghafourian. Traditional and modern cell culture in virus diagnosis. *Osong Public Health and Research Perspectives*, 7(2):77–82, 2016.

87 Parth Malik, Srirupa Mukherjee, and Tapan Kumar Mukherjee. Microbial contamination of mammalian cell culture. In Tapan Kumar Mukherjee, Parth Malik, and Srirupa Mukherjee, editors, *Practical Approach to Mammalian Cell and Organ Culture*, pages 187–231. Springer Nature, Singapore, 2023.

88 Ibrahim Ahmed Hamza, Lars Jurzik, Klaus Überla, and Michael Wilhelm. Methods to detect infectious human enteric viruses in environmental water samples. *International Journal of Hygiene and Environmental Health*, 214(6):424–436, November 2011.

89 Alexander A Dolskiy, Irina V Grishchenko, and Dmitry V Yudkin. Cell cultures for virology: Usability, advantages, and prospects. *International Journal of Molecular Sciences*, 21(21):7978, October 2020.

90 Mark W LeChevallier, George D Di Giovanni, Jennifer L Clancy, Zia Bukhari, Shan Bukhari, Jeffrey S Rosen, Jose Sobrinho, and Michelle M Frey. Comparison of method 1623 and cell culture-PCR for detection of *Cryptosporidium* spp. in source waters. *Applied and Environmental Microbiology*, 69(2):971–979, 2003.

91 Eugene V Koonin. Does the central dogma still stand? *Biology Direct*, 7(1):27, 2012.

92 Arup Panda and Tamir Tuller. Determinants of associations between codon and amino acid usage patterns of microbial communities and the environment inferred based on a cross-biome metagenomic analysis. *npj Biofilms and Microbiomes*, 9(1):5, 2023.

93 Xinzhu Wei and Jianzhi Zhang. On the origin of compositional features of ribosomes. *Genome Biology and Evolution*, 10(8):2010–2016, August 2018.

94 Søren M Karst, Morten S Dueholm, Simon J McIlroy, Rasmus H Kirkegaard, Per H Nielsen, and Mads Albertsen. Retrieval of a million high-quality, full-length microbial 16s and 18s rRNA gene sequences without primer bias. *Nature Biotechnology*, 36(2):190–195, 2018.

95 Matthew D Berg and Christopher J Brandl. Transfer RNAs: Diversity in form and function. *RNA Biology*, 18(3):316–339, March 2021.

96 Stephanie A Thatcher. DNA/RNA preparation for molecular detection. *Clinical Chemistry*, 61(1):89–99, 2015.

97 Brandon C Iker, Kelly R Bright, Ian L Pepper, Charles P Gerba, and Masaaki Kitajima. Evaluation of commercial kits for the extraction and purification of viral nucleic acids from environmental and fecal samples. *Journal of Virological Methods*, 191(1):24–30, 2013.

98 Marina Sabatier, Antonin Bal, Grégory Destras, Hadrien Regue, Grégory Quéromès, Valérie Cheynet, Bruno Lina, Claire Bardel, Karen Brengel-Pesce, Vincent Navratil, et al. Comparison of nucleic acid extraction methods for a viral metagenomics analysis of respiratory viruses. *Microorganisms*, 8(10):1539, October 2020.

99 Hanliang Zhu, Haoqing Zhang, Ying Xu, Soňa Laššáková, Marie Korabečná, and Pavel Neužil. PCR past, present and future. *BioTechniques*, 69(4):317–325, October 2020.

100 Yann Karlen, Alan McNair, Sébastien Perseguers, Christian Mazza, and Nicolas Mermod. Statistical significance of quantitative PCR. *BMC Bioinformatics*, 8(1):131, December 2007.

101 Grace Adams. A beginner's guide to RT-PCR, qPCR and RT-qPCR. *The Biochemist*, 42(3):48–53, 2020.

102 Stephen A Bustin, Vladimir Benes, Jeremy A Garson, Jan Hellemans, Jim Huggett, Mikael Kubista, Reinhold Mueller, Tania Nolan, Michael W Pfaffl, Gregory L Shipley, et al. The MIQE Guidelines: Minimum information for publication of quantitative real-time PCR experiments. *Clinical Chemistry*, 55(4):611–622, April 2009.

103 Mark A Borchardt, Alexandria B Boehm, Marc Salit, Susan K Spencer, Krista R Wigginton, and Rachel T Noble. The environmental microbiology minimum information (EMMI) guidelines: qPCR and dPCR quality and reporting for environmental microbiology. *Environmental Science & Technology*, 55(15):10210–10223, August 2021.

104 Alexander A Morley. Digital pcr: A brief history. *Biomolecular Detection and Quantification*, 1(1):1–2, 2014.

105 Stephen J Salipante and Keith R Jerome. Digital PCR—An emerging technology with broad applications in microbiology. *Clinical Chemistry*, 66(1):117–123, 2020.

106 The dMIQE Group and Jim F Huggett. The digital MIQE guidelines update: Minimum information for publication of quantitative digital PCR experiments for 2020. *Clinical Chemistry*, 66(8):1012–1029, August 2020.

107 P Monis, R Yang, A Paparini, and U Ryan. Comparison of next-generation droplet digital PCR (ddPCR) with quantitative PCR (qPCR) for enumeration of *Cryptosporidium* oocysts in faecal samples. *International Journal for Parasitology*, 44(14):1105–1113, 2014.

108 Warish Ahmed, Wendy J M Smith, Suzanne Metcalfe, Greg Jackson, Phil M Choi, Mary Morrison, Daniel Field, Pradip Gyawali, Aaron Bivins, Kyle Bibby, et al. Comparison of RT-qPCR and RT-dPCR platforms for the trace detection of sars-cov-2 rna in wastewater. *ACS Es&T Water*, 2(11):1871–1880, 2022.

109 Rachael Lappan, Thanavit Jirapanjawat, Deborah A Williamson, Sigrid Lange, Steven L Chown, and Chris Greening. Simultaneous detection of multiple pathogens with the taqman array card. *MethodsX*, 9:101707, 2022.

110 Alexandra S Whale, Jim F Huggett, and Svilen Tzonev. Fundamentals of multiplexing with digital pcr. *Biomolecular Detection and Quantification*, 10:15–23, Dec 2016.

111 Gouthami Rao, Drew Capone, Kevin Zhu, Abigail Knoble, Yarrow Linden, Ryan Clark, Amanda Lai, Juhee Kim, Ching-Hua Huang, Aaron Bivins, and Joe Brown. Simultaneous detection and quantification of multiple pathogen targets in wastewater. *PLOS Water*, 3(2):e0000224, February 2024.

112 C R Woese and G E Fox. Phylogenetic structure of the prokaryotic domain: The primary kingdoms. *Proceedings of the National Academy of Sciences of the United States of America*, 74(11):5088–5090, November 1977.

113 David A W Soergel, Neelendu Dey, Rob Knight, and Steven E Brenner. Selection of primers for optimal taxonomic classification of environmental 16S rRNA gene sequences. *The ISME Journal*, 6(7):1440–1444, 2012.

114 Francesca De Filippis, Manolo Laiola, Giuseppe Blaiotta, and Danilo Ercolini. Different amplicon targets for sequencing-based studies of fungal diversity. *Applied and Environmental Microbiology*, 83(17):e00905-17, 2017.

115 N P Mthethwa, I D Amoah, P Reddy, F Bux, and S Kumari. A review on application of next-generation sequencing methods for profiling of protozoan parasites in water: Current methodologies, challenges, and perspectives. *Journal of Microbiological Methods*, 187:106269, 2021.

116 Mariah K Taylor, Evan P Williams, Thidathip Wongsurawat, Piroon Jenjaroenpun, Intawat Nookaew, and Colleen B Jonsson. Amplicon-based, next-generation sequencing approaches to characterize single nucleotide polymorphisms of orthohantavirus species. *Frontiers in Cellular and Infection Microbiology*, 10:565591, 2020.

117 Qing Wang, Xu Wang, Pheobe S Tang, Grace M O'leary, and Ming Zhang. Targeted sequencing of both DNA strands barcoded and captured individually by RNA probes to identify genome-wide ultra-rare mutations. *Scientific Reports*, 7(1):3356, 2017.

118 James M Heather and Benjamin Chain. The sequence of sequencers: The history of sequencing DNA. *Genomics*, 107(1):1–8, 2016.

119 Heena Satam, Kandarp Joshi, Upasana Mangrolia, Sanober Waghoo, Gulnaz Zaidi, Shravani Rawool, Ritesh P Thakare, Shahid Banday, Alok K Mishra, Gautam Das, et al. Next-generation sequencing technology: Current trends and advancements. *Biology*, 12(7):997, 2023.

120 Yong-Xin Liu, Yuan Qin, Tong Chen, Meiping Lu, Xubo Qian, Xiaoxuan Guo, and Yang Bai. A practical guide to amplicon and metagenomic analysis of microbiome data. *Protein & Cell*, 12(5):315–330, May 2021.

121 Alena M Fricker, Daniel Podlesny, and W Florian Fricke. What is new and relevant for sequencing-based microbiome research? A mini-review. *Journal of Advanced Research*, 19:105–112, 2019.

122 N J Ashbolt. Environmental (Saprozoic) pathogens of engineered water systems: Understanding their ecology for risk assessment and management. *Pathogens*, 4(2):390–405, June 2015.

123 Laura Moreno-Mesonero, María Antonia Ferrús, and Yolanda Moreno. Determination of the bacterial microbiome of free-living amoebae isolated from wastewater by 16s rRNA amplicon-based sequencing. *Environmental Research*, 190: 109987, 2020.

124 Christopher Quince, Alan W Walker, Jared T Simpson, Nicholas J Loman, and Nicola Segata. Shotgun metagenomics, from sampling to analysis. *Nature Biotechnology*, 35(9):833–844, 2017.

125 Jacob Friedrich Hess, T A Kohl, M Kotrová, K Rönsch, T Paprotka, V Mohr, T Hutzenlaub, M Brüggemann, R Zengerle, S Niemann, et al. Library preparation for next generation sequencing: A review of automation strategies. *Biotechnology Advances*, 41:107537, 2020.

126 Karrie K K Ko, Kern Rei Chng, and Niranjan Nagarajan. Metagenomics-enabled microbial surveillance. *Nature Microbiology*, 7(4):486–496, April 2022.

127 Emily Garner, Benjamin C Davis, Erin Milligan, Matthew Forrest Blair, Ishi Keenum, Ayella Maile-Moskowitz, Jin Pan, Mariah Gnegy, Krista Liguori, Suraj Gupta, et al. Next generation sequencing approaches to evaluate water and wastewater quality. *Water Research*, 194:116907, April 2021.

128 Sandra Martínez-Puchol, Marta Rusiñol, Xavier Fernández-Cassi, Natàlia Timoneda, Marta Itarte, Cristina Andrés, Andrés Antón, Josep F Abril, Rosina Girones, and Sílvia Bofill-Mas. Characterisation of the sewage virome: Comparison of NGS tools and occurrence of significant pathogens. *Science of the Total Environment*, 713:136604, 2020.

129 Tulio M Fumian, Julia M Fioretti, Jennifer H Lun, Ingrid A L dos Santos, Peter A White, and Marize P Miagostovich. Detection of norovirus epidemic genotypes in raw sewage using next generation sequencing. *Environment International*, 123(2):282–291, February 2019.

130 Davida S. Smyth, Monica Trujillo, Devon A. Gregory, Kristen Cheung, Anna Gao, Maddie Graham, Yue Guan, Caitlyn Guldenpfennig, Irene Hoxie, Sherin Kannoly, et al. Tracking cryptic SARS-CoV-2 lineages detected in NYC wastewater. *Nature Communications*, 13(1):635, 2022.

131 Charles N Haas. Quantitative microbial risk assessment and molecular biology: Paths to integration. *Environmental Science & Technology*, 54(14):8539–8546, July 2020.

132 Fiona Fouhy, Adam G Clooney, Catherine Stanton, Marcus J Claesson, and Paul D Cotter. 16S rRNA gene sequencing of mock microbial populations- impact of DNA extraction method, primer choice and sequencing platform. *BMC Microbiology*, 16(1):123, June 2016.

133 Samuel M Nicholls, Joshua C Quick, Shuiquan Tang, and Nicholas J Loman. Ultra-deep, long-read nanopore sequencing of mock microbial community standards. *GigaScience*, 8(5):giz043, May 2019.

134 Francesco Durazzi, Claudia Sala, Gastone Castellani, Gerardo Manfreda, Daniel Remondini, and Alessandra De Cesare. Comparison between 16s rRNA and shotgun sequencing data for the taxonomic characterization of the gut microbiota. *Scientific Reports*, 11(1):3030, 2021.

135 Joshua G Harrison, W John Calder, Bryan Shuman, and C Alex Buerkle. The quest for absolute abundance: The use of internal standards for DNA-based community ecology. *Molecular Ecology Resources*, 21(1):30–43, 2021.

136 Ezra Orlofsky, Maya Benami, Amit Gross, Michelle Dutt, and Osnat Gillor. Rapid MPN-Qpcr screening for pathogens in air, soil, water, and agricultural produce. *Water, Air, & Soil Pollution*, 226(9):303, 2015.

137 Christine M Carey, Hung Lee, and Jack T Trevors. Comparison of most probable number-pcr and most probable number-foci detection method for quantifying infectious Cryptosporidium parvum oocysts in environmental samples. *Journal of Microbiological Methods*, 67(2):363–372, 2006.

138 Sanjiv R Shah, Staci R Kane, Maher Elsheikh, and Teneile M Alfaro. Development of a rapid viability RT-PCR (RV-RT-PCR) method to detect infectious SARS-CoV-2 from swabs. *Journal of Virological Methods*, 297:114251, 2021.

139 Andreas Nocker, Priscilla Sossa-Fernandez, Mark D Burr, and Anne K Camper. Use of propidium monoazide for live/dead distinction in microbial ecology. *Applied and Environmental Microbiology*, 73(16):5111–5117, Aug 2007.

140 V V Mabasa, W B van Zyl, M B Taylor, and J Mans. Quantification and potential viability of human noroviruses in final effluent from wastewater treatment works in Pretoria, South Africa. *Food and Environmental Virology*, 16(2):200–215, 2024.

141 B Reyneke, M Waso, T Ndlovu, T Clements, B Havenga, S Khan, and W Khan. EMA- versus PMA-amplicon-based sequencing to elucidate the viable bacterial community in rainwater. *Water, Air, & Soil Pollution*, 233(4):103, 2022.

142 Yu Yang, Yu Deng, Xianghui Shi, Lei Liu, Xiaole Yin, Wanwan Zhao, Shuxian Li, Chao Yang, and Tong Zhang. QMRA of beach water by nanopore sequencing-based viability-metagenomics absolute quantification. *Water Research*, 235:119858, 2023.

Part II

Quantifying Exposure

7

Sources of Pathogens

One of the critical requirements in evaluating risks to people exposed to pathogens is the assessment of exposure (Section 3.1). One aspect of exposure assessment is quantifying the contribution of pathogen sources to an environment, which will be a focus of this chapter. In subsequent chapters, potential interventions, fate and transport processes, and ultimately characterizing risk will be discussed. To understand sources, the potential reservoirs and ports of entry, as denoted in Figure 2.13, need to be assessed.

7.1 Generic Approach

The objective of quantifying sources is to estimate the number of (viable) organisms entering an environmental compartment per unit of time. This, in general, has two components:

- The concentration (#/volume or #/mass) of organisms entering the environmental compartment. In Chapter 6, methods to quantify pathogen numbers were discussed in detail. Estimating the concentration in various sources requires understanding the particular analytical method(s) used and their limitations.
- The bulk flow (mass/time or volume/time) contains pathogens entering the environmental compartment from the source.

The two quantities mentioned earlier can be multiplied to obtain the flux of pathogens entering an environment (#/time). If there is variability in the bulk flow and/or concentration over time, the product may yield an instantaneous flux, and integration can produce a time-weighted average. These flux terms serve as inputs to exposure assessment, as will be discussed in Chapter 13.

7.2 Human Excreta

Excretion is the biological process by which an organism eliminates metabolic waste products and other materials that are not useful [1]. Collectively, these are called excreta.

Fecal and urine production rates vary across populations and individuals' dietary patterns (particularly in fecal output, which is influenced by dietary fiber consumption). This has been systematically reviewed, and a fecal median production of 128 g/capita-day (wet-weight basis) was computed [2], with individual ranges of 51–796, all in healthy adults.

Median urinary output was 1.42 L/capita-day, with a range of 0.6–2.4, and was strongly dependent on fluid intake [2].

It has long been known that feces contain many pathogens. Until recently, it had been thought that normal human urine is sterile. However, this has now been overtaken by new data showing spectra of bacterial flora [3]. In addition, individuals with urinary tract infections (bacteria, fungi) or infected with some viruses may also excrete these agents. Some bacterial populations may exist in viable but nonculturable forms and thus may not be enumerated except by special methods [4].

A broad spectrum of pathogens may be present in excreta. Some of these are enumerated in Table 7.1. In addition, fungal agents of concern can be excreted.

Not all these pathogens have been systematically surveyed in human excreta. Some examples where a more robust data set exists are:

Shigella are gastrointestinal pathogens that are excreted in feces. It has been estimated that symptomatic individuals can excrete 10^5–10^9 *Shigella* per gram of wet feces and that asymptotic individuals can excrete 100–10^6 per gram [5].

Norovirus are a widespread pathogen transmitted by various water routes, person-to-person direct contact, fomites, and subsequent hand-to-mouth exposure. Atmar et al. exposed human volunteers to this virus and found that positive individuals could excrete the virus in their feces for up to 56 days post-exposure. The median peak amount of virus shedding, as measured by rt-qPCR, was 95×10^9 gene copies per gram of feces (range 0.5–1640 $\times10^9$) [6]. A similar set of values, but more detailed, was obtained through a careful study of individuals across multiple Norovirus outbreaks [7]. *Giardia* and *Cryptosporidium* are protozoan gastrointestinal pathogens that both humans and animals can excrete. Measurement has been difficult, and thus, the databases are sparse. Daniels et al. [8] reported on a survey of 85 individuals with diarrhea in villages in coastal India. 12% each was positive for *Giardia* and *Cryptosporidium* with geometric means of 2.6 (range 0.1–64) and 31.9 (range 1–491) organisms per gram of feces, respectively. The wide range may reflect the intrinsic propensity of different individuals, or of individuals at different time points during infection, to excrete pathogens.

Table 7.1 Bacterial, viral, protozoal, and helminth pathogens in human excreta.

Viruses	Bacteria	Protozoa	Helminths
Adenoviruses	*Aeromonas*	*Balantidium coli*	*Cestodes*
Hepatitis A	*Arcobacter*	*Blastocystis*	*Diphyllobothriidae*
Hepatitis E	*Campylobacter*	*Cyclospora cayetanensis*	*Echinococcus* spp.
Norovirus	*Pathogenic Escherichia coli*	*Cryptosporidium* spp.	*Taenia* spp.
Caliciviruses	*Shigella*	*Entamoeba histolytica*	*Nematodes*
Papillomavirus	*Helicobacter pylori*	*Giardia duodenalis*	*Ascaris* spp.
Polioviruses	*Leptospira*	*Microsporidia*	*Hookworms*
Enteroviruses	*Salmonella*	*Toxoplasma gondii*	*Toxocara* spp.
Polyomavirus	*Vibrio cholerae*		*Trichuris trichiura*
Rotavirus			*Trematodes*
Astrovirus			*Liver flukes*
			Intestinal flukes
			Paragonimus spp.
			Schistosoma spp.

Source: Modified from https://www.waterpathogens.org/toc.

The variability in both excreta production rates and pathogen concentrations among individuals means that incorporating this variability is necessary for a full consideration of the source terms for exposure. As populations contributing waste increase, the relative variability of the overall composition becomes less.[1] Several examples illustrate this.

Example 7.1 Contribution from a Single Individual

What is the rate of contribution of Norovirus from the feces of a single infected individual at peak excretion?

Solution:

If an average value at peak excretion is 10^9 gene copies per gram, and a mean fecal output of 128 g/d is used, then the rate becomes:

$$128 \times 10^9 = 1.28 \times 10^{11} \text{gene copies/d}$$

But both the concentration and the fecal output have ranges, so if the ranges used are (for concentration) $0.5 - 1650 \times 10^9$ and for fecal output (using a factor of 2 range) 64–256 g/d.[2]

In this case, using the upper and lower bounds of each of the inputs, we can bracket the output by:

$$\text{Lower bound} = \left(0.5 \times 10^9\right) \times 64 = 3.2 \times 10^{10} \text{ per day}$$

$$\text{Upper bound} = \left(1640 \times 10^9\right) \times 256 = 4.2 \times 10^{14} \text{ per day}$$

As this example indicates, the contribution can be quite extreme if a single individual is a source. However, if there is a population, the aggregate may be more consistent, even if each source may be highly uncertain or variable. Considering this in more detail requires characterizing the source contributions beyond specifying ranges, and also knowing the probability distributions of their occurrence. Such probabilistic descriptions will be discussed in Chapter 13.

7.3 Exhalation

Breathing rates vary with gender, age, activity level, and health status [9]. Even with the variation due to these factors, there is intrinsic person-to-person variability. A long-term average value across gender for adults aged 11–70 years is 16 $\frac{m^3}{d}$. The bulk of this flow is inhaled and exhaled, albeit with different compositions, e.g., the oxygen concentration is greater during inspiration than during exhalation, and vice versa for carbon dioxide concentration. Exhaled air can contain other metabolic products and microorganisms (or their gene fragments).

---

1 This is a general concept worth keeping in mind – as the number of individuals contributing to a source increases, even though there may be individual to individual variability, as long as their contributions are statistically independent, the overall variability will be diminished.

2 For this calculation, it is assumed that the concentration in feces is statistically independent of the fecal output. This is unlikely in symptomatic cases, where diarrhea is a result of the infection. Therefore, input correlation is needed, changing the total input statistics. The treatment of correlated inputs from variables with uncertainty and variability will be discussed in Section 14.2.

During exhalation, liquid particles are produced over a size spectrum originating from various locations in the respiratory and oral tracts. Production has been measured in studies of healthy human volunteers engaged in different activities.

Eiche and Kuster [10] measured liquid released in a small sample ($N = 2$ or 4) of individuals engaged in breathing, speaking, and speaking in a raised voice. Their key findings are abstracted in Table 7.2.

Using a larger sample size ($N = 33$ for breathing and speaking, and $N = 8$ for singing), Archer et al. [11] measured aerosol liquid generation rate, which is summarized in Table 7.3

The emissions rates in Tables 7.2 and 7.3 are similar when it is considered that for pure water, the density would be 1000 $\frac{ng}{nL}$, and that the liquid in a produced aerosol would likely be slightly denser due to contained solutes. In addition, when emitted, evaporation results in a reduction in size and increase in density of the particles once they are expelled (Section 8.4.1). It should also be noted that there is wide variability in the emission of particles during speaking, with some individuals potentially being consistent "super emitters" [12] perhaps due either to their speech patterns or differences in physiological characteristics.

The size of particles emitted from various activities results from their generation at different locales in the oral and respiratory tracts. Smaller ($< 10\ \mu m$) particles are generated primarily in the bronchioles and larynx, while larger particles emanate directly from the mouth [13]. Various pathogenic microorganisms have been isolated from exhaled air (during normal breathing, as well as during coughing or sneezing) [14]. However, only recently, in a few cases, have exhaled breath emissions rates of pathogens been quantified. Parenthetically, it may not be the case that the concentration of pathogens within aerosols as freshly exhaled is the same as might exist within the respiratory system, since the production of the aerosol requires the formation of particles from a surface, and it is well known that, due to the hydrophobicity of many microorganisms, interfacial accumulation occurs [15].

The exhaled breath of patients reporting flu symptoms was collected, and bacteria present in the condensation from their breath were sampled, yielding between about 5000 and 50,000 $\frac{cfu}{m^3}$ [16].

Table 7.2 Emission rate (nL/h) of aerosols from individuals engaged in different activities.

Activity	Mean emissions	Standard deviation
Breathing	0.92	0.90
Speaking	2.30	1.43
Raised voice	6.11	3.55

Source: Data from [10].

Table 7.3 Emission rate (ng/h) of aerosols from individuals engaged in different activities.

Activity	Mean emissions	Interquartile range
Breathing	20.5	0.61–61
Speaking	432	256–720
Singing	1116	288–1440

Source: Data from [11].

Table 7.4 PCR gene copies of SARS-CoV-2 detected in respiratory aerosols of patients during different activities.

Activity	Mean	Interquartile range
Breathing (30 min)	63.5	0–277.6
Talking (15 min)	477.9	234.5–1356.5
Singing (15 min)	713.6	135.2–1216.1

Source: Data from [18].

Milton et al. [17] devised an innovative breath collection system and found that influenza patients (confirmed by swab) exhaled both rt-qPCR detectable influenza genomes and culturable virus. Concentrations were more significant in the fine ($< 5\mu m$) fraction than in the coarse fraction, with a median of about 100 gene copies emitted in 30 minutes of breathing. Some subjects had very high emissions (in excess of 20,000), and viable viruses could be cultured in these.

In a similar work, Coleman et al. [18] measured the rt-PCR detectable SARS-CoV-2 in exhaled aerosols from diagnosed and clinically positive patients. Talking and singing generated substantially greater virus emissions than breathing (Table 7.4). This study also found that during talking and singing, most emissions were in the fine fraction of the aerosol, whereas breathing had the most emissions in the coarse fraction.

As noted above, loud talking and singing can increase the emission of exhaled particles. In individuals with respiratory infections, coughing and sneezing may produce additional aerosol, which could be laden with infectious agents. Orchestral instruments, especially brass and woodwind, can create a larger flow of aerosols, which can have a greater velocity and thus be transported further, compared to singing [19].

Example 7.2 Estimate for SARS-CoV-2 Concentration in Aerosol

Use the information in Table 7.4 and the information in Table 7.3 to estimate the SARS-CoV-2 concentration in aerosols emitted during singing.

Solution

For 15 minutes of singing, 713.6 gene copies are emitted. During the same time, 1116/4 ng of aerosol are emitted. If the density of the aerosol is approximately that of water (actually it will likely be somewhat denser), i.e., 1 $\frac{ng}{nL}$, then the concentration is given by:

$$\frac{713.6}{\frac{1116}{4}} = 2.56\frac{gc}{nL}$$

Episodic emissions during sneezing [20] and coughing can produce short-term high emission events, which can be expelled at higher velocity [21]. Such episodic emissions over long-duration exposure may pose a significant risk, especially for those close to the source [22].

Lindsley et al. [23] measured the cough emissions from nine individuals having symptomatic influenza both during their symptomatic illness and after recovery. Their key findings were:

- The average production of aerosol particles per cough was 38.3 pL, with a standard deviation of 43.7 pL. Each cough emitted an average of 75,400 particles, with a standard deviation of 97,300.

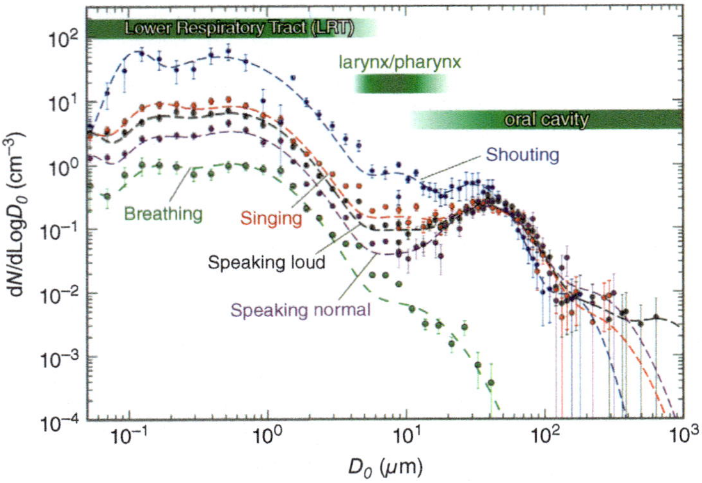

Figure 7.1 Particle size distribution of emitted aerosols vs. activity. *Source:* [24]/Elsevier/CC BY 4.0/https://www.sciencedirect.com/science/article/pii/S0021850222001380?ref=pdf_download&fr=RR-2&rr=9cba8daf2d9c59db/last accessed Feb. 12, 2026.

- After recovery, cough aerosol production decreased to an average of 26.4 pL with a standard deviation of 45.6 pL. Each post-recovery cough emitted an average of 52,200 particles, with a standard deviation of 98,600.

There is inter-individual variability in both the volume and number of particles produced during the symptomatic phase and recovery phases, and a substantial number of particles in all cases were in the respirable range.

When particles are exhaled or emitted, they have a diversity of sizes. This will affect their fate and transport, to be discussed in Chapter 8. Respiratory activity will affect the size distribution, which in turn influences the proportion of particles originating from different portions of the respiratory tract [24]. This is shown in Figure 7.1. Smaller, more numerous particles originate in the lower respiratory tract, whereas larger particles originate from the larynx/pharynx and the oral cavity. Speaking loudly and shouting increases aerosol production, but most notably, the finer particles from the lower respiratory tract. Note that, physically, these finest particles (0.5 μm) can still be large enough, theoretically, to contain an infectious agent (especially for viruses, but also for some bacteria).

7.4 Other Bodily Discharges

Other bodily discharges can be sources of pathogens into the environment. Principal among these are nasal discharges, phlegm and sputum, vomit, and blood. Often, such discharges accompany infection with pathogens. Depending on the fluid and hygiene practices, this may result in contamination of the liquid (wastewater), the production of contaminated fomites, or air emissions.

A study of nasal discharge fluids in children with runny noses showed [25] between 10^7 and 10^8 bacteria per mL. Using 16S mRNA sequencing, most samples were dominated by one or more of the following pathogenic bacteria: *Haemophilus aegyptius/influenzae, Moraxella catarrhalis/nonliquefaciens*, or *Streptococcus pneumoniae*. During the early SARS-CoV-2 pandemic, a study of adults coming in for voluntary testing who had symptomatic runny noses found a broad diversity of respiratory pathogens (viral as well as bacterial) in nearly 90% of subjects [26] using PCR.

Phlegm (sputum) is an accumulated mixture of respiratory cells, mucus, and other materials expelled via coughing and can be provoked by an infectious agent [27].

The presence of Mycobacterium tuberculosis in sputum from infected individuals has long been known [28]. This was the motivation behind many "anti-spitting" campaigns of the late 19th and early 20th centuries [29].

More extensive studies have been conducted using modern molecular techniques. Wolff et al. measured a spectrum of bacterial and viral pathogens in the sputum of children and adult patients diagnosed with community-acquired pneumonia [30]. At least one pathogen was detectable in 93% of subjects ($N = 236$), and positivity was greater in sputum than in nasopharyngeal or oropharyngeal swabs.

In human saliva, many pathogens can be isolated, with these primarily those associated with dental and oral cavity diseases [31]. However, few were capable of transmission via an intermediate environmental reservoir, as opposed to direct contact.

Many gastrointestinal infections induce vomiting as a symptom. Exposure to vomit from infected individuals has not been widely noted as a risk factor for inducing secondary cases. A notable exception is Norovirus, which has been cited as a risk factor in numerous case reports, e.g., [32, 33]. Kirby et al. [34] measured the presence of norovirus genomes by PCR in experimental volunteers infected with the pathogen and found between 10^4 and 10^6/mL in vomit (which appeared to be more frequent than diarrhea as a symptom).

Some pathogens can be present at significant concentrations in their hosts' blood. There is a broad spectrum of pathogens that can be present, including many that are (usually) in geographically defined locations[3] [35]. Protection of blood is essential in blood donor, transfusion, and organ donor programs. Generally, risk is mitigated by excluding at-risk donors, testing, and deliberate treatment.

If blood enters the environment and the pathogen persists until it is exposed to another host, a risk may arise. This condition appears to be highly restrictive to many of the agents found in blood. Most bloodborne pathogens are transmitted by direct contact between the host and susceptible individual or by direct spray or spatter from the host onto the mucus membranes of a susceptible individual. In healthcare settings, these transmission routes are often deliberately intercepted by containment rooms and the use of appropriate personal protective equipment by medical workers.

Studies that have quantified pathogen concentrations in the blood include:

- In individuals infected with human immunodeficiency virus (HIV), virus levels in the blood have been reported to average 30–3500 $TCID_{50}$/mL [36].
- Individuals infected with the Ebola virus in outbreak situations have been observed to have virus levels (measured by rt-qPCR) of 10^6–10^9/mL in their blood [37].
- Patients with severe bacterial bloodstream infections (bacteremia) have reported concentrations between 1 and 10^4 CFU/mL in blood [38].

In infected individuals, blood may be contained in other fluids, such as feces, vomit, or sputum. Therefore, pathogens ordinarily considered purely blood-borne could pose hazards associated with dissemination via these different vehicles, provided they remain viable.

3 For example, filoviruses such as Ebola.

There is strong documentation of infection to healthcare workers by needlestick injuries[4] [39]. However, there are only sporadic cases and outbreak reports of other exposure via contaminated blood-containing fluids.

A sampling of air for hemoglobin-containing aerosols during dental, surgical, and autopsy procedures documents the (at least theoretical) potential for dissemination of pathogens that might be present in blood [40]. In a small study of patients, some of whom were hepatitis B carriers, undergoing dental procedures, PCR-positive measurements were observed in occasional aerosol samples.

There is significant evidence that some filoviruses, which are ordinarily bloodborne, may be transmissible via aerosol routes (presumably including blood splatter) [41]. The significance of this route in natural circumstances is controversial [42], although several filoviruses have been of concern as bioweapons or bioterrorist agents.

7.5 Skin-associated Pathogens

The skin is the largest organ in the human body. It has a complex structure, including hair follicles, sweat glands, and a circulatory system. The topmost layer of skin, the epidermis, is in contact with the environment. When intact, it serves to protect the host (and lower layers of the skin from physical, chemical, and biological disturbances [43]. It also hosts a distinct microbiome. In normal healthy individuals, the epidermis is ordinarily a hostile environment for growth and survival of many organisms, but does contain an indigenous characteristic microbiome [44]:

- Bacteria dominated by those in the phyla *Actinomycetota, Bacillota, Bacteroidota,* and *Pseudomonadota*
- Fungi, often dominated by the genus *Malassezia*
- Mites of the genus *Demodex* (including their associated microflora)
- Unstable viral occupants, primarily double-stranded DNA, often including *Polyomaviridae* and *Papillomaviridae*

Disturbances, injuries, or disease processes may alter the skin flora, and characteristic bacterial, fungal, and viral pathogens may then proliferate [45].

Particular host factors and environmental exposures may influence the composition of the skin microbiome. These include [46, 47]:

- Degree of host sweating
- Presence of skin folds
- Host age and genetic factors
- Ethnicity
- Occupation
- Clothing practices
- Use of cosmetic and hygiene products, soaps, and moisturizers
- Climate and geography

4 Needlestick injuries are cases where a sharp instrument, such as a scalpel or a hypodermic needle, that has been in contact with contaminated blood penetrates protective barriers and breaks the skin of an uninfected person, causing direct subcutaneous and possibly direct blood exposure.

The skin may be viewed as a set of diverse habitats for microorganisms, supplying nutrients (via skin lipids and proteins) in diverse niches with respect to moisture, pH and temperature [48]. Aerobic bacteria can reach $10^7/cm^2$ in moist areas, and $100/cm^2$ in drier areas. Anaerobic bacteria can also be present at levels up to $10^6/cm^2$.

In healthy individuals, there is a normal process of epidermal loss and replacement by newly synthesized cells. This process, termed desquamation [49], has been estimated to be 1000 cells per cm^2 per hour [50]. A skin area of 1.8 m^2 [46] translates into a loss of 1.8×10^7 cells per hour.

Once shed, the skin and any adherent microorganisms (including pathogens) in its microbiome can serve as a source of multiple environmental compartments, specifically water and air. The skin with adherent microorganisms may be shed into the indoor air, some fraction may settle onto surfaces as dust, and some may be shed onto clothing. Some desquamation (with any associated microorganisms) may be washed off of the skin with normal showering or bathing, and thus represent a potential pathogen source in wastewater.

Qian et al. [51] measured microbial emission rates using 16S rt-qPCR (for bacteria) and 18S rt-qPCR (for fungi) in classrooms occupied by university students. They found emission rates of 37×10^6 and 7.3×10^6 gene copies per person hour, respectively, for bacteria and fungi. From the genetic analysis, 18% were from taxa associated with the human skin microbiome (the remainder from other human inputs). Additionally, the modal particle diameter containing either fungi or bacteria was 3 μm, indicating that it could have a high potential to remain airborne and be respirable.

Another skin-related microbial impact is the transference of microorganisms from the skin to clothing. Measurements of clothing worn by healthy volunteers [52] found total bacterial counts between 10^4 and 10^7 cfu/garment and fungal levels (measured by qPCR) between 30 and 1000 / garment. Sequencing showed a broad diversity of taxa, including numerous pathogenic bacterial and fungal types.

It has long been known that transmission of pathogens from worn clothing can serve as a mode of transmission of infectious organisms to others. Duguid et al. [53] found that mild activity could release microbial particles from clothing, which the use of a dustproof covering could mitigate. A substantial fraction of particles liberated from clothing can be in the 2–10μm range and remain suspended for considerable times [54]. Routine handling of soiled hospital linens has been demonstrated to release microorganisms, as assessed by qPCR and sequencing [55].

7.6 Our Biological Cohabitants

People interact with numerous other biota. These other biota, and their environments, may harbor or amplify human pathogens. Animals may exhale or excrete wastes with pathogenic microorganisms. People may be exposed to these agents by contact with animals, plants, or their environment. The contact may occur in various venues, including residences, offices, recreational sites, and agricultural settings.

A variety of insects can also inhabit indoor environments [56]. However, since diseases transmitted by them involve either mechanical conveyance to water, dust, or surfaces, or direct blood-borne infection by biting, they will not be discussed further.

7.6.1 Plants

Over 80% of the US households have one or more plants [57].

Soil, both agricultural and in use for houseplants, can serve as a medium for the survival and persistence of a diverse spectrum of bacterial and fungal pathogens [58]. *Legionella*, especially but not only *L. longbeachae* in soils, especially potting soils and compost, has been studied [59]. There have been outbreaks of legionellosis associated with exposure to potting soils [60].

It has been established that enteric bacterial pathogens can survive and multiply on plants grown and harvested for food [61]. There have been numerous reported outbreaks associated with the consumption of pathogen-contaminated produce [62]. However, there has not been a systematic consideration of the occurrence and significance of pathogens in house and ornamental plants.

It should be noted that other venues for the occurrence and growth of pathogens on plants exist beyond domestic and agricultural contexts. Green walls (interior and exterior to buildings) and green roofs are increasingly being deployed to achieve multiple objectives [63]. While it is understood that such systems can foster complex ecological communities, including microorganisms [64], the extent to which they may enable pathogen survival and proliferation and subsequent exposure remains an issue that requires further study.

7.6.2 Pets

Pets are frequent cohabitants of households with people. Although dogs and cats are the most common (Table 7.5), there is a broad diversity of other pet species.

Pets possess their own microbial communities, which can be influenced by exchange with their human "friends." Some have asserted that such interchanges can benefit people by increasing immune system priming [66]. However, there are clear instances in which human exposure to pathogens has been increased via pet exposure.

In principle, pets can transmit infectious agents to people via many of the same routes involved in environmentally mediated person-to-person transmission. These mechanisms can include (depending on the pet):

- Contact of pet fur, skin, and hair with surfaces and subsequent fomite transmission, or direct contact with those animal surfaces and subsequent touching of portals of entry
- Contamination of surfaces from saliva and nasal discharges
- Human handling of animal excreta
- Exhalation of infectious air particles

Table 7.5 Pet ownership statistics in the US (2024).

Pet	Number of households (in millions)
Dog	65,100,000
Cat	46,500,000
Freshwater fish	11,100,000
Bird	6,100,000
Small animal	6,700,000
Reptile	6,000,000
Horse	2,200,000
Saltwater fish	2,200,000

Source: Data from [65].

Another route of exposure associated with pet care is the occurrence of pathogens in pet food, particularly enteric bacterial pathogens. This has been reviewed, and data on microbial levels and examples of outbreaks resulting, by Lambertini et al. [67].

Sequencing of the microbiome of the skin, fur, and hair of cats and dogs shows the diversity of microorganisms as complex as in humans [68]. This includes genera that can be human and animal pathogens. Evidence of cross-transmission has emerged from a detailed study of methicillin-resistant *Staphylococcus aureus* transmission between humans and pet cats and dogs, as assessed using PCR and gel electrophoresis typing [69].

Contact with animal wastes, either during routine pet management or from the cleanup of spills, etc., may result in transmission of illnesses (Table 7.6).

There are other routes of exposure to pathogens harbored by pets, principally through direct contact. Some examples of these routes are:

- Biting and scratching (e.g., Toxoplasma from cats [75], rabies)
- Insect vectors (pathogens transmitted from infected pets to humans by intermediate insects)
- Direct human contact with pets, their fur, or licking behavior, including pets sleeping in beds [76].

The increasing popularity (and variety) of exotic pets presents new opportunities for emergence and transmission of infectious diseases [77].

7.6.3 Agricultural Animals

The cultivation of animals in agriculture has a tremendous impact on the environment. This includes the potential for this activity to be a source of pathogens into various environmental

Table 7.6 Example pet-related infectious diseases.

Agent(s)	Routes of exposure	Animal hosts
Protozoa		
Toxoplasma	Ingestion of soil contaminated by animals and fomites	Cats
Cryptosporidium	Contact with animal feces	Dogs, cats, ferrets, rodents
Giardia	Contact with animal feces	Dogs, cats, ferrets, rodents
Bacteria		
Chlamydophila psittaci	Inhalation of aerosolized bird droppings	Disease is known as psittacosis ("parrot fever"). Organism was previously genus *Chlamydia*
Salmonella	Contact with surfaces contaminated with feces	Dogs, cats, birds, turtles and other reptiles, birds (also see [74])
Leptospira	Contact with water and soil on which animal feces have deposited	Dogs, rats
Campylobacter	Contact with contaminated surfaces on which animal waste has been contacted	Cats, dogs, birds
Mycobacterium	Contact with contaminated water	Fish
Viruses		
Norovirus	Contact with food or environmental surfaces contaminated with animal feces	Dogs

Source: Compiled and adapted from [70–73].

compartments. In terms of total numbers, the inventory of animals under cultivation far exceeds the human population (Table 7.7).

In animal agricultural operations, pathogens can enter the environment through multiple routes and ultimately become a source of human exposure. This may result in occupational exposure among workers tending to or processing the animals, as well as exposure to the broader community. These may include water pollution, particularly associated with manure management and runoff, and direct air emissions [78].

Although contributions from individual animals or animals in small farms can be significant, particularly important sources of pollution, including pathogens, can come from animal feeding operations (AFOs). There are over 1 million locations in the United States with livestock and poultry [79].

An AFO is defined as "a lot or facility where animals are kept confined and fed or maintained for 45 or more days per year, and crops, vegetation, or forage growth are not sustained over a normal growing period" [80].

Concentrated animal feeding operations (CAFOs) are AFOs that meet thresholds of having a certain number of specific types of animals, and facilities so designated are required to have waste discharge permits under the Clean Water Act.

Regardless of classification as a small farm, an AFO or a CAFO, animal agriculture may be a source of environmental pathogens. There have been reports in which such facilities have served to transmit infectious diseases to people, some of which are described below.

It is well established that manure and wastes from animal operations contain a broad spectrum of viruses, bacteria, protozoa, and other organisms that may be human pathogens [81].

The amount of manure generated by livestock can be considerable. Fecal output of some commercially important animal species is summarized in Table 7.8.

Table 7.7 Inventory of major animal groups in agriculture. Data abstracted from USDA (2017).

Animal	Inventory (1000s)
Cattle	94,399
Hogs and pigs	73,230
Chickens (slaughtered for meat)[a]	9,050,716
Chickens (layers)	375,800

Source: https://www.aphis.usda.gov/sites/default/files/demographics2017.pdf.
[a] USDA reports 9.05×10^9 chickens slaughtered per year. The number at inventory is calculated assuming an average age at slaughter of 8 weeks.

Table 7.8 Fecal output of different livestock.

Animal	Average weight (kg)	Fecal waste (kg/d)
Dairy or beef calf	50	1.65
Dairy heifers	399	9.8
Lactating dairy cow	614	25.3
Beef cattle	360	14.4
Pigs	61	2.7
Chickens	1.8	0.12

Source: Data from [82].

Table 7.9 Bacterial pathogens in dairy manure.

Bacteria	Parameter	Range
Salmonella	Proportion positives (%)	5–8
	Concentration[a]	2.6–4.6
Campylobacter	Proportion positives (%)	5–38
	Concentration[a]	1.8–4.5
E. coli O157:H7	Proportion positives (%)	9.7–28
	Concentration – $\log_{10}$ mean	3.08
	$\log_{10}$ standard deviation	1.49

Source: Data from [83].
[a] Concentrations are $\log_{10}$ culturable bacteria per wet gram of manure.

Jahne et al. [83] have summarized pathogenic bacteria in dairy manure. The data are shown in Table 7.9. Management of such wastes to reduce the risk associated with human exposure, either directly, or after dispersal in the environment (air, water, land) is essential. There may also be direct risks to workers from animal handling, milking, etc., which can be mitigated by handwashing and the use of personal protective equipment [84].

Historically, handling manure at or near the facility generating the waste has been common. However, especially for AFOs and CAFOs, the intensity of manure production makes this infeasible, and therefore, a variety of management practices (i.e., treatment prior to disposal via land or water) have emerged [85], although these do not totally obviate pathogen risks. Therefore, even with treatment, off-site impacts from manure management or associated with the manure management practices themselves are expected.

Land spraying of liquid manure has been used as a management practice. In the course of such application, there is the potential for medium-range dispersal of pathogens in droplets or aerosols from any spraying devices. Burch et al. measured bacteria and viruses in the air as a function of distance from the point of spraying using both PCR and culture-based methods. Positive signals could be detected (including some by culture) at distances up to 210 m from the point of application [86].

Haack et al. used PCR to assess contamination in 19 small watersheds downstream of animal agricultural operations [87]. They were able to detect signals from *Shigella, Salmonella, Campylobacter*, pathogenic *E. coli,* and several fecal indicator organisms.

Other studies have found that watersheds downstream of animal agricultural applications can be contaminated with protozoans, including *Cryptosporidium* oocysts and *Giardia* cysts [88]. Loads could be increased during rainfall periods, when the accumulated land deposited manure washed into the watershed. Water contamination with viruses from manure and agricultural operations appears to have been less widely studied, although evidence for such contributions from a variety of human-pathogenic viruses exists [89].

During the preparation of animals for sale as food, waste is generated at slaughterhouses and wet markets where the animals are processed. This can result in the dissemination of pathogens from animals into the environment via air or water emissions [90].

There have been case studies that indicate that animal agriculture can be a source of adverse human health effects from pathogens.

From 2007 to 2010, there was a major outbreak of *Coxiella burnetii*, the causative agent of Q fever, in the Netherlands. This resulted in over 4000 cases. There was a clear association with sources on sheep and goat dairy farms. Van Leuken et al. [91] found a strong relationship between downwind (from farms containing infected animals) case occurrence and the predicted airborne concentration from atmospheric dispersion models, modulated by terrestrial factors such as vegetation and land use.

The farm environment is one that often provides extensive opportunities for interchange of infectious agents between animal species, e.g., cattle, swine, and birds, and between birds and people. This provides opportunities for the rearrangement of pathogen traits and adaptation to new hosts, with potential for "breakout" into the human population. This has been most documented in the case of influenza virus [92]. It is not surprising that human cases of "bird flu" have emerged in workers who directly handle and manage livestock [93].

7.6.4 Wild Animals

Just as animals kept as pets and reared for agriculture can represent sources of pathogens to the environment, wild animals can also be sources. This has been particularly important in the case of watersheds that serve as sources of drinking water and as recreational water resources. Therefore, many of the issues discussed earlier, such as manure management, arise in the context of wild animals.

Several documented studies indicate the importance of these sources to environmental pathogen loading.

Campylobacter jejuni, a significant cause of gastroenteritis, has been commonly found in New Zealand's recreational rivers and streams. A study, using gel electrophoresis serotyping of strains six New Zealand recreational found *Campylobacter* spp. and *C. jejuni* in 33.2% and 20.4% of 509 samples, respectively. The bacteria were present under both low and high river flows, with a higher likelihood of detection in winter than in summer after adjusting for river-flow effects. A diverse range of *C. jejuni* multilocus sequence types was found. While most strains are associated with wild birds and are rarely linked to human disease, some strains associated with poultry and ruminants that were also detected in human infections were also detected, indicating a potential public health risk [94].

Using a similar approach in waters in Luxembourg and the Netherlands, attribution of strains of *Campylobacter jejuni* and *Campylobacter coli* could be made. In Luxembourg, water strains were attributed to wild birds (61.0%), followed by poultry (18.8%), ruminants (15.9%), and pigs (4.3%), whereas in the Netherlands, water strains were mainly attributed to poultry (51.7%), wild birds (37.3%), ruminants (9.8%), and pigs (1.2%) [95]. This indicates that while farmed animals were significant sources, wild birds were also substantial sources in both countries, and in Luxembourg, they were dominant.

Protozoan (as well as indicator) organism contamination of waters used as drinking water supplies has been examined in multiple cases, since elevated levels of *Cryptosporidium* and *Giardia* may necessitate more intensive water treatment. In some cases, the dominant sources may be wildlife, while in other cases, domestic animals may be dominant, and wildlife may be a secondary source.

Cox et al. [96] sampled animal feces from wild and domesticated animals in the Sydney, Australia watershed. They measured indicators by MPN, protozoa by microscopic count, and viruses by MPN-CPE assay. Excerpts of their data for protozoa are shown in Table 7.10. Although some wildlife sources were found, the authors concluded that dominant sources were domestic (agricultural and pets).

Table 7.10 Protozoans in animal fecal samples in Sydney, Australia watershed.

Animal domestic	Cryptosporidium oocysts			Giardia cysts		
	Positive/total samples	Median concentration	Range	Positive/total samples	Median concentration	Range
Calf	4/7	9	0–183	7/9	133	0–533
Adult cattle	2/9	0	0–10	7/9	68	0–293
Horse	1/9	0	0–5	2/9	0	0–8
Pig	7/9	367	0–6000	5/9	11	0–>16,000
Poultry	0/7	0	0	2/9	0	0–67
Cat	3/7	0	0–17	1/7	0	0–>7143
Wildlife						
Dog	2/8	0	0–>5000	5/8	835	0–>6061
Brushtail possum	2/2	54	20–89	0/2	0	0
Kangaroo	4/11	0	0–1257	0/11	0	0
Rat	0/4	0	0	0/4	0	0
Rabbit	1/2	38	0–77	0/2	0	0

Source: Data from [96].
Concentrations are #/gram wet weight of feces.

In the New York City watershed, fecal samples from small mammals were collected. These were processed via PCR for the 18S rRNA gene, and the products were sequenced to determine the genotypes of *Cryptosporidium* that could be obtained. The protozoan sequences could be isolated from 13 animal species. While a number of the sequences were presumptively host-specific for particular animal hosts, potentially human-infective strains were also observed [97].

A similar study was conducted in the Portland, Oregon watershed [98]. While *Cryptosporidium* sequences could be detected in 13.8% of the wildlife fecal samples examined ($N = 506$), only a minority of the sequences corresponded to strains believed to be of human health significance.

The studies in Portland, New York City, and elsewhere point to the value of sequencing. In addition, this highlights the need to quantify below the genus level to detect and assess the public health significance of pathogens.

7.7 Waste Management Practices

Facilities for managing waste from people can themselves be sources of pathogen contamination of the environment. This may be an intrinsic property of the facility or result from uncontrolled operations. However, in general, they are proximate, not ultimate, sources in that pathogens must have been introduced by other sources, such as human or animal waste. Certain bacteria, however, such as those that have been termed "saprozoic" [99], may multiply under favorable conditions. Other bacteria, including *Salmonella* and *Listeria,* can regrow in warm, moist, organic-rich conditions such as compost piles [100]. Regardless of whether regrowth occurs, waste management practices may allow pathogens to spread more widely in the environment.

Previously, the issue of agricultural manure management was discussed. This can result in off-site migration by air dispersal, by seepage or runoff into water, or by carriage via animal or insect vectors.

7.7.1 Sanitary Landfills

The disposal of municipal solid waste, comprising domestic and non-hazardous commercial waste, was historically done either by incineration or open dumping. When problems with waste migration from open dumps became more understood, the concept of a sanitary landfill was developed. Modern municipal sanitary landfills are constructed with liners (consisting of clay and/or geomembranes), leachate collection systems, and, at closure, capped with an upper membrane to minimize water intrusion into the landfill. They also contain a gas collection system to remove methane and other gases produced by the decomposition of organic material in the waste. Typically, during operations, the landfill is used as a series of "lifts," in which a portion of the landfill is filled with waste from a daily operation, then immediately covered with clean soil (a "daily cover"). The daily cover minimizes the dispersal of solid waste by wind and the conveyance of contamination by animal vectors (birds, insects). After multiple days of operation, a "cell" of the landfill may be closed off with additional cover material.

Parenthetically, it should be noted that more recently, some landfills have been designed as bioreactors. In this concept [101], rather than aiming to exclude water from the landfill, the idea is to control the recirculation of water *through* the system to encourage biological decomposition and possibly to create additional volumetric capacity for new waste to be disposed of within.

In classical or bioreactor landfills, the membrane systems may be imperfect or sustain damage from abrasive waste materials, burrowing animals, or tree roots. As this occurs, the potential exists for leachate to percolate outside the landfill, thereby conveying contaminants from the landfill.

Pathogens (and indicator organisms) may enter the solid waste stream from a variety of sources. Gerba et al. inventoried these, including food wastes, pet wastes, biosolids,[5] hygienic products, and absorbent products. They concluded that the dominant sources of the indicator fecal coliforms come from food waste, while the dominant sources of *Salmonella*, enteroviruses and pathogenic protozoans, come from pet feces. The dominant source of norovirus comes from biosolids. Absorbent materials in waste and hygienic products did not dominate as sources for any of the pathogen groups [102].

While modern landfill design incorporates multiple levels of containment, leachate collection, and reduction of potential sources of infiltration [103], as in any engineered system, complete elimination is not feasible. Pathogens and indicator organisms can survive, to some extent, within the landfill and be mobilized by leachate [104]. In the event of imperfections, these materials can be transported through liner systems into adjacent groundwater and may pose a risk to those using such aquifers for a water supply [105]. The degree to which such transport occurs and poses a risk depends on the groundwater flow gradient, aquifer hydrogeology, the separation distance with potential users, and chemical factors that might promote or diminish attenuation [106] and resulting risk.

The handling of solid waste during emplacement and prior to the application of daily cover material, as well as the manipulation of the waste stream, such as in resource recovery facilities (e.g., source separation), may result in airborne emissions containing indicator organisms or pathogens [107]. This may pose a risk to onsite workers and potentially to individuals in the near vicinity offsite.

7.7.2 Biosolids Application to Land

Biosolids are the semisolid residuals from wastewater treatment. They may consist (depending on the design of the treatment plant) of a treated mixture of the following:

Primary sludge resulting from the simple sedimentation of raw wastewater before further treatment.

Secondary sludge resulting from the biological flocs produced during wastewater treatment processes such as activated sludge or trickling filters, wherein a portion of the organic material removed from the wastewater produces new biomass.

This sludge mixture is collected and treated further to reduce the concentration of putrescible organic material and to reduce concentrations of pathogens that are present.[6] Upon treatment, this sludge mixture is formally termed a biosolid. Under the US EPA regulatory framework, biosolids are divided into Class A and Class B [108].

With respect to pathogen and indicator organisms, Class B biosolids are sludges that have been treated with a process for significant reduction of pathogens (PFRP) or have been treated to achieve a fecal coliform concentration below 2×10^6/g solids at the time of disposal. Examples of PFRPs are aerobic digestion, air drying, anaerobic digestion, composting, and lime stabilization. There is also a requirement for treatment to achieve a reduction in vector attraction. Class B biosolids may only be applied to land with restrictions, and may not be otherwise distributed to the public.

---

5 Which are the partially treated residuals from wastewater treatment.

6 Pathogens in biosolids result from their removal from the wastewater during treatment by sorption and occlusion within the organic flocs that form by biological wastewater treatment.

Class A biosolids are suitable for public distribution and for unrestricted land application. They must have been treated with a process for further reduction of pathogens (PSRP). Examples of PSRPs are composting, heat drying, heat treatment, thermophilic aerobic digesting, beta ray, gamma ray irradiation and pasteurization. Class A performance may also be achieved by monitoring microbial levels and demonstrating that each of the following is attained:

- *Salmonella* < 3 MPN/4g
- Fecal coliforms < 1000 MPN/g
- Enteric virus < 1 PFU/4g
- Viable helminth ova < 1/4g

Depending on the physical state of the biosolids and the land use receiving them, they may be land-applied differently. If the biosolids are still flowable, generally less than 10% solids, they can be sprayed or injected subsurface. Biosolids of greater concentrations have a cake-like consistency and can be applied via a truck-like manure spreader or by use of a "slinger" [109]. The application of biosolids using any of these approaches may result in the production of airborne particles (solid or liquid/semisolid) that can be dispersed in the atmosphere.

Viau and Pecci used PCR and qPCR to measure pathogens in Class A and Class B biosolids. There is a clear reduction in prevalence (and concentration) of pathogens associated with more intensive treatment (Table 7.11).

While the presence of pathogens in biosolids, especially Class B material, might pose a risk to those offsite, measurements of aerosol concentrations of pathogens and indicators suggest that such risks are insubstantial [111], and the enforcement of separation distances can further reduce the risk.

7.7.3 Wastewater Treatment

As noted previously, raw sewage can contain pathogens. The conveyance of raw sewage, either in open channels [112] or in sewers with head space [113], can release aerosols containing these pathogens.

Wastewater treatment generally involves biological processes — two of the most common are activated sludge and trickling filters. The activated sludge process involves contacting the (settled)

Table 7.11 Pathogen percent positivity in biosolids receiving various treatments (determined by PCR).

Pathogen	Class B (MAD)	TPAD	COM	MH
Total samples	16	8	10	2
Adenovirus	88	75	70	100
Legionella pneumophila	31	25	50	0
Staphylococcus aureus	19	50	0	0
Clostridioides difficile	25	38	0	0

Source: Modified from [110].
MAD - Mesophilic anaerobic digestion
TPAD - Thermophilic anaerobic digestion (Class A)
COM - MAD plus composting (Class A)
MH - Composting plus heat pelletization (Class A)
Mean detection limits: 347–3801 gene copies/gram dry weight of solids

Figure 7.2 Activated sludge aeration tank. *Source:* Marcos von Sperling/Public domain/https://commons.wikimedia.org/wiki/File:Activated_sludge_sewage_treatment_plant_aeration_tank.png/last accessed Feb. 10, 2026.

Figure 7.3 Photograph of a trickling filter. *Source:* From https://www.geograph.org.uk/photo/1450096 CC-By-2.

wastewater and vigorously mixing (Section 7.7.3) or introducing fine air bubbles from below to allow the degradation of organic material in the presence of oxygen. This can result in emitted droplets or aerosols at rates depending on the specific mixing or aeration process (Figure 7.2).

Trickling filters contain a bed of crushed gravel or synthetic media over which a fine spray of wastewater is allowed to flow, typically using a rotating arm application as in Figure 7.3. The splash of the liquid on the media (as well as drift from the spray arms themselves) may result in dispersal of microbial aerosol from the system.

Other potential sources for microbial-containing aerosol generation at wastewater treatment plants include (in addition to the above) points at which the flow from pipes goes into open tanks, resulting in splashing (such as influents to equalization or to a combination of flows in a flow channel), in-plant conveyance channels, and hydraulic jumps used for mixing or aeration.

Table 7.12 Partition factors (air/water) for microbial groups in biological wastewater treatment.

Treatment process	Microorganism	Partition coefficient
Activated sludge	Mesophilic bacteria	1.7×10^{-10}
Activated sludge	Mesophilic fungi	7.5×10^{-6}
Activated sludge	Thermotolerant fungi	1.5×10^{-6}
Trickling filter	Mesophilic bacteria	1.3×10^{-9}
Trickling filter	Mesophilic fungi	2.7×10^{-7}
Trickling filter	Thermotolerant fungi	1.1×10^{-7}

Source: Data from [117].

Pathogens or indicator organisms in the air at wastewater treatment plants have been detected, for example:

- Heinonen-Tanski et al. detected viable coliforms and coliphages in aerosols of full-scale treatment plants in Finland [114]. Greater concentrations (up to 100 cfu or pfu/m^3) were observed in the vicinity of preliminary and primary treatment than in the latter stages.
- Masclaux et al. used qPCR to detect adenovirus, norovirus, and the hepatitis E virus in air samples at 31 wastewater treatment plants in the canton of Zurich, Switzerland [115]. Adenovirus gene copies were present in all summer samples and 97% of winter samples (at levels up to 2.3×10^6 gene copies/m^3. Norovirus was infrequently present (3/123 samples), and hepatitis E was absent. The authors suggested that these aerosols might be responsible for gastrointestinal symptoms in workers. A limitation, however, is that gene copy isolation via qPCR does not necessarily indicate the presence of viable viruses.
- Itarte et al. [116] did intensive sampling at one wastewater treatment plant in Spain using qPCR and sequencing. They detected adenovirus in almost all aerosol samples at levels up to 100 gene copies/m^4. Sequencing detected a broad spectrum of other viral taxa in aerosol samples.

Bauer et al. [117] measured viable bacteria and fungi counts in activated sludge, trickling filter water, and in the air immediately above these processes. They characterized their results as a partition coefficient, i.e., CFU/m^3 in the air divided by the CTU/m^3 in the water. Results are given in Table 7.12.

During the formation of aerosols from gas bubbles in a liquid, microorganisms can accumulate at the interface due to surface tension effects. The more hydrophobic microorganisms will be preferentially accumulated. Ji et al. [118] have reviewed the basic physics of these processes. Kataki et al. have comprehensively reviewed the literature on bioaerosolization in wastewater treatment [119].

7.8 Other Sources via Aerosolization of Water

In addition to waste management systems, microorganisms from the water environment can be emitted into the environment as aerosols. This can occur for organisms deposited into water from other sources, such as human or animal waste, or for other microorganisms that may either passively persist in the water environment or actively grow.

It has long been known that in aquatic environments, microorganisms, depending on their hydrophobicity, can accumulate at the air-water interface. Interfacial turbulence can result in the ejection of droplets that are emitted to the overlying air [120]. More recently, in the context of the highly polluted Tijuana River, the emission of pathogens as aerosols from the river has been documented, with sampling indicating transport of these organisms some distance from the source [121, 122]. This route is also supported by a meta-analysis of recreational water epidemiological studies, which show the significance of increased respiratory illnesses [123]. Clearly, emission rates will be highly site-specific, depending on water turbulence, meteorological factors, and water quality.

Cooling towers are widely used at the building scale (for air conditioning) and the industrial scale (for both air conditioning and for power generation). These can result in the emission of water droplets and aerosols, depending on the design and mode of operation. Some bacteria, such as *Legionella*, can multiply at warm temperatures when biocides are not continuously maintained and may thence trigger outbreaks [124]. The emission of the particles from these towers can be coupled with atmospheric dispersion models to assess the potential for transport to receptors [125].

In the indoor environment, microbial-containing aerosols can be emitted from a diverse spectrum of uses, including toilets [126], sinks [127], and showers [128].

7.9 Discussion Questions and Problems

1 Using the data from Table 7.4, if an auditorium contains 100 people of whom 10% are talking, what is the estimate for the generation of gene copies per minute in the entire room?

2 Given Kirby's [34] estimate of $10^4 - 10^6 \frac{virus}{mL}$ of norovirus in vomit, if a person has one event expelling 250 mL of vomit, how much norovirus is released. If 1% of this is aerosolized, how much is dispersed?

3 A dairy herd consists of 20 dairy heifers. Given data in this chapter:
   a How much manure (kd/d) is produced?
   b What is the estimated quantity of *E. coli* O157:H7 produced per day?

4 The US EPA has a Pathogen Equivalency Committee to designate new processes as PSRPs or PFRPs (https://www.epa.gov/biosolids/pathogen-equivalency-committee). What is a recent process designated as a PFRP? And write a two-page summary of how it works and key design parameters.

5 A certain activated sludge plant uses $10 \frac{m^3}{d}$ of air and has been found to have a concentration of a certain pathogenic bacterium of five organisms per liter in the aeration tank. Estimate the total flux (number of organisms per time) emanating from this process.

References

1 John E Hall and Michael E Hall. *Guyton and Hall Textbook of Medical Physiology E-Book: Guyton and Hall Textbook of Medical Physiology E-Book*. Elsevier Health Sciences, 2020.

2 C Rose, A Parker, B Jefferson, and E Cartmell. The characterization of feces and urine: A review of the literature to inform advanced treatment technology. *Critical Reviews in Environmental Science and Technology*, 45(17):1827–1879, 2015.

3 Michael I Kogan, Yulia L Naboka, Khalid S Ibishev, Irina A Gudima, and Kurt G Naber. Human urine is not sterile - shift of paradigm. *Urologia Internationalis*, 94(4):445–452, March 2015.

4 Anderson Marc, Bollinger Daphne, Hagler Ashley, Hartwell Hadley, Rivers Bryan, Ward Kristie, and Todd R. Steck. Viable but nonculturable bacteria are present in mouse and human urine specimens. *Journal of Clinical Microbiology*, 42(2):753–758, 2004.

5 Steven L Percival and David W Williams. Chapter 11- Shigella. In Steven L Percival, Marylynn V Yates, David W Williams, Rachel M Chalmers, and Nicholas F Gray, editors, *Microbiology of Waterborne Diseases (Second Edition)*, pages 223–236. Academic Press, London, second edition, 2014.

6 Robert L Atmar, Antone R Opekun, Mark A Gilger, Mary K Estes, Sue E Crawford, Frederick H Neill, and David Y Graham. Norwalk virus shedding after experimental human infection. *Emerging Infectious Disease*, 14(10):1553–1557, October 2008.

7 P F M Teunis, F H A Sukhrie, H Vennema, J Bogerman, M F C Beersma, and M P G Koopmans. Shedding of norovirus in symptomatic and asymptomatic infections. *Epidemiology and Infection*, 143(8):1710–1717, 2015.

8 Miles E Daniels, Arpit Shrivastava, Woutrina A Smith, Priyadarshi Sahu, Mitsunori Odagiri, Pravas R Misra, Pinaki Panigrahi, Mrutyunjay Suar, Thomas Clasen, and Marion W Jenkins. *Cryptosporidium* and *Giardia* in humans, domestic animals, and village water sources in rural India. *American Journal of Tropical Medicine and Hygiene*, 93(3):596, 2015.

9 United States Environmental Protection Agency. Exposure factors handbook. *Technical Report* EPA/600/R-09/052, United States Environmental Protection Agency, Washington, D.C., 2011.

10 Thomas Eiche and Martin Kuster. Aerosol release by healthy people during speaking: Possible contribution to the transmission of SARS-CoV-2. *International Journal of Environmental Research and Public Health*, 17(23):9088, December 2020.

11 Justice Archer, Lauren P McCarthy, Henry E Symons, Natalie A Watson, Christopher M Orton, William J Browne, Joshua Harrison, Benjamin Moseley, Keir E J Philip, James D Calder, et al. Comparing aerosol number and mass exhalation rates from children and adults during breathing, speaking and singing. *Interface Focus*, 12(2): 20210078, April 2022.

12 Sima Asadi, Anthony S Wexler, Christopher D Cappa, Santiago Barreda, Nicole M Bouvier, and William D Ristenpart. Aerosol emission and superemission during human speech increase with voice loudness. *Scientific Reports*, 9(1):2348, 2019.

13 Lidia Morawska, Giorgio Buonanno, Alex Mikszewski, and Luca Stabile. The physics of respiratory particle generation, fate in the air, and inhalation. *Nature Reviews Physics*, 4(11):723–734, 2022.

14 Kevin P Fennelly. Particle sizes of infectious aerosols: Implications for infection control. *The Lancet Respiratory Medicine*, 8(9):914–924, 2020.

15 Abigail C Dommer, Rommie E Amaro, and Kimberly A Prather. Understanding erosol-mediated disease transmission. *ACS Central Science*, 11(12):2319–2328, December 2025.

16 Zhenqiang Xu, Fangxia Shen, Xiaoguang Li, Yan Wu, Qi Chen, Jie Xu, and Maosheng Yao. Molecular and microscopic analysis of bacteria and viruses in exhaled breath collected using a simple impaction and condensing method. *PLOS ONE*, 7(7): 1–8, 2012.

17 Donald K Milton, M Patricia Fabian, Benjamin J Cowling, Michael L Grantham, and James J McDevitt. Influenza virus aerosols in human exhaled breath: Particle size, culturability, and effect of surgical masks. *PLoS Pathogens*, 9(3):e1003205, 2013.

18 Kristen K Coleman, Douglas Jie Wen Tay, Kai Sen Tan, Sean Wei Xiang Ong, The Son Than, Ming Hui Koh, Yi Qing Chin, Haziq Nasir, Tze Minn Mak, Justin Jang Hann Chu, et al. Viral load of severe acute respiratory syndrome coronavirus 2 (SARS-CoV-2) in respiratory aerosols emitted by patients with coronavirus disease 2019 (COVID-19) while breathing, talking, and singing. *Clinical Infectious Diseases*, 74(10):1722–1728, May 2022.

19 Lingzhe Wang, Tong Lin, Hevander Da Costa, Shengwei Zhu, Tehya Stockman, Abhishek Kumar, James Weaver, Mark Spede, Donald K. Milton, Jean Hertzberg, et al. Characterization of aerosol plumes from singing and playing wind instruments associated with the risk of airborne virus transmission. *Indoor Air*, 32(6):e13064, 2022.

20 Z Y Han, W G Weng, and Q Y Huang. Characterizations of particle size distribution of the droplets exhaled by sneeze. *Journal of The Royal Society Interface*, 10(88):20130560, November 2013.

21 Meg VanSciver, Shelly Miller, and Jean Hertzberg. Particle image velocimetry of human cough. *Aerosol Science and Technology*, 45(3):415–422, February 2011.

22 Rachael M Jones, Yoshifumi Masago, Timothy Bartrand, Charles N Haas, Mark Nicas, and Joan B Rose. Characterizing the risk of infection from Mycobacterium tuberculosis in commercial passenger aircraft using quantitative microbial risk assessment. *Risk Analysis*, 29:355–365, 2009.

23 William G Lindsley, Terri A Pearce, Judith B Hudnall, Kristina A Davis, Stephen M Davis, Melanie A Fisher, Rashida Khakoo, Jan E Palmer, Karen E Clark, Ismail Celik, et al. Quantity and size distribution of cough-generated aerosol particles produced by influenza patients during and after illness. *Journal of Occupational and Environmental Hygiene*, 9(7): 443–449, 2012.

24 Gholamhossein Bagheri, Oliver Schlenczek, Laura Turco, Birte Thiede, Katja Stieger, Jana M Kosub, Sigrid Clauberg, Mira L Pöhlker, Christopher Pöhlker, Jan Moláček, et al. Size, concentration, and origin of human exhaled particles and their dependence on human factors with implications on infection transmission. *Journal of Aerosol Science*, 168:106102, 2023.

25 Kaoru Haro, Midori Ogawa, Mitsumasa Saito, Koichi Kusuhara, and Kazumasa Fukuda. Bacterial composition of nasal discharge in children based on highly accurate 16s rRNA gene sequencing analysis. *Scientific Reports*, 10(1):20193, 2020.

26 Elandri Fourie, Eva S Bording, Sjoerd M Euser, Paul Badoux, Mildred E Haverkort, and Marlies A van Houten. Nasal discharge: A unique enhanced alternative for the detection of respiratory pathogens in adults. *Journal of Medical Virology*, 95(8):e29047, 2023.

27 S Farzan. Chapter 38: Cough and sputum production. In H K Walker, W D Hall, and J W Hurst, editors, *Clinical Methods: The History, Physical, and Laboratory Examinations*. Butterworths, 1990.

28 Masahiro Shirai, Hiroshi Hayakawa, Hiroshi Uchiyama, Kingo Chida, and Hirotoshi Nakamura. Clinical significance of potential pathogenic microorganisms of sputum in patients with pulmonary tuberculosis. *Respirology*, 6(4):311–315, 2001.

29 Patrick J O'Connor. Spitting positively forbidden: The anti-spitting campaign, 1896-1910. *Master's Thesis*, University of Montana, 2015.

30 Bernard J. Wolff, Anna M. Bramley, Kathleen A. Thurman, Cynthia G. Whitney, Brett Whitaker, Wesley H. Self, Sandra R. Arnold, Christopher Trabue, Richard G. Wunderink, Jon McCullers, Kathryn M. Edwards, Seema Jain, and Jonas M. Winchell. Improved detection of respiratory pathogens by use of high-quality sputum with TaqMan array card technology. *Journal of Clinical Microbiology*, 55(1):110–121, 2024/05/15 2016.

31 Jørgen Slots and Henrik Slots. Bacterial and viral pathogens in saliva: Disease relationship and infectious risk. *Periodontology 2000*, 55(1):48–69, February 2011.

32 P R Chadwick and R McCann. Transmission of a small round structured virus by vomiting during a hospital outbreak of gastroenteritis. *Journal of Hospital Infection*, 26(4):251–259, 1994.

33 P Godoy, M Alsedà, R Bartolomé, D Clavería, I Módol, P Bach, G Mirada, and À Domínguez. Norovirus gastroenteritis outbreak transmitted by food and vomit in a high school. *Epidemiology and Infection*, 144(9):1951–1958, July 2016.

34 Amy E Kirby, Ashleigh Streby, and Christine L Moe. Vomiting as a symptom and transmission risk in norovirus illness: Evidence from human challenge studies. *PLoS One*, 11(4):e0143759, 2016.

35 Bjørg Marit Andersen. *Prevention and Control of Infections in Hospitals: Practice and Theory*. Springer International Publishing, Cham, 2019.

36 Richard W Johnson, Ernest R Blatchley III, and Dana R Mason. HIV and the bloodborne pathogen regulation: Implications for the wastewater industry. *Water Environment Research*, 66(5):684–691, 1994.

37 Jonathan S Towner, Pierre E Rollin, Daniel G Bausch, Anthony Sanchez, Sharon M Crary, Martin Vincent, William F Lee, Christina F Spiropoulou, Thomas G Ksiazek, Mathew Lukwiya, et al. Rapid diagnosis of Ebola hemorrhagic fever by reverse transcription-PCR in an outbreak setting and assessment of patient viral load as a predictor of outcome. *Journal of Virology*, 78(8): 4330–4341, 2004.

38 O Opota, A Croxatto, G Prod'hom, and G Greub. Blood culture-based diagnosis of bacteraemia: state of the art. *Clinical Microbiology and Infection*, 21(4):313–322, 2015.

39 Haldo Treviño and Minerva Angélica Romero Arenas. Systematic review of blood-borne pathogen exposure rates among medical students. *Journal of Surgical Research*, 255:66–70, 2020.

40 F Perdelli, A M Spagnolo, M L Cristina, M Sartini, R Malcontenti, M Dallera, G Ottria, R Lombardi, and P Orlando. Evaluation of contamination by blood aerosols produced during various healthcare procedures. *Journal of Hospital Infection*, 70(2):174–179, 2008.

41 Elizabeth K Leffel and Douglas S Reed. Marburg and Ebola viruses as aerosol threats. *Biosecurity and Bioterrorism: Biodefense Strategy, Practice, and Science*, 2(3):186–191, 2004.

42 Berhanu Mekibib and Kevin K Ariën. Aerosol transmission of Filoviruses. *Viruses*, 8(5):148, 2016.

43 Hani Yousef, Mandy Alhajj, and Sandeep Sharma. Anatomy, skin (integument), epidermis. https://www.ncbi.nlm.nih.gov/books/NBK441980/, January 2024.

44 Katarzyna Adamczyk, Agnieszka A Garncarczyk, and Paweł P Antończak. The microbiome of the skin. *Dermatology Review/Przegląd Dermatologiczny*, 105(2):285–297, 2018.

45 Keisha Findley and Elizabeth A Grice. The skin microbiome: A focus on pathogens and their association with skin disease. *PLOS Pathogens*, 10(11):1–3, November 2014.

46 Elizabeth A Grice and Julia A Segre. The skin microbiome. *Nature Reviews Microbiology*, 9(4):244–253, April 2011.

47 Krzysztof Skowron, Justyna Bauza-Kaszewska, Zuzanna Kraszewska, Natalia Wiktorczyk-Kapischke, Katarzyna Grudlewska-Buda, Joanna Kwiecińska-Piróg, Ewa Wałecka-Zacharska, Laura Radtke, and Eugenia Gospodarek-Komkowska. Human skin microbiome: Impact of intrinsic and extrinsic factors on skin microbiota. *Microorganisms*, 9(3):543, March 2021.

48 David N Fredricks. Microbial ecology of human skin in health and disease. *Journal of Investigative Dermatology Symposium Proceedings*, 6(3):167–169, 2001.

49 Leonard M Milstone. Epidermal desquamation. *Journal of Dermatological Science*, 36(3):131–140, 2004.

50 Chiara Fantinato, Ane Elida Fonneløp, Øyvind Bleka, Magnus Dehli Vigeland, and Peter Gill. The invisible witness: air and dust as DNA evidence of human occupancy in indoor premises. *Scientific Reports*, 13(1):19059, 2023.

51 J Qian, D Hospodsky, N Yamamoto, W W Nazaroff, and J Peccia. Size-resolved emission rates of airborne bacteria and fungi in an occupied classroom. *Indoor Air*, 22:339–351, January 2012.

52 Kelly Whitehead, Jake Eppinger, Vanita Srinivasan, M Khalid Ijaz, Raymond W Nims, and Julie McKinney. Microbiome of clothing items worn for a single day in a non-healthcare setting. *Microbiology Research*, 2023.

53 J P Duguid and A T Wallace. Air infection with dust liberated from clothing. *The Lancet*, 2(6535):845–849, November 1948.

54 D Licina and W W Nazaroff. Clothing as a transport vector for airborne particles: Chamber study. *Indoor Air*, 28(3):404–414, May 2018.

55 Alina Handorean, Charles E Robertson, J Kirk Harris, Daniel N Frank, Natalie Marie Hull, Cassandra Vogel Kotter, Mark J Stevens, Darrel Baumgardner, Norman R Pace, and Mark Hernandez. Microbial aerosol liberation from soiled textiles isolated during routine residuals handling in a modern health care setting. *Microbiome*, 3:11116277, 2015.

56 Misha Leong, Matthew A Bertone, Amy M Savage, Keith M Bayless, Robert R Dunn, and Michelle D Trautwein. The habitats humans provide: Factors affecting the diversity and composition of arthropods in houses. *Scientific Reports*, 7(1):15347, 2017.

57 Melinda Knuth, Hayk Khachatryan, and Charlie Hall. Consumer houseplant purchasing report 2021. *Technical Report*, Floral Marketing Fund, 2021.

58 Dennis J Baumgardner. Soil-related bacterial and fungal infections. *The Journal of the American Board of Family Medicine*, 25(5):734–744, 2012.

59 S Casati, A Gioria-Martinoni, and V Gaia. Commercial potting soils as an alternative infection source of *Legionella pneumophila* and other *Legionella species* in Switzerland. *Clinical Microbiology and Infection*, 15(6):571–575, 2009.

60 Stephen T Chambers, Sandy Slow, Amy Scott-Thomas, and David R Murdoch. Legionellosis caused by non-*Legionella pneumophila* species, with a focus on *Legionella longbeachae*. *Microorganisms*, 9(2):291, 2021.

61 Jeong-A Lim, Dong Hwan Lee, and Sunggi Heu. The interaction of human enteric pathogens with plants. *Plant Pathology Journal*, 30(2):109–116, 2014.

62 S D Bennett, S V Sodha, T L Ayers, M F Lynch, L H Gould, and R V Tauxe. Produce-associated foodborne disease outbreaks, USA, 1998–2013. *Epidemiology and Infection*, 146(11):1397–1406, 2018.

63 Maria Manso and João Castro-Gomes. Green wall systems: A review of their characteristics. *Renewable and Sustainable Energy Reviews*, 41:863–871, 2015.

64 Kanchane Gunawardena and Koen Steemers. Living walls in indoor environments. *Building and Environment*, 148:478–487, 2019.

65 Michelle Hegna. Pet ownership statistics 2024. https://www.forbes.com/advisor/pet-insurance/pet-ownership-statistics/, January 2024.

66 Mariana C Salas Garcia, Ashley R Schorr, Wyatt Arnold, Na Fei, and Jack A Gilbert. Pets as a novel microbiome-based therapy. In M Ramiro Pastorinho and Ana Catarina A Sousa, editors, *Pets as Sentinels, Forecasters and Promoters of Human Health*, pages 245–267. Springer International Publishing, Cham, 2020.

67 Elisabetta Lambertini, Robert L Buchanan, Clare Narrod, and Abani K Pradhan. Transmission of bacterial zoonotic pathogens between pets and humans: The role of pet food. *Critical Reviews in Food Science and Nutrition*, 56(3):364–418, 02 2016.

68 J Scott Weese. The canine and feline skin microbiome in health and disease. *Veterinary Dermatology*, 24(1):137–e31, 2013.

69 J S Weese, H Dick, B M Willey, A McGeer, B N Kreiswirth, B Innis, and D E Low. Suspected transmission of methicillin-resistant staphylococcus aureus between domestic pets and humans in veterinary clinics and in the household. *Veterinary Microbiology*, 115(1):148–155, 2006.

70 Peter M Rabinowitz, Zimra Gordon, and Lynda Odofin. Pet-related infections. *American Family Physician*, 76(9):1314–1322, 2007.

71 I Ghasemzadeh and S H Namazi. Review of bacterial and viral zoonotic infections transmitted by dogs. *Journal of Medicine and Life*, 8(Spec Iss 4):1, 2015.

72 Peter Damborg, Els M Broens, Bruno B Chomel, S Guenther, Frank Pasmans, Jaap A Wagenaar, J Scott Weese, Lothar H Wieler, Ulrika Windahl, Daisy Vanrompay, et al. Bacterial zoonoses transmitted by household pets: State-of-the-art and future perspectives for targeted research and policy actions. *Journal of Comparative Pathology*, 155(1):S27–S40, 2016.

73 Jason W Stull, Jason Brophy, and J S Weese. Reducing the risk of pet-associated zoonotic infections. *Cmaj*, 187(10):736–743, 2015.

74 R H McCoy and Ramon J Seidler. Potential pathogens in the environment: Isolation, enumeration, and identification of seven genera of intestinal bacteria associated with small green pet turtles. *Applied Microbiology*, 25(4):534–538, 1973.

75 Sarah Legge, Pat L Taggart, Chris R Dickman, John L Read, and John C Z Woinarski. Cat-dependent diseases cost Australia AU$6 billion per year through impacts on human health and livestock production. *Wildlife Research*, 47(8):731, 2020.

76 Lucie A Zanen, Johannes G Kusters, and Paul A M Overgaauw. Zoonotic risks of sleeping with pets. *Pathogens*, 11(10):1149, 2022.

77 Bruno B Chomel, Albino Belotto, and François-Xavier Meslin. Wildlife, exotic pets, and emerging zoonoses. *Emerging Infectious Disease*, 13(1):6, 2007.

78 Patricia D Millner. Bioaerosols associated with animal production operations. *Bioresource Technology*, 100(22):5379–5385, November 2009.

79 Claudia Clopeland. Animal waste and water quality: EPA regulation of concentrated animal feeding operations (CAFOs). *Technical Report* RL31851, Congressional Research Service, 2010.

80 Carrie Hribar. Understanding concentrated animal feeding operations and their impact on communities. Report, National Association of Local Boards of Health; National Center for Environmental Health (U.S.). Environmental Health Services Branch, Bowling Green, OH, 2010.

81 Mark D Sobsey, L A Khatib, Vincent Hill, E Alocilja, and Suresh D Pillai. Pathogens in animal wastes and the impacts of waste management practices on their survival, transport and fate. In *Animal Agriculture and the Environment:*, National Center for Manure and Animal Waste Management White Papers., pages 609–666. American Society of Agricultural and Biological Engineers, St. Joseph, Michigan, 2006.

82 Delia Grace Edward R. Atwill, Xunde Li and Victor Gannon. Zoonotic waterborne pathogen loads in livestock. In Robert Bos Al Dufour, Jamie Bartram and Victor Gannon, editors, *Animal Waste, Water Quality and Human Health*, chapter 3, pages 73–114. IWA Publishing, 2012.

83 Michael A Jahne, Shane W Rogers, Thomas M Holsen, Stefan J Grimberg, and Ivan P Ramler. Emission and dispersion of bioaerosols from dairy manure application sites: Human health risk assessment. *Environmental Science & Technology*, 49(16):9842–9849, 2015.

84 Alexis L Mraz, Nandini Mutyala, Seana Cleary, and Brenda F Seals. Is personal protective equipment worth the hassle? Annual risk of cryptosporidiosis to dairy farmers and how personal protective equipment and handwashing can mitigate it. *Microorganisms*, 11(10):2413, September 2023.

85 Kerry A Hamilton, Warish Ahmed, Eleanor Rauh, Channah Rock, Jean McLain, and Rebecca L. Muenich. Comparing microbial risks from multiple sustainable waste streams applied for agricultural use: Biosolids, manure, and diverted urine. *Current Opinion in Environmental Science & Health*, 14:37–50, April 2020.

86 T R Burch, S K Spencer, J P Stokdyk, B A Kieke, R A Larson, A D Firnstahl, A M Rule, and M A Borchardt. Quantitative microbial risk assessment for spray irrigation of dairy manure based on an empirical fate and transport model. *Environmental Health Perspectives*, 125:087009, 2017.

87 Sheridan K Haack, Joseph W Duris, Dana W Kolpin, Michael J Focazio, Michael T Meyer, Heather E Johnson, Ryan J Oster, and William T Foreman. Contamination with bacterial zoonotic pathogen genes in U.S. streams influenced by varying types of animal agriculture. *Science of The Total Environment*, 563-564:340–350, September 2016.

88 G Wilkes, T A Edge, V P J Gannon, C Jokinen, E Lyautey, N F Neumann, N Ruecker, A Scott, M Sunohara, E Topp, and D R Lapen. Associations among pathogenic bacteria, parasites, and environmental and land use factors in multiple mixed-use watersheds. *Water Research*, 45(18):5807–5825, November 2011.

89 Gauthami Penakalapati, Jenna Swarthout, Miranda J Delahoy, Lydia McAliley, Breanna Wodnik, Karen Levy, and Matthew C Freeman. Exposure to animal feces and human health: A systematic review and proposed research priorities. *Environmental Science & Technology*, 51(20):11537–11552, October 2017.

90 Adel Al-Gheethi, Nyuk Ling Ma, Parveen Fatemeh Rupani, Naznin Sultana, Maizatul Azrina Yaakob, Radin Maya Saphira Radin Mohamed, and Chin Fhong Soon. Biowastes of slaughterhouses and wet markets: An overview of waste management for disease prevention. *Environmental Science and Pollution Research*, 30(28):71780–71793, September 2021.

91 J P G Van Leuken, A N Swart, J Brandsma, W Terink, J Van de Kassteele, P Droogers, F Sauter, A H Havelaar, and W Van der Hoek. Human Q fever incidence is associated to spatiotemporal environmental conditions. *One Health*, 2:77–87, December 2016.

92 Jenny L Mace and Andrew Knight. Influenza risks arising from mixed intensive pig and poultry farms, with a spotlight on the United Kingdom. *Frontiers in Veterinary Science*, 10:1310303, 2023.

93 Hinh Ly. Highly pathogenic avian influenza H5N1 virus infections of dairy cattle and livestock handlers in the United States of America. *Virulence*, 15(1):2343931, 12 2024.

94 Rima D Shrestha, Anne C Midwinter, Jonathan C Marshall, Julie M Collins-Emerson, Eve J Pleydell, and Nigel P French. *Campylobacter jejuni* strains associated with wild birds and those causing human disease in six high-use recreational waterways in New Zealand. *Applied and Environmental Microbiology*, 85(24):e01228-19, September 2019.

95 L Mughini-Gras, C Penny, C Ragimbeau, F M Schets, H Blaak, B Duim, J A Wagenaar, A de Boer, H M Cauchie, J Mossong, et al. Quantifying potential sources of surface water contamination with Campylobacter jejuni and Campylobacter coli. *Water Research*, 101:36–45, 2016.

96 P Cox, M Griffith, M Angles, D Deere, and C Ferguson. Concentrations of pathogens and indicators in animal feces in the Sydney watershed. *Applied and Environmental Microbiology*, 71:5929–5934, 2005.

97 Peter E Ziegler, Susan E Wade, Stephanie L Schaaf, Yung-Fu Chang, and Hussni O Mohammed. *Cryptosporidium* Spp. from small mammals In the New York City watershed. *Journal of Wildlife Diseases*, 43(4):586–596, October 2007.

98 Xunde Li, Tran Nguyen, Chengling Xiao, Ann Levy, Yone Akagi, Sarah Silkie, and Edward R Atwill. Prevalence and genotypes of *Cryptosporidium* in wildlife populations co-located in a protected watershed in the Pacific Northwest, 2013 to 2016. *Microorganisms*, 8(914):15, 2020.

99 N J Ashbolt. Environmental (saprozoic) pathogens of engineered water systems: Understanding their ecology for risk assessment and management. *Pathogens*, 4(2):390–405, June 2015.

100 N Paniel, S Rousseaux, P Gourland, M Poitrenaud, and J Guzzo. Assessment of survival of *Listeria monocytogenes, Salmonella Infantis* and *Enterococcus faecalis* artificially inoculated into experimental waste or compost. *Journal of Applied Microbiology*, 108(5):1797–1809, May 2010.

101 C H Benson, M A Barlaz, D T Lane, and J M Rawe. Practice review of five bioreactor/recirculation landfills. *Waste Management*, 27(1):13–29, 2007.

102 Charles P Gerba, Akrum H Tamimi, Charles Pettigrew, Anne V Weisbrod, and Vijay Rajagopalan. Sources of microbial pathogens in municipal solid waste landfills in the United States of America. *Waste Management*, 29(8):781–789, 2011.

103 Syed R Qasim. *Sanitary Landfill Leachate: Generation, Control and Treatment*. Routledge, 2017.

104 R S Engelbrecht, M J Weber, P Amirhor, D H Foster, D LaRossa, J F Malina, and B P Sagik. Biological properties of sanitary landfill leachate. In *Symposium Virus Survival In Water And Wastewater Systems, Water Resources*, volume 7, pages 201–217. University of Texas, 1974.

105 Charles N Haas, Jin Anotai, and Richard S Engelbrecht. Monte Carlo assessment of microbial risk associated with landfilling of fecal material. *Water Environment Research*, 68(7):1123–1131, November 1996.

106 Rui Xiang, Ya Xu, Yu-Qiang Liu, Guo-Yuan Lei, Jing-Cai Liu, and Qi-Fei Huang. Isolation distance between municipal solid waste landfills and drinking water wells for bacteria attenuation and safe drinking. *Scientific Reports*, 9(1):17881, 2019.

107 Agnieszka Kalwasińska and Aleksandra Burkowska. Municipal landfill sites as sources of microorganisms potentially pathogenic to humans. *Environmental Science: Processes & Impacts*, 15:1078–1086, 2013.

108 US Environmental Protection Agency. Title 40—protection of environment chapter i—environmental protection agency subchapter o —sewage sludge part 503 standards for the use or disposal of sewage sludge. https://www.ecfr.gov/current/title-40/part-503, ND.

109 J P Brooks, B D Tanner, K L Josephson, C P Gerba, C N Haas, and I L Pepper. A national study on the residential impact of biological aerosols from the land application of biosolids. *Journal of Applied Microbiology*, 99:310–322, 2005.

110 E Viau and J Peccia. Survey of wastewater indicators and human pathogen genomes in biosolids produced by class A and class B stabilization treatments. *Applied and Environmental Microbiology*, 75(1):164–174, January 2009.

111 Benjamin D Tanner, John P Brooks, Charles N Haas, Charles P Gerba, and Ian L Pepper. Bioaerosol emission rate and plume characteristics during land application of liquid class B biosolids. *Environmental Science & Technology*, 39(6):1584–1590, March 2005.

112 Olivia Ginn, Lucas Rocha-Melogno, Aaron Bivins, Sarah Lowry, Maria Cardelino, Dennis Nichols, Sachchida Nand Tripathi, Freddy Soria, Marcos Andrade, Mike Bergin, et al. Detection and quantification of enteric pathogens in aerosols near open wastewater canals in cities with poor sanitation. *Environmental Science & Technology*, 55(21):14758–14771, November 2021.

113 Charles N Haas, Taylor Rycroft, Kyle Bibby, and Leonard Casson. Risks from ebolavirus discharge from hospitals to sewer workers. *Water Environment Research*, 89(4):357–368, 2017.

114 Helvi Heinonen-Tanski, Tiina Reponen, and Jari Koivunen. Airborne enteric coliphages and bacteria in sewage treatment plants. *Water Research*, 43:2558–2566, 2009.

115 Frédéric G. Masclaux, Philipp Hotz, Drita Gashi, Dessislava Savova-Bianchi, and Anne Oppliger. Assessment of airborne virus contamination in wastewater treatment plants. *Environmental Research*, 133(8):260–265, August 2014.

116 Marta Itarte, Miquel Calvo, Lola Martínez-Frago, Cristina Mejías-Molina, Sandra Martínez-Puchol, Rosina Girones, Gertjan Medema, Sílvia Bofill-Mas, and Marta Rusiñol. Assessing environmental exposure to viruses in wastewater treatment plant and swine farm scenarios with next-generation sequencing and occupational risk approaches. *International Journal of Hygiene and Environmental Health*, 259:114360, June 2024.

117 Heidi Bauer, M Fuerhacker, F Zibuschka, H Schmid, and H Puxbaum. Bacteria and fungi in aerosols generated by two different types of wastewater treatment plants. *Water Research*, 36(16):3965–3970, 2002.

118 Bingqiang Ji, Amrit Singh, and Jie Feng. Water-to-air transfer of nano/microsized particulates: enrichment effect in bubble bursting jet drops. *Nano Letters*, 22(13): 5626–5634, July 2022.

119 Sampriti Kataki, Rupam Patowary, Soumya Chatterjee, Mohan G Vairale, Sonika Sharma, Sanjai K Dwivedi, and Dev Vrat Kamboj. Bioaerosolization and pathogen transmission in wastewater treatment plants: Microbial composition, emission rate, factors affecting and control measures. *Chemosphere*, 287:132180, January 2022.

120 Duncan C Blanchard. The ejection of drops from the sea and their enrichment with bacteria and other materials: A review. *Estuaries*, 12(3):127, September 1989.

121 Matthew A Pendergraft, Pedro Belda-Ferre, Daniel Petras, Clare K Morris, Brock A Mitts, Allegra T Aron, MacKenzie Bryant, Tara Schwartz, Gail Ackermann, Greg Humphrey, et al. Bacterial and chemical evidence of coastal water pollution from the Tijuana River in sea spray aerosol. *Environmental Science & Technology*, 57(10):4071–4081, March 2023.

122 Matthew A Pendergraft, Derek J Grimes, Sarah N Giddings, Falk Feddersen, Charlotte M Beall, Christopher Lee, Mitchell V Santander, and Kimberly A Prather. Airborne transmission pathway for coastal water pollution. *PeerJ*, 9:e11358, June 2021.

123 A Mannocci, G La Torre, A Spagnoli, A G Solimini, C Palazzo, and M De Giusti. Is swimming in recreational water associated with the occurrence of respiratory illness? A systematic review and meta-analysis. *Journal of Water and Health*, 14(4):590–599, August 2016.

124 Sandra M Walser, Doris G Gerstner, Bernhard Brenner, Christiane Höller, Bernhard Liebl, and Caroline E W Herr. Assessing the environmental health relevance of cooling towers – A systematic review of legionellosis outbreaks. *International Journal of Hygiene and Environmental Health*, 217(2):145–154, 2014.

125 Kerry A Hamilton, Mark T Hamilton, William Johnson, Patrick Jjemba, Zia Bukhari, Mark LeChevallier, and Charles N Haas. Health risks from exposure to Legionella in reclaimed water aerosols: Toilet flushing, spray irrigation, and cooling towers. *Water Research*, 134:261–279, 2018.

126 Madison P Goforth, Stephanie A Boone, Justin Clark, Priscilla B Valenzuela, Julie McKinney, M Khalid Ijaz, and Charles P Gerba. Impacts of lid closure during toilet flushing and of toilet bowl cleaning on viral contamination of surfaces in United States restrooms. *American Journal of Infection Control*, 52(2):141–146, February 2024.

127 K Cole and J E Talmadge. Mitigation of microbial contamination from waste water and aerosolization by sink design. *Journal of Hospital Infection*, 103(2):193–199, October 2019.

128 Kerry A Hamilton, Mark T Hamilton, William Johnson, Patrick Jjemba, Zia Bukhari, Mark LeChevallier, Charles N Haas, and P L Gurian. Risk-based critical concentrations of *Legionella pneumophila* for indoor residential water uses. *Environmental Science & Technology*, 53(8):4528–4541, February 2019.

8

Transport, Growth, and Decay in the Environment

8.1 Chapter Overview

As stated previously (Section 3.2), one of the key paradigms of environmental engineering is the importance of describing quantitatively the source → fate and transport → receptor continuum. In Chapter 7, the sources for pathogens in the environment were enumerated. This chapter aims to quantitatively describe the relationship between those sources and the concentrations of pathogens to which a potential susceptible individual (the receptor) may be exposed.

In a 1933 lecture, Albert Einstein said:

> "It can scarcely be denied that the supreme goal of all theory is to make the irreducible basic elements as simple and as few as possible without having to surrender the adequate representation of a single datum of experience." [1]

A version of this idea underlies the application of quantitative models in engineering. Engineering models of a system can be constructed at different levels of complexity. Choosing the appropriate complexity level is part of the art of engineering and is driven by the application(s) or decision(s) that need to be made. There are trade-offs in the choice of complexity, including:

- Simple models generally have fewer parameters and data requirements and can be executed more rapidly on smaller computational platforms.
- Simple models generally are less capable of describing finer levels of detail at spatial or temporal scales. They may treat concentrations as of uniform composition and/or time-invariant.
- More complex models generally require greater details of the physical setting being modeled.
- Formulation and implementation of more complex models require greater sophistication of those developing and using such models.

8.2 Taxonomy of Models

Models used to describe the fate and transport of contaminants in the environment can be distinguished by whether they have 0, 1, 2, or 3 spatial dimensions and whether they assume a steady-state (no temporal variation) or unsteady-state (time-varying) approach. Regardless of these specifications, model formulation proceeds through a series of steps.

The first step in model development is to define the system of interest being modeled. This depends on the desired use of the model's output, and the system's definition may be used to simplify the later computational steps. Some examples of system definitions for problems relevant to environmental pathogen engineering are:

- All the air within a single room
- All the air in multiple rooms in a building and in the ventilation system that might connect them
- Water within the plumbing of a building
- Water within an aquifer, lake, or river being used as a source of drinking water
- The top meter of soil on an agricultural plot receiving and including applied Class B biosolids
- The air at or below an elevation of 25 m overlying a community

Regardless of how a system is defined, there may be fluxes of pathogens into and out of the system, and internal processes that result in pathogen growth, decay, and transformation (some of which may arise from interactions with other constituents in the system). It may also be desirable to model pathogens in a system based on their differential properties (which could impact the estimated risk to receptors or the persistence of the pathogens themselves). In this latter case, there may be transformation processes that convert pathogens with one set of properties into those with another. Examples of properties that might merit inclusion (in more complex models) are:

- Presence of a dormant state, such as spores, viable but nonculturable (VBNC) forms, cysts, or oocysts
- Occurrence of organisms in singles and in aggregates containing multiple organisms. Organisms may shift between these several types of occurrences.
- Occurrence of organisms in occluded solids of various forms
- Organisms in different physiological states, which may affect susceptibility to decay or ability to grow
- Exchange of genetic elements among organisms may result in differentially expressed traits. An example of this is exchange of resistance plasmids, which will not be discussed further.

In any system, either for the total population of a pathogen of interest or for the total in any one of the states, the application of a principle of mass conservation leads to the verbal equality in Equation 8.1.

$$\begin{aligned}&\text{Rate of change of organisms in system} = \\ &\qquad\qquad \text{Rate of input} - \text{Rate of output} \\ &\qquad\qquad\qquad \pm\text{Rate(s)of reaction}\end{aligned} \tag{8.1}$$

This verbal statement will be translated into mathematical forms in subsequent portions of this chapter.

There are several types of inputs or outputs to/from a system so defined.

- Bulk movement of the fluid containing microorganisms transported in or out of the system is termed advection. Therefore, the bulk motion of the fluid must be described to describe this.
- Exchanges across the boundaries not associated with the bulk flow of the medium. There are several types of these, including:

Diffusion/dispersion: If there are spatial differences in concentrations of pathogens (or of pathogens in a particular state), then the random motion of the pathogens or small packets of fluid in which they are contained (termed "eddies") will serve as a driving force to act against such gradients. It can be described as an apparent flux.

Fluxes due to external forces: Pathogens may be, or may be contained within particles that are denser than the fluid in which they occur, or may have electrical charges or magnetic properties. The presence of an external field (gravitational, electrostatic, or magnetic) can cause pathogens or particles to move along the field. The effect of this is to impose an additional flux.

Fluxes due to transference with an external phase: The system may be in contact with an external environment containing pathogens. If that external environment is a separate phase, there could be a driving force for pathogens move between that external phase and the system. Examples include the atmosphere as an external phase of a system defined as water, or a surface in contact with a system defined as water or air.

When transport includes advection and diffusion/dispersion, the overall system is described by an advection–dispersion process. If there is one or more reactions, as noted in the following, the system is defined as an advection–dispersion–reaction (ADR) process.

Reactions are processes within the system that result in an increase or decrease in the number of pathogens (or in some particular state of the pathogens). These may occur intrinsically or via interaction with other (chemical, physical, or biological) components of the system. Some examples of these are:

- There may be internal sources of pathogens, such as exhalations from an infected individual or dislodgement/secretion/excretion from pets.
- Organisms may be inactivated by spontaneous decay or via interaction with inhibitory factors such as light, chemical disinfectants, or antagonistic biota.
- Organisms may grow if conditions for their multiplication are favorable.
- When subpopulations are being modeled, organisms in one subpopulation may be changed into another.

8.2.1 Batch and Simple Flow Systems

The simplest system that can be analyzed is a closed zero-dimensional system. Denoting it as zero-dimensional is tantamount to noting the absence of spatial gradients within the system, i.e., it is completely mixed.

If there is a zero-dimensional system of constant volume (V) and pathogen concentration (C), and there are no inputs, outputs or internal sources or sinks, but possibly reactions, then the verbal mass balance of Equation 8.1 can be written as;

$$V\frac{dC}{dt} = rV \tag{8.2}$$

where r is the sum of all reaction rates that may occur in the volume. With initial conditions and the expression of the reaction rate in terms of C, this differential equation can be solved. Simple rate laws for decay reactions are shown in Table 8.1.

The reaction rates in Table 8.1 are written as spontaneous and unimolecular, meaning that they are not the result of interaction with any other constituents (biotic or abiotic).

An example of combining a rate into the batch mass balance is shown in Example 8.1.

Table 8.1 Examples of simple reaction rates for decay.

	Expression for *r*
Zero-order	$-k$
First-order	$-kC$
Second-order	$-kC^2$
"x" order	$-kC^x$

Example 8.1 Simple Decay Reaction in Closed Volume

A pathogen at initial concentration C_0 in a closed box decays at an xth order reaction with reaction rate k. Derive the expression for concentration (C) versus time. For arbitrary values of k and x, derive the expression for 90% decay (i.e., 1 log).

Solution:
The starting point is substituting the xth order rate expression from Table 8.1 into Equation 8.2 to yield:

$$V\frac{dC}{dt} = -kC^xV$$

V can be canceled from both sides, and this equation can be directly integrated (with the given initial condition) as in:

$$\int_0^t dt = \int_{C_0}^{C} \frac{dC}{kC^x}$$

Evaluation of this integral (providing that $x \neq 1$) yields:

$$t = \frac{C^{1-x} - C_0^{1-x}}{k(1-x)}$$

This gives the following expression for time to 90% decay:

$$t_{90\%} = \frac{0.1^{1-x} - C_0^{1-x}}{k(1-x)}$$

Note that the time to one-log decay will depend on the initial concentration. If $x = 1$, i.e., the decay process is first order, concentration decay is exponential, and the time to one log decay is independent of initial concentration.

Note that for a pure decay process, as this example illustrates, there is only a trivial steady state approached at long time, when $C \to 0$. For more complex processes, or when there is an internal source (as well as reactions and other sinks), the steady state for a simple well-mixed volume, even without inputs and outputs, may be nontrivial.

Even for a closed box system, there may be multiple components of interest. Consider that a pathogen may exist in viable, injured, and killed states. The concentration of these can be written as C_v, C_i and C_k. There may be a first-order rate of conversion from one state to another, as examples given in Table 8.2.

Table 8.2 Example reactions and rates for a viable to injured to killed process.

Process	Reaction	Loss rate
Conversion from viable to injured	$C_v \rightarrow C_i$	$-k_{vi}C_v$
Conversion from injured to killed	$C_i \rightarrow C_k$	$-k_{ik}C_i$

With such processes and reactions in Table 8.2, a mass balance can be written for each of the three species. Note, by virtue of the reaction, a viable organism that is injured produces an injured organism (resulting in a positive rate for the injured organism mass balance), and similarly for an injured organism that is killed. Therefore, the following equations result:

$$\begin{aligned}\frac{dC_v}{dt} &= -k_{vi}C_v \\ \frac{dC_i}{dt} &= k_{vi}C_v - k_{ik}C_i \\ \frac{dC_k}{dt} &= k_{ik}C_i\end{aligned} \tag{8.3}$$

These equations form a system of linear first-order differential equations and, therefore, can be solved analytically [2]. It can also be solved numerically, and this is presented since such an approach is helpful for more complex problems.

Example 8.2 Numerical Solution of System in Equation 8.3

Compute the numerical solution to the system defined by Equation 8.3 with the following conditions:

- $@t = 0, C_v = 1000, C_i = 0, C_k = 0$
- $k_{vi} = 0.1, k_{ik} = 0.03$

Solution:

The following listing in Python uses the built-in numerical initial value ODE solver to solve the system of equations under defined conditions. The parameters are coded in the user-defined function.

```
import scipy.integrate as numint
import numpy as np
import matplotlib.pyplot as plt
def batch_three_component(t,C):
    kvi=0.1
    kik=0.03
    Cv=C[0];Ci=C[1];Ck=C[2]
    dy_v=-kvi*Cv
    dy_i=kvi*Cv-kik*Ci
    dy_k=kik*Ci
    return[dy_v,dy_i,dy_k]

t_span=[0,100]
y0=[1000,0,0]
results=numint.solve_ivp(batch_three_component, t_span, y0,dense_output=True)
print(results.y)
plt.rcParams.update({'font.size': 18})
plt.plot(results.t,results.y[0],'r--',results.t,results.y[1],'bs',
         results.t,results.y[2],'g^')
plt.legend(['viable','injured','killed'])
plt.xlabel('Time')
```

(Continued)

Example 8.2 (Continued)

The results of the calculation are shown in Figure 8.1. Note the build-up of the intermediate of injured organisms (C_i) and the subsequent inactivation of this subpopulation.

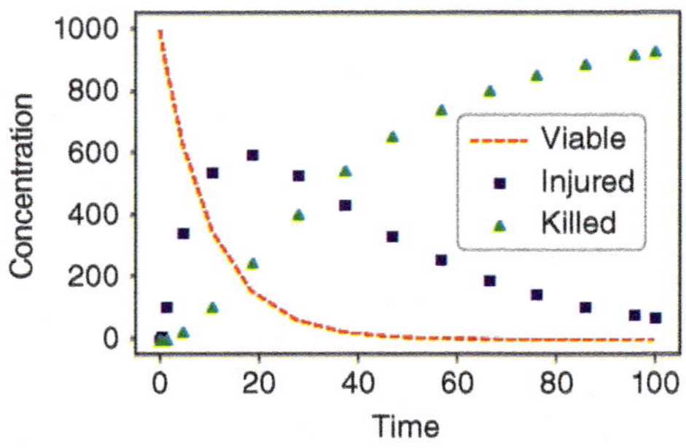

Figure 8.1 Output of solution to Example 8.2.

Note that this system only has a trivial steady-state (at $t \to \infty$) where all the initial organisms are killed. This approach can be used for more complex inactivation and subpopulation processes.

There are other source and sink terms that can arise in addition to intrinsic decay. These will be discussed in more detail in Section 8.4. Microorganisms (or the particles with which they are associated) may settle by gravity, or may be deposited to surfaces at the side of a space (and some may be resuspended). They may collide with each other and agglomerate or break up, changing their effective size. They may be inactivated by chemical or physical agents. They may grow under favorable conditions.

In general, real systems outside a laboratory are not closed. One generalization of the model is a completely mixed volume with both inlet and outlet flows. These systems are often referred to as a "completely mixed flow reactor" (CMFR) or a "completely stirred tank reactor" (CSTR). The definition of a CSTR or CMFR is that the concentrations within the volume are identical throughout and that the exit concentration is equal to this mixed internal concentration. To achieve this, a strong degree of mixing is needed.

How effective mixing is achieved in such volumes is beyond the scope of this book. However, standard references are available for the design and analysis of mixing conditions, e.g., [3]. Mixing can be verified in flow-through systems using tracer tests [4].

Many physical systems of interest can be (approximately) described as well-mixed volumes with flow. Some examples include:

- Air in a well-ventilated room
- A small lake, pond, or reservoir
- A tank used in water or wastewater treatment

Even where a volume is not completely mixed, it can often be described (approximately) as a system of multiple mixed volumes in series.

For a well-mixed volume with volumetric flow given by Q, inlet concentration of C_{in}, and potential internal sources with a total strength E (in units of number of organisms per unit time), the mass balance can be written as in Equation 8.4.

$$V\frac{dC}{dt} = rV + QC_{in} - QC + E \tag{8.4}$$

If Equation 8.4 is divided throughout by V, and the parameter θ is defined as $\frac{V}{Q}$, with dimensions of time, this equation can be written as:

$$\frac{dC}{dt} = r + \frac{C_{in} - C}{\theta} + \frac{E}{V} \tag{8.5}$$

For a non-reactive system ($r = 0$) with no internal sources ($E = 0$) if the initial concentration in C_0 and there is no input of contaminant (i.e., pathogen, $C_{in} = 0$), Equation 8.5 can be integrated to yield Equation 8.6.

$$C = C_0 \exp\left(-\frac{t}{\theta}\right) \tag{8.6}$$

The parameter θ is referred to as the "mean residence time," which denotes the average time that material spends within the volume.[1] The experiment described in the development of Equation 8.6 is a tracer experiment, which can be used to determine the average residence time of a system, and the entire residence time distribution.[2] Equation 8.6 is plotted in Figure 8.2. Note that the dimensionless group $\frac{t}{\theta}$ can be used for generalizability.

For this system, it is clear that there are significant portions of material that exit at times less than (and also greater than) the mean residence time.

In the presence of a reaction rate, such as those in Table 8.2 and a source term, Equation 8.5 can be solved, either at steady state (with $\frac{dC}{dt} = 0$) or dynamically by numerical integration. Note that as in the closed system case, if there are multiple subpopulations or interacting populations, then a mass balance equation in the form of Equation 8.5 needs to be written, and the resulting algebraic (at steady state) or differential (for dynamic systems) equations need to be solved simultaneously.

For example, if there is a first-order decay rate at a steady state, then Equation 8.7 can be derived.

$$C = \frac{C_0 + \theta\left(\frac{E}{V}\right)}{1 + k\theta} \tag{8.7}$$

From this equation, for example, if the inlet concentration, the source strength ($\frac{E}{V}$), the mean residence time, and the decay rate constant are given, then the steady-state concentration can be computed. This would be an important input parameter in a risk assessment, and engineering interventions might be considered to reduce exposure, such as increasing residence time or adding processes to increase inactivation.

---

1 In the field of indoor air, and elsewhere, it is common to use the reciprocal of θ as a characteristic. In indoor air, this is called the air-exchange rate, or, with units of hours, the air changes per hour (ACH). In continuous culturing of biological systems, particularly under controlled laboratory conditions, this is called the dilution rate.

2 There are several types of tracer experiments. The one described here is a washout experiment in which the system contains material, the influent is devoid of that material, and the effluent concentration is measured versus time. It is also possible to conduct a tracer experiment in which the system initially has zero concentration, and the influent has a constant concentration—this is termed a step-up experiment. A third approach is to start with the system at zero concentration and "instantaneously" add a known amount of material—this is termed a pulse experiment. Analysis of the results of a step-up or pulse experiment requires using Equation 8.5 and integrating it with the different appropriate initial conditions.

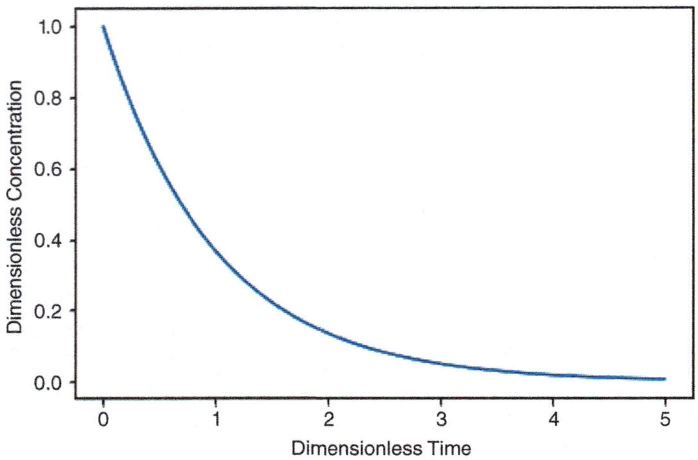

Figure 8.2 Dimensionless concentration ($\frac{C}{C_0}$) versus dimensionless time ($\frac{t}{\theta}$) for a CSTR washout tracer experiment.

This is illustrated by an example in which Equation 8.6 is modified for dynamic inlet conditions and both reaction and a (dynamic) internal source are considered.

Example 8.3 Complete Mix Flow Reactor with Time Variable Input, Reaction, and Internal Source

Consider a completely mixed volume (e.g., a room) with inlet concentrations that are sinusoidally variable time with a period of 12 hours. There is a first-order decay rate with $k = 1\ h^{-1}$ and an internal source that is sinusoidally variable with a period of 0.5 hours (perhaps a coughing individual). The mean residence time is 0.333 h. Compute the internal concentration versus time.

Solution:

Equation 8.6 can be used, setting $r = -1.0\ hr^{-1} \times C$. The source term is described as sinusoidal with a range between 50 and 100 organisms/hr according to the following equation:

$$\frac{E}{V} = 10\left[1 + \cos\left(\frac{2\pi t}{12}\right)\right]$$

The inlet concentration is also sinusoidal with a range from 10 to 20 organisms/L.

With these substitutions, and with an initial concentration (of organisms in the volume), the differential equation can be set up for solution. The code for the solution is given as follows.

```
import scipy.integrate as numint
import numpy as np
import matplotlib.pyplot as plt
def dynamicCSTR(t,C):
    # note - time units in hours
    k=1; C_in=10.0*(1+np.cos(2*np.pi*t/(12)))
    theta=0.333
    EoverV=50.0*(1+np.sin(2*np.pi*t/0.5))
    return -k*C+(C_in-C)/theta + EoverV

t_span=np.array([0.0,24.0])
```

```
y0=np.array([0.])
results=numint.solve_ivp(dynamicCSTR, t_span, y0,dense_output=True,
first_step=1,rtol=1e-8,atol=1e-12,method='DOP853')
print(results.y)
plt.rcParams.update({'font.size': 18})
plt.plot(results.t,results.y[0],'r--')
plt.xlabel('Time')
plt.ylabel('Concentration')
```

The output of this calculation is shown in Figure 8.3. The low-frequency fluctuation, in this case, represents the time-variant input, while the high-frequency fluctuations reflect the internal dynamic source.

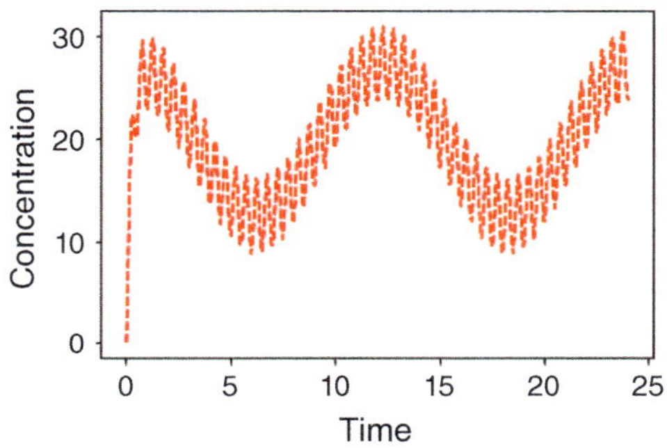

Figure 8.3 Output of complete mix flow system with first-order reaction and time-varying inputs and source term.

The CMFR/CSTR approach represents one extreme in which incoming material is instantaneously mixed with the contents of the system. A diametrically opposite approach is a flow system in which all of the entering material remains with the material in which it entered. This is termed a plug flow system. From the conceptual idea that if such a system occurred in a pipe, one could imagine each entering element of material traveling through the pipe in a plug, without mixing with material that entered earlier or later. This can alternatively be imagined as a conveyor belt of (very small) containers each containing an increment of material and progressing in isolation until they are discharged, as shown schematically in Figure 8.4. It should be immediately apparent that the discharge/effluent concentrations can be given by solving the appropriate equations for closed batch systems.

Real systems generally have mixing conditions that are neither ideally plug flow nor completely mixed. More complex frameworks are needed to describe the conditions in such intermediate systems.

8.2.2 Modeling Approaches for Complex Flow Systems

If the system flow is neither perfectly mixed nor plug flow, several approaches are available for quantitative description. These are broadly described as box models and as continuum models [5]. In either case, while the fluid (water, air, etc.) in motion can be described as a continuum, the microbial contaminants and particles on which they might be conveyed can also be treated in a continuum fashion (i.e., characterized by concentrations)—this is termed an Eulerian approach or

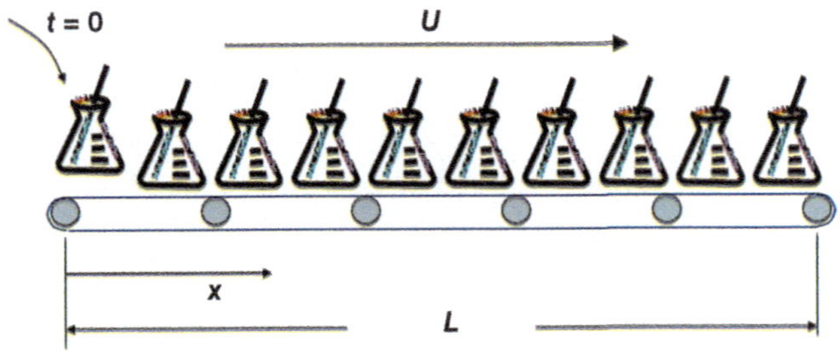

Figure 8.4 Analogy of a plug flow system as a conveyor belt of discrete closed volumes. U is the velocity of the conveyor, L is the length, and thus $\theta = \frac{L}{U}$.

by describing their individual discrete behavior—this latter approach is termed Lagrangian, and also a particle-tracking approach. Broadly speaking, this gives the following potential alternatives:

- Box models describing both particles and the containing fluid as concentrations—Eulerian–Eulerian approach.
- Continuum models describing both particles and their containing fluid being transported in a volume of interest—broadly termed Eulerian–Eulerian advection–dispersion–reaction (ADR) models.
- Models in which the fluid being transported is described by a concentration model, while particles are tracked individually, termed an Eulerian–Lagrangian approach. The fluid may be modeled either using a box model approach or an ADR relationship.

The details of these three approaches are discussed in Section 8.3 and subsequent sections.

In the context of pathogen fate and transport, Lagrangian or particle-tracking models can describe the stochastic variability associated with small numbers of contaminants, which is significant (Figure 4.3).

8.3 Quantitative Description of Transport

The objective of this section is to outline the approaches to modeling pathogen fate and transport in various settings. As noted above, central to the analysis of this is an understanding of the flow patterns of fluid in which microorganisms may be contained. This can be characterized by the use of tracer curves as in the single well-mixed volume (CSTR) in Figure 8.2. However, even tracer curves and residence time distributions provide incomplete information in and of themselves to characterize reactions occurring in flow-through systems. This older approach is systematically described by Nauman [6].

8.3.1 Box Models

The simplest box model, with an Eulerian–Eulerian framework, is one single well-mixed volume, as was depicted in Figure 8.2 for a washout tracer experiment. The equation for the washout tracer (Equation 8.6) can be made dimensionless by dividing by the initial concentration, C_0, and functionally is a complementary cumulative distribution. When taking the complement of that, the functional form is referred to as an "F" curve, i.e., for the single mixed volume:

$$F(t) = 1 - \exp\left(\frac{t}{\theta}\right) \tag{8.8}$$

This represents the fraction of material that exits the system on or before time t. The derivative of the F curve $\frac{df}{dt}$ is the "E" curve, which is a formal probability density function, termed the residence time distribution [7]. Hence, for the single mixed volume system, the following results:

$$E(t) = \frac{dF}{dt} = \frac{1}{\theta}\exp\left(-\frac{t}{\theta}\right) \tag{8.9}$$

This is formally an exponential probability distribution with mean equal to θ [8]. If there are two identical complete mixed volumes in series, perhaps two consecutive tanks, or two rooms with connected ventilation,[3] as in Figure 8.5, then a differential equation for the mass balance can be written for each of the two compartments. With no internal source and no internal reaction, and with inflow concentration to the first compartment of C_0, this results in the system of equations:

$$\begin{aligned}\frac{dC_1}{dt} &= \frac{C_0 - C_1}{\theta}\\ \frac{dC_2}{dt} &= \frac{C_1 - C_2}{\theta}\end{aligned} \tag{8.10}$$

The coupled system in Equation 8.10 can be solved (for a step-up experiment) with the conditions that at $t = 0$, $C_1 = C_2 = 0$. This results in the analytical solution for the F curve in Equation 8.11, where θ is the mean residence time for *each* of the compartments.

$$F = 1 - \left(\frac{t}{\theta} + 1\right)\exp\left(-\frac{t}{\theta}\right) \tag{8.11}$$

Equation 8.11 can be differentiated to yield the E curve for the two-compartment system:

$$E(t) = \left(\frac{t}{\theta^2}\right)\exp\left(-\frac{t}{\theta}\right) \tag{8.12}$$

This can be generalized to a system of N identical mixed volumes in series, each with a mean residence time of θ. The E curve for such a cascade is:

$$E(t) = \frac{t^{N-1}}{(N-1)!\theta^N} \exp\left(-\frac{t}{\theta}\right) \tag{8.13}$$

This is the equation for the Erlang distribution [9], which is a special case of the gamma distribution for integer parameter N. Note that when $N = 1$, this reduces to the exponential distribution. The mean of this distribution is $N\theta$. The mean residence time is given as $N\theta$ and the variance is $N\theta^2$.

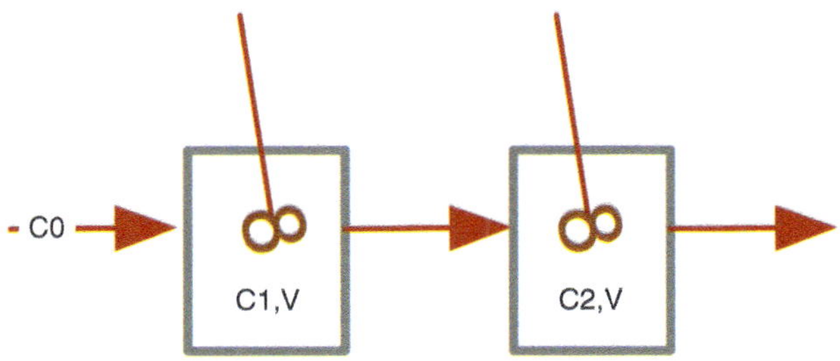

Figure 8.5 Schematic of two well-mixed volumes in series.

3 It is assumed that the time in the connecting pipe or duct is negligible.

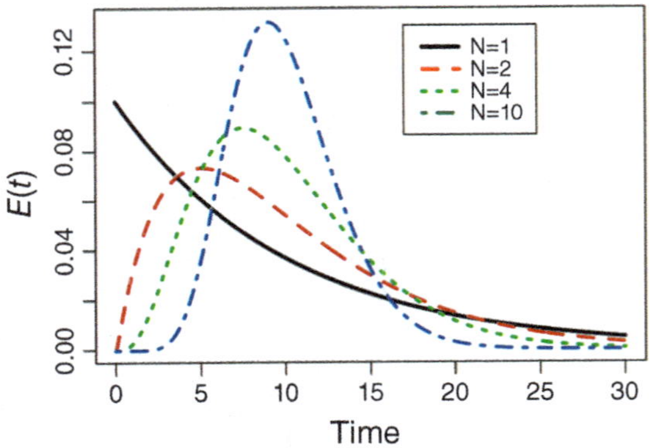

Figure 8.6 E curves for multiple well-mixed systems in series with mean residence time = 10.

Figure 8.6 plots the $E(t)$ curve versus time for a system with an overall mean residence time of 10, from 1 to 10 volumes in series. Note that as the number of volumes in series increases, the spread of the distribution decreases (for example, the standard deviation decreases) and ultimately (at greater than 10 volumes in series) tends toward greater symmetry.

If there are reactions and/or internal sources, then the system in Equation 8.10 can be generalized. Depending on the reaction order, an analytic solution may be possible. The following example illustrates a numeric solution.

To develop the numerical solution, the two-compartment model (without reaction or internal sources) in Equation 8.13 can be extended by adding the reaction rate and/or source terms as shown in Equation 8.5. Depending on circumstances, there may be either or both reaction and source terms in one or both of the two connected compartments. Starting from initial conditions, the system can be numerically integrated using the approach shown in Example 8.3.

In the realm of buildings, the spaces may be considered as a complex network of mixed volumes connected by ducts (perhaps with intermediate air treatment devices). The National Institute of Standards and Technology has developed software, CONTAM, to formulate and execute such models, including transport of contaminants [10] using this continuum box approach.

8.3.2 Box Models with Particle Tracking

Box models with particle tracking treat each microorganism or particle containing microorganisms as a discrete entity and analyze the dynamics of particle numbers in a system over time. In this manner, the stochastic variability associated with low particle numbers can be explicitly computed, and the persistence (or disappearance) of low numbers of organisms from a system can be ascertained.

There are three types of processes that can affect particle numbers in a volume. These will first be described, and then a series of example computations will be presented.[4]

---

4 In this subsection, the explicit analysis of a system on an individual particle basis will be used. There is also an approach in which the ensemble of particles is examined to study the time and spatial evolution of the mean concentration and higher-order moments. But, in the author's opinion, this is somewhat less useful if the objective is ultimately to connect fate and transport to risk. Furthermore, with the rapidly increasing power of direct computational simulation, detailed theoretical analysis is diminishing in utility.

8.3.2.1 Case 1—No Source or Reaction

The first situation, which is present in all flow-through systems, is the loss of particles associated with the flow itself. In a complete mixed volume, if the mean residence time is θ, then in any interval Δt, each particle present has a probability of exiting equal to $\frac{\Delta t}{\theta}$. If each particle behaves identically and independently, and if at the beginning of an interval, there are N_t particles, then the probability distribution for particles at the successive time interval $(t + \Delta t)$ may be described by a binomial. But it is just as easy to take a brute force approach to computation with the following logic:

1. Initialize with N_0 particles.
2. Compute a uniform random number over the interval $< 0, 1 >$ for each particle occurring at time t.
3. For a Δt, count the number of particles for which its random number is $> \frac{\Delta t}{\theta}$. Set the $N_{t+\Delta t}$ equal to this count.
4. Set $t = t + \Delta t$
5. If the desired time interval is not exceeded and particles remain, return to step 2.

Python code for this case is below for a single mixed volume with a mean residence time of 0.5 hours. The time step is kept small (0.01 hours) relative to the residence time, and a small number of initial particles ($N = 20$) are added at the initial time. The simulation is repeated several times due to stochastic variability.

```
import numpy as np
import matplotlib.pyplot as plt
import pandas as pd

Initial_N=20
delta_t=1/100     # work in time units of hours
theta=0.5        # mean residence time
endtime=6*theta
runs=10
tableau=pd.DataFrame()

trial=0
while trial<runs:
    simtime=0
    N=Initial_N
    counts=np.array([N])
    while simtime<=endtime:
        random_vector=np.random.rand(N)
        N=(delta_t/theta < random_vector).sum()
        counts=np.append(counts,N)
        simtime=simtime+delta_t
    trial=trial+1
    tableau[chr(trial)]=counts
plt.plot(np.arange(0,endtime+2*delta_t, delta_t),tableau,'|')
plt.rcParams.update({'font.size': 18})
plt.grid()
plt.xlabel('Time (hours)')
plt.ylabel('Particles Remaining')
```

The output of the 10 runs are shown in Figure 8.7. Due to the small number of particles, stochastic variability is clearly evident, especially at longer times. Note that for the 10 simulations, the "extinction" time (time at which the last particle exists) ranges from one to more than three hours (i.e., from two to more than 6 times the mean residence time). This simulation can be repeated for many more trials to get confidence limits on the extinction time.

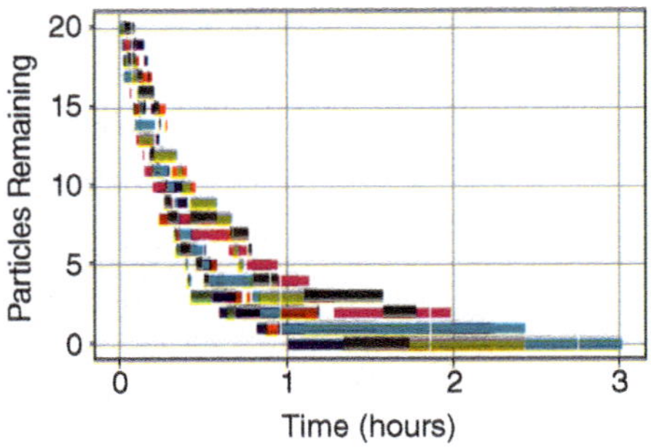

Figure 8.7 Trajectories of 10 runs for a complete mix volume initialized with 20 particles.

8.3.2.2 Case 2–Concomitant Loss or Reaction

If there are also loss processes in the form of reaction rates, then these are additive to the loss process in Case 1. On an instantaneous basis, the reaction rates (e.g., from Table 8.1 can be divided by the instantaneous microbial concentration (C) to give a term that has units of $time^{-1}$. For example, for the xth order reaction, a pseudo first-order term can be written as:

$$k^* = kC^{x-1} \tag{8.14}$$

This can be added to the loss from flow to yield a net probability of loss of a particle from the system at Δt to equal:

$$p_{\text{loss}} = \left(\frac{1}{\theta} + k^*\right)\Delta t \tag{8.15}$$

Note that k^* is a function of C, and so at each time step it would need to be evaluated based on the total particles (organisms) at that time step.

8.3.2.3 Case 3–Internal Sources or Growth Terms

If there are internal sources, such as emission from an infected individual, resuspension from surfaces, etc., that can be quantified as an emission rate, E (number of organisms per unit time), then the expected number emitted in a small time, Δt, would be $E\Delta t$. This number can be regarded as the mean in a Poisson distribution, and at each time increment, a random number of emitted organisms is generated.

The growth of a population can be described using a concept of instantaneous growth rate, μ with units of time^{-1}. For pure exponential growth, this will be a constant, while for more complex growth relationships μ may be a function of other conditions, including the concentration of microorganisms present in a system.

The presence of growth and/or an internal source term can allow the persistence of microorganisms for a substantial period of time, even in a flowing system. The following example is a modification of case 1, but adding a growth term with $\mu = 0.05h^{-1}$ and $E = 5\frac{\text{organisms}}{h}$. The code for this example follows.

```
import numpy as np
import matplotlib.pyplot as plt
import pandas as pd

Initial_N=20
delta_t=1/100     # work in time units of hours
```

```
theta=0.5          # mean residence time
mu=0.05             # first order growth rate
E=5                # emission rate per unit time
endtime=10*theta
runs=10
tableau=pd.DataFrame()

trial=0
while trial<runs:
    simtime=0
    N=Initial_N
    counts=np.array([N])
    while simtime<=endtime:
        random_vector=np.random.rand(N)
        survivors=(delta_t/theta < random_vector).sum()
        new_from_growth=np.random.poisson(mu*N*delta_t)
        new_from_source=np.random.poisson(E*delta_t)
        newcounts=survivors+new_from_growth+new_from_source
        counts=np.append(counts,newcounts)
        N=newcounts
        simtime=simtime+delta_t
    trial=trial+1
    tableau[chr(trial)]=counts
plt.plot(np.arange(0,endtime+2*delta_t, delta_t),tableau,'|')
plt.rcParams.update({'font.size': 18})
plt.grid()
plt.xlabel('Time (hours)')
plt.ylabel('Particles Remaining')
```

The output for this simulation (10 runs) is shown in Figure 8.8. Note that the time axis is extended to 5 hours, i.e., 10 residence times. Even at that prolonged time, in many runs there is the persistent presence of organisms resulting from the growth and source terms—essentially a non-zero steady state is reached in a proportion of trajectories.

8.3.3 Advection Reaction Processes–Eulerian–Eulerian Approach

To describe a flowing system with advection and reaction, it is necessary to combine models of fluid flow with those for contaminant transport and reaction. The description of the flow of fluid requires the solution of transport equations for momentum as well as overall conservation of mass. The solution of these bulk transport equations is beyond the scope of this book. For an introduction

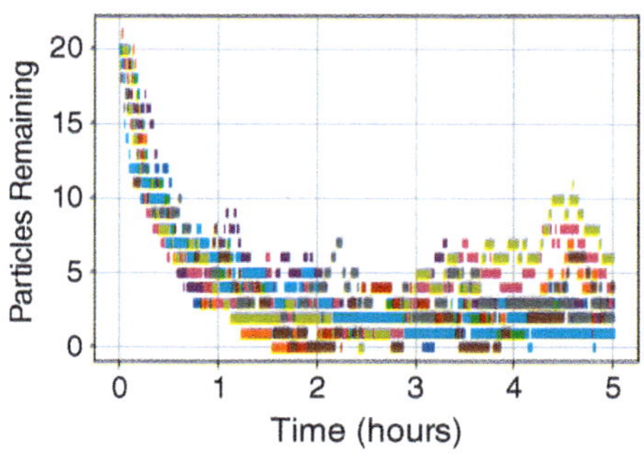

Figure 8.8 Trajectories of 10 runs for a complete mix volume initialized with 20 particles, with growth rate and internal source.

to the topic, consult Logan [11]. In many cases, the use of computer intensive solutions (computational fluid dynamics), abbreviated as CFD [12], may be required.

8.3.3.1 Brief Outline of Fluid Flow Modeling

A brief outline of the underlying concepts for describing flow in systems is presented in this subsection. This may be useful for those who choose to use CFD models to solve problems.

If a fluid, such as air or water, is moving in a system, in any volume of the system, the conservation of mass must be satisfied as well as the conservation of momentum (functionally equivalent to a conservation of energy relationship where there is no other source or sink for energy).

In a system, at any point and time, there is a local velocity, $\vec{v}$, which is a vector (denoted by the overbar arrow) having Cartesian components $< v_x, v_y, v_z >$. For steady, incompressible flow, the mass balance may be written in one of two equivalent forms[5]:

$$\frac{\partial v_x}{\partial x} + \frac{\partial v_y}{\partial y} + \frac{\partial v_z}{\partial z} = 0$$
$$\nabla \cdot \vec{v} = 0 \tag{8.16}$$

For incompressible flow of fluid with constant velocity, moving with even, laminar flow, the conservation of momentum may be written as Equation 8.17, which is called the Navier–Stokes Equation.

$$\rho \frac{D'\vec{v}}{D't} = \rho g - \nabla P + \mu \nabla^2 \vec{v} \tag{8.17}$$

This is a vector equation (i.e., with Cartesian components in the three coordinates) and notation defined as follows:

- $\nabla = \langle \frac{\partial}{\partial x}, \frac{\partial}{\partial y}, \frac{\partial}{\partial z} \rangle$
- $\nabla^2 = \frac{\partial^2}{\partial x^2} + \frac{\partial^2}{\partial y^2} + \frac{\partial^2}{\partial z^2}$
- $\frac{D'}{D't} = \frac{\partial}{\partial t} + v_x \frac{\partial}{\partial x} + v_y \frac{\partial}{\partial y} + v_z \frac{\partial}{\partial z}$

The Navier–Stokes equation in this form depicts flow that is perfectly smooth, termed laminar, as depicted at the bottom of Figure 8.9. It was observed as early as Leonardo Da Vinci [13] that flow often had parcels, termed "eddys" as depicted at the top of Figure 8.9 that did not move smoothly or regularly.

The research identified that, with increasing flow velocity, eddy formation dominated. Reynolds [14] quantified when this onset occurred and defined it by a dimensionless number which has come to be referred to in his honor as the Reynolds number. In the current nomenclature, this is defined by

$$Re = \frac{\rho v L}{\mu} \tag{8.18}$$

In this equation, μ is the dynamic viscosity, with dimensions of mass/(length × time), v is a characteristic velocity, and L is a characteristic length. For pipes, L is the diameter. For other channels, it is the hydraulic radius.[6] For flow in pipes and conduits, the transition to turbulence typically occurs with Re greater than 2000.

---

5 This is derived by considering a differential cube of sides *dx*, *dy*, and *dz*, and applying conservation of mass to that cube.

6 Defined as the cross-sectional area of the flow divided by the perimeter of the conveyance in contact with that flow.

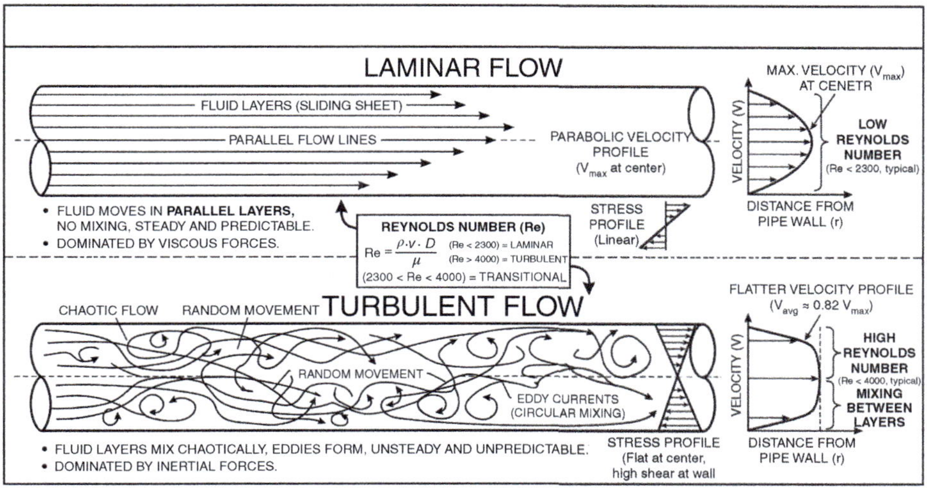

Figure 8.9 Laminar and turbulent flow in pipes.

With the onset of turbulence, as first articulated by Reynolds [14], the Navier–Stokes applies only to average conditions at any point in space. At any location in a system, there may be a particular *average* velocity vector[7] $\boldsymbol{v}$ and a vector of mean velocity *fluctuations* associated with movement of the fluid eddies. The spatial and temporal properties of these fluctuations will be governed by parallel equations, which are in turn coupled to the equations for mean velocity and which require additional assumptions to model the interaction between the mean velocities and the velocity fluctuations—this is referred to as the turbulence closure problem.

The velocity fluctuations will further affect the transport of contaminants such as microorganisms. The random motion of eddies will superimpose (and often dominate) the transport of material contained therein via the introduction of a dispersive flux.

8.3.3.2 Contaminants in a Flowing System

Of most significance to this work is how flow affects the transport of microorganisms and other contaminants. Fick [15] was the first to propose that materials are transported by a flux that is inversely linearly proportional to their concentration gradients. This is analogous to various other linear proportionality relationships in engineering, such as Hooke's law (for force on a mechanical spring), Ohm's law (for current in the presence of a voltage difference), and Fourier's law (for flow of heat in the presence of a temperature gradient). In basic mathematical terms, for a one-dimensional system that is uniform and isotropic,[8] this can be written mathematically as:

$$J_x = -D\frac{\partial C}{\partial x} \tag{8.19}$$

7 In Cartesian coordinates, it may have components in each of the three—x, y, and z—directions.

8 Uniformity means that the diffusion coefficient, or more generally the dispersion coefficient, is constant throughout the system of interest. Isotropy , as opposed to anisotropy, means that the transport property of diffusion is the same in all coordinate directions. Anisotropic behavior can occur in multiphase systems, such as groundwater flow, where the presence of a surface influences the mechanical properties of the medium over the physical range of interest.

Where J is the flux in the x direction, and D is the diffusion coefficient for the material being transported.[9]

In a more general form, a Fick's law flux can be added into the 3D mass balance, under homogeneous and isotropic conditions, to form the general relationship in Equation 8.20. The second derivatives in this equation result from the fact that J is a spatial function, and thus itself may have derivatives in the three coordinate axes.

$$\frac{dC}{dt} + \left(v_x\frac{\partial C}{\partial x} + v_y\frac{\partial C}{\partial y} + v_z\frac{\partial C}{\partial z}\right) - \epsilon_C\left(\frac{\partial^2 C}{\partial x^2} + \frac{\partial^2 C}{\partial y^2} + \frac{\partial^2 C}{\partial z^2}\right) - r = 0 \tag{8.20}$$

The solution of this equation requires specification of the geometry of the system of interest as well as initial and boundary conditions. An examination of the terms in this equation notes that specific simplifications may occur based on:

Reaction: Whether or not decay (or formation) processes exist (if not, $r = 0$).
Dimensionality: One, two, or three spatial dimensions.
Time dependence: Whether the system is at steady state ($\frac{\partial C}{\partial t} = 0$) or not.
Boundary conditions: Whether or not there is some constant input (i.e., at a particular spatial coordinate, such as x=0, the concentration is fixed), or perhaps time varying.

A further generalization of Equation 8.20 could be the inclusion of internal source terms, such as E in Equation 8.4, which may be a function of the physical location of the internal source(s).

An exhaustive treatment of analytical solutions is available in the literature [16], and for more complex situations, numerical solutions may be needed. Some special cases of analytical solutions will be presented as follows.

8.3.3.2.1 One-dimensional Systems

One-dimensional systems may be typified by flow in a duct, pipe, or river. In such a system, Equation 8.20 can be simplified to

$$\frac{\partial C}{\partial t} = -v_x\frac{\partial C}{\partial x} + \epsilon\frac{\partial^2 C}{\partial x^2} - r \tag{8.21}$$

In the absence of reaction, the final term, r becomes zero. This is a second-order partial differential equation and requires an initial condition and two boundary conditions for solution. For a tracer experiment, the initial condition is $C = 0$ at all x. There are alternative boundary conditions used in this solution, which may introduce differences, particularly for short or highly dispersive systems (ϵ large). However, a particularly useful and robust analytical solution has been proposed by Nauman and Buffham [17]. For a pulse tracer of mass M introduced instantaneously into a flow Q, the equation becomes [4]:

$$C = \frac{M}{Q\sigma\sqrt{2\pi}}\left(\frac{\theta}{t}\right)^{\frac{3}{2}}\exp\left[-\frac{(t-\theta)^2}{2\left(\frac{t}{\theta}\right)\sigma^2}\right] \tag{8.22}$$

9 The symbol "D" will be used strictly for molecular, or Brownian diffusion, while the symbol "ϵ" will be used for overall dispersion which also includes transport in eddies of a flowing fluid.

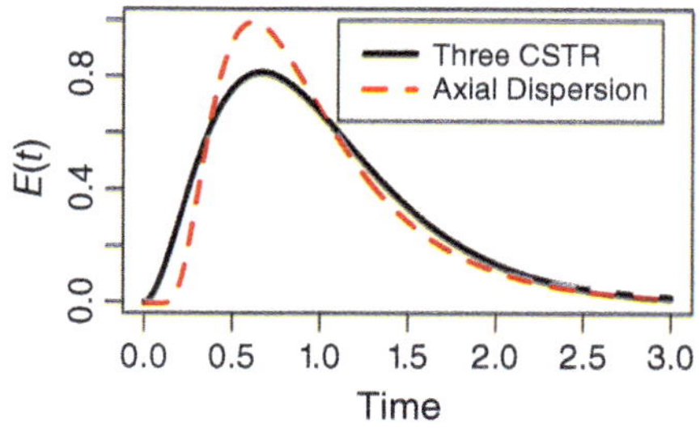

Figure 8.10 Comparison of axial dispersion to compartment in series model for mean residence time of 1.0 and variance equal to three compartment model.

In this equation, $\theta = \frac{v_x}{L}$, where L is the flow path in the x direction. This equation is of the functional form of an inverse Gaussian distribution [8]. Note that the mean (θ) and variance (σ^2) appear explicitly in this equation. Variance can be written in terms of the dispersion via the following:

$$\sigma^2 = \frac{2\theta^3 \varepsilon}{t^2} \tag{8.23}$$

If a tracer experiment is done, the mean and the variance (of arrival times at the outlet) of the tracer can be calculated. These are sufficient to describe either the compartment in series model or the dispersion model; however, the fitted distributions based on these first two moments may differ. This is shown in Figure 8.10, where the functions for a mean residence time (θ) and a variance equivalent to that of the three compartments in the series model are used. Note that the axial dispersion model is more "peaked" than the three compartments in series model.

Fitting Tracer Data and Its Limitations

A full discussion of performing tracer experiments is beyond the scope of this work, and there is nothing uniquely pertinent to microorganisms. However, several issues should be noted. First, the Erlang (gamma) model for residence time (Equation 8.13) and the Inverse Gaussian (axial dispersion) equation (Equation 8.22) are formally probability density functions with support over the positive real line. There are a large number (presumably, in theory, an infinite number) of potential density functions [8]. While they might be (partially) specified by matching the first two moments (mean and variance), they differ in higher moments and, as exemplified by Figure 8.10, in shape.

Second, while the method of moments has often been used for the analysis of tracer data, e.g., [18], it is clear that this can lead to bias and fail to use all the information present in the data. A nonlinear regression approach is superior and naturally leads to more systematic comparisons of alternative models when a sufficiently rich data set is available [4, 19].

Third, the ability of relationships derived from fundamentally different approaches, such as the compartments-in-series versus the axial-dispersion frameworks, to describe the same data leads to the conclusion that the fit of tracer data alone is insufficient to deduce the internal structure of a flowing system.

Finally, in reacting systems, the residence time distribution characterizes only the system's end state with respect to the tracer. A critical piece of information for predicting the performance of a system with reacting constituents is **when** mixing occurs to produce the

(Continued)

Fitting Tracer Data and Its Limitations (Continued)

effluent mixture. In the "conveyer belt" model (Figure 8.4), fluid remains with its entering neighbors until the last possible moment before being discharged (and mixed) to comprise the effluent—this is generally termed the segregated flow approach. An alternative approach is to assume that mixing occurs at the earliest possible moment, termed complete micromixing [20]. In general, these approaches will yield different estimates of a system's performance [21, 22]. This will be discussed in more detail in Section 11.2.2.

Under steady-state conditions, where $\frac{\partial C}{\partial t} = 0$, Equation 8.21 becomes an ordinary differential equation. If the reaction rate (r) is specified, this can be solved analytically (in some cases) or numerically. For the particular case of a first-order decay ($r = -kC$), the analytical solution is given by Equation 8.24.

$$C = C_0 \quad \exp\left[\frac{\upsilon_x x}{2\epsilon}\left(1 - \sqrt{\frac{4k\epsilon}{\upsilon_x^2} + 1}\right)\right] \tag{8.24}$$

In general, for other rate laws, a numerical solution of the steady state differential equation is necessary. If the reaction rate is power law, of order "w," then Equation 8.21 can be simplified to the following:

$$\epsilon\frac{d^2C}{dx^2} - \upsilon_x\frac{dC}{dx} - kC^w = 0 \tag{8.25}$$

Since this is a second-order differential equation, it requires two conditions for numerical solution. As Nauman and Buffham [17] note, depending on whether the inlet or outlets are treated as open or closed, different conditions apply. In the case of systems with both closed inlets and outlets, such as in a pipe or some sort of engineered reaction system, the following inlet and outlet conditions in Equation 8.26 are appropriate. The exit boundary condition specifies that there is no dispersive flux at the exit, and only advection.

$$\begin{aligned} C &= C_0 \quad @x = 0 \\ \frac{dC}{dx} &= 0 \quad @x = L \end{aligned} \tag{8.26}$$

In the case of open systems, which might be pertinent in rivers and streams (although if such systems are long, the differences from the closed boundary condition solutions are minor), the proper formulation of both "inlet" and "outlet" conditions becomes more difficult, and essentially requires the numerical solution of Equation 8.25 from distances far upstream of the system of interest to far downstream of the system of interest.

Even with the closed boundary conditions, the numerical solution of Equation 8.25 is a mixed boundary value problem. This can be solved in Python, using the package `solve_bvp` in the SciPy library. In Matlab, `bvp4c` can be used.

8.3.3.2.2 Three-dimensional Systems

For many cases of interest, numerical solutions of the governing equations are needed if a Eulerian–Eulerian approach is needed. However, many special cases have tractable analytical solutions. Herein, two are presented.

In the first example, a number of pathogens are released instantaneously at coordinates (0,0,0) in an infinite 3-D system with a velocity in the x-direction of υ_x.In such systems, dispersion in the x-direction can be neglected [23], and the temporal and spatial evolution of concentration is described by Equation 8.27.

$$C = \frac{M}{(4\pi t)^{3/2}\sqrt{\epsilon_x \epsilon_y \epsilon_z}} \exp\left[-\frac{\left(\frac{(x-v_x t)^2}{\epsilon_x} + \frac{y^2}{\epsilon_y} + \frac{z^2}{\epsilon_z}\right)}{4t}\right] \tag{8.27}$$

The relationship in Equation 8.27 forms a basis for the description of contaminant transport in the atmosphere as described in Section 8.6.1. Note also that the dispersion coefficients ($\epsilon_x, \epsilon_y, \epsilon_z$) may be different in the coordinate directions. The estimation of these dispersion coefficients is discussed in Section 8.3.4.

A second example is the continuous introduction of organisms per unit time at coordinates (0,0,0) into an infinite 3D space with a first-order decay reaction. This results in the solution given in Equation 8.28.

$$C = \left(\frac{m}{4\pi x\sqrt{\epsilon_y \epsilon_z}}\right) \exp\left[-\frac{kx}{v_x} - \frac{v_x}{4x}\left(\frac{y^2}{\epsilon_y} + \frac{z^2}{\epsilon_z}\right)\right] \tag{8.28}$$

8.3.4 *A Priori* Estimation of Dispersion

A central data need for the use of advection–dispersion models is the estimation of the dispersion coefficient(s), given by the (possibly subscripted) ϵ's. In a perfectly quiescent system, the primary dispersion mechanism is the Brownian diffusion of the microorganisms (or the particles to which they may be attached) themselves. This purely intrinsic (to the microorganisms or particles) diffusion coefficient can be computed via the Stokes–Einstein relationship [24] given in Equation 8.29. This assumes the particles are significantly larger than the molecules of the medium in which they diffuse (actually greater than the mean free path of those molecules), that they behave as rigid spheres, and that other interactions (electrostatic, magnetic, etc.) can be neglected. For many microorganisms, these may be only approximate.

$$D_i = \frac{\kappa_B T}{3\pi \mu d_i} \tag{8.29}$$

Where:

κ_B is the Boltzmann constant
T is the absolute temperature
μ is the viscosity
d_i is the diameter of the diffusing species

The characteristics of the bulk medium in which diffusion occurs have a significant effect on the diffusivity. For small molecules in water, the room temperature molecular diffusivity is order of magnitude $10^{-9}\frac{m^2}{s}$. For small molecules in air, the room temperature diffusivity is $10^{-6} - 10^{-5}\frac{m^2}{s}$.

Example 8.4 Estimation of Diffusivity Using Stokes Einstein Equation
Estimate the diffusivity of an aerosolized particle containing pathogens if its diameter is 3 μm and the temperature is 20 °C.

(Continued)

Example 8.4 (Continued)

Solution:

If SI units are used, $k_B = 1.38x10^{-23}\frac{J}{K}[or\frac{kgm^2}{s^2K}]$, $\mu = 1.8x10^{-5}\frac{kg}{m \cdot s}$ and $d_i = 3x10^{-6}m$.

Substituting this into Equation 8.29 yields:

$$D = \frac{(1.38 \times 10^{-23})(293)}{3\pi(1.8 \times 10^{-5})(3 \times 10^{-6})} = 7.945 \times 10^{-12}\frac{m^2}{s}$$

Note that this is substantially lower than that for small molecules in air, as expected.

One significant concept is that even when fluid motion is very small, some energy is dissipated via the formation of small-scale eddies, which have random motion of their own. This random motion of eddies gives rise to a dispersion (termed eddy dispersion) that may be (many) orders of magnitude more than the pure diffusion of microorganisms or small particles.

If a CFD approach is used for analyzing fluid motion *and* contaminant transport, then intrinsically the motion of eddies and their contained contents are described.

Alternatively, the contaminant mass balance equation (Equation 8.20) or its simplifications can be used apart from simulating turbulent fluid flow, if correlations for the dispersion (ϵ_c) are available.[10]

Such correlations depend on the fluid-flow characteristics and are typically interrelationships among several dimensionless variables. The Reynolds Number (*Re*) was defined in Equation 8.18. Correlations for turbulent diffusivity often involve relationships between *Re*, the Schmidt Number (*Sc*), and the Peclet Number (*Pe*). These relationships arise because there is an intrinsic connection between transport of mass (i.e., contaminants) and transport of momentum/energy. The intrinsic Schmidt number is defined as the ratio of momentum diffusion to mass diffusion given by Equation 8.30:

$$Sc = \frac{\mu}{D\rho} \tag{8.30}$$

With turbulent flow, the fluid and its contaminants have transport properties that are a function of the turbulent flow, and so a turbulent Schmidt number (S) can be defined by Equation 8.31, using turbulent viscosity (μ_t) and dispersion (ϵ).

$$S = \frac{\mu_t}{\epsilon\rho} \tag{8.31}$$

Momentum transport and mass transport (dispersion) are influenced by interactions between the moving fluid and the surrounding boundaries in which it is contained—in particular, the roughness of the boundaries may influence both of these quantities. Studies indicate that S is often in the range of 0.5–1.5 [25, 26].

A third dimensionless quantity of use is the Peclet number (*Pe*). This is the ratio between advective flow and dispersive flow and can be given by

$$Pe = \frac{Lv_x}{D} \tag{8.32}$$

Under turbulent conditions, D in Equation 8.32 is replaced by ϵ to get a turbulent Peclet number (P)

10 Or if dispersion differs with different directions, ϵ_x, ϵ_y, and ϵ_z

Specific correlations for different physical situations have been reported in the literature. Examples of these follow.

Sheriff and O'Kane [26] measured turbulent diffusivity of tracer gas (nitrous oxide) in air flowing through circular ducts and reported the following correlation:

$$\frac{\epsilon\rho_{\text{air}}}{\mu_{\text{air}}} = 25.27\frac{Re}{10,000} - 3.82 \tag{8.33}$$

Example 8.5 Estimate of Dispersion in an Air Duct

Using the correlation of Sheriff and O'Kane [26], estimate the dispersion of contaminants in an air duct that is 0.3 m in diameter with an air velocity of 10 $\frac{m}{s}$ at 20°C.

Solution:

First, note that the diameter of the particles does not enter the calculation—it is assumed that their diameter is less than that of the eddies in which they are conveyed. To compute Re, the viscosity and density of air are needed. At the stated temperature,[11] these are (in consistent SI units):

$$\rho_{\text{air}} = 1.204\frac{kg}{m^3}$$

$$\mu_{\text{air}} = 18.13 \times 10^{-6}\frac{N \cdot s}{m^2}$$

Using this information, Re can be calculated:

$$Re = \frac{v_x \cdot d \cdot \rho_{\text{air}}}{\mu_{\text{air}}} = \frac{10 \cdot 0.3 \cdot 1.204}{18.13 \times 10^{-6}} = 200{,}667$$

Note that this is very strongly in the turbulent region. This can be substituted into Equation 8.33 to yield:

$$\frac{\epsilon\rho_{\text{air}}}{\mu_{\text{air}}} = 25.27\frac{Re}{10{,}000} - 3.82$$

$$\frac{\epsilon\rho_{\text{air}}}{\mu_{\text{air}}} = 25.27\frac{200{,}667}{10{,}000} - 3.82 = 503$$

$$\therefore \epsilon = \frac{503 \cdot 18.13 \times 10^{-6}}{1.204\,\text{kg/m}^3} = 7.58 \times 10^{-3}\frac{\text{m}^2}{\text{s}}$$

Note that this is orders of magnitude greater than the diffusivity of molecules or small particles, as predicted from the Stokes–Einstein relationship. Hence, turbulent dispersion is the dominant mechanism vis-à-vis molecular or small particle diffusion. In general, this will hold for nearly all environmental problems of interest.

11 It is tacitly assumed that the pressure within the duct is 1 atm.

A number of situations arise in which microorganism flow occurs through fixed beds packed with solid material. Examples of these settings include permeation through soil, engineered filtration processes (air or water), adsorption processes, and analytical equipment. Rastegar and Gu [27] provide a correlation for dispersion in such situations (Equation 8.34).

$$P = \frac{0.18 + 0.008R^{0.59}}{d_p \cdot \eta} \tag{8.34}$$

In this relationship, d_p is the diameter of particles in the packed bed, R is the Reynolds number based on the packed bed porosity, and η is the packed bed porosity.

A general approach to dispersion in channels, which could include pipes, canals, etc. has been relating the dispersion to the shear velocity at the surface of the containment [11]. This relationship is given by Equation 8.35:[12]

$$\epsilon = 10.1R_h \cdot \upsilon_{sh} \tag{8.35}$$

In this relationship, R_h is the hydraulic radius (defined as the cross-sectional area of the flow divided by the length of the cross-section with which it is in contact. The parameter υ_{sh} is the shear velocity, which is related to the Darcy-Weisbach friction factor (f) by:

$$u_{sh} = \sqrt{\frac{f}{8}} \tag{8.36}$$

The friction factor in turn is defined by the implicit Colebrook equation (Equation 8.37) [28].

$$\frac{1}{\sqrt{f}} = -2\log_{10}\left(\frac{\kappa}{3.7} + \frac{2.51}{Re\sqrt{f}}\right) \tag{8.37}$$

where κ is the relative roughness of the pipe or channel (defined as the ratio of the absolute roughness to the diameter or to four times the hydraulic radius). Classically, a nomograph, called a Moody diagram, was used to solve this equation. However, it can also be readily solved iteratively on a computer.

In rivers, streams and natural channels, the shear velocity is related to the slope of the hydraulic gradient (S) and R_h, with g being the gravitational acceleration via [11]:

$$\upsilon_s h = \sqrt{gSR_h} \tag{8.38}$$

S can be predicted from the Manning equation, using the discharge flow rate and the effective surface roughness (which is governed by characteristics of the stream bed or channel, and vegetation), as in a similar approach [29].

8.3.5 Advection Reaction Processes—Eulerian–Lagrangian Approach

With the Eulerian–Lagrangian approach, the continuity equation for the bulk fluid (Equation 8.16) and the Navier–Stokes equation (Equation 8.17) are solved with a suitable turbulent closure assumption (using computational fluid dynamics) to obtain the velocity field at all points within space—this field consists of the average velocity vector and also the vector of velocity fluctuations

12 There does not seem to be a reason why this relationship would not apply both in water and air, though estimation of dispersion using this approach in air does not appear to be widely used.

at each point in space. This may be (in steady-state problems) a constant field or, in transient problems, a time-varying field. Since the transport of microorganisms is not expected to affect the transport of bulk fluid, it is possible to describe the former separately once the velocity field is known.

Guha provides an overview of the approaches that have been used in these simulations [30]. On any particle in the system, Newton's law can be used to relate the particle velocity to the forces on the particle Equation 8.39, recognizing that both velocity $\vec{v}$ and forces ($\vec{F}_i$) are vectors.

$$\frac{d\vec{v}}{dt} = \sum_i \vec{F}_i \tag{8.39}$$

Multiple forces can act on particles, including the following:

- Drag force
- Gravity
- Lift
- Thermophoretic force (in the presence of a temperature gradient)
- Electrical forces (in the presence of an electrical field and particles that are charged or can be induced to carry a charge)
- Magnetic forces

In calculating the individual particle motion, the stochastic fluctuations of the fluid's velocity need to be considered. By tracking many particles through a system, the ensemble result produces the distribution of particle transport. One way to evaluate the turbulent velocity fluctuations is at each time and location, for each particle, add a random vector whose components are from an uncorrelated normal distribution with zero mean and standard deviation equal to $\sqrt{\frac{2\kappa}{3}}$, where κ is the turbulence kinetic energy (usually obtainable from the CFD solver) [31].

The Eulerian–Eulerian and Eulerian–Lagrangian methods were compared to describe particle motion in a simulated indoor air environment [32]. The results were comparable. Under steady-state conditions, the computational effort for the two approaches was similar, whereas under transient conditions, the Eulerian–Lagrangian approach was more computationally efficient.

It is possible to conduct the Lagrangian portion of the computation in parallel if the behaviors of individual particles are independent.[13] In such cases, a large number of particles can be introduced concurrently, and their individual trajectories computed through the velocity field by application of Equation 8.39.

Zoka et al. used an Eulerian–Lagrangian approach to model dispersion of SARS-CoV-2 particles released from an individual breathing [33], coughing or sneezing in a room. An example of their calculations is shown in Figure 8.11.

With particle-tracking models, the physical-chemical changes in particles and the rate processes that affect pathogen persistence can be addressed. In the Eulerian–Eulerian approach, such factors would be included with the rate process term in the governing equation. In the Eulerian–Lagrangian context, the behavior of each particle containing microorganisms, or every single microorganism, can be modeled.

Li et al. [31] incorporated the evaporation of expelled respiratory particles within a room, considering the evaporation of water. This required consideration of the evaporative cooling of the particles and their reduction in size due to water loss. The diminution in size, in turn, increases the rate of persistence in the bulk indoor air.

---

13 This might not be the case if there is the possibility for particle agglomeration and/or breakup during their transit.

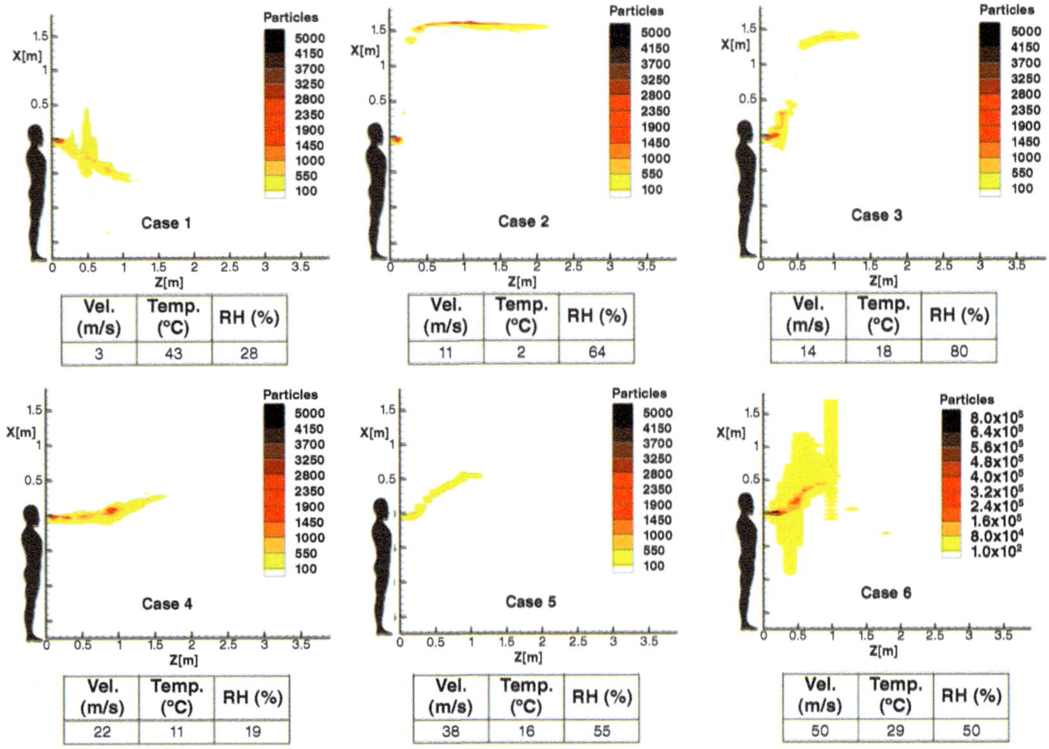

Figure 8.11 Dispersion of particles from an infector in a room for various scenarios. *Source:* [33] / MDPI / CC BY 4.0 / https://www.mdpi.com/2073-4433/12/8/986 / last accessed Feb. 12, 2026

If disinfection processes are used, the Eulerian–Lagrangian approach can be modified (as can obviously the Eulerian–Eulerian approach via a local reaction rate). Details of rate processes are discussed in the next section. Disinfection is discussed in more detail in Chapter 11. Zhang et al. have reviewed the use of computational fluid dynamics for modeling disinfection, primarily in the context of water [34].

In the presence of disinfection, there will be a rate process describing the rate of inactivation of viable organisms over time, i.e., r_d, in units of # organisms per volume per time. Consider a single microorganism traversing a space that has a chemical disinfectant or a physical disinfection field (e.g., ultraviolet radiation) present. From a single particle point of view, its inactivation can be described by:

$$\frac{dN}{dt} = -r(x(t), y(t), z(t)) \tag{8.40}$$

where N is the fraction of particles following that trajectory that are viable. Hence, during the course of a simulation of a particle, this derivative can be evaluated by summing up the reactions based on the coordinate position of the particle[14] and the proportion of particles that are viable at the exit can be obtained. This approach has been used for UV disinfection of water [35], chemical disinfection of water [36], chemical decontamination of indoor air [37], and UV disinfection of indoor air [38].

14 This is necessary since chemical disinfectant concentration or physical disinfectant field intensity may vary with space and time.

8.4 Quantitative Description of Rate Processes

From the previous material in this chapter, it is clear that reaction rates for various processes are important to characterize and describe microbial fate and transport adequately. The concept of a (batch) curve of the phases of microbial growth and decay was already encountered in Chapter 6 (Figure 6.4). This section will provide a general tour of many of the processes that can result in microbial growth or loss, with the objective of providing rate functions (r) that can be incorporated into mass balance relationships. Rate processes to be considered can be broadly classified according to Figure 8.12 and will be discussed in turn.

8.4.1 Physical

Physical factors can cause the decay (inactivation) of pathogens, as discussed further as follows. Aside from this, other rate processes arising from physical interactions can affect the fate and transport of pathogens.

One key factor is the size of particles in or on which pathogens may be contained. In indoor [39] and outdoor air [40], there is a broad spectrum of sizes in which pathogens may occur. This size can be dynamic because processes in the air will influence the effective size of the containing particles [41]. For microorganisms in airborne particles, the effective particle size influences both fate and transport and pathogenicity, since the size of the inhaled particle influences its propensity to deposit in different respiratory tract sites.[15]

In contaminated aquatic systems, pathogens may exist associated with a spectrum of particle sizes. This will depend both on the nature of the initial contamination as well as any intervening treatment processes [43]. In aquatic systems, particle size will also substantially influence the fate and transport of associated pathogens and can be dynamic.

8.4.1.1 Processes Unique to Air

A unique set of processes for particles emitted to the air, whether by direct exhalation from infected individuals, or from spray generated from aqueous systems (such as open water environments or waste treatment), is the tendency for the size of the particles to decrease with equilibration with the surrounding atmosphere. Vejerano and Marr [44] summarized some of these issues for the case of aerosolized particles from virus-infected individuals:

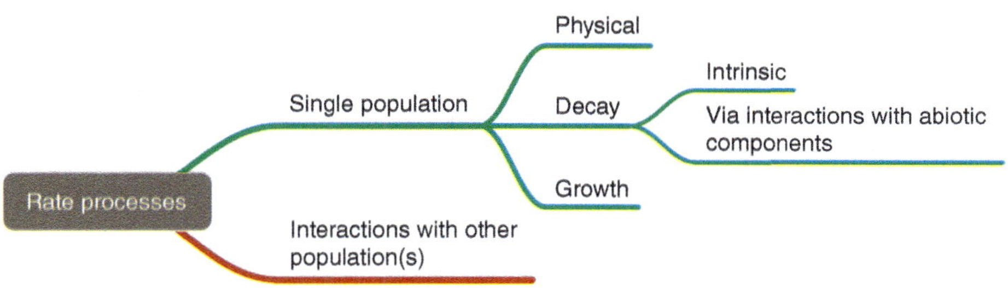

Figure 8.12 Types of rate processes considered.

15 Historically, pathogens emitted via the respiratory tract, by breathing, speaking, coughing, etc., were differentiated into droplets versus aerosols. There was also a belief that the larger droplets (classically > $5 \mu m$) would settle rapidly in indoor air. However, scholarship in light of the COVID-19 pandemic indicates that such distinctions were based on a misunderstanding of both basic epidemiology and air physics and transport [42]. Therefore, such differentiation is avoided in this book.

"A 60 μm droplet has a volume of approximately 0.1 nl and would contain approximately 1 virion if the virus concentration in the bulk liquid were 10^7 ml^{-1}, if the bulk concentration also applied to the droplet. Assuming a simplified model of respiratory fluid composition, the droplet would contain approximately 1 ng of salt, approximately 1 ng of total protein and approximately 0.06 ng of surfactant. The mass of each component is at least five orders of magnitude larger than the mass of the virion...

When expelled from the respiratory system, droplets undergo an abrupt change in RH from approximately 100% to ambient. In all but the most humid environments, the droplets evaporate quickly to approximately half their initial diameter, and concomitantly, the concentrations of salts, proteins and other components increase by nearly an order of magnitude due to the loss of water."

Therefore in describing the behavior of such particles in the atmosphere, it is necessary to consider several processes operating simultaneously:

- There will be evaporation of water from the particle, with a flux depending on temperature, relative humidity, and relative velocity (influencing mass transfer coefficients).
- The driving force for evaporation will be a function of the vapor pressure of water from the particle, which in turn will depend on the water activity of the water in the particle. This water activity will depend on the concentration of internal solids and the particle diameter (affecting surface tension).
- With time, the particle shrinks, solutes will concentrate, and the particle cools due to loss of enthalpy from the vaporized water.
- Evaporation will decrease when an equilibrium state is reached.

The physical changes in the particles containing microorganisms may also influence their decay rate over time.

Lighthart [45] has outlined an approach to quantifying these processes. This approach assumes that the microbial particles contain, as primary solutes, inorganic salts (e.g., NaCl). There are three equations that need to be solved simultaneously. The differential equation describes the time dependency of particle diameter in Equation 8.41. The evaporative cooling of the particle surface is represented by the differential equation in Equation 8.42. As temperature changes, as evaporation concentrates the remaining solids concentration in the particle, and as the particle curvature increases, there is a change in effective vapor pressure given by Equation 8.43. The symbols in these equations are defined in Table 8.3.

$$\frac{dd_p}{dt} = \frac{4D_w M}{R\rho_p d_p}\left(\frac{p_\infty}{T_\infty} - \frac{p_d}{T_d}\right)\left[\frac{2\lambda + d_p}{d_p + 5.33\left(\frac{\lambda^2}{d_p}\right) + 3.42\lambda}\right] \tag{8.41}$$

$$\frac{dT_d}{dt} = \frac{12}{d_p^2\rho_p C_p}\left[\frac{D_w M_w H}{R}\left(\frac{p_d}{T_d} - \frac{p_\infty}{T_\infty}\right) - k(T_\infty - T_d)\right] \tag{8.42}$$

$$p_d = \frac{p_s}{1 + \left(\frac{6imM_w}{M_s\rho\pi d_p^3}\right)} exp\left(\frac{4\gamma M_w}{\rho RTd_p}\right) \tag{8.43}$$

If the internal concentration of solute increases to the solubility limit, the shrinking of emitted particles and evaporation will decrease. Convection currents in outdoor or indoor environments may also increase evaporative losses beyond those predicted by the Lighthart equations.

Table 8.3 Definition of variables in Equations 8.41–8.43.

Symbol	Definition
C_p	Specific heat of water vaporization
D_w	Molecular diffusivity of water vapor in air
d_p	Particle diameter
H	Latent enthalpy of fusion of water
i	Number of ions formed per molecule of solute dissolved
k	Thermal conductivity of air
M_s	Molecular weight of solute
M_w	Molecular weight of water
m	Mass of dissolved solute
T_d	Temperature at particle surface
T_∞	Ambient air temperature
R	Gas constant
p_d	Saturation vapor pressure of water at the particle surface
p_s	Saturation vapor pressure of pure water at ambient temperature
p_∞	Ambient water vapor pressure
ρ	Density of water
ρ_p	Density of the particle at $t = 0$
λ	Mean free path of air molecules

While the Lighthart approach may be suitable for particles emitted from environmental compartments, such as sprays from water surfaces, it may be of limited use except as a qualitative framework when particles are emitted from sources with high concentrations of organic materials. Such materials may alter vapor pressure, surface tension, and other properties, and this could occur in emissions from waste treatment facilities and directly via breathing, coughing, and sneezing by infected individuals.

In the specific context of particles emitted from people, Stiti et al. [46] used a modified approach from that of Lighthart to model the shrinkage of emitted particles from breathing and coughing of individuals. They compared the results with experimental measurements of particle shrinkage over time. Within seconds to minutes, even large particles are shown to shrink down to sizes given by the residual solids in the remaining particle, at which point they can remain readily suspended in the environment.

8.4.1.2 Generally Applicable Processes

Multiple other physical processes affect the fate and transport of microorganisms (and particles that contain them) in water or air. Perhaps the most significant is the collision of particles to form larger particle aggregates.[16]

The classical modeling of flocculation kinetics derives from early work of Smoluchowski [47]. For two classes of particles with number concentration of n_i and n_j and with diameters d_i and d_j,

16 In the water treatment field, the term "coagulation-flocculation" is used to denote a two-step process in which chemicals are added to reduce repulsive forces between particles (destabilization) and the collisions are encouraged to occur via mechanical forces, including Brownian motion, resulting in aggregation. The term flocculation will be used here to denote the overall process of aggregation, regardless of whether there is a previous step of destabilization.

the rate of flocculation, i.e., the rate at which particles of these two classes collide to produce a successful aggregate is given by Equation 8.44

$$r_{\text{floc}}(i, j) = \alpha\beta(i, j)n_i n_j \tag{8.44}$$

where α is the collision efficiency factor, i.e., the proportion of collisions that produce a successful aggregate, and $\beta(i, j)$ is the collision frequency between particles of the two classes. Collision efficiency is a function of any particle pre-conditioning, or the intrinsic surface chemistry of the particles, as well as interparticle forces upon collision [48]. The quadratic dependence on particle number means that as the initial concentration of particles that can flocculate increases, the initial half-life decreases.

There are two principal drivers of collision rate ($\beta(i, j)$). One driver is the random motions of small particles due to the thermal kinetic energy of the fluid in which they are contained. This is termed perikinetic flocculation and is generally the exclusive mechanism for particles in air. The rate constant for perikinetic flocculation is given by Equation 8.45.

$$\beta_{\text{perikinetic}}(i, j) = \left(\frac{2kT}{3\mu}\right)\left(\frac{1}{d_i} + \frac{1}{d_j}\right)(d_i + d_j) \tag{8.45}$$

Note that the perikinetic rate, particularly in the case of monodisperse particles ($d_i = d_j$) is *independent of particle diameter.* The second driver is the collision of particles conveyed by randomly moving fluid eddies, resulting from energy dissipation (perhaps from imparted mechanical action) into the fluid. This is important in water and other liquid systems, especially for larger particles. Especially in the water literature, this is termed orthokinetic flocculation. The rate constant for orthokinetic flocculation is given by Thomas et al. [47] (Equation 8.46).

$$\beta(i, j)_{\text{orthokinetic}} = \left(\frac{G}{6}\right)(d_i + d_j)^3 \tag{8.46}$$

Computation of the RMS velocity gradient (G) is illustrated by Example 8.6. Note that, unlike the case of perikinetic flocculation, the rate of orthokinetic flocculation *increases* with the cube of the particle diameter.

Example 8.6 RMS Velocity Gradient in a Room

Consider a room with an air change (AC) rate of 5 h^{-1} $\left(= \frac{5}{3600} s^{-1}\right)$, and a pressure drop of 0.01 inches of water across the room. To compute the "G" value for orthokinetic flocculation, the power per unit volume needs to be computed. Per Camp [49]:

$$G = \sqrt{\frac{P}{\mu V}} \tag{8.47}$$

where P is the power and V is the volume of the room. Furthermore, P can be expressed in terms of pressure drop (Δp) via:

$$P = Q\Delta p \tag{8.48}$$

with Q being the flow rate through the room. Dividing 8.48 by V and substituting the definition of air change rate ($=Q/V$), the following results:

$$G = \sqrt{\frac{AC\Delta p}{\mu}} \tag{8.49}$$

with units conversion, at 20 °C with air viscosity of $18.13 \times 10^{-6} Pa \cdot s$, rhia yields $G = 13.8 s^{-1} G = 13.8 s^{-1}$.

The overall flocculation rate between particles of diameter i and j is then equal to $\beta_{\text{perikinetic}}(i, j) + \beta_{\text{orthokinetic}}(i, j)$.

This analysis has a number of simplifying assumptions including [47]:

1. Fluid motion in the vicinity of the colliding particles is laminar.
2. Particles are spherical, and remain so after collision.
3. Collisions only involve two particles.
4. No breakup of flocs occurs.
5. Particles are monodisperse

Perhaps limitations (1) and (2) are the most problematic. In the realm of water flocculation, Han and Lawler have developed a more complex framework to improve predictions [50] particularly when orthokinetic flocculation is dominant.[17] Item (3), to some extent, is also handled by the Han and Lawler formulation and is expected to be significant only at high particle concentrations. The issue of floc particle breakup and polydisperse populations will be discussed below. Despite its limitations, the Smoluchowski/Camp and Stein approach provides a useful approximate framework. Several examples show this.

First, by equating the perikinetic rate (Equation 8.45) and orthokinetic rate (Equation 8.46) the condition where both are equal can be derived, given in Equation 8.50, with perikinetic dominating at lesser diameters and orthokinetic dominating at greater diameters.

$$d_{p=o} G^{\frac{1}{3}} = \left(\frac{kT}{2\mu}\right)^{\frac{1}{3}} \tag{8.50}$$

Table 8.4 shows the conditions at which orthokinetic and perikinetic rates are equal. So for the situation in Example 8.6, with a G value of 13.8 s^{-1} in air, the orthokinetic and perikinetic rates are equal at:

$$d_{p=o} = \left(\frac{1.308 \times 10^{-23} \cdot 293}{2 \cdot 1.813 \times 10^{-5} \cdot 13.8}\right)^{\frac{1}{3}} = 1.97 \times 10^{-6} m$$

Therefore, for these conditions, particle flocculation for particles less than about two microns will be dominated by perikinetic flocculation, while particles greater than this diameter will be more affected by orthokinetic flocculation.

Table 8.4 Condition for perikinetic and orthokinetic rates to be equal at 20 °C.

Medium	Viscosity (20 °C, Pa-s)	$d_{p=o} G^{\frac{1}{3}}, m^{\frac{1}{3}}$
Water	1.002×10^{-3}	1.559×10^{-6}
Air	1.813×10^{-5}	5.937×10^{-6}

17 In aqueous flocculation, a third mechanism for collisions can be significant, termed sedimentation, in which larger and/or denser particles having a greater settling velocity overtake and collide with smaller particles, leading to some additional agglomeration.

Example 8.7 Particle Flocculation Example

Consider a well-mixed room with an air change (AC) rate of 5 h^{-1}. Inlet air has 100 5 μm particles per m^3. It is desired to have flocculation of these (as the only process operative in the room), reducing the primary particle concentration by 90%. What should the overall flocculation rate be? How much power per volume is needed from the pressure drop in the room plus any internal mixing (e.g., fans)? Assume a temperature of 20 °C and a collision efficiency of 1.0.

Solution:

The starting points are Equation 8.44 for the flocculation rate and Equation 8.5 (which, for steady state, $\frac{dc}{dt} = 0$, and recognizing that $\theta = \frac{1}{AC}$). Given only one type of particle ($n_i = n_j$), this results in the following quadratic equation in n:

$$-\beta n^2 + (n_{\text{in}} - n) + n_{\text{in}} AC = 0$$

This can be explicitly solved for β:

$$\beta = \frac{(n_{\text{in}} - n) \cdot AC}{n^2}$$

After substituting given values, it can be found that:

$$\beta = \frac{(100 - 10) \cdot \frac{5}{3600}}{10^2} = 2.3 \times 10^{-3} \frac{m^3}{\text{particle} \cdot s}$$

$\beta_{\text{perikinetic}} = 5.937 \times 10^{-6} \beta_{\text{orthokinetic}} = 2.3 \times 10^{-3}$

Rearranging Equation 8.46, the G value can be solved for:

$$G = \frac{6 \cdot \beta_{\text{orthokinetic}}}{(2 \cdot 5 \times 10^{-6})^3} = \frac{6(2.3 \times 10^{-3})}{(10^{-5})^3} = 1.38 \times 10^{13} \quad s^{-1}$$

On the face of it, this is so high as to be impractical, which becomes obvious by application of Equation 8.47,

$$\frac{P}{V} = G^2 \mu = (1.38 \times 10^{13})^2 \cdot (1.813 \times 10^{-5}) = 3.45 \times 10^{21} \frac{W}{m^3}$$

Hence, reliance on flocculation for the substantial reduction of micron-sized particles in air under usual conditions is unrealistic. For larger particles, with a higher initial particle density and/or in water, flocculation via the orthokinetic mechanism may be more important.

When two particles form an agglomerate, their effective size increases. The degree of increase depends on the individual particle geometry and on its composition. In a simple case of the agglomeration of two spherical fluid particles of diameter d_i and d_j, if the agglomerate is also a fluid sphere, and the total volume is conserved, then simple geometry yields:

$$d_{i+j} = \sqrt[3]{d_i^3 + d_j^3} \tag{8.51}$$

In another simplified example, if the collision is between two spherical particles that retain their shape and merely form a stable attachment, the child particle has a complex shape with a characteristic length (depending on the angle) between the diameter of the larger particle and the sum of the diameters of the two particles.

In even more complex situations, chains or aggregates of particles with fractal properties may be formed [51]. It may then be more appropriate to characterize a system as having an evolving distribution of particle diameters.

It is possible to describe the evolution of particle sizes during aggregation by population balances, in which particles are subdivided into size ranges. The aggregation rates (whether by perikinetic or orthokinetic mechanisms) are used to define the loss of smaller particles by aggregation and the formation of larger particles. This can be used to account for the "shearing" of larger particles by fluid motion, which may be important on longer time scales. Jeldres et al. describe such population balance models, which will not be discussed further [48].

In general, this aggregation process will result in only minor loss, except for large particles, long times, or when (in water) vigorous mixing is used to promote the orthokinetic mechanism.

Another major set of physical processes is those which result in deposition or settling to bounding surfaces, with potential resuspension. Since these are interphase processes, they will be discussed in Section 8.5.2.

8.4.2 Decay

A general process common to all microorganisms is decay or death. A natural starting point needs to consider the definition of death for microorganisms. This has long been known to be problematic. As noted by Rahn [52], in the context of bacteria, but which generalizes to all microorganisms of concern here:

> "A study of death should logically begin with a definition of death, but right there, difficulties arise. Death cannot be defined by positive criteria; it can be characterized only by the absence of some property which is essential to life. But different groups of biologists differ in their views on the most essential properties of living organisms. The death of a cell is not always determined by the same method; the loss of motility, of respiration or of other enzyme activities, the increase in permeability which makes plasmolysis impossible and causes the absorption of dyes, are variously used as criteria of death. These criteria do not always appear simultaneously ... Thus, a cell may be alive according to one definition, and dead according to another.
>
> The bacteriologist has no choice of definition because bacteria are too small to permit an easy study of any of the above criteria. The criterion almost universally used by him to define death is the loss of reproduction."

As was shown in Chapter 6, there are many ways to detect and quantify pathogens. Each of these measures implicitly defines viability in a different manner, e.g., the ability to grow in culture, (bacterial) ability to exclude certain dyes, the ability of (a portion) of the genome to be duplicated, etc. Therefore, the concept of decay needs to be conditioned on the measurement used. For pathogens, the gold standard is whether the organism can infect and ultimately cause an adverse effect in the host. Ultimately, forces leading to decay stem from the concept that microorganisms, like any form of life, are highly organized structures and therefore have reduced entropy. The laws of thermodynamics indicate that isolated systems proceed toward a state of maximum entropy. Therefore, without a continuous input of energy, microorganisms will decay via their own or their hosts' metabolism. The various environmental conditions under which microorganisms exist will influence the rate at which this decay occurs.

The kinetics of microorganism inactivation can be simple or complex. In approaching the question of what model to use, the maxim that all models are wrong but some are useful [53]. From an engineering point of view, the simplest model that best fits the data is preferable. A baseline simple model is that of first-order decay, i.e., the rate of inactivation of microorganisms ($r_d = -k_d N$), where k_d is the decay constant, and N is the concentration of viable organisms. In a closed, batch system, this is an exponential decay which can be depicted as a straight line on a semilog plot. For a simple exponential decay, only one parameter, k_d, is sufficient to describe the system.[18]

It has long been known that simple first-order decay is often an insufficient model, and that more complex behavior is noted. Some examples of this more complex behavior in batch systems are noted in Figure 8.13. With these models, more than a single parameter is required to describe the decay process. As an aside, in this section only intrinsic decay, as perhaps affected by general environmental variables, is the focus, and later in the book the behavior of decay during interventions using chemical or physical agents is discussed. Living organisms are often more complex than simple textbook examples, or as has been stated, "The Bugs Don't Read The Books" [54].

Decay with shoulders, tailing off, and the other deviations shown in Figure 8.13 may arise from multiple types of models. In fact, a decay curve, such as in this figure, can be regarded as a complementary cumulative probability distribution with support over the non-negative real axis. There are, in principle, an infinite number of such functions, and a large number that have been the subject of study [8].

Hence, the consistency of experimental decay curves with a model, in and of itself, does not provide proof of the correctness of the model assumptions. If a model derivation is based on a reaction rate approach, and if the same reaction rates also provide good agreement with a second type of reactor (e.g., a CSTR) then this is strong supportive evidence for the underlying assumptions.

The focus of this discussion is the description of the decay process as arising from one or more reactions, each with a rate constant. Alternatively, the decay of a population of microorganisms can be viewed as the result of a "hazard rate" imposed on populations of varying ages or fitnesses. This former approach will first be presented, and then the hazard rate approach will be presented.

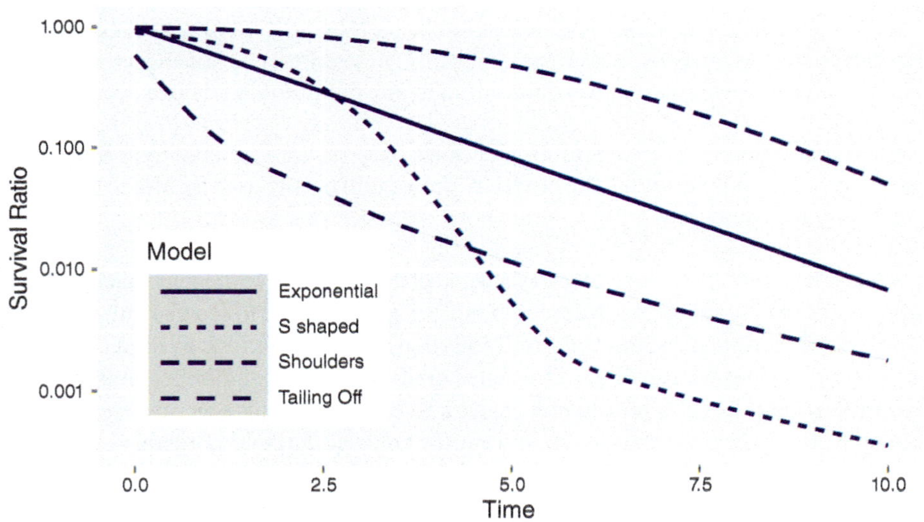

Figure 8.13 Schematic of exponential and more complex decay relationships in closed batch systems.

18 Or equivalently, the half life—i.e., time for 50% reduction, the decimal reduction time—i.e., time for 90% reduction, or some similar parameter may be used.

8.4.2.1 Reaction Rate Approach

There are several approaches to modeling the more complex decay behaviors using a rate based approach. These include the following:

- Intrinsic heterogeneity of subpopulations
- Simultaneous reactions to alter the susceptibility of individuals in a population
- Diffusional limitations
- Requirement for multiple events before decay occurs
- Impact of quorum sensing on decay

These approaches will be discussed in turn.

The most straightforward complexification of the exponential model is to consider the population of microorganisms undergoing decay to be composed of subpopulations, each with an intrinsic decay rate. In the case of two populations with initial concentrations N_1^o and N_2^o, having two different decay rates,[19] $k_{d,1}$ and $k_{d,2}$, the following two differential equations describe the system:

$$\begin{aligned}\frac{dN_1}{dt} &= -k_{d,1}N_1\\ \frac{dN_2}{dt} &= -k_{d,2}N_2\end{aligned} \tag{8.52}$$

If $N_1^o = x \cdot N^o$ and $N_2^o = (1-x) \cdot N^o$, then the result is:

$$\frac{N}{N^o} = x \cdot \exp(-k_{d,1}t) + (1-x) \cdot \exp(-k_{d,2}) \tag{8.53}$$

This model can describe a tailing-off behavior, where both the initial rapid and the latter slower decays are approximately linear on a semilog plot. This approach can clearly be generalized to more than two populations. In the extreme, there may be a probability distribution for the decay rates among a heterogeneous population. If $f(k_d)$ is a density function for the distribution of decay rates, then overall survival at time t can be given by Equation 8.54.

$$\frac{N}{N^o} = \int_0^\infty f(\kappa)e^{-\kappa t}d\kappa \tag{8.54}$$

It is important to note that the subpopulations may arise from multiple causes. They may be genetically (and thus phenotypically) different. They may be organisms in different physiological states and degrees of vulnerability. Or they may be in various physical states (perhaps some are partially occluded or protected by other constituents in the environment).

A second set of causes of nonexponential decay could arise if, during the course of inactivation, parallel reaction(s) occur which either predispose or act to protect from decay. A simple modification to a first-order decay can illustrate this by allowing the rate constant to be time-variant (Equation 8.55).

$$\begin{aligned}\frac{dN}{dt} &= -kN\\ \frac{dk}{dt} &= \alpha k\end{aligned} \tag{8.55}$$

where $k = k_0$ at $t = 0$

19 For convenience, it is assumed that $k_{d,1} > k_{d,2}$

The integrated solution of Equation 8.55 can be shown to be:

$$ln\left(\frac{N}{N_0}\right) = -\frac{[k_0(\exp(\alpha t) - 1)]}{\alpha} \quad (8.56)$$

Depending on whether α is positive or negative, this equation results in curves with different shapes, i.e., shoulders or tails, respectively,

A third set of causes of non-exponential decay results from requiring multiple events to occur before death arises. This can arise via several different mechanisms:

- There may be multiple organisms (bacteria, virus, etc.) in relatively stable aggregates that remain stable during the viable enumeration process—and *all* organisms in the aggregate must be inactivated for that aggregate to be considered inactivated. Figure 8.14 shows an example of stable aggregates (and the distribution of aggregate size) of the Delta variant of SARS-CoV-2 produced from cell culture [55].
- An individual viable organism may have multiple copies of a critical biomolecule and for death to occur *all* copies must be destroyed. This is the case, for example, with DNA, where both strands of DNA (in the case of prokaryotic or eukaryotic cells or double-stranded DNA virus) must be damaged.
- If there are repair mechanisms that can (at least partially) contravene the degradation resulting in death, then prolonged exposure to the adverse conditions may be required.

The concept of multiple hits being required before inactivation occurs leads to a classical model, termed the multi-hit model [56], which can be written as in Equation 8.57 [57].

$$\frac{N}{N_0} = 1 - \frac{\Gamma(\eta, k \cdot t)}{\Gamma(\eta)} \quad (8.57)$$

In this equation, η is the number of hits required, $\Gamma(x)$ is the mathematical gamma function, and $\Gamma(x, y)$ is the incomplete gamma function. This equation reduces to the simple exponential if $\eta = 1$, and for $\eta > 1$, it produces shoulders. It is also possible to convolute this with a statistical distribution for required hits (for example, see the right-hand panel of Figure 8.14) to produce a more complex survival curve.

Another potential mechanism for non-exponential survival, although attempts to incorporate it into quantitative survival models have been limited, is quorum sensing in bacteria (and possibly higher microorganisms). The metabolism of bacteria results in the release of small molecules that can serve as a signal to the population to regulate individual cell responses [58]. This has the

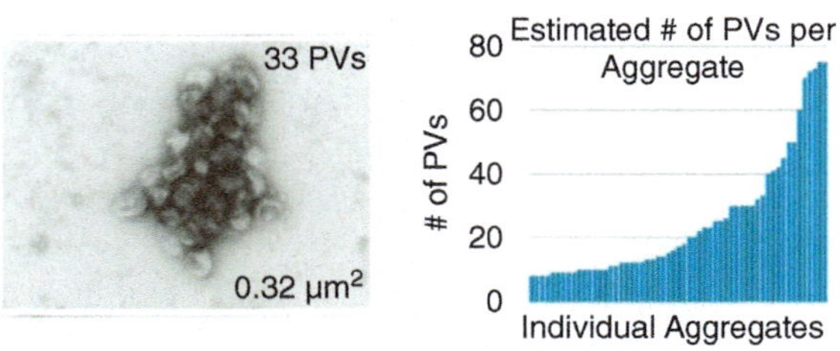

Figure 8.14 Electron micrograph (left) of aggregate of delta variant of SARS-CoV-2 grown on cell culture, and histogram of aggregate size (right). *Source:* From [55]/MDPI/CC BY 4.0/https://www.mdpi.com/1999-4915/14/5/1024/last accessed Feb. 12, 2026.

potential to accelerate or modulate decay. With some chemical disinfection processes, this may have been manifested as an effect of the initial population density on the rate of inactivation of the population [59].

Multiple combinations of these rate-based processes can occur, leading to a broader set of potential decay curve shapes. As noted already, the consistency of the decay curve's shape with a model does not prove the validity of the assumptions used to derive that model.

8.4.2.2 Hazard Rate Approach

Another approach to developing decay relationships is to view a survival curve as a complementary cumulative probability distribution. A starting point is defining the survival ratio (also termed the survival function), S(t), as $S(t) = \frac{N(t)}{N_0}$. S can vary between 1 (at t=0) and 0 (at long times).

The hazard rate is defined as the instantaneous death rate of the survivors remaining at time t, and denoted as $h(t)$. This is mathematically defined by Equation 8.58.

$$h(t) = -\frac{1}{S(t)} \cdot \frac{d(S(t))}{dt} \tag{8.58}$$

As an example, for the simple exponential decay, $S(t) = \exp(-kt)$, mathematical substitution results in $h(t) = k$, i.e., there is a constant hazard rate. There are many other survival models with non-constant hazard rates [60], a few of which are presented below in detail. The mathematical forms of the survival functions and the hazard functions are given in Table 8.5.

The Weibull, Gamma, and Gompertz model all simplify to the exponential as a special case. Depending on the value of the second (non-rate) parameter, these three models can give a hazard rate that increases with time (shoulders on the survival curve) or one that decreases with time (tailing off). The lognormal model gives a sigmoid survival curve, with a hazard function that initially increases then decreases.

There is often a biological interpretation based on the behavior of the hazard rate over time. The constant hazard rate of the exponential implies no change in the instantaneous sensitivity of organisms versus time. If the hazard rate increases with time, this could reflect the accumulation of damage that eventually results in inactivation. If the hazard rate decreases with time, this could reflect the presence of multiple subpopulations in which the more fragile are inactivated first, or a parallel process in which there is "hardening" of survivors [61]. However, caution in firmly asserting a

Table 8.5 Common two parameter survival distributions and hazard functions.

Survival function	$S(t)$	$h(t)$	Parameter range
Weibull	$exp[-(kt)^m]$	$mk(kt)^{m-1}$	$k, m > 0$
Gamma[a]	$1 - \frac{\Gamma(kt, \beta)}{\Gamma(\beta)}$	$\frac{(kt)^\beta}{t\Gamma(\beta)} \cdot \frac{\exp(-kt)}{1 - \Gamma(kt, \beta)}$	$k, \beta > 0$
Gompertz	$\exp\left[-\frac{\gamma}{k}(e^{kt} - 1)\right]$	γe^{kt}	$k, \gamma > 0$
Lognormal[b]	$1 - \Phi\left[\frac{ln(a \cdot t)}{\sigma}\right]$	$\frac{\exp\left[-\frac{(ln(at))^2}{2\sigma^2}\right]}{1 - \Phi\left(\frac{ln(at)}{\sigma}\right)} \cdot \frac{1}{t\sigma\sqrt{2\pi}}$	$a, \sigma > 0$

[a] $\Gamma(a, b)$ is the incomplete gamma function, and $\Gamma(a)$ is the usual gamma function.

[b] $\Phi()$ is the standard cumulative normal distribution.

mechanism based solely on the hazard function's temporal structure, in the absence of additional information, is risky.

It should be noted that, in addition to the two-parameter models, there are survival models with three (and more) parameters, for example, a three-parameter extension of the Weibull model [62]. However, distinguishing between the two-parameter models based on data fit is often problematic, and small data sets may not provide sufficient power to justify the more complex models. Such higher-order models will describe more complex, non-monotonic behavior of the hazard function. The fitting of experimental data to survival models will be discussed later in this subsection.

The survival functions in Table 8.5 are written for well-mixed batch systems. To be useful, it would be desirable to derive the corresponding reaction-rate expressions for each function. These could then be used in mass balance expressions for different system configurations, e.g., complete mix, axial dispersion, etc. The process is illustrated for the Weibull survival function in Example 8.8.

Example 8.8 Rate Expression Derivation for Weibull Survival Model

It is desired to determine the reaction rate which, when integrated in a batch mass balance yields the Weibull survival function from Table 8.5.

Solution:

Recognize that the survival function is for $\frac{N}{N_0}$. This can be differentiated with respect to N to yield:

$$\frac{dN}{dt} = -mNk^m t^{m-1} \tag{8.59}$$

The survival function in Table 8.5 can be rearranged to solve for t:

$$t = \frac{1}{k}\left[-\ln\left(\frac{N}{N_0}\right)\right]^{\frac{1}{m}} \tag{8.60}$$

Using Equation 8.60 to substitute for t in Equation 8.59, the following is obtained:

$$\frac{dN}{dt} = -kmN\left[-\ln\left(\frac{N}{N_0}\right)\right]^{m-1} \tag{8.61}$$

The right-hand side of Equation 8.61 can be interpreted as the decay reaction rate and only involves state variables (i.e., N and N_0).

A similar approach, though not necessarily yielding simple analytical expressions, can be used for the other survival functions.

8.4.2.3 Phenomenological and Empirical Models

There is a broader class of models that have been developed on a purely empirical basis. However, each of them conceptually can be considered a survival distribution, and a (perhaps not analytically tractable) hazard function derived. Mitchell and colleagues reviewed data on viruses in untreated groundwater [63] and indicators and pathogens in natural surface waters [64]. They fit data to a suite of empirical decay models, and compared the adequacy of fit and uncertainty of parameter estimates. Table 8.6 summarizes the principal models other than the ones discussed previously in this chapter.

These empirical functions can also depict various types of nonexponential survival curves. In a review of viral persistence in groundwater, the Juneja and Marks models and the exponential damped model emerged as the superior alternatives to the exponential [63].

Table 8.6 Empirical survival functions.

Name	$\left(\frac{N}{N_0}\right) =$	Parameter restrictions
Logistic	$\frac{2}{1+\exp(-k_1 t)}$	$k_1 > 0$
Exponential damped	$\exp(-k_1 t \cdot e^{-k_2 t})$	$k_1, k_2 > 0$
Juneja and Marks 1	$1-(1-e^{-k_1 t})^{k_2}$	$k_1, k_2 > 0$
Juneja and Marks 2	$[1+\exp(k_1 + k_2 ln(t))]^{-1}$	
Gompertz-Makeham	$\exp\left[-k_3 t - \frac{k_1}{k_2}(e^{k_2 t} - 1)\right]$	$k_1, k_2, k_3 > 0$

Source: Abstracted from [63, 64].

Table 8.7 Burr Type XII and III complementary cumulative distributions expressed as survival functions.

Name	$\frac{N}{N_0} =$
Burr Type III	$1-\left[1+\left(\frac{a}{t}\right)^{-c}\right]^{-k}$
Burr Type XII	$\left[1+\left(\frac{t}{a}\right)^{c}\right]^{-k}$

It should also be noted that since survival functions have the mathematical form of a complementary cumulative distribution function, any such function that has a range of support (i.e., argument) over the positive real line can be a plausible survival function. There are a variety of families of such functions that may be promising alternatives. These include the Type XII and Type III Burr distributions discussed by Tadikamalla [65]. Depending on parameter values, these can cover a wide range of skewness and kurtosis values and have a number of other distributions as special cases or can nearly approximate such distributions. These distributions have simple analytical function forms, summarized in Table 8.7.

8.4.2.4 Fitting Decay Rate Models to Data

Given a spectrum of potential models and data on the persistence/decay of a pathogen in a particular environment, it is desired to determine the most suitable model(s) to describe the data, the best estimates for the parameters, and the uncertainty in those estimates.

If the number of experimental data points are small, and if they span a narrow range of survivals, there may not be a possibility to differentiate adequacy of fit statistically, however if results are extrapolated, different models may yield quite disparate results. This is illustrated in Figure 8.15, where the parameters of three models are fixed using the times for one log (t_{90}) and two logs (t_{99}) inactivation to determine the parameters of three models. Note within the 0 to three log inactivation ratio range, there is only a small difference among these models, but at three logs inactivation (i.e., below a survival ratio of 0.001), there is substantial and increasing divergence among the models.

In Section 6.2, the principle of maximum likelihood was used to analyze the statistics of simple count data. This can be extended to analyze decay experiments. Ideally, data sets would have raw count data (at known volumes) versus time, and the mean concentration could be modeled as a function of time with a particular decay model to determine decay model parameters. The layout of such an experiment might be as in Table 8.8. This table is given in "tidy" form in which each

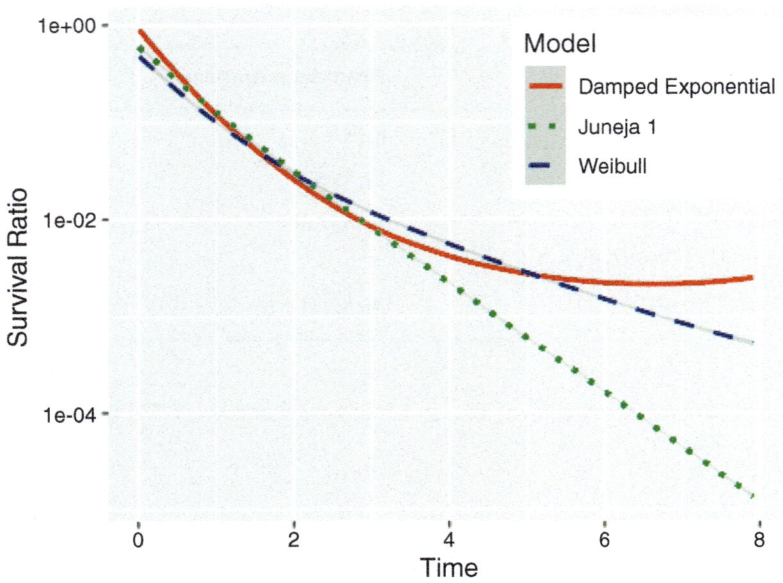

Figure 8.15 Comparison of different decay models with similar values for t_{90} and t_{99}.

Table 8.8 Schema of decay experiment using count data presented in "tidy" form.

Time	Amount of sample	Viable organisms (#)
t_1	V_1	x_1
t_2	V_2	x_2
t_3	V_3	x_3
…		
t_i	V_i	x_i

row represents a single data point (count) with the associated descriptive data [66]. This is a conventional and convenient form for statistical software analysis.

A starting point for analyzing such data is a generalization of Equation 6.3 for multiple sample amounts (V). Using the schema in Table 8.8, and assuming underlying Poisson statistics, an overall likelihood function can be

$$L = \left[\frac{(\mu_1 V_1)^{x_1}}{x_1!} e^{-\mu_2 V_2}\right]\left[\frac{(\mu_3 V_3)^{x_2}}{x_2!} e^{-\mu_4 V_4}\right] \dots \left[\frac{(\mu_i V_i)^{x_i}}{x_i!} e^{-\mu V_i}\right] \tag{8.62}$$

The values for the mean concentrations (μ_i) are given using the initial concentration of organisms (μ_0), which may be an unknown to be estimated from the data and the decay model to be evaluated, denoted by $f()$, which is a function of unknown parameters in the decay model to be evaluated, denoted by the vector $\vec{\Theta}$ as shown in:

$$\mu_i = \mu_0 f(\vec{\Theta}|t_i) \tag{8.63}$$

The optimum values of the parameters in the decay model can be obtained by finding the parameters $\vec{\Theta}$, which maximize L or equivalently minimize− $\log(L)$. If there are n observations, then the dimensionality of $\vec{\Theta}$, i.e., the number of parameters estimated (p) from the data must be less than n.

The best possible model is one that perfectly fits the data. In this case, the counts for sample i, x_i are presumed to be exactly predicted, and so a "null" likelihood can be written (denoted as L_0) in Equation 8.64.

$$L_0 = \left[\frac{(x_1)^{x_1}}{x_1!}e^{-x_1}\right]\left[\frac{(x_2)^{x_2}}{x_2!}e^{-x_2}\right]\dots\left[\frac{(x_i)^{x_i}}{x_i!}e^{-x_i}\right] \tag{8.64}$$

Essentially, the null likelihood is formulated with a model containing $p=n$ parameters. This is also called a saturated model.

It is expected that $L < L_0$. It is useful to take logarithms and, furthermore, to examine the differences between the logs of the likelihoods for the fitted model and the null. Conventionally, this is done using the deviance (D) defined by Equation 8.65.

$$D = -2(\log(L) - \log(L_0)) \tag{8.65}$$

The factor of 2 is used to make the deviance comparable to the residual sum of squares in an ordinary regression model. There are several uses of the deviance[20]:

- The value of deviance can be used as a test of the goodness of fit of a model. If $D < \chi^2_{df=(n-p)}$ at the desired upper percentile, then the null hypothesis that the chosen model provides a statistically acceptable fit cannot be rejected.
- If there are two models (denoted by subscripts 1 and 2) in which model 1 is a special case of model 2, and $p_1 < p_2$ (an example of this is the exponential decay model versus the Weibull model) then the improvement provided by model 2 can be tested by comparing $D_1 - D_2$ to $\chi^2_{df=p_2-p_1}$. If the difference in deviances is greater than the critical χ^2, then the equivalence of fit is rejected, and model 2 can be regarded as superior. If model 1 is not a special case of model 2, then an alternative approach must be used—AIC (Akaike information criterion) or BIC (Bayesian information criterion) described later should be used.[21]

An example of count data from a decay experiment is given in Table 8.9. The fit of different decay models can be tested using the maximum likelihood approach.

For this data, the fitting of an exponential, Weibull, lognormal and Gompertz–Makeham model is done using the code in R shown below.[22] In executing this fit, the best values for the survival

Table 8.9 Data for a hypothetical decay experiment with count data.

Time (h)	Volume (mL)	Counts
0	0.1	53
10	0.3	120
30	1	140
90	2	50
180	2	25
360	5	35

20 There is a similarity in the discussion here to the goodness of fit and comparison of alternative models presented in Section 6.2.1.

21 AIC or BIC can also be used when the models are nested, and many argue that this can be a more robust approach than the use of the χ^2 criterion.

22 Previous coding was done in Python, which is highly suitable for mathematical simulation. In the author's opinion, currently, R is more suitable for data analysis and statistics, so it is preferable in this context.

models, *as well as the initial microorganism concentration* μ_0 are estimated—the initial concentration should generally be treated as a fitting parameter rather than a fixed value, since in general the measurement of microorganisms at time zero may be no less subject to experimental variability than measurements at later times The input data to this code is an Excel sheet with three columns of data—time, volume, and counts. This is read into the program as input.

```
library(readxl)
library(optimx)
data<-read_excel('CountData.xlsx')
times<-data$'time (hr)'
volumes<-data$'volume (mL)'
observed_counts<-data$counts
observed_mu<-observed_counts/volumes
#---------------------------------
# Exponential Decay
mu_Exponential<-function(theta,t){
  mu0=theta[1]
  k=theta[2]
  y=exp(-k*t)
  return(y*mu0) }

# Weibull Decay
mu_Weibull<-function(theta,t){
  mu0=theta[1];k=theta[2];m=theta[3]
  y=exp(-(k*t)^m)
  return(y*mu0) }

# LogNormal Decay
mu_LogNormal<-function(theta,t){
  mu0=theta[1];a=theta[2];sigma=theta[3]
  argument=(log(a*t))/sigma
  y=1-pnorm(argument)
  return(y*mu0)}

# Gompertz Makeham (or other three parameter)
mu_GompertzMakeham<-function(theta,t){
  mu0=theta[1];k1=theta[2];k2=theta[3];k3=theta[4]
  A=k3*t
  B=k1/k2
  C=exp(k2*t)-1
  y=exp(-A-B*C)
  return(y*mu0) }
#---------------------------------
# Deviance
deviance<-function(theta,function_name,observedN,times,volumes){
  predicted<-do.call(function_name,list(theta=theta,t=times))
  predictedN<-predicted*volumes
  predictedLogL<-observedN*log(predictedN+1e-300)
   -predictedN-lfactorial(observedN)
  predictedLogL<-sum(predictedLogL)
  NullLogL<-observedN*log(observedN+1e-300)
   - observedN - lfactorial(observedN)
  NullLogL<-sum(NullLogL)
  return(-2*(predictedLogL-NullLogL)) }
#---------------------------------
# Driver
model<-'mu_GompertzMakeham'
if (model=='mu_Exponential'){
  initial_params<-c(300,0.02225)
} else if (model=='mu_Weibull'){
  initial_params<-c(300,1,1)
```

```
} else if (model=='mu_LogNormal'){
  initial_params<-c(300,1,1)
} else if (model=='mu_GompertzMakeham'){
  initial_params<-c(1000,-1,-1,.1)
}

initial_deviance<-deviance(initial_params,model,
  observed_counts,times,volumes)
print(initial_deviance)

# USE OPTIMR TO CALL DEVIANCE AND OPTIMIZE
best<-optimr(initial_params,deviance,gr=NULL, hess=NULL, method='Nelder-Mead', lower=-
     Inf, upper=Inf, hessian=FALSE, control=list(trace=1),model,observed_counts,times,
     volumes)

bestparam=(best$par)
plottimes=seq(0,400,0.1)
bestmu<-do.call(model,list(theta=bestparam,t=plottimes))
bestdeviance=(best$value)
print(bestparam)
print(bestdeviance)
```

The results of fitting these models are shown in Table 8.10—the optimized deviances as well as the best fitted parameter values are presented. Note that the exponential fit has two parameters, the Weibull and lognormal have three parameters, and the Gompertz–Makeham model has four parameters—including the value of μ_0, which is estimated from the data. Also note that for this problem, the number of data points (n) is six.

As discussed in Section 6.2, alternative selection criteria among models include the AIC and BIC statistics, including when models are not special cases of other models [67]. The statistics for deviance were presented in Equations 6.10 and 6.9. The values of AIC and BIC for the four models used are shown in Table 8.11.

From this fitting, several conclusions are apparent immediately:

Table 8.10 Results of fitting data in Table 8.9 to candidate survival distributions.

Model	Deviance	Best μ_0	Best parameters		
Exponential	197.92	238.83	0.0141		
Weibull	57.81	699.44	0.0970	0.4655	
Lognormal	51.33	374.37	0.0518	1.2642	
Gompertz–Makeham	167.03	529.84	0.8881	−0.9279	0.0129

Table 8.11 AIC and BIC for different models fitting data in Table 8.9 to candidate survival distributions.

Model	AIC	BIC
Exponential	205.92	201.50
Weibull	75.81	63.18
Lognormal	69.33	56.71
Gompertz–Makeham	215.03	174.20

- The exponential is a special case of the Weibull survival distribution (with one additional degree of freedom) and of the Gompertz–Makeham survival distribution with two degrees of freedom. The values of the upper percentile of χ^2 for one and two degrees of freedom are 3.84 and 5.99, respectively (Table 8.12). The difference in deviances exceeds this critical percentile, and hence the more highly parameterized functions are preferred to the exponential.
- Based on AIC and BIC, the lognormal survival function is the one preferred to the other alternatives (Table 8.11).
- For the best fitting survival function, the lognormal, the residual deviance (51.33) is greater than the critical χ^2 at 3 (=6-3) degrees of freedom, hence there is still a lack of fit to the data. There could be several reasons for this that could be explored:
  - The likelihood function assumes Poisson variability for counts, and there may be greater variability as discussed in Section 6.2.1. This could be tested by formulating the likelihood function using, e.g., the negative binomial distribution.
  - An alternative, or more complex (containing more than two parameters for the survival function) may provide a better fit that reduces the deviance and does not excess a goodness of fit criterion.

Rather than data on counts, data may only be available on concentrations. Perhaps these are computed by dividing colony counts by equivalent volume examined, or by one of the other methods discussed in Section 6.2. The use of a Poisson or other discrete distribution to construct a likelihood function is no longer appropriate, but concentrations must be regarded as observations suitable for modeling by a continuous distribution, with a lognormal being the most frequently appropriate.[23]

An example of such data is the results from Easton et al. [68], where batch laboratory experiments were conducted to assess the persistence of *E. coli* O157:H7 in creek water (Table 8.13). As an aside, it is possible for such data to contain an observation with a concentration of "0" or with an observation of TNTC (too numerous to count). This must be treated using special methods for regression with censored data [69, 70]— discussion of these methods is beyond the scope of coverage here.[24]

Table 8.12 Upper percentiles of the χ^2 distribution.

Degrees of freedom	Upper 95th percentile
1	3.84
2	5.99
3	7.82
4	9.49

23 or alternatively the log concentrations can be modeled by a normal distribution.

24 It is generally good practice for those conducting experiments on persistence to report the equivalent volumes for all samples analyzed. In such cases, a "0" concentration is really less than or equal to the minimum number of organisms that can be quantified. Similarly, the maximum number of organisms that can be counted and the equivalent volume examined should be recorded, and a TNTC observation is a concentration greater than the maximum number of quantifiable organisms. Therefore, censored data analysis constructs the full likelihood equation, but for lower-censored observations, the lower tail of the CDF should be used, and for upper-censored observations (i.e., TNTC), the upper tail of the CDF should be used. This problem has often been referred to as a Tobit analysis, and packages are available for its solution, such as the `CensReg` package in R [71].

Table 8.13 Survival of *E. coli* O157:H7 in creek water.

Time (days)	Cells per mL
0	5.47×10^7
1.15	4.924×10^6
3.16	9.18×10^5
7.3	1.32×10^5
10.6	8.98×10^4
14.7	1.99×10^4
22	4.41×10^3

Source: Data from [68].

Table 8.14 Fit of data in Table 8.13 to candidate decay models.

Model	Deviance	Best μ_0	Best parameters		
Exponential	22.25	$4.66 \cdot 10^6$	0.206		
Weibull	13.31	$1.10 \cdot 10^7$	4.58	0.336	
Lognormal	10.00	$5.76 \cdot 10^7$	46.1	2.36	
Gompertz–Makeham	23.72	$3.85 \cdot 10^6$	$-4.84 \cdot 10^4$	$-1.69 \cdot 10^5$	0.212

With these caveats, if the symbol "y" is the log-transform of concentrations, and f is the survival function with parameters $\vec{\theta}$ then the deviance can be written as[25] which should be minimized is given by Equation 8.66.

$$D = \sum_{i=0}^{n} \left[y_i - \ln\left(N_0 f(\vec{\theta}, t_i) \right) \right]^2 \tag{8.66}$$

The fitting software is modeled so that concentrations, rather than colony counts, are used, and the result of fitting candidate survival functions is shown in Table 8.14.

Using the AIC and BIC criteria, the preferred model using either approach is the lognormal decay (Table 8.15). It is not possible to do a strict goodness of fit test as it was in the case of count data.[26] However, a plot of the fit and a detailed examination of the residuals may be useful for assessing systematic shortcomings in the fit.

In Figure 8.16, the observed concentrations are plotted along with the best fit to the lognormal decay relationship. There is no visual indication of systematic discrepancies between the model and data (note that the y-axis is on a logarithmic scale). While formal tests of the residuals, e.g., for bias, correlation, and heteroscedasticity, could be performed, given the small number of data points, such tests could only detect gross deviations.

A third type of data (in addition to counts and concentrations) could arise from the use of quantal methods to assess pathogen survival. For example, the use of an MPN test for bacteria, a presence–absence method for viruses, or digital PCR is all quantal, as noted in Section 6.2. Analysis of this

25 Note that the null model would be one in which the observations themselves are the model, and thus the null likelihood would be equal to zero.

26 If there were replicate experiments at each time, it would be possible to assess goodness of fit by comparing the residuals from the regression fit with the replication error. This is a standard problem in non-linear regression, and is well covered in the literature [72] and will not be discussed further.

Table 8.15 AIC and BIC criteria for analysis of concentration decay experiment of Eaton et al.

Model	AIC	BIC
Exponential	30.25	25.84
Weibull	31.31	18.68
Lognormal	28.00	15.37
Gompertz–Makeham	71.72	30.89

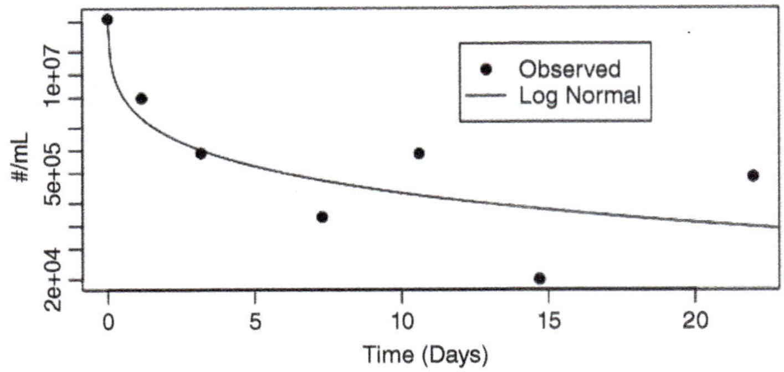

Figure 8.16 Fit of decay data from Table 8.13 to the lognormal model.

type of data in a survival experiment is akin to the analysis of dose–response experimental data, which will be covered in Section 14.1.

8.4.2.5 Modulators of Decay Rates and Data Sources

Decay rates and kinetics can be influenced by the conditions in which microorganisms occur. Such conditions can include deliberate changes to increase decay, which will be covered in subsequent chapters. However, even in the absence of such interventions, the nature of the surrounding environment can influence fate.

As with any reaction, increasing the temperature increases the reaction rate, i.e., decay rate. In natural systems, on surfaces, and in the air, as temperature increases, pathogen persistence decreases. The precise magnitude of this effect will depend on the specific environment and other factors. The impact of freezing on organisms will be complex and will depend on the rate of freezing, the humidity (which may result in the formation of ice crystals), and the specific pathogen. Freezing may provide an additional mechanism for pathogen inactivation [73].

Any environmental system will be in equilibrium with some form of water. In aqueous (whether natural or human-altered) systems, an increase in salinity or total dissolved solids will result in a decrease in water activity (a_w). This decrease will reduce the propensity for pathogen growth, among those that can grow in environmental conditions. With extremely low activity, under drying conditions, actual decay can occur.[27]

---

27 Although some pathogens can form resistant stages that can survive desiccation, such as spores of bacteria and fungi, and cysts and oocysts of protozoa.

Relationship Between Water Activity and Relative Humidity

Water activity and relative humidity are interrelated concepts. The relationship is given by Raoult's law. If x_w is the mole fraction of water in liquid (with 1.0 being pure water) and if the other constituents in the solution are nonvolatile, the vapor pressure of water above the solution is given by:

$$P_w = P_w^o \cdot x_w \quad (8.67)$$

where P_w^o is the pure water vapor pressure at the system temperature.

In the atmosphere, the term relative humidity, abbreviated as RH is the ratio of the actual water vapor pressure to the pure water vapor pressure, i.e., $RH = \frac{P_w}{P_w^o}$. Conventionally, relative humidity is expressed as a percentage rather than as a fraction.

A number of natural environments have low water activity, such as marine systems, saline and hypersaline groundwaters and hot springs. Certain industrial brines, including those from hydraulic fracturing of oil and gas deposits, are hypersaline. There are numerous bacterial taxa that are specialized for growth and persistence in such systems, although few are pathogens. Another class of low water activity environments is those that have high sugar content or are partially or completely dried [74]. Typically, water activity below 0.85 is inhibitory for the growth of prokaryotes (except for the specialized nonpathogenic groups), while certain yeasts and molds can grow with about 0.7 water activity.

There is a broad range of persistence for bacteria, even non-spore formers, under dried conditions on surfaces and soils. This range is partially explained by differences in cellular biochemistry and structure and ranges from survival times exceeding centuries (in permafrost) of actinomyetes and Gram-positive spore-forming bacilli to years (e.g., *Listeria* to months from many bacteria) [75].

It is not surprising that there have been outbreaks of bacterial diseases attributed to the persistence and growth of pathogens in low-moisture, i.e., low-water-activity, foods [76]. Although food pathogens are not the focus in the book, there are a number of outbreaks involving organisms in such environments that suggest the potential for risk in other (non-food) low water activity circumstances, such as:

- *Clostridium*
- *Cronobacter*, previously classified as *Enterobacter sakazakii*
- *Listeria*
- *Salmonella*
- *Staphylococcus aureus*

Exposure of pathogens under low-moisture conditions could occur via the fomite route, contact with soil, or wind-borne dispersal of microorganisms in dust or soil.

For viruses, the effect of relative humidity on persistence in the air is more complex than for bacteria. In bacteria, the principal effect of low water activity is disruption of osmoregulation. For viruses, it has long been known [77] that the relationship between decay rate and RH is non-monotonic, with maximum decay occurring at intermediate humidity. This is reinforced in the review of the survival of enveloped viruses [78].

It has also been found that some viruses can be excreted and contained within vesicles from the cells that they have infected [79]. These virions, often referred to as "membrane cloaked" may thereby be somewhat protected from decay by natural processes or added disinfectants in the environment [80].

It is possible to modulate the sensitivity of organisms to environmental inactivation through surface coatings. Virus naturally occurring in silica-rich environments, when dispersed by air, can exhibit enhanced persistence, and this can be mimicked by deliberate silica coating prior to dispersal [81]. It is known that in the 2001 anthrax terrorism incidents , spores of the bacterium had been "weaponized" via an external coating, which presumably affected their persistence and their ability to be retained by deposition on surfaces.

Another master variable that could influence decay is pH. In aqueous systems, pH influences the chemical speciation of substances such as ammonia, which are typically more toxic in their unionized forms. In addition, the osmoregulatory system in vegetative organisms (bacteria, protozoa, fungi) maintains a relatively constant internal pH, and excursions of the external pH from neutrality will result in additional stress that could adversely affect persistence.

Sunlight, particularly near visible UV and shorter-wavelength UV, may enhance the decay of pathogens in the air and on the surfaces of water, soil, and other exposed systems. In older literature, it was known that pathogens in air decay more rapidly in the open (outdoor) air than in rooms, and this was due to an "open air factor" [82].

It is now clear that the effect of sunlight (and artificially produced indoor germicidal light) may arise via at least two mechanisms [83]. There can be a direct photo-induced inactivation. There can also be inactivation from an indirect mechanism in which organic matter may be excited by irradiation to produce free radicals (such as the hydroxyl radical), which then can produce additional decay even in the absence of direct illuminance.

A final potential modulator of significance is the environmental microbiome in which pathogens occur or are present. This has been recognized as a potential intervention location [84]. Other microorganisms co-occurring with pathogens may enhance pathogen persistence by excreting growth factors or removing inhibitory substances, or may diminish persistence through antibiosis, predation, or competition for resources.

Ideally, it would be useful to predict microbial decay (and where relevant, growth) parameters as a function of environmental conditions. In the field of food safety, controlled experiments have been practical to develop a predictive microbiology for pathogens.

Predictive Microbiology

The field of predictive microbiology emerged in the 1990s as a way to systematically predict the growth and decay of (primarily) bacterial pathogens in food [85]. In that application, time, temperature, pH, water activity, and the presence of preservatives are the key drivers of concern, in addition to the type of foodstuff itself.

By conducting many experiments under controlled conditions, parameters of growth and decay were determined as functions of those conditions. From such experiments, online databases have been developed that are useful for assessing food safety risks, for example, COMBASE [86], which is accessible via a web browser at url https://combase.errc.ars.usda.gov/.

While no such database has emerged for predictive microbiology of pathogens in the environment, some reviews and meta-analyses have been done. Examples of these point to the feasibility of a long-term goal of predicting environmental decay rates. Several are summarized as follows.

- Dean et al. [63] re-analyzed data on the persistence of bacteriophage and enterovirus in groundwaters. They fit individual experiments to a broad set of decay models and developed classification trees based on water quality to select specific persistence models.
- Boehm et al. [87] conducted a meta-analysis of decay experiments for mammalian viruses and bacteriophages in surface waters. They noted a range of decay rates over more than an order of magnitude, with some apparent systematic differences between groups. Decay rates were generally higher when experiments were done at high temperatures, in the presence of sunlight and under estuarine conditions. An exponential decay model was used as an underlying basis.
- Dean and Mitchell [64] did a meta-analysis of the persistence of fecal indicator bacteria, bacteriophage, and pathogenic bacteria, viruses, and protozoa in natural surface waters. They examined multiple decay models to assess the most appropriate descriptors for T_{90} and T_{99}, i.e., time for one log and two log inactivation. In general, most of the time, models other than the exponential were most suitable as determined by BIC. Temperature and predation were significant determinants of persistence, along with less frequently reported factors such as pH and turbidity.
- Silverman and Boehm [88] performed a systematic meta-analysis of decay of enveloped viruses in natural waters and wastewaters. They used the exponential decay model and found an order-of-magnitude range in decay across their data, with significant effects of virus type, enumeration method, water type, temperature, and sterility. In general, the enveloped viruses had higher decay rates than those previously reported for non-enveloped viruses.

8.4.3 Growth

The increase of microbial numbers in a system may occur by multiple mechanisms. The first is pure multiplication, in which one viable microorganism becomes more than one organism.[28]

An ultimate aim of growth modeling is to describe all phases of the growth curve—lag, exponential, stationery, and decline (Figure 6.4). Simpler models may only capture the details of one or two of these phases; however, with a sufficiently complex model, all phases of the microbial life cycle can be described.

The most straightforward classes of models are termed autonomous. They may be written as differential equations that do not include either time or the system's temporal history as variables when modeling population increase. The most basic case is the exponential relationship, where μ is generally referred to as the instantaneous growth rate (Equation 8.68). Recall from a batch mass balance that the right-hand side of this equation can be defined as a reaction rate, in this case for growth, which can be written as r_g.

$$\frac{dN}{dt} = \mu N \tag{8.68}$$

The exponential relationship depicts only the exponential stage of growth, not the other stages.

---

28 For bacteria and most protozoa, this is typically one microorganism that becomes two in the course of one generation. However, for viruses, an increase alone may result in more new virions being produced per generation in a host (termed the burst size). While these are mechanistic differences, the mathematical treatment of the different processes can be approached similarly. In some situations, particularly artificially produced in the laboratory, synchronous growth can occur , in which all members of a microbial population are at the same stage of their life cycle at any one time, and the overall growth becomes more of a step function rather than a continuous smooth curve. In such cases, more complex special-case models may be useful, which will not be discussed further [89].

There have been many ways to generalize the exponential to describe the other phases of growth. One of the earliest generalizations was the simple logistic developed by Verhulst [90] using the concept that growth continues only until saturation (where $N=K$) is approached. This can be written in differential and integrated form as in Equation 8.69.

$$\frac{dN}{dt} = \mu N\left(1 - \frac{N}{K}\right)$$
$$N = \frac{N_0 \cdot K}{(K - N_0)\exp(-\mu t) + N_0} \tag{8.69}$$

The simple logistic equation can describe exponential and stationary growth phases, but not lag or decay. As a result, a number of expansions of the logistic equation have been devised. Few of these have analytical solutions. In Table 8.16, these generalizations of the growth rate are presented; to obtain the curve for N versus t, numerical integration of the batch equation is needed.

Using one of the logistic models, or any model that can be written using a reaction rate that incorporates only state variables, is advantageous because the reaction rate itself can be used in the mass balance of a system other than a closed batch reactor. For example, consider a single CSTR at steady state with a mean hydraulic residence time of Θ. If there is an influence concentration of microorganisms of N_{inf}, then the resulting mass balance if the theta-logistic model applied would be:

$$N_{\text{inf}} - N + \mu N\left[1 - \left(\frac{N}{K}\right)^{\theta_1}\right]\Theta = 0 \tag{8.70}$$

Thus, if the kinetic parameters (μ, K, θ_1) are known, this implicit equation can be solved for N as a function of Θ.

Another class of growth models is non-autonomous, but only considers a state variable of microorganism concentration. An initial approach to this class of models was proposed by Ludwig in 1929, as recapitulated by Wang and Gao [94]. The basic exponential model is considered to be modified by a time-dependent decay as in Equation 8.71, which can describe the growth phase and a stationary phase.

$$\frac{dN}{dt} = \mu N \exp(-\omega t) \tag{8.71}$$

The integrated form of this equation is:

$$N = N_0 \exp\left[\frac{\mu}{\omega}(1 - e^{-\omega t})\right] \tag{8.72}$$

By introducing a time shift and redefining variables, Zwietering et al. [95] developed a widely used modification of this basic model, which is termed the Gompertz in the form of Equation 8.73.

Table 8.16 Generalizations of the logistic growth rate expressions.

Name	Reference	Expression for r_g	Comments
Theta-logistic	[91]	$\mu N\left[1 - \left(\frac{N}{K}\right)^{\theta_1}\right]$	Growth phase may be curvilinear and not exponential.
Two-parameter extension	[92]	$\mu N^{\theta_2}\left[1 - \left(\frac{N}{K}\right)^{\theta_1}\right]$	Describes a density-dependent lag phase.
Three-parameter extension	[93]	$\mu N^{\theta_2}\left[1 - \left(\frac{N}{K}\right)^{\theta_1}\right]^{\theta_3}$	Can give good fidelity to lag, growth, and stationary phases.

This model can describe lag, exponential, and stationary phases; however, because it is a non-autonomous derived relationship, it cannot readily be used to develop a rate function for systems other than batch systems.

$$ln\left(\frac{N}{N_0}\right) = a \cdot \exp\left[-e^{(b-ct)}\right] - a \cdot \exp\left(-e^{b}\right) \tag{8.73}$$

The Gompertz and the multiparameter logistic models can often give similar-shaped curves. As described in Haas et al. [96], growth of *Listeria monocytogenes* in a sample of pasteurized milk could be fit to both the Gompertz and the multiparameter logistic models. In Figure 8.17, a comparison of the best-fit curves is shown. For the particular data in question, using the AIC criterion, the Gompertz model was preferable. However, a qualitative examination of the fitted curves shows quite similar behavior.

The development of the Gompertz formulation can be generalized in a number (perhaps infinite) number of ways. Essentially Equation 8.71 can be viewed as a product of a rate parameter (μN) and an "adjustment function" of time [97]. A diverse suite of growth models that describe the lag, growth, and stationary phases can be represented by using an alternative adjustment function. However, this approach retains the characteristic of the Gompertz as being non-autonomous and, hence, difficult to use outside of a batch system.

Another set of approaches is to model the population of a pathogen that is growing not only by a single parameter, i.e., concentration, but also by characterizing some properties of the population that contribute to its growth, survival, and decay. These could include interactions with the population's environment, such as nutrient consumption, excretion of by-products, and perhaps inhibition or inactivation by those by-products. Broadly speaking, these are classified as structured modeling approaches.

An early example of this was that of Young and Bungay, who described an "internal" growth rate of a microorganism that exhibited a first-order lag relative to that determined from the external nutrient composition [98]. This was able to model a lag phase, and as substrate depletion occurs, a stationary phase. A simple way to describe this loose coupling between external nutrients and internal growth rate was devised by Rodin et al. [99], who introduced an externally driven rate (μ_e) directly given by a Monod function of substrate, and an internal growth rate with a first-order lag.

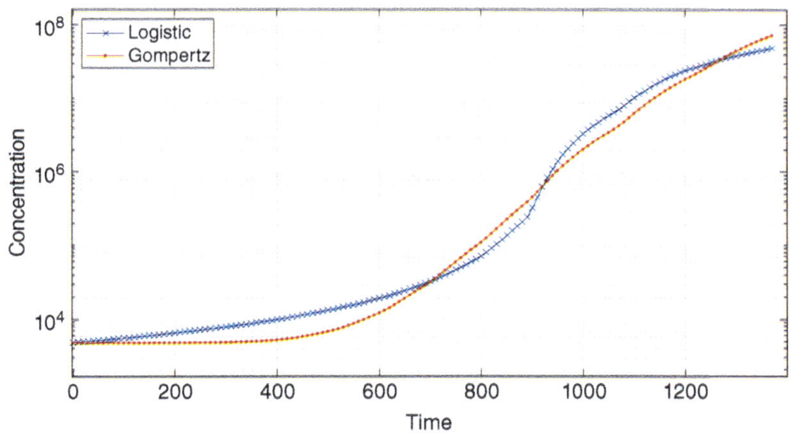

Figure 8.17 Comparison of Gompertz vs. logistic curves.

These effects are shown below, where α is the adjustment rate between the internal growth rate and the external target.

$$\frac{dN}{dt} = \mu N \tag{8.74}$$

$$\mu_e = \frac{\mu_{\max} S}{S + K_S} \tag{8.75}$$

$$\frac{d\mu}{dt} = \alpha(\mu_e - \mu) \tag{8.76}$$

$$\frac{dS}{dt} = (S > 0) \cdot \frac{-\mu N}{Y} \tag{8.77}$$

This model can describe a lag phase, and as substrate is consumed, a stationary phase. The Boolean expression in Equation 8.77 prevents the substrate from going negative, although due to the lab inherent in Equation 8.76, there is some residual growth even when the external substrate level goes to zero.

Possibly the earliest effort to develop a structured model for microbial growth was that of Williams [100]. He depicted a microbial population as comprising synthetic and genetic compartments. The rate of increase of the synthetic compartment is modeled as proportional to the nutrient concentration (and its increase consumes nutrients proportionately), and the rate of increase of the genetic compartment (which is, in effect, the population number concentration) is proportional to the product of its concentration and the synthetic compartment concentration. This model, when mathematically operationalized, yielded a logistic growth curve. It could also model the effects (on growth lag) of inoculating with populations of different readiness, as depicted by the ratio of synthetic to genetic compartments.

Esener et al. [101] elaborated and validated this approach, in effect, by considering the synthetic compartment to be proportional to bacterial RNA. They demonstrated that such a two-compartment model, incorporating Monod substrate kinetics, fitted data obtained in batch and chemostat experiments.

Monod Kinetics

It is often desirable to relate rates of a process, such as microbial growth rate, to the nutrient concentrations on which those rates depend. In many situations, the following equation, first proposed by Monod [102], has been found to be useful, where μ is the growth rate and S is the nutrient ("Substrate") concentration:

$$\mu = \frac{\mu_{\max} S}{S + K_S} \tag{8.78}$$

$\mu_{\max}$ and K_S are kinetic constants for the model. When $S << K_S$, this is first order, and when $S >> K_S$, the growth rate reaches the asymptote $\mu_{\max}$.

Another phenomenon occurring during microbial growth, particularly for vegetative organisms, is the excretion of products, some of which themselves could be inhibitory. This could lead to a reduction in growth as products accumulate in a batch system, which will also occur with nutrient depletion.

Structured models for microbial growth can be complex, though they have generally been developed only for industrial organisms important in biosynthesis, rather than pathogens. An early example of this is the modeling of the growth of the industrially important yeast, *Saccharomyces cerevisiae* [103]. This structured model included 26 different rates, including uptake and metabolism of carbon and nitrogen sources, synthesis of RNA, cell envelope, and DNA, and respiration and death. Such highly structured models, while capable of describing fine details of microbial population growth and decay, also require the specification of numerous parameters to execute. Hence, in general, more parsimonious and lumped models are frequently used for pathogen modeling.

There are numerous approaches to these intermediate-complexity models that could be used not only in the context of batch growth but also in flowing systems.

The salient features of the microbial batch growth curve can be depicted by a simple modification of the model just presented. If growth occurs simultaneously with the excretion of an inhibitor (I), which causes microbial inactivation,[29] then the model can be modified as in Equations 8.79–8.83.

$$\frac{dN}{dt} = \mu N - k_d I N \tag{8.79}$$

$$\mu_e = \frac{\mu_{\max} S}{S + K_S} \tag{8.80}$$

$$\frac{d\mu}{dt} = \alpha(\mu_e - \mu) \tag{8.81}$$

$$\frac{dS}{dt} = (S > 0) \cdot \frac{-\mu N}{Y} \tag{8.82}$$

$$\frac{dI}{dt} = k_I N \tag{8.83}$$

Execution of this model is shown in Figure 8.18 for a particular set of parameters. Note qualitatively, this depicts lag, exponential, stationary (albeit short), and decay phases. Decay is initiated as the substrate is depleted and the inhibitor builds up. Because of the lag between the growth rate and external substrate (Equation 8.81), some growth continues even after the external substrate is exhausted, potentially from accumulated intra-organism resources. Since this model consists of a set of autonomous differential equations (for a batch system), the corresponding rates could describe behavior in other systems, such as a completely mixed system.

An alternative approach, which has found application in many situations in food microbiology, has been proposed by van Impe and colleagues [104, 105]. They start with the concept of a time-dependent adjustment function introduced by Baranyi [97], which depicts the physiological "readiness" of a population and describes an initial lag phase. By adding factors that correspond to substrate depletion or inhibitor buildup,[30] they are also able to depict the stationary phase. Adding a decay-rate term would enable modeling the full growth curve. From these concepts, they define "P" models and "S" models, in which there is product inhibition or substrate depletion.

The "P" model, with an added decay coefficient, is given by Equations 8.84–8.86.

$$\frac{dN}{dt} = \mu_{\max}\left(\frac{\eta}{1+\eta}\right)\left(1 - \frac{P}{K_P}\right)N - k_d N \tag{8.84}$$

$$\frac{d\eta}{dt} = \mu_{\max}\eta \tag{8.85}$$

$$\frac{dP}{dt} = \mu_{\max} Y_P\left(\frac{\eta}{1+\eta}\right)\left(1 - \frac{P}{K_P}\right)N \tag{8.86}$$

---

29 Note that, in general, all microbial populations will have a spontaneous constant decay rate. The inhibitor decay rate exceeds that, and so it is ignored in this model. The functional form of the decay relationship is similar to the disinfection rates seen in Chapter 11. Also, this model assumes that the inhibitor production rate is proportional to the viable number of organisms; an alternative framework could be that the production rate of I is proportion to the organism *growth rate*.

30 They assume the inhibitors act solely to diminish growth rather than to inactivate agents.

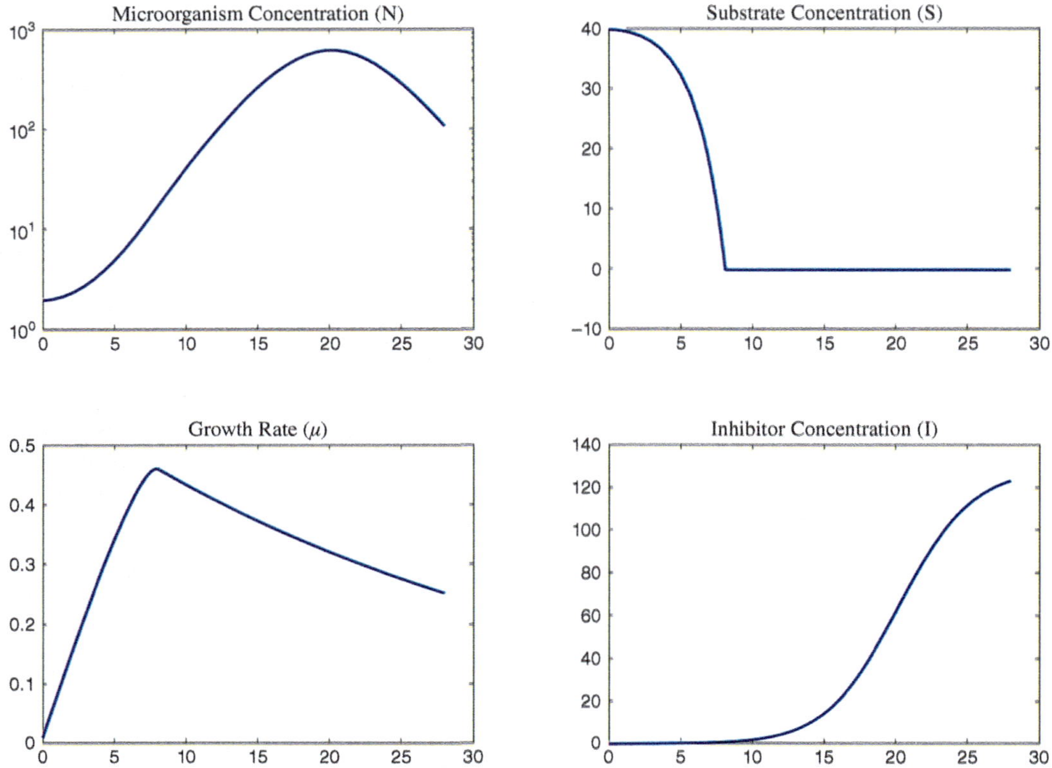

Figure 8.18 Model output of growth with lag and inhibitor excretion. Parameters: $\mu_{max} = 4$, $K_S = 25$, $\alpha = 0.03$, $Y = 0.4$, $k_i = 0.02$, $k_d = 0.005$.

This system has initial conditions @t=0, $N = N_0$, $\eta = \eta_0$ and $P = P_0$. If data exist only on N versus t, then there are five adjustable parameters, μ_{max}, K_P, k_d, η_0, and Y_P.

The "S" model, with added decay coefficient, is given by equations

$$\frac{dN}{dt} = \mu_{max}\left(\frac{\eta}{1+\eta}\right)SN - k_d N \tag{8.87}$$

$$\frac{d\eta}{dt} = \mu_{max}\eta \tag{8.88}$$

$$\frac{dP}{dt} = \frac{\mu_{max}}{Y_S}\left(\frac{\eta}{1+\eta}\right)SN \tag{8.89}$$

The initial conditions for this model are $N = N_0$, $\eta = \eta_0$ and $S = S_0$. Like the P model, there are five adjustable parameters, μ_{max}, Y_S, k_d, S_0 and η_0.

Testing of the P and S models against actual data indicates that they provide an equally satisfactory fit to experimental data [106], and therefore, additional information, such as measurements of substrate or product, would be needed to distinguish them from each other.

Other processes can lead to increasing numbers of (measured) viable organisms. As noted in Section 6.5, some bacteria can enter into a VBNC state. Under suitable conditions, these organisms can recover culturability [107]. More work is needed to elucidate the kinetics of these processes. Some bacteria can form dormant spores, and protozoans can form dormant cysts or oocysts. The germination from dormancy may form another rate for increase in viability, and the use of models describing the transition from dormancy to viability, while at an early stage, shows promise [108].

8.4.4 Stochastic Treatment of Growth and Decay

The previous discussion on growth described the dynamics of microorganisms using a continuous framework. Analogous to the earlier presentations, it is also possible to analyze growth using a discrete framework that focuses on individual organisms. This allows more detailed consideration of heterogeneity within a population as well as the variability in the projection of the prediction of growth and/or decay dynamics.

Williams and colleagues [109, 110] were perhaps the first to consider microorganism multiplication in a stochastic framework. Using terminology consistent with prior models presented here, they consider that if there is a population of N organisms at time t, then *for each organism,* one of the following three events must occur by $t + \Delta t$:

- An organism can divide into two progeny with probability $\mu\Delta t$
- An organism can die with probability $k\Delta t$
- An organism can remain unchanged, by the law of total probability with a probability of $1 - \mu\Delta t - k\Delta t$

In this basic birth–death model, the behavior of a single organism is independent of the behavior of any other organism at that time (i.e., the probabilities are independent and identical). With this approach, there is a finite (perhaps small) probability of a small initial inoculum growing to large numbers or, likewise, a large inoculum decaying to extinction.

Python code for this model is shown as follows. Each run is initialized with a Poisson-distributed random integer with a mean of 10 organisms. At each time interval, uniform random numbers are drawn to determine whether there is a death or a new birth. This is repeated for a number of restarts, and the set of trajectories is assembled to determine the mean and standard deviation at each time point.

```
import numpy as np
import matplotlib.pyplot as plt
tableaux = []
for reps in range(0, 100):
    N0 = np.random.poisson(10)
    k = 0.18
    mu = 0.25
    N = N0
    t = 0
    del_t = 0.03
    birth_prob = mu * del_t
    death_prob = k * del_t
    tableau = []
    times=[]
    while t < 100:
        t += del_t
        R = np.random.rand(N)
        born = np.sum(R < birth_prob)
        dead = np.sum((1 - R) < death_prob)
        N += born - dead
        tableau.append(N)
        times.append(t)
    tableaux.append(tableau)
```

The output from this simulation is shown in Figure 8.19. This analysis could be readily extended, for example, by sampling a larger number of trajectories to get the probability of extinction before any particular time.

The basic birth–death model is an example of a Markov Chain model, in which the state at the next time increment of a set of random variables is a function only of the conditions at the prior state [111], i.e., there is no "memory" of the time history of the system. This basic approach can be generalized in several ways, including discarding this basic assumption. Some potential extensions may include:

- The parameters k and μ might be regarded as random variables from a particular distribution, and for each time step a new set of these parameters sampled from that distribution.
- During growth, there may be excretion of product and/or consumption of substrate. The values of k and μ may then be made to be functions of the extant values of P and S.
- At the "birth" of a new organism, a "clock" might be started, and the values for the birth and death parameters given as functions of the age of each organism.
- With viruses, each generation may result in the production of a large number of progeny rather than (as is generally the case with bacteria) binary fission. Therefore, the burst size may be modeled as a random integer with a distribution over the duration of each generation.

For further generalizations, using the framework of agent-based models for microbial populations shows promise. They can incorporate heterogeneity within space, among members of a population, interactions among multiple populations, and variability in physiological state [112, 113]. However, as with any modeling effort, increasing complexity requires more input parameters (which may not all be available) and greater computational effort.

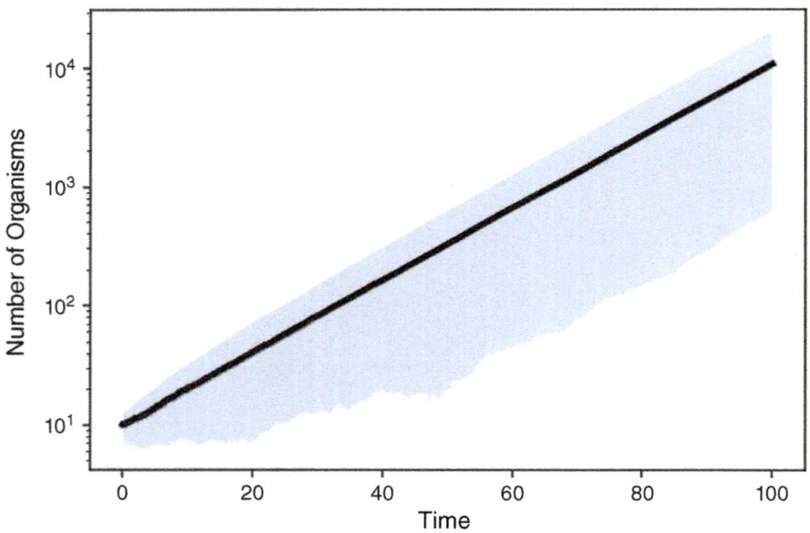

Figure 8.19 Williams birth–death model output. Solid line is mean of 100 trials and cloud shows range within one standard deviation. Conditions: N_0 drawn from Poisson with mean of 10, $\Delta t = 0.03$, $k = 0.18$, $\mu = 0.25$.

8.4.5 Predator–Prey and Other Biotic Interactions

A single population of microorganisms, whether pathogenic or not, rarely exists in isolation. Instead, multiple populations can occur in the same space and time, although spatial and temporal variability can arise. Detailed studies using molecular methods (Chapter 6 have shown the rich diversity of microorganisms, for example, in the indoor environment [114] and in different drinking water systems [115].

Given the co-occurrence of different populations, there is potential for direct or indirect interaction. Broadly speaking, these interactions may be neutral, i.e., no impact of one population on another; conflictual, with some adverse impact; or beneficial, with some positive impact [116]. These are further characterized into different forms, described in the box.

Types of Population Interactions Summarized From [116]

Conflictual Interactions

Parasitism/Predation: One organism (the predator or the parasite) benefits and the other is negatively affected (prey or host).

Antibiosis: One organism is adversely affected, while the other (the excretory/secretor) may not be impacted (or may benefit from the inhibition of the target population). In a more generic sense, this can also be called **amensalism**.

Competition: Both populations adversely affected by the other through a their need to use common resources such as nutrients.

Beneficial Interactions

Cometabolism: The metabolism of interacting populations is complementary—perhaps one population excretes a growth factor or essential nutrient for another, or perhaps one population uses a product that is inhibitory for the other.

Mutualism: This is the classical interaction understood by the lay public as symbiosis, where each population benefits from the other.

Commensalism: One population benefits from the interaction with no effect on the other.

There are several well-documented examples of interacting populations involving pathogens (or indicator organisms) in environmental settings.

Ashbolt has used the term "saprozoic" pathogen to describe bacteria that can persist and multiply in decaying organic material. Specifically, *Legionella*, *Mycobacterium*, and potentially other organisms can colonize within vegetative amoeba that graze in biofilms present on wetted surfaces [117]. A schematic is shown in Figure 8.20 from [118]. The association appears to benefit the bacteria by increasing resistance to water disinfection and providing a favorable environment for persistence and multiplication. The ecological benefits to the protozoan hosts are less clear. This association has been found to occur in many venues where water is piped, as well as in other locations such as ornamental architectural water features, spas, and misters/sprayers.

Pathogens, including viruses, naturally decay within waste treatment systems and wastewater. One mechanism for this decay is the excretion of exoenzymes and other inhibitory factors to damage virus components [119].

Bdellovibrio are a genus of Gram-negative bacteria that can prey upon other bacteria. They have been found frequently in waste treatment systems [120]. The predation of pathogenic bacteria can

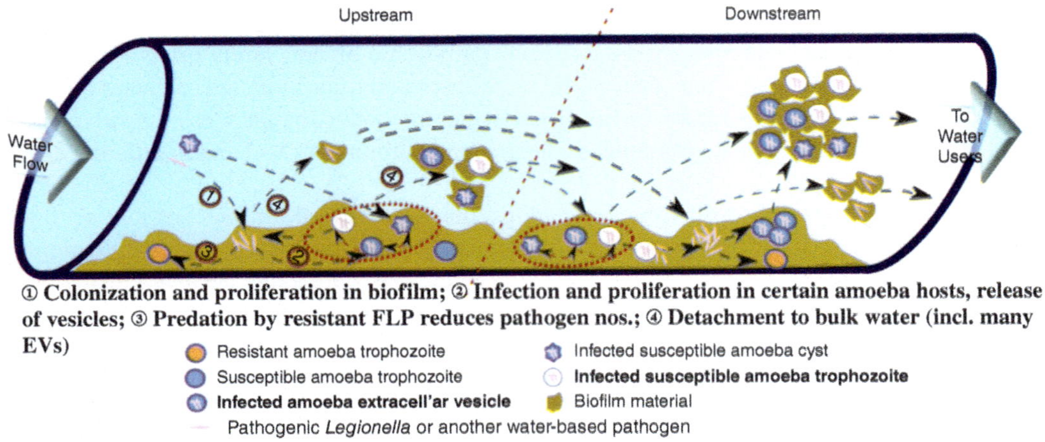

Figure 8.20 Conceptual model of *Legionella* colonization of biofilm on the side of a pipe wall. *Source:* From [118]/Frontiers Media S.A./CC BY 4.0/https://www.frontiersin.org/journals/cellular-and-infection-microbiology/articles/10.3389/fcimb.2023.1200478/full/last accessed Feb. 12, 2026.

be measured in controlled laboratory experiments [121], although more careful work is needed to ascertain under what circumstances this could be a dominant removal mechanism.

Discharging wastewater effluent into a stream introduces a potential source of prey, in the form of pathogenic and indicator bacteria, into an environment where protozoan predators may naturally occur. This could result in a (desired) natural decay of the released bacteria, which may be disrupted if the effluent also contains a disinfectant residual toxic to the protozoa. This can be described by a predator-prey relationship modified by the toxicity of residual disinfectant [122]. Under certain conditions, the toxicity of residual disinfectant may produce a paradoxical effect in which bacteria in the receiving water regrow in the presence of the residual. At the same time, they may decay in the absence of disinfection.[31]

The quantitative modeling of multiple populations interacting requires formulating mass balances for each population and specifying rates of interaction between populations. The basic concepts were outlined by Lotka and Volterra [123, 124]. Davis has formulated a generic basis of the interaction relationships that can be used to describe the multitude of relationships that can exist between populations [125], given in Equations 8.90 and 8.91.

$$\frac{dN_1}{dt} = N_1(\alpha_1 + \beta_{11}N_1 + \beta_{12}N_2) \tag{8.90}$$

$$\frac{dN_2}{dt} = N_2(\alpha_2 + \beta_{21}N_1 + \beta_{22}N_2) \tag{8.91}$$

With suitable combinations of coefficients, these equations can describe all of the interactions as summarized in Table 8.17.

This two-population system can readily be solved numerically and its stability properties examined, as shown previously (e.g., [122]). The β_{ii} terms, if negative, produce growth versus time relations that are qualitatively similar to a logistic. This can readily be extended to more than two populations, which would require more interaction parameters.

31 Although in the absence of disinfection of the effluent, the bacterial concentration from the starting condition would be greater than for a disinfected effluent.

Table 8.17 Combinations of coefficients in Equations 8.90–8.91 describing two population interactions.

Type of interaction	β_{ij}	β_{ji}	Comments
Neutralism	0	0	Populations behave independently
Mutulism	>0	>0	Both populations benefit
Commensalism	>0	0	One benefits, no effect on other. metabolic product is a nutrient for another organism (possibly *Legionella* colonization of protozoa
Predation/parasitism	>0	<0	Bacteriphage infecting host bacterium
Amensalism	0	<0	Secretion of inhibitory or antibiotic factors to other organism
Competition	<0	<0	Use of same food or electron acceptor

Source: Modified from [125].

Other modifications can be made to this framework to suit particular systems. Examples of such modifications are:

- If population 1 is prey and population 2 is predator, then the growth rate of the prey (β_{11}) may depend on substrate concentrations, perhaps via Monod kinetics.
- In a predator–prey or parasite–host situation, the growth rate of the predator (parasite) and the consumption rate of the prey (host) may be a function of the prey/host concentration, perhaps in a Monod fashion.
- The α's may function as decay (if they are negative) as well as growth terms, and the decay could depend on other properties of the environment, such as the occurrence of a toxicant or disinfectant.[32]
- Different populations could exist in different physical spaces, in which case the fluxes between those spaces would need to be defined and incorporated in modified equations—which would either be ODEs (if the different spaces are modeled as mixed compartments) or PDEs if there are definable concentration gradients.

8.5 Quantitative Description of Other Flux Processes

Other external fields or forces may impose additional flux terms on pathogen movement in the environment.

8.5.1 External Fields

8.5.1.1 Gravitational Force

Microorganisms generally have densities different from those of the fluids (air, water) in which they are conveyed. Therefore, in the presence of gravity, they will tend to move in the direction of the gravitational field (if they are denser than the fluid in which they are carried, which is usual) or opposite to the direction of the field (if they are less dense).

Pure bacterial cultures have densities (in liquid) of 1.035–1.093 g/mL. Fungal hyphae have densities of 1.035–1.093 g/mL [126]. Resistant stages, e.g., spores, may be expected to be denser.

32 Which, if it was time variant might require another differential equation for its simulation, and the relevant α could be replaced by a function of the toxicant.

Viruses tend to be denser due to the lower water content of their virions. Siegl and Frosner [127] measured the density of hepatitis A to be 1.34 g/mL.

Also, in bacteria and higher microorganisms, internal structures called gas vesicles or gas vacuoles may form and be metabolically controlled [128]. This can reduce the effective density of particles in which such organisms are present and, in turn, reduce (or even reverse) their sedimentation velocity.

In the environment, microorganisms are often associated with other particles or adsorbing materials. Therefore, their response to gravitational (and other) forces is the response of the aggregate in which they are found. Furthermore, microorganisms may have a variety of shapes; for example, see the diversity of viral shapes in Figure 2.2 and the diversity of bacterial shapes in Figure 2.6. Therefore, while there is a relationship between particle size and sedimentation velocity under idealized conditions, for complex geometries given by Stokes' law, it is generally not possible to directly ascribe a particle size based on sedimentation behavior for environmental particles containing pathogens.

Particles in water to which microorganisms are attached or in which they may be enmeshed have a variety of densities. Common mineral particles, such as quartz, exhibit a density of approximately 2.65 g/cm^3. These particles dominate inorganic sediments in aquatic environments. Organic matter, including plant debris, animal waste products, and microbial cells, typically exhibits densities ranging from 1.1 to 1.5 g/cm^3. Organic particles are less dense than mineral particles due to their chemical composition and structure.

The organic and inorganic matter may combine in the aqueous environment to form flocs that also engulf a mass of water, thereby further reducing the effective density to a value closer to that of water.

As noted in Section 8.4.1, for bioaerosols, particle density may be time varying since the initial emission from a source (whether human or environmental) due to evaporation of fluid. Typically, then, the density of bioaerosols increases with time.

In the case of particles either in air or water, in a gravitational field, there will be a tendency for the particles to move in the direction of a gravitational field.[33]

If a particle is independently settling within a gravitational field, there is a balance between three forces, as shown in Figure 8.21. From Newton's Law, the gravitational force is given by:

$$F_g = mg \tag{8.92}$$

where g is the gravitational acceleration, i.e., at the surface of the Earth, 980.6 m/s^2. From Archimedes principle, the (upward) buoyant force is equal in magnitude to the gravitational force exerted on a mass of fluid that is displaced by the volume of the particle. If the particle is a sphere of diameter d_p, and the fluid has a density of ρ_f then the buoyant force is equal to:

$$F_b = -\rho_f \, g\left(\frac{d_p^3}{6}\right) \tag{8.93}$$

The gravitational and buoyant forces act in the same direction. The third force, that of drag, may be vector-valued based on the velocity of the particle.

---

33 If the particle is less dense than the fluid, it will actually move in opposition due to the gravitational field, since the denser fluid will displace the particles. However, in natural settings this is uncommon; in engineered processes such as flotation separation, the goal is to remove particles by decreasing their density, typically by making conditions suitable for them to adhere to gas bubbles.

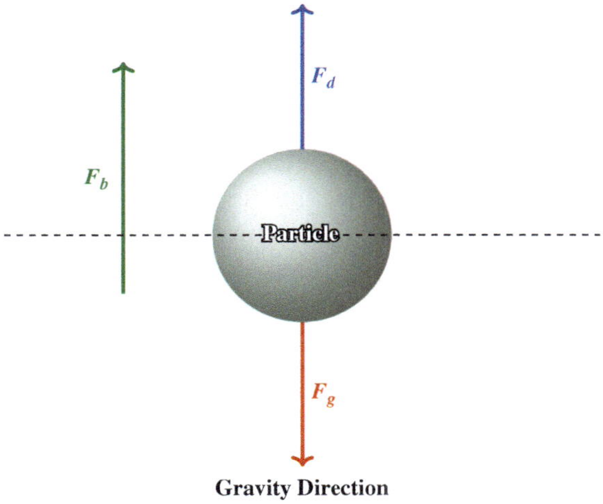

Figure 8.21 Force balance on particle moving in gravitational field.

The drag force has a directional vector *inverse* to the velocity vector, and has a magnitude ($\| F_d \|$) given by[34]:

$$\| F_d \| = \frac{\pi}{8} d_p^2 \rho_f \| v_p \| C_d \tag{8.94}$$

where $\| v_p \|$ is the magnitude of the velocity and C_d is a dimensionless drag coefficient.

The drag coefficient is an empirically derived dimensionless quantity that has been found to be related to the particle Reynolds number R defined analogously to Equation 8.18:

$$R = \frac{\rho_f \, v_p \, d_p}{\mu} \tag{8.95}$$

Goossens [129] reviewed extensive experimental data and prior correlations for drag coefficients on rigid spherical particles and concluded that the best relationship is given by Equation 8.96, at least up to R of 380,000.

$$C_d = \frac{24}{R} + 0.44 \tag{8.96}$$

At small values of the Reynolds number, the first term in this equation is dominant, and this is the regime termed Stokes' law. At large (>1000) Reynolds numbers, the second term is dominant, and this is termed the Newtonian regime.

In the case of non-spherical and/or non-rigid (and this deformable) particles, more complex relationships incorporating other properties need to be used, which require more detailed characterization of the particle properties [130].

Using the definition of the Reynolds number, at low R, which would be most frequent for problems of interest here, the drag force can be written as Equation 8.97.

$$\| F_d \| = 3\pi\mu d_p \| v_p \| \tag{8.97}$$

34 From basic vector operations, recall that for a vector, $\vec{A}$, its magnitude is given in terms of its components by: $\| A \| = \sqrt{A_x^2 + A_y^2 + A_z^2}$

At high R, the drag force can be expressed via Equation 8.98, which is termed Newton's law.

$$\| F_d \| = 0.173\rho_f d_p^2 \| v_p \|^2 \tag{8.98}$$

Note that at low Reynolds' number, the drag force depends on viscosity, while this is not the case at high Reynolds' number. In the intermediate case, the full expression of Equation 8.96 needs to be combined with the definition of Reynolds' number and substituted into Equation 8.94.

For very fine particles, especially in air, an additional modification is needed, called the Cunningham correction factor. This accounts for non-continuum effects as the particle size approaches the fluid molecules' mean free path. It can be expressed as [23]:

$$C_c = 1 + \frac{\lambda_f}{d_p}\left[2.51 + 0.80\ \exp\left(-\frac{0.55 d_p}{\lambda_f}\right)\right] \tag{8.99}$$

The drag force should be divided by C_c to incorporate this correction. In this relationship, λ_f is the mean free path of molecules in the fluid. For air under standard conditions, it is 0.066 μm [23].

Parenthetically, it should be noted that the drag force should be considered whenever a particle is moving relative to the fluid in which it is contained—it is a vector in 180° opposition to the direction of motion. The gravitational force is always in the direction of the gravitational field, and the buoyant force is always in opposition to the gravitational field.

The expressions for the gravitational, drag, and buoyant forces can be used directly as forces in particle balance equations, or they can be solved to get velocities and used as flux terms in the continuity equation approach. In the latter case, combining the three forces and solving for the particle settling velocity under low Reynolds number conditions yields Equation 8.100, which is often termed Stokes' law.

$$v_p = \left[\frac{C_c g d_p^2}{18}\left(\frac{\rho_p - \rho_f}{\mu}\right)\right] \tag{8.100}$$

Given a settling velocity, then the presence of a gravitational field induces a flux (assuming particles are denser than the fluid) downwards in the amount of $v_p C$, where C is the concentration of particles in the liquid.

This formulation assumes smooth, rigid spheres. Correction factors are available for nonspherical geometries [131]. For non-rigid, i.e., deformable particles, the situation is more complex since the movement of the particle may result in particle deformation, which is also going to be a function of the mechanical properties of the particle itself. It is possible to take material samples and directly measure settling velocities of particles (solids) under quiescent conditions. This can then be used directly to calculate gravitational flux.

Relationship Between Settling Velocity and Flux

If particles in a volume are at a concentration of C (in units of number or mass per volume), and have an effective settling velocity of v_x, then the gravitational flux in units of number or mass per area (normal to gravity) per time is given by:

$$J_x = v_x \cdot C \tag{8.101}$$

Note that the concepts of settling velocity can be used in the context of centrifugation. Under such circumstances, the gravitational and buoyant forces on particles need to be replaced by the centrifugal force on a particle (and the fluid that it displaces). The centrifugal force (F_c) on a particle of mass m in a centrifuge rotating at a rate of ω and located at a distance of r_{axis} from the axis of rotation is given by Equation 8.102.

$$F_c = m_p \omega^2 r_{\text{axis}} \tag{8.102}$$

8.5.1.2 Electrical Force

A system can be subject to external electrical fields. Some particles containing microorganisms, or some microorganisms themselves, may have net charges or may have inducible charges. The nature and magnitude of these charges can be influenced by the chemistry of the surrounding fluid, and by the motion of the particles themselves.

The interaction between particles, their charges, the surrounding fluid, and its chemistry is shown schematically in Figure 8.22. When particles move, they are surrounded by some tightly bound fluid which may have ions that partially (or completely) neutralize the intrinsic surface charge on the particle itself. The degree of this neutralization is influenced by the ionic composition of the surrounding fluid—higher ionic strengths will tend to increase neutralization and compress the layer of fluid moving with the particle. Fluid pH may directly alter the chemistry of functional groups on the surface. Operationally, as shown in this figure, what is directly measurable is a quantity called the ζ (zeta) potential, which is the effective voltage relative to the ground at the surface of the surrounding fluid that moves with the particle in a field. It is this potential at the surface of the shear layer (and the size including the shear layer) that governs the properties of a charged particle in an electric field.

From basic physics, if a particle carries a net charge of "z" ions, and it is in a DC electric field with a strength E (in units of potential per length, e.g., V/m), then the force on that particle is F_e given

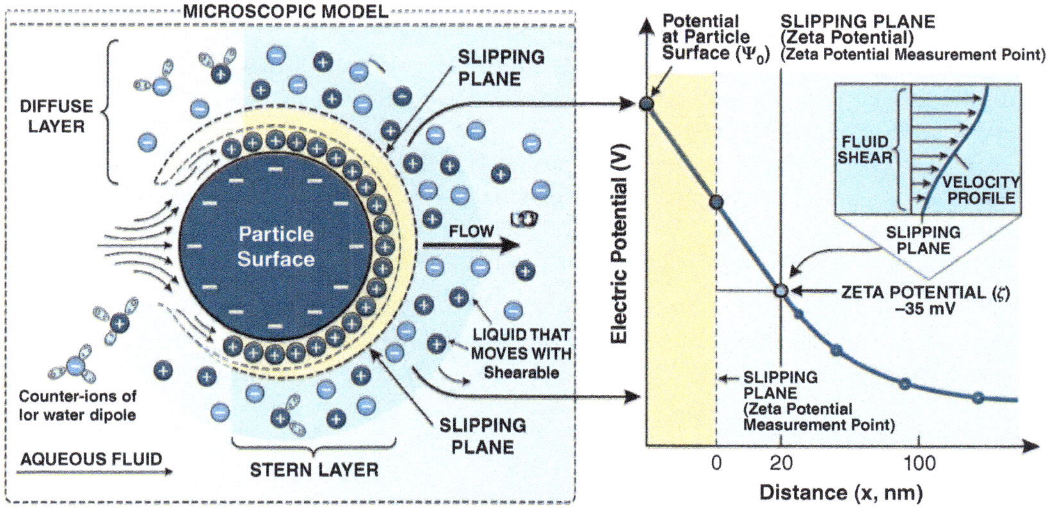

Figure 8.22 Schematic of charged particle moving through a fluid, showing slip plane and zeta potential.

by Equation 8.103. The symbol e denotes the physical constant charge per ion in the SI system, $1.602 \cdot 10^{-19}$ Coulomb.

$$F_e = z \cdot eE \tag{8.103}$$

Particles may have effective charges derived from several mechanisms [45]. As suggested by Figure 8.22, ionizable groups at the surface of the particle may be partially dissociated and incompletely neutralized by the fluid that moves along with the particle. The formation of droplets in detaching from a bulk fluid, or being formed (under laboratory conditions) may acquire a charge from that process. When a charged particle passes through an electric field, it can be induced to acquire additional charge. Collisions may result in the transference of charges between particles. As charged particles move through a fluid, depending on the particle capacitance and the fluid's conductivity, some charge dissipation may occur.

Charges on particles are rarely measured directly but are instead inferred from measurements of their motion under defined conditions. Classical techniques involved direct microscopic examination of particles in an electric field, but more modern methods use light scattering or electroacoustic methods [132].

Generally, gravitational and buoyant forces can be neglected, and particles of interest are in the low Reynolds number range. Therefore, at mechanical equilibrium, the electrostatic force given by Equation 8.103 can be equated to the drag force given by Equation 8.97. With the incorporation of Cunningham's correction, the electrophoretic mobility (U) defined as the ratio of particle velocity (v_p) to electric potential gradient (E) can be written as Equation 8.104.

$$U = \frac{v_p}{E} = \frac{z\ e\ C_c}{3\pi\mu d_p} \tag{8.104}$$

From dimensional analysis, U (in SI units) will have dimensions of $\frac{m^2}{V \cdot s}$.

In practice, even individual microorganisms from a pure culture [133] in fluid suspension or in an aerosol from a pure suspension [134] have a distribution of velocities and hence of electrophoretic mobilities due to their heterogeneity.

There is an effect of the chemistry of the surrounding fluid (or in air, the potential electrostatic dissipation) on electrophoretic mobility of individual microorganisms. There may not be substantial differences among different taxonomic groups, however; for example one study of three species (gram positive and gram negative) of bacteria found a range of electrophoretic mobilities from -1.57 to $-2.83 \cdot 10^{-8} \frac{m^2}{V \quad s}$ at low ionic strength [135]. If microorganisms are on or within other particles, then their electrophoretic mobility will depend on the adherent or carrier material.

The electrophoretic mobility can be related to the zeta potential (ζ) via Equation 8.105 [132].

$$U = \frac{2\zeta F(\kappa d_p)}{3\mu_f} \tag{8.105}$$

In this equation, μ_f is the viscosity of the fluid medium, d_p is the actual particle diameter, κ is the Debye length of the medium, which accounts for charges within the fluid, and F is the Henry's function. Henry's function accounts for the apparent additional thickness of any counterions from the fluid associated with the particle.

The Debye length is given by Equation 8.106:

$$\kappa = \sqrt{\frac{e \sum_i z_i^2 C_i}{\epsilon kT}} \tag{8.106}$$

where

- z_i and C_i are the charges on and the bulk concentrations of ionic species in the medium
- ϵ is the permittivity of the bulk medium
- k is the Boltzmann constant

Then the function $F(d_p\kappa)$ can be evaluated using an analytical approximation[35] provided by Swan and Furst [136] given in Equation 8.107.

$$F(y) = \frac{16 + 18y + 3y^2}{16 + 18y + 2y^2} \tag{8.107}$$

Once the zeta potential has been computed, in principle, the surface charge on particles can be obtained; however, to do so requires extensive knowledge of the particle's surface chemistry and the surrounding media.

What is established, however, is that if a particle (either microorganisms themselves or the solids with which microorganisms are associated) electrophoretic mobility (or zeta potential) is measured as a function of bulk media pH, there can be a value of pH that results in zero electrophoretic mobility (and zero zeta potential). This is called the isoelectric point, or zero point of charge.

Perhaps the earliest systematic study of bacterial isoelectric points (IEP) using electrophoretic methods was that of Harden and Harris [137]. They showed a broad range of IEP among bacteria (pH 1.75–4.15), although there is no evidence of systematic differences between Gram-positive and Gram-negative taxa. The low IEP indicates that, under relatively neutral conditions, organisms of all taxa tend to be negatively charged.

Michen and Graule [138] conducted a broad review of measured IEP values for viruses, including human, animal, and plant viruses and bacteriophages. Among these were both enveloped and non-enveloped taxa. They concluded that:

> "IEPs of viruses are found in pH range from 1.9 to 8.4; most frequently, they are measured in a band of 3.5 < IEP < 7. However, the data appear to be scattered widely within single virus species."

Hence, under neutral conditions, most but not all viruses are expected to be negatively charged.

Pure mineral phases can exhibit a wide range of IEPs, from acidic to alkaline. A number of such measurements are shown in Table 8.18.

Low organic soils and clays have been found to have a range of IEPs between 2.3 and 8.4 measured by different techniques [140].

8.5.1.3 Magnetic Force

In any system, the natural magnetic field of the Earth imposes a magnetic gradient on a system. This might be enhanced or contravened by deliberately introduced magnetic fields. While mineral particles may have magnetic properties that favor orientation in a magnetic field, only a small number of bacteria (and perhaps some of their predators) exhibit magnetic susceptibility. To date, none of these are known to be pathogenic [141]. However, some of these magnetotactic bacteria are motile and will move in the presence of a magnetic field.

---

35 The exact form involves an exponential integral.

Table 8.18 Isoelectric points for selected minerals in water.

Mineral	Isoelectric point
Alpha aluminum oxide (Al_2O_3)	8–9
Gamma aluminum oxide (Al_2O_3)	7–8
Iron (III) oxide (maghemite) (Fe_2O_3)	3.3–6.7
Iron (II,III) oxide (magnetite) (Fe_3O_4)	6.5–6.8
Magnesium oxide (MgO)	9.8–12.7
Manganese (IV) oxide (MnO_2)	4–5
Silicon dioxide (SiO_2)	1.7–3.5

Source: Excerpted from [139].

8.5.2 Interphase Transfer Processes

Interphase transfer processes occur when microorganisms in one phase (either free organisms or organisms associated with other materials) can be transported to another phase. To be rigorous, the concept of a phase needs to be defined, which is not a simple problem. Antonoff [142] developed a definition founded on thermodynamic concepts:

> "Mechanically separable parts of a system are called phases if one of them, or some constituents thereof, may become part of another in a reversible way."

With this definition, transfer between surfaces and air or water, and between air and water, can occur in several contexts. These have been documented and quantified using different approaches.

As the phase definition implies, material transfer between phases can generally be reversible. In other words, microorganisms or particles on which they occur can be transported from phase A → phase B and phase B → phase A.

In generic terms, the net flux (which may be negative) in amount per unit time per unit area (of contact between the phases) of contaminant (microorganisms or particles containing them) from phase A to phase B can be written as a linear driving force model in the form of Equation 8.108.

$$J_{AB} = \mathcal{K}_{\mathcal{AB}}(C_A - C_{A|B}) \tag{8.108}$$

Where $\mathcal{K}_{\mathcal{AB}}$ is an interphase transport coefficient (typically with dimensions of length/time), C_A is the concentration of the material of interest in phase A, and $C_{A|B}$ is the concentration *that would exist* in phase A if it was in equilibrium with the existing concentration in phase B. In many cases, the flux only in one direction is considered, and $C_{A|B}$ is treated as zero. This relationship is particularized in the case of specific phase types between which material is transferred.

8.5.2.1 Indoor Environments

Application of models in previous sections of this chapter (either box models or advection-dispersion models) in the indoor environment is sufficient for determining the flux of microorganisms *from the bulk indoor air to surrounding surfaces*, providing that the particle force balance terms are included. The mass transfer coefficient from air to room surfaces is generally termed the deposition velocity and depends on the particle sizes. This results in an effective first-order loss rate that depends on particle size. For typical rooms with 1–10 μm-diameter particles, it has been measured to range from 0.4 to 10 h^{-1} [143].

However, accumulated particles and microorganisms on these surfaces are not fully immobilized, and there may be a countervailing source that returns to the bulk air.

Mechanically, when particles contact a surface, they retain residual kinetic energy that may exceed the energy of attraction to the surface materials, and therefore, they rebound. This may be characterized by a "coefficient of restitution," which is the fraction of particles contacting a surface that rebound from the contact. Detailed estimation of this depends on the material and mechanical properties of both the particles and surfaces, and the energy with which the impact occurs [144].

Once deposited on a surface, mechanical agitation can result in additional resuspension. This was highlighted during the 2001 anthrax spore incident, where resuspension of spores during ordinary office activities was observed [145].

Qian et al. [146] systematically reviewed walking-induced particle resuspension in indoor environments. The analysis reveals that resuspension is a significant source of indoor particulate matter, particularly for particles sized 0.7–10 μm, with increasing mass flux observed for larger particles. The loading on the surface during activity will influence the emission flux, and the authors suggested the need for standardized methods and terminology in comparing studies. The observed resuspension fractions reported in the literature, including studies using total mass and microbiological measurements, ranged from 10^{-7} to 10^{-3}.

The primary mechanisms driving resuspension are controlled by adhesive and hydrodynamic forces, with rolling detachment identified as the predominant process. Surface roughness, particle size, and environmental variables, including relative humidity and ventilation, significantly influence resuspension rates. Biological particles often demonstrate higher resuspension fractions due to their physicochemical properties.

Infants (and presumably pets) can cause particles to be resuspended from floors. Using careful measurements, this has been quantified [147, 148]. This and other resuspension forces may result in greater concentrations of microbial particles in the near-floor environment, representing the breathing zone for infants and children, and thus could pose an enhanced risk.

Microbial particles may be captured on the clothing of occupants in an interior space. There is some indirect evidence from the 2001 anthrax incident of this being an exposure route, and subsequent controlled experiments show the potential for resuspension of respirable material that has been deposited on clothing [149]. Clothing that has been worn in a contaminated environment onto which microbial particles have deposited would also represent a means of transport to another environment, where those particles can be resuspended [150].

8.5.2.2 Outdoor Land Air Exchange

Pathogens can be found on soil originating from several mechanisms as discussed in Chapter 7. Human or animal waste, including sewage sludges (Section 7.7.2), may be deposited on land. There may be dry or wet deposition of bioaerosol particles from the atmosphere to land. While on land, pathogens may pose an exposure risk to people via fomites. Some deposited material may become resuspended in the air, resulting in additional exposure via this route. This has been demonstrated from studies on reaerosolization of *Bacillus* spores from soils, including material that had been resident on land for decades [151]. It is recognized that wind-blown dust may significantly influence atmospheric composition and local-to-global changes in biogeochemical cycles [152].

Wind can be a vehicle to "lift" bioaerosols and particles with adhering microorganisms from land surfaces. Quantitative modeling is complex and highly dependent on the nature of the bioaerosol or adhering particle, the land surface, and meteorological conditions. Fundamentally, the force of the wind must be sufficient to overcome the net of attractive forces, which include electrostatic,

London-van der Waals and capillary adhesive forces [153]. Hence, most estimates of resuspension come from direct measurements of emissions.

Korcz et al. estimated the total wind-blown dust flux from all of Europe. They derived an overall emission factor of PM_{10} of 0.139 $\frac{Mg}{km^2 y}$, with average flux in natural, agricultural, and anthropogenic areas of 0.021, 0.157, and 0.118, respectively [154]. PM_{10} is the concentration of particles with diameters less than or equal to 10 µm. If the target microorganism concentration is known, then this can be used to assess the flux. Furthermore, these authors tabulate estimated emission rates by soil type and land use, which may be used for more detailed estimation.

Jahne et al. conducted a bacterial risk assessment of airborne exposure from manure application sites, reviewing the literature on emission rates. For *E. coli* and enterococcus, the emission from sites actively receiving manure application or Class B wastewater sludge application corresponded to 0.002% of the applied bacterial loadings [155].

8.5.2.3 Water Sediment Exchange

Solids containing pathogens may settle via gravity in pipes, streams, lakes, and other aquatic environments. However, sufficient velocity over such settled sediments can resuspend the sediment and associated pathogens. This may be related to the shear forces associated with the flow, as has been documented in natural streams [156]. A more mechanistic analysis of these processes on regular surfaces may be performed using detailed force estimates on particles. However, the importance of particle–particle and surface forces must be considered [157].

8.5.2.4 Liquid Air Exchange

Turbulence at the water/air interface may result in the formation of small particles detached from the water and containing microorganisms that can be transported in the air. These rates are partly governed by mechanical forces from wind and tidal forces leading to particle generation at the surface [158].

Complicating estimates of fluxes are differential compositions of material at the air/water interface compared to the bulk water. A schematic of the processes that may occur within this microlayer is shown in Figure 8.23. The accumulation of particles and organics within this microlayer can also alter the physical properties near the surface, such as viscosity, which can, in turn, influence the extent to which material exchange with the atmosphere occurs. The fragmentation of material near the surface due to the action of wind and tidal forces can be a major mechanism for such production of potentially air-conveyed particles, in addition to bubble rise through the water column and bursting at the surface [159].

The accumulation of microorganisms within the surface microlayer will depend on their association with particles, and also the hydrophobicity of the organisms.[36] There is variability in the hydrophobicity of pathogens. The effective hydrophobicity of pathogens will be in part a function of their composition, particularly of their outer layers (capsids and envelopes for virus, cell membrane and cell envelope for bacteria, etc.) and also the chemistry of the water surrounding the organism(s) [161].

As suggested in Figure 8.23, there is strong evidence that many viruses can accumulate in the surface layer of water, which Rahlff has termed the "virion neuston" [160]. Bacteria vary in their hydrophobicity, principally based on the lipid content of their membranes or cell envelopes, for

36 The organism hydrophobicity will also influence its propensity to associate with solid surfaces

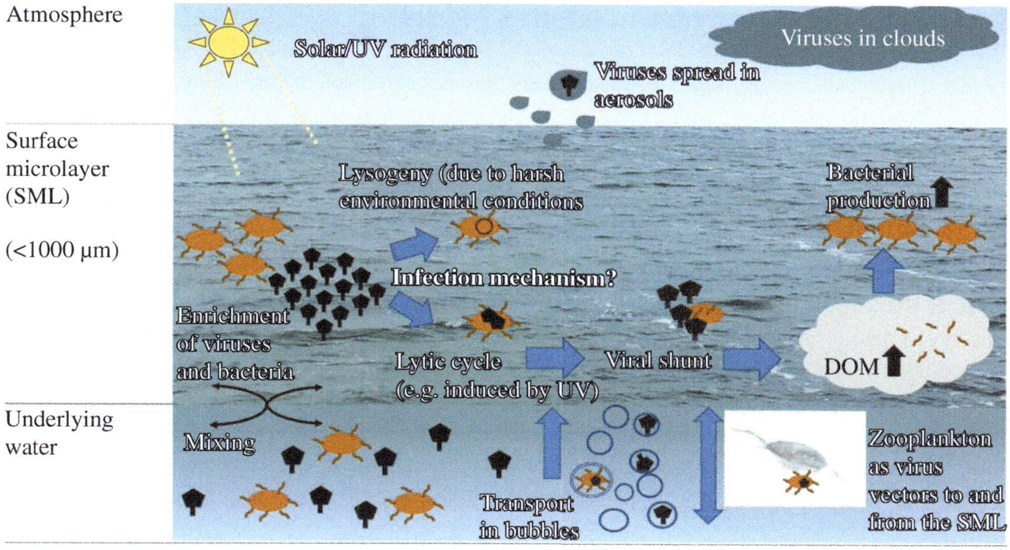

Figure 8.23 Microbial and chemical interactions in surface microlayer. *Source:* From [160]/MDPI/CC BY 4.0/https://www.mdpi.com/1999-4915/11/2/191/last accessed Feb. 12, 2026.

example with *Mycobacterium* being known for strong hydrophobicity. As a result, the selective accumulation of bacteria at air/water interfaces has been observed [162].

While it is not yet possible to estimate *a priori* the flux of microbial-containing particles into the atmosphere from a water body, there are some field studies that have demonstrated the occurrence of such transport. The following are particular examples:

- Pendergraft et al. conducted an experiment in which they released a dye tracer into the water off Imperial Beach, Ca [163]. During the first two days post-release, in two of three experiments, they detected air concentrations of the dye over land at a distance up to 668 m inland and 720 m downwind of the ocean. This demonstrates that ocean-to-air transfer can occur.
- In subsequent work, this group analyzed air inland for organic chemicals and for bacteria (using 16S RNA sequencing) [164]. The authors concluded that this demonstrates the contribution of coastal water pollution, such as from sewage outfalls to atmospheric contamination and then to possible human exposure to both chemical pollutants and biological agents.
- Abdool-Ghany sampled sea surface microlayers and aerosol samples on beaches in Miami, Florida, during a sewage spill incident. They used 16S RNA to conduct bacterial community analysis [165]. They observed the occurrence of potential pathogens *Vibrio, Corynebacterium,* and *Staphylococcus*, as well as other organisms, in the overlying aerosol, often with a proportional increase relative to their presence in the water. Buoyancy effects of the lower-density sewage plume relative to the underlying seawater may have aided in this enrichment.
- There are several polluted urban rivers with high organic loading, and in-stream aeration has been implemented to mitigate oxygen depletion. However, a study in the heavily polluted Newton Creek in New York City found, using 16S sequencing, that genera from potential pathogen-containing taxa could be found in aerator foam samples, as well as the water column [166].

8.6 Longer-range and More Complex Models

The individual processes discussed in this chapter may be combined to describe pathogen fate and transport, in particular, larger and more complex situations. It is impossible to give an exhaustive list of these applications. However, particular examples will be described.

8.6.1 Atmospheric Dispersion Models

Contaminants, including pathogens and particles on or in which pathogens are found, when released to the atmosphere, can be transported and dispersed in the air. Modeling this dispersion can involve consideration of multiple factors, including meteorology, particle size and density, reactions, terrain and topography, and the relative positions of the source(s) and receptor(s). Classical textbook Gaussian plume models are based on continuous steady releases over smooth terrain. However, substantial extensions and elaborations of this basic approach have been developed, although they cannot be reviewed here [167].

The basic Gaussian plume model for a continuous point source of contamination can be written as Equation 8.109 [168].

$$C(x,y,x) = \frac{Q}{2\pi\sigma_y\sigma_z\bar{u}} \exp\left[-\frac{(y-y_s)^2}{2\sigma_y^2}\right] \times \left\{\exp\left[-\frac{(z-z_p)^2}{2\sigma_z^2}\right] + \exp\left[-\frac{(z+z_p)^2}{2\sigma_z^2}\right]\right\} \tag{8.109}$$

In this equation, Q is the continuous emission rate from the source (quantity per time), C is the concentration at coordinates (x,y,z), $\bar{u}$ is the average wind speed, and σ_y and σ_z are the transverse and vertical dispersion coefficients. The source is located at x=0 (the direction of the wind), y=y_s (the transverse direction), and z=z_p relative to the ground surface.[37]

The dispersion coefficients in Equation 8.109 are functions of meteorological conditions. There are multiple approaches to estimating these depending on available data. A common, approximate approach is to define them as correlations based on stability class [169] defined by Table 8.19.

The stability classes can then be used to estimate σ_y and $sigma_z$ by use of Equations 8.110 and 8.111 with the parameters for these equations given in Table 8.20. Units for x, the I, J, and K coefficients and σ's are all in meters.

$$\sigma_y(x) = \exp\left\{I_y + J_y ln(x) + K_y[ln(x)]^2\right\} \tag{8.110}$$

$$\sigma_z(x) = \exp\left\{I_z + J_z ln(x) + K_z[ln(x)]^2\right\} \tag{8.111}$$

Meselson et al. analyzed the 1979 cluster of anthrax in the vicinity of a bioweapons laboratory in Sverdlovsk, USSR [171]. They concluded that the geographic distribution of cases was consistent with atmospheric dispersion from an accidental point source instantaneous release of *Bacillus anthracis*, and using a basic Gaussian dispersion model. Using rough estimates for a dose-response relationship, they estimated the initial release could have been in the milligram-to-gram range, which would have been plausible.

Jahne et al. [155] assessed the flux of pathogens to the atmosphere from land application of animal manure. They then used an EPA atmospheric dispersion model, AERMOD [172], a steady-state

37 This is a combination of the physical height of the source and any plume rise that occurs. There are various approaches to estimating plume rise [168].

Table 8.19 Stability classes based on meterological conditions.

Surface wind speed (m/s)	Daytime insolation			Night-time conditions	
	Strong	Moderate	Slight	Thin overcast or >4/8 low cloud	<=4/8 cloudiness
<2	A	A–B	B	E	F
2–3	A–B	B	C	E	F
3–5	B	B–C	C	D	E
5–6	C	C–D	D	D	D
>6	C	D	D	D	D

Source: Data from [169].

Table 8.20 Coefficients for atmospheric dispersion correlations as a function of stability class.

Coefficient	A	B	C	D	E	F
I_y	−1.104	−1.634	−2.054	−2.555	−2.754	−3.143
J_y	0.9878	1.0350	1.0231	1.0423	1.0106	1.0148
K_y	−0.0076	−0.0096	-0.0076	−0.0087	−0.0064	−0.0070
I_z	4.679	−1.999	−2.341	−3.186	−3.783	−4.490
J_z	−1.7172	0.8752	0.9477	1.1737	1.3010	1.4024
K_z	0.2770	0.0136	−0.0020	−0.0316	−0.0450	−0.0540

Source: Data from [170].

version of the Gaussian dispersion model useful for describing near-source impacts. This was then coupled with pathogen dose-response models to assess human risk. The authors concluded that the risk 100 m off-site for many pathogens was greater than those levels commonly accepted in the context of drinking water exposures.

Van Leuken et al. [173] did a systematic review of atmospheric dispersion models used for pathogen exposure and risk assessment. They outlined attributes of the many available models, including algorithm (Gaussian, Lagrangian transport, Eulerian transport), source (continuous versus instantaneous), inclusion of settling and other loss terms, and particle size distribution. They presented an annotated bibliography of numerous prior studies. They noted that:

> "A majority of the reviewed studies, however, lacked quantitative analyses and application of full quantitative microbial risk assessments (QMRA)... [in the future] we recommended determining well-quantified emission and inactivation rates and applying dosimetry and dose–response models to estimate infection probabilities in the population at risk."

Stated simply, there is often insufficient underlying information to take full advantage of atmospheric dispersion models in a risk assessment context.

8.6.2 Indoor Air Models

Since people spend much of their time indoors rather than outdoors (Figure 2.14), several approaches have been developed to model exposure to pathogens in indoor air. There are numerous potential sources of particles that could be generated and dispersed within the interior air of rooms. These include water fixtures generating drops, aerosols and mists such as showers, toilets (Figure 8.24). misters, humidifiers and architectural water features. In commercial and industrial sources, numerous other sources can exist, such as floor cleaners [174], dental equipment in dentist offices [175], etc. Given such sources, the concentration (and concentration time) profile can be estimated using box models or one of the more sophisticated transport models introduced in Section 8.3.3 or Section 8.3.5.

Higham et al. [177] performed detailed measurements of the aerosol generation rates from toilets, including particle size distribution characteristics. They used this in a microbial risk framework to assess the secondary transmission risk from shared toilet facilities and tested alternative design features, including ventilation, to mitigate that risk.

Hamilton et al. used box modeling approaches to develop guidelines for exposure to *Legionella pneumophila* to achieve acceptable risk levels from aerosols generated by showers and toilets [178].

For full-room, multi-room, or building-scale dispersion of contaminants, the program CONTAM, developed by the (US) National Institute of Standards and Technology (NIST), has been widely used [10]. This program can model individual rooms, or zones within rooms, as completely mixed compartments connected by flows (ventilation systems), and considers sources and reactions. This has been explicitly extended to consider bioaerosols [179]. This framework enables consideration of various mitigation strategies,[38] including room UV irradiation, air filtration, use of room air purifiers, etc. The types of processes considered are illustrated schematically in Figure 8.25.

Yan et al. applied this model to the fate and transport of SARS-CoV-2 aerosols in exemplar buildings of several types: Medium Office, Large Office, Small Hotel, Stand-alone Retail, and Secondary School [180]. They tested various mitigation strategies, including masking, and concluded “for

Figure 8.24 Production of aerosols immediately after toilet flushing, visualized using fluorescent dyes. *Source:* From [176]/Springer Nature/CC BY 4.0/https://www.nature.com/articles/s41598-022-24686-5/last accessed Feb. 12, 2026.

38 The characterization and performance of various mitigation strategies will be discussed in Chapters 9–11.

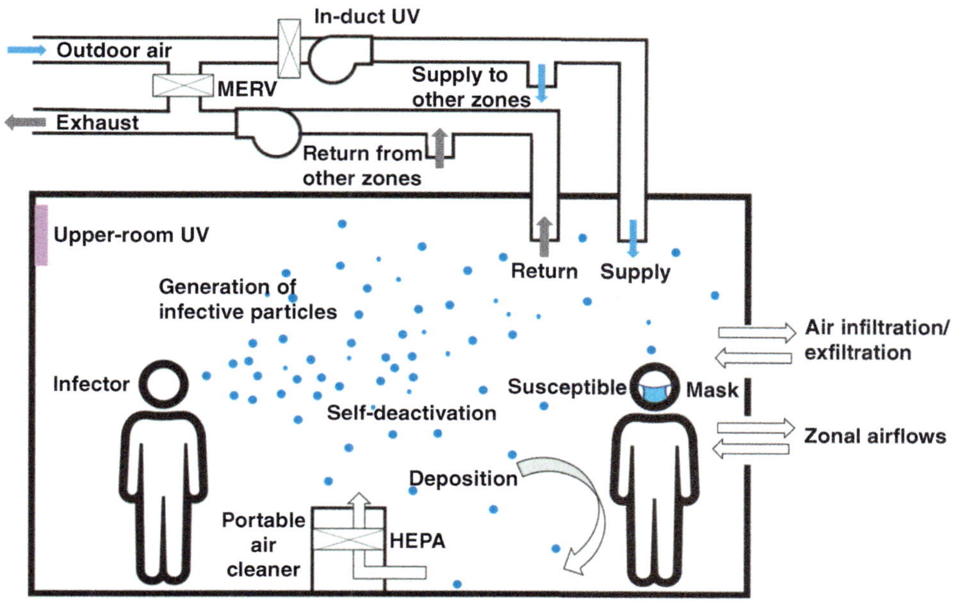

Figure 8.25 Conceptual model of some processes in modeling fate and transport in a single mixed room. *Source:* From [180]/MDPI/CC BY 4.0/https://www.mdpi.com/2075-5309/13/1/102?utm_campaign=releaseissue_buildingsutm_medium=emailutm_source=releaseissueutm_term=titlelink_cover/last accessed Feb. 12, 2026.

small spaces, in-duct air treatment would be insufficient for mitigating infection risks and additional in-room treatment devices would be needed."

8.6.3 Water Quality Models

Pathogen fate and transport in lakes, reservoirs, rivers, and streams is important because exposure to pathogens may occur by consumption of (treated or untreated) drinking water or inadvertent ingestion or inhalation/aspiration of water during recreation. There are many potential sources, as delineated in Chapter 7. Within the water body, many individual processes could affect pathogen dynamics, other than intrinsic growth and/or decay, including [181]:

- Attachment and detachment to particles
- Sedimentation and resuspension
- Advective and dispersive transport
- Inactivation by UV light penetrating the water column
- Growth and/or decay may be influenced by vertical gradients in temperature, dissolved oxygen, oxygen-reduction potential, and total dissolved solids. This is especially true in deeper lakes or reservoirs with seasonal thermoclines[39] and turnovers.

39 Any lake or reservoir of moderate or large size in temperate locations will undergo thermal stratification. In summer locations, warm water will lie at the top of the water body, and more dense cold water will remain at the bottom. There will typically be a relatively thin zone in which the temperature changes rapidly with depth—this is called the thermocline and forms a relative impediment to vertical transport and mixing. This phenomenon was first described in the late 19th century [182].

There will also potentially be multiple sources of pathogens in a surface water body, many of which were enumerated in Chapter 7. These can include inputs from surface water and groundwater tributaries, overland flow from land containing wastes (urban, suburban, agricultural), direct deposition of atmospheric particulates, and excreta from migratory and resident birds [183] and other fauna. Direct discharge of treated wastewater and flow from septic tank leach fields may be sources. Improper discharges from boat toilets may also be sources of pathogens. It is often desirable to ascertain the relative magnitude of multiple diverse sources, and water quality models may be useful tools in this task. Two case studies are illustrative.

Hoyer et al. [184] evaluated the potential for pollution from beaches (due to *Cryptosporidium* oocysts) to adversely affect drinking water intakes in Lake Tahoe, NV.[40] To assess this, they constructed a classical hydrodynamic model of flows in the lake and modeled pathogen persistence using a particle tracking, Lagrangian approach in which they considered UV inactivation (and consequently light penetration through the water column) as well as dispersion and advection. They conducted their simulations in a back-tracking mode, in which ensembles of organisms at the drinking water intake were simulated in reverse time to track the probability that they might have originated at the recreational beach sites. The analysis concluded that there was a low (0.02%) probability that viable *Cryptosporidium* at the drinking water intakes could have originated from the recreational area, which tended to alleviate concern (at least from a microbial contamination view) that the beach areas posed a substantive risk to the quality of the drinking water.

Tolouei et al. [186] modeled the fate and transport of *E. coli*, *Cryptosporidium* and *Giardia* in Lake Ontario to determine the potential impacts on the water intake for Toronto. They used a fully Eulerian approach for hydrodynamics and for pathogen fate and transport and considered as sources two rivers, three creeks, and two wastewater treatment plants under both dry weather and wet weather (storm) conditions. They were then able to construct probabilistic models for the pathogen concentrations at the drinking water intake for various scenarios. Based on this analysis, they concluded that the dominant sources of *Giardia* were the wastewater treatment plants, while the rivers were the dominant sources for *E. coli* and *Cryptosporidium*. Notwithstanding these findings, they noted that the projected risks from the pathogens remained within acceptable levels.

Groundwater transport of pathogens can be of concern since a substantial proportion of water from these systems have been found to contain pathogens [187], and there have been waterborne outbreaks [188] associated with the use of groundwater (which is often untreated).

The physical nature of groundwater coupled with the ability of low, single exposures to provoke an adverse effect leads to the need to model the precise transport dynamics. As noted by Hunt and Johnson [189]:

> "understanding pathogen processes requires ... the focus is on the leading edges of plumes and preferential flow paths, even if such features move only a very small fraction of the aquifer flow. Moreover, the relatively short survival times of pathogens in the subsurface require greater attention to very fast (<10 year) flow paths."

These authors further note the importance of pathogen/subsurface solid interactions, the presence of filtration mechanisms (which do not occur for solutes), adsorption (which may or may not be reversible), and portions of the subsurface liquid volume that may be inaccessible to pathogens due to their size. In addition, as with any medium, pathogen decay and/or (for vegetative organisms) growth may occur.

40 This is a deep, high-altitude alpine lake, which has long been known for its water clarity [185].

The aquifer medium may be heterogeneous and anisotropic, requiring more complex models of fluid flow. In addition, as illustrated in Figure 8.26, there may be complex macro structures, such as impermeable layers, infiltration to or exfiltration from surface waters, withdrawal via vegetative transpiration, and the existence of injection or withdrawal wells.

The fundamental quantitative theory of microbial fate and transport in the subsurface was reviewed by Bradford et al. [190]. The authors note that several aspects of the underlying processes are incompletely understood. Furthermore, due to the unique characteristics of individual groundwater systems, models must be reconfigured for each application. The presence of fracturing, including in karst aquifers, may produce complex flow pathways that can cause pathogens to be transported differently from soluble materials [191].

One critical engineered water infrastructure system is used for the distribution and conveyance of treated drinking water. This comprises pipes, pumps, storage tanks, fire hydrants, and treated water sources (product water for treatment plants or pumping from groundwater wells—with or without treatment).

The need to simulate hydraulic flow patterns and transport of physical, chemical, and biological contaminants in these systems has led to the development of the EPANET software package. This was initially developed by the US EPA but has been released as open-source. The repository for

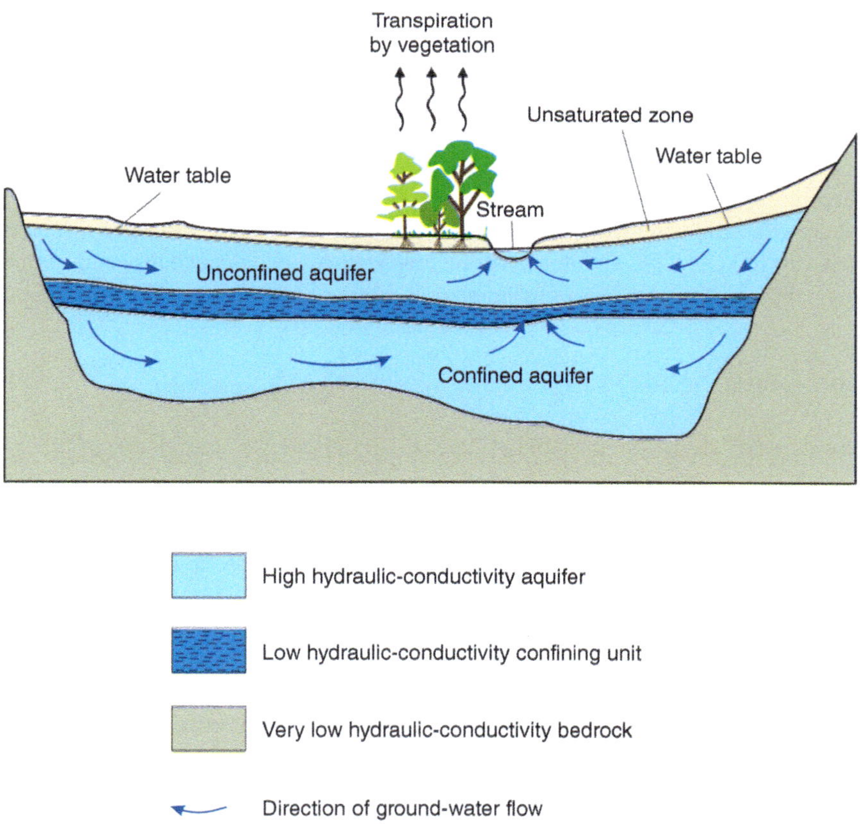

Figure 8.26 Schematic of an aquifer. *Source:* Hans Hillewaert/CC0 1.0/https://en.wikipedia.org/wiki/Aquifer#/media/File:Aquifer_en.svg/last accessed Feb. 10, 2026.

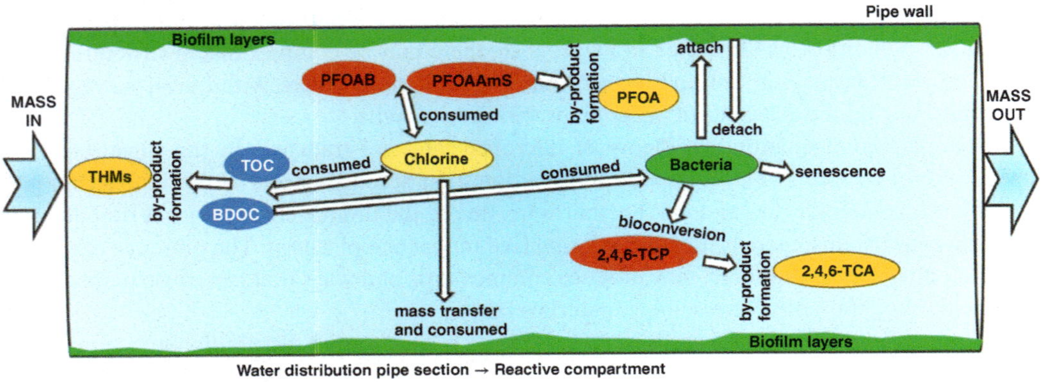

Figure 8.27 Processes included in EPANET-C model. *Source:* [192]/MDPI/CC BY 4.0.

the software is at https://github.com/OpenWaterAnalytics/EPANET. As a result, there have been a number of elaborations and "forks" of the underlying model to consider, in addition to flow and pressure, the persistence of disinfectant residual, decay and growth of microorganisms, attachment and detachment to pipe wall surfaces, and formation of disinfection by-products.

In these models, axial dispersion is generally neglected because of variable flow velocity caused by fluctuations in system supply and demand. However, transport to and from the pipe walls and potential reactions on the surfaces are considered.

One fork of the base EPANET model, termed EPANET-C, was developed by Abhijith and Ostfeld [192]. They considered multiple chemical and biological components, as depicted in Figure 8.27, including bacterial growth within the water phase and at the pipe wall. They developed an interface for the model to run in MATLAB, and demonstrated its performance on a trial system. More chemical and biological processes are being considered for addition to this framework.

8.7 Discussion Questions and Problems

1 Consider a closed, well-mixed system with pathogens that decay via a second-order process. The initial concentration is 1000 organisms/L. If the rate constant is $1\frac{L}{hr - \text{organism}}$, how long will it take to the organisms to decay to 1 organism/L?

2 Revise the model given by Equation 8.3 to include the possibility that "injured" organisms recover to viability with a rate constant of 0.05. Modify the computer code and plot the subpopulations versus time.

3 A single well-mixed tank receives water containing an average of 10 viruses per 100 L, with diurnal sinusoidal fluctuations between 0 and 20 per 100 L. The flow is constant with $\Theta = 20$ hours. This is treated with a disinfectant that inactivates the virus according to a first-order reaction (with virus). What should be the inactivation rate (assuming a constant disinfection process) such that the virus concentration is *always* below 0.1 organisms per 100 L. This computation can be done analytically or numerically.

4 Derive the rate expression corresponding to the gamma survival distribution in Table 8.5. Note that some experience with special functions or computer algebra programs may be needed.

5 Start with the basic differential equation for the Gompertz growth equation (Equation 8.73). If a negative power law function, $(t + t_0)^{-q}$ is used, what will the shape of the resulting growth curve look like and how does this compare with the Zweitering version of the Gompertz?

6 Consider the process described by the lag and autoinhibitor growth in equations 8.79–8.83, with the parameters from the batch output in Figure 8.18. Apply this to a single CSTR receiving an influent concentration of N_{inf} organisms, and write the algebraic equations that would predict the steady state effluent concentrations.

7 Legnani et al. [193] present multiple data sets on growth of *Pseudomonas aeruginosa* in mineral water over a long (5 year) time period. Use the Van Impe "*P*" and "*S*" models to fit several of these data sets. Note that this will require considerable coding expertise, in which the numerical solution of the differential equation system is embedded within a nonlinear curve-fitting routine.

8 An infectious individual expels bioaerosols of 10μm diameter, with a density of 1.2 $\frac{g}{cm^3}$. How long will it take to reach the floor if it is emitted at a height of 1.5 m. For this example, the following simplifying assumptions may be made (which, from the prior discussion in this chapter, is known not to be strictly true):
   - There is no evaporation of material from the particle.
   - The room is assumed to be quiescent at a temperature of 25°C, and thermal currents can be neglected.

9 The electrophoretic mobility of *Brevibacillus brevis*, formerly termed *Bacillus brevis* was measured between −3 and +3 $\frac{\mu m \cdot cm}{V \cdot s}$ [137]. Given the conditions reported in that paper, estimate the bacterium's zeta potential range.

10 An agricultural field is being used for the disposal of biosolids. Wind produces 1 $\frac{kg}{d}$ of dust that contains a total viable organism count of 100 gm^{-1}. If the wind blows toward a receptor at 2.5 $\frac{m}{s}$ at night with thin overcast, draw the concentration profile versus distance from 100 m to 1 km downwind.

References

1 Alice Calaprice. *The Ultimate Quotable Einstein*. Princeton University Press, Princeton, NJ, 2010.

2 Charles William Gear and Linda Ruth Petzold. ODE methods for the solution of differential/algebraic systems. *SIAM Journal on Numerical analysis*, 21(4):716–728, 1984.

3 D S Dickey and J B Fasano. *Mechanical Design of Mixing Equipment*, chapter 21, pages 1247–1332. John Wiley & Sons, Ltd, 2003.

4 Charles N Haas, Josh Joffe, Mark S Heath, and Joseph Jacangelo. Continuous flow residence time distribution function characterization. *Journal of Environmental Engineering*, 123(2):107–114, February 1997.

5 M S Saidi, M Rismanian, M Monjezi, M Zendehbad, and S Fatehiboroujeni. Comparison between Lagrangian and Eulerian approaches in predicting motion of micron-sized particles in laminar flows. *Atmospheric Environment*, 89:199–206, June 2014.

6 E Bruce Nauman. Residence time theory. *Industrial Engineering Chemistry Research*, 47:3752–3766, 2008.

7 H Scott Fogler. *Elements of Chemical Reaction Engineering*. Prentice Hall, 5th edition, 2016.

8 Norman L Johnson, Samuel Kotz, and N Balakrishnan. *Continuous Univariate Distributions*, volume 1. Wiley Interscience, New York, 2nd edition, 1994.

9 Nick T. Thomopoulos. *Statistical Distributions*. Springer International Publishing, 2017.

10 National Institute of Standards and Technology. Contam. https://www.nist.gov/services-resources/software/contam, January 2021.

11 Bruce E Logan. *Environmental Transport Processes*. John Wiley & Sons, 2012.

12 Atul Sharma. *Introduction to Computational Fluid Dynamics: Development, Application and Analysis*. Springer Nature, 2021.

13 Wolfgang Rodi. Turbulence modeling and simulation in hydraulics: A historical review. *Journal of Hydraulic Engineering*, 143(5):03117001, May 2017.

14 Osborne Reynolds. IV. On the dynamical theory of incompressible viscous fluids and the determination of the criterion. *Philosophical Transactions of the Royal Society A*, 186:123–164, 1895.

15 Adolf Fick. Ueber diffusion. *Annalen der Physik*, 170(1):59–86, 1855.

16 John Crank. *The Mathematics of Diffusion*. Oxford University Press, 1979.

17 E Bruce Nauman and B A Buffham. *Mixing in Continuous Flow Systems*. John Wiley and Sons, New York, 1983.

18 Susan M Teefy and Philip C Singer. Performance and analysis of tracer tests to determine compliance of a disinfection scheme with the SWTR. *Journal of the American Water Works Association*, 82(12):88–98, 1990.

19 Charles N Haas. Moment analysis of tracer experiments. *Journal of Environmental Engineering*, 122:1121–1123, 1996.

20 T N Zwietering. The degree of mixing in continuous flow systems. *Chemical Engineering Science*, 11:1–15, 1959.

21 Charles N Haas. Micromixing and dispersion in chlorine contact chambers. *Environmental Technology Letters*, 9:35–44, 1988.

22 Stephen A Craik. Effect of micro-mixing conditions on predictions of *Cryptosporidium* inactivation in an Ozone contactor. *Ozone: Science & Engineering: The Journal of the International Ozone Association*, 27:487–494, 2005.

23 William W Nazaroff and Lisa Alvarez-Cohen. *Environmental Engineering Science*. John Wiley & Sons, 2000.

24 Christina Cruickshank Miller. The Stokes-Einstein law for diffusion in solution. *Proceedings of the Royal Society of London. Series A, Containing Papers of a Mathematical and Physical Character*, 106(740):724–749, 1924.

25 I B Goldman and J M Marchello. Turbulent Schmidt numbers. *International Journal of Heat and Mass Transfer*, 12(7):797–802, 1969.

26 N Sheriff and D J O'Kane. Eddy diffusivity of mass measurements for air in circular duct. *International Journal of Heat and Mass Transfer*, 14(5):697–707, 1971.

27 Seyed Omid Rastegar and Tingyue Gu. Empirical correlations for axial dispersion coefficient and Peclet number in fixed-bed columns. *Journal of Chromatography A*, 1490:133–137, March 2017.

28 D J Zigrang and N D Sylvester. A review of explicit friction factor equations. *Journal of Energy Resources Technology*, 107(2):280–283, June 1985.

29 George J Arcement and Verne R Schneider. Guide for selecting Manning's roughness coefficients for natural channels and flood plains. Water supply paper, US Geological Survey, 1989.

30 Abhijit Guha. Transport and deposition of particles in turbulent and laminar flow. *Annual Review of Fluid Mechanics*, 40(1):311–341, January 2008.

31 Xiangdong Li, Yidan Shang, Yihuan Yan, Lin Yang, and Jiyuan Tu. Modelling of evaporation of cough droplets in inhomogeneous humidity fields using the multi-component Eulerian-Lagrangian approach. *Building and Environment*, 128:68–76, 2018.

32 Z Zhang and Q Chen. Comparison of the Eulerian and Lagrangian methods for predicting particle transport in enclosed spaces. *Atmospheric Environment*, 41(25): 5236–5248, August 2007.

33 Hamid Motamedi Zoka, Mohammad Moshfeghi, Hadi Bordbar, Parham A Mirzaei, and Yahya Sheikhnejad. A CFD approach for risk assessment based on airborne pathogen transmission. *Atmosphere*, 12(8):986, 2021.

34 Jie Zhang, Andrés E Tejada-Martínez, and Qiong Zhang. Developments in computational fluid dynamics-based modeling for disinfection technologies over the last two decades: A review. *Environmental Modelling & Software*, 58:71–85, 2014.

35 Joel J Ducoste, Dong Liu, and Karl Linden. Alternative approaches to modeling fluence distribution and microbial inactivation in ultraviolet reactors: Lagrangian versus Eulerian. *Journal of Environmental Engineering*, 131(10):1393–1403, October 2005.

36 B A Wols, J A M H Hofman, W S J Uijttewaal, L C Rietveld, and J C Van Dijk. Evaluation of different disinfection calculation methods using CFD. *Environmental Modelling & Software*, 25(4):573–582, April 2010.

37 Shamia Hoque, Bakhtier Farouk, and Charles N Haas. Development of artificial neural network based metamodels for inactivation of anthrax spores in ventilated spaces using computational fluid dynamics. *Journal of the Air and Waste Management Association*, 61(9):968–982, 2011.

38 V D'Alessandro, M Falone, L Giammichele, and R Ricci. Eulerian–Lagrangian modeling of cough droplets irradiated by ultraviolet–C light in relation to SARS-CoV-2 transmission. *Physics of Fluids*, 33(3):031905, March 2021.

39 J Qian, D Hospodsky, N Yamamoto, W W Nazaroff, and J Peccia. Size-resolved emission rates of airborne bacteria and fungi in an occupied classroom. *Indoor Air*, 22:339–351, January 2012.

40 Marcus Clauß. Particle size distribution of airborne microorganisms in the environment-a review. *Landbauforsch Applied Agricultural and Forestry Research*, 65(2):77–100, 2015.

41 Lidia Morawska, Giorgio Buonanno, Alex Mikszewski, and Luca Stabile. The physics of respiratory particle generation, fate in the air, and inhalation. *Nature Reviews Physics*, 4(11):723–734, 2022.

42 K Randall, E T Ewing, L C Marr, J L Jimenez, and L Bourouiba. How did we get here: what are droplets and aerosols and how far do they go? A historical perspective on the transmission of respiratory infectious diseases. *Interface Focus*, 11:20210049, October, 2021.

43 Frank J Loge, Robert W Emerick, Tim R Ginn, and Jeannie L Darby. Association of coliform bacteria with wastewater particles: Impact of operational parameters of the activated sludge process. *Water Research*, 36(1):41–48, 2002.

44 Eric P Vejerano and Linsey C Marr. Physico-chemical characteristics of evaporating respiratory fluid droplets. *Journal of The Royal Society Interface*, 15(139):20170939, February 2018.

45 Bruce Lighthart. Physics of microbial bioaerosols. In *Atmospheric Microbial Aerosols: Theory and Applications*, pages 5–27. Springer, 1994.

46 Mehdi Stiti, Guilaume Castanet, Andrew Corber, Marcus Alden, and Edouard Berrocal. Transition from saliva droplets to solid aerosols in the context of COVID-19 spreading. *Environmental Research*, 204: 112072, 2022.

47 D N Thomas, S J Judd, and N Fawcett. Flocculation modelling: A review. *Water Research*, 33(7):1579–1592, May 1999.

48 Ricardo I Jeldres, Phillip D Fawell, and Brendan J Florio. Population balance modelling to describe the particle aggregation process: A review. *Powder Technology*, 326:190–207, February 2018.

49 Thomas R Camp. Velocity gradients and internal work in fluid motion. *Journal of the Boston Society of Civil Engineers*, 30:219–230, 1943.

50 Mooyoung Han and Desmond F Lawler. The (relative) insignificance of G in flocculation. *Journal-American Water Works Association*, 84(10):79–91, 1992.

51 John Gregory and Charles R O'Melia. Fundamentals of flocculation. *Critical Reviews in Environmental Control*, 19(3):185–230, January 1989.

52 Otto Rahn. *Injury and Death of Bacteria by Chemical Agents*. Biodynamica Monographs, Normandy, MO, 1945.

53 George E P Box. Science and statistics. *Journal of the American Statistical Association*, 71(356):791–799, December 1976.

54 Vincent P Olivieri. Oral remarks. direct quotation, ca 1985.

55 Jennifer D Petersen, Jianming Lu, Wendy Fitzgerald, Fei Zhou, Paul S Blank, Doreen Matthies, and Joshua Zimmerberg. Unique aggregation of retroviral particles pseudotyped with the delta variant SARS-CoV-2 spike protein. *Viruses*, 14(5):1024, 2022.

56 J Maynard Smith. *Mathematical Ideas in Biology*. Cambridge University Press, London, 1968.

57 D Bettega, P Calzolari, A Ottolenghi, and L Tallone Lombardi. Criteria and techniques for analysing cell survival data. *Radiation and Environmental Biophysics*, 30(1):53–70, March 1991.

58 Rajesh E Kannan and Supreet Saini. Mathematical modelling of quorum sensing in bacteria. *INAE Letters*, 3(3):175–187, 2018.

59 Charles N Haas and Baris Kaymak. Effect of initial microbial density on inactivation of Escherichia coli by monochloramine. *Journal of Environmental Engineering and Science*, 7:237–245, 2008.

60 E T Lee. *Statistical Methods for Survival Data Analysis*. John Wiley & Sons, 2003.

61 Andrew F Brouwer, Marisa C Eisenberg, Justin V Remais, Philip A Collender, Rafael Meza, and Joseph N S Eisenberg. Modeling biphasic environmental decay of pathogens and implications for risk analysis. *Environmental Science & Technology*, 51(4):2186–2196, February 2017.

62 Tieling Zhang and Min Xie. Failure data analysis with extended Weibull distribution. *Communications in Statistics - Simulation and Computation*, 36(3):579–592, May 2007.

63 Kara Dean, Austin Wissler, J Sebastian Hernandez-Suarez, A Pouyan Nejadhashemi, and Jade Mitchell. Modeling the persistence of viruses in untreated groundwater. *Science of The Total Environment*, 717:134599, May 2020.

64 Kara Dean and Jade Mitchell. Meta-analysis addressing the implications of model uncertainty in understanding the persistence of indicators and pathogens in natural surface waters. *Environmental Science & Technology*, 56(17):acs.est.1c07552, August 2022.

65 Pandu R Tadikamalla. A look at the Burr and related distributions. *International Statistical Review / Revue Internationale de Statistique*, 48(3):337–344, December 1980.

66 Hadley Wickham. Tidy data. *Journal of Statistical Software*, 59(10):1–23, 2014.

67 J B Johnson and K S Omland. Model selection in ecology and evolution. *Trends in Ecology & Evolution*, 19(2):101–108, February 2004.

68 J E Easton, Melinda Lalor, Joseph J Gauthier, and R E Pitt. In-situ die-off of indicator bacteria and pathogens. In *Proceedings: AWRA's 1999 Annual Water Resources Conference-Watershed Management to Protect Declining Species*, 1999.

69 Charles N Haas and Joseph Jacangelo. Development of regression models with below detection limit data. *Journal of Environmental Engineering*, 119(2):214–230, 1993.

70 Charles N Haas and B Heller. Averaging TNTC counts. *Applied and Environmental Microbiology*, 54:2069–2072, 1988.

71 Arne Henningsen. Estimating censored regression models in R using the censReg package. *R Package Vignettes*, 5:12, 2010.

72 Harvey J Motulsky and Lennart A Ransnas. Fitting curves to data using nonlinear regression: A practical and nonmathematical review. *The FASEB Journal*, 1(5):365–374, 1987.

73 Yukie Miyamoto-Shinohara, Takashi Imaizumi, Junji Sukenobe, Yukie Murakami, Sugio Kawamura, and Yasuhiko Komatsu. Survival rate of microbes after freeze-drying and long-term storage. *Cryobiology*, 41(3):251–255, 2000.

74 W D Grant. Life at low water activity. *Philosophical Transactions of the Royal Society of London. Series B: Biological Sciences*, 359(1448):1249–1267, August 2004.

75 Malcolm Potfis. Desiccation tolerance of prokaryotes. *Microbiology Review*, 58(4):755–805, 1994.

76 Larry R Beuchat, Evangelia Komitopoulou, Harry Beckers, Roy P Betts, François Bourdichon, Séamus Fanning, Han M Joosten, and Benno H Ter Kuile. Low-water activity foods: Increased concern as vehicles of foodborne pathogens. *Journal of Food Protection*, 76(1):150–172, January 2013.

77 J E Benbough. Some factors affecting the survival of airborne viruses. *Journal of General Virology*, 10(3):209–220, March 1971.

78 Kaisen Lin and Linsey C Marr. Humidity-dependent decay of viruses, but not bacteria, in aerosols and droplets follows disinfection kinetics. *Environmental Science & Technology*, 54(2):1024–1032, January 2020.

79 Marianita Santiana, Sourish Ghosh, Brian A Ho, Vignesh Rajasekaran, Wen-Li Du, Yael Mutsafi, Dennise A De Jesús-Diaz, Stanislav V Sosnovtsev, Eric A Levenson, Gabriel I Parra, et al. Vesicle-cloaked virus clusters are optimal units for inter-organismal viral transmission. *Cell Host Microbe*, 24(2):208–220, August 2018.

80 Mengyang Zhang, Sourish Ghosh, Mengqiao Li, Nihal Altan-Bonnet, and Danmeng Shuai. Vesicle-cloaked rotavirus clusters are environmentally persistent and resistant to free chlorine disinfection. *Environmental Science & Technology*, 56(12):8475–8484, June 2022.

81 James R Laidler, Jessica A Shugart, Sherry L Cady, Keith S Bahjat, and Kenneth M Stedman. Reversible inactivation and desiccation tolerance of silicified viruses. *Journal of Virology*, 87(24):13927–13929, December 2013.

82 R A Hobday. The open-air factor and infection control. *Journal of Hospital Infection*, 103(1):e23–e24, September 2019.

83 Kara L Nelson, Alexandria B Boehm, Robert J Davies-Colley, Michael C Dodd, Tamar Kohn, Karl G Linden, Yuanyuan Liu, Peter A Maraccini, Kristopher McNeill, William A Mitch, et al. Sunlight-mediated inactivation of health-relevant microorganisms in water: A review of mechanisms and modeling approaches. *Environmental Science: Processes & Impacts*, 20:1089–1122, 2018.

84 Fiona P Brennan, Beatrix W Alsanius, Ana Allende, Catherine M Burgess, Helena Moreira, Gro S Johannessen, Paula M L Castro, Mieke Uyttendaele, Pilar Truchado, and Nicola J Holden. Harnessing agricultural microbiomes for human pathogen control. *ISME Communications*, 2(1):44, May 2022.

85 Tom McMeekin, June Olley, David Ratkowsky, Ross Corkrey, and Tom Ross. Predictive microbiology theory and application: Is it all about rates? *Food Control*, 29(2):290–299, 2013.

86 József Baranyi and Mark L Tamplin. Combase: A common database on microbial responses to food environments†. *Journal of Food Protection*, 67(9):1967–1971, 2004.

87 Alexandria B Boehm, Andrea I Silverman, Alexander Schriewer, and Kelly Goodwin. Systematic review and meta-analysis of decay rates of waterborne mammalian viruses and coliphages in surface waters. *Water Research*, 164(1):114898, 2019.

88 Andrea I Silverman and Alexandria B Boehm. Systematic review and meta-analysis of the persistence of enveloped viruses in environmental waters and wastewater in the absence of disinfectants. *Environmental Science & Technology*, 55(21):14480–14493, 2021.

89 Stephan Noack, Wolfgang Klöden, and Thomas Bley. Modeling synchronous growth of bacterial populations in phased cultivation. *Bioprocess and Biosystems Engineering*, 31(5):435–443, 2008.

90 P F Verhulst. Notice sur la loi que la population suit dans son accroissement. *Correspondance Mathematique et Physique Publiee par A*, 10:113–129, 1838.

91 J V Ross. A note on density dependence in population models. *Ecological Modelling*, 220:3472–3474, 2009.

92 M A Savageau. Growth equations: A general equation and a survey of special cases. *Mathematical Biosciences*, 48:267–278, 1980.

93 A Tsoularis and J Wallace. Analysis of logistic growth models. *Mathematical Biosciences*, 179:21–55, 2002.

94 Jianlong Wang. The Gompertz model and its applications in microbial growth and bioproduction kinetics: Past, present and future. *Biotechnology Advances*, 72:108335, 2024.

95 M H Zwietering, I Jongenburger, F M Rombouts, and K Van't Riet. Modeling of the bacterial growth curve. *Applied and Environmental Microbiology*, 56:1875–1881, 1990.

96 Charles N Haas, Joan B Rose, and Charles P Gerba. *Quantitative Microbial Risk Assessment*. John Wiley & Sons, Hoboken, NJ, 2nd edition, 2014.

97 J Baranyi, T A Roberts, and P McClure. A non-autonomous differential equation to model bacterial growth. *Food Microbiology*, 10(1):43–59, 1993.

98 T B Young, D F Bruley, and H R Bungay III. A dynamic mathematical model of the chemostat. *Biotechnology and Bioengineering*, 12(5):747–769, 1970.

99 J B Rodin, G K Lyberatos, and S A Svoronos. A simple model to describe transient differences between cell number and biomass growth rates of Escherichia coli. *Biotechnology and Bioengineering*, 37:127–132, 1991.

100 F M Williams. A model of cell growth dynamics. *Journal of Theoretical Biology*, 15(2):190–207, May 1967.

101 A A Esener, T Veerman, J A Roels, and N W F Kossen. Modeling of bacterial growth; Formulation and evaluation of a structured model. *Biotechnology and Bioengineering*, 24(8):1749–1764, August 1982.

102 J Monod. The growth of bacterial cultures. *Annual Review of Microbiology*, 3:371–394, 1949.

103 D E Steinmeyer and M L Shuler. Structured model for *Saccharomyces cerevisiae*. *Chemical Engineering Science*, 44(9):2017–2030, 1989.

104 J Vanimpe, F Poschet, A Geeraerd, and K Vereecken. Towards a novel class of predictive microbial growth models. *International Journal of Food Microbiology*, 100(1-3):97–105, April 2005.

105 Kristel Bernaerts, Els Dens, Karen Vereecken, Annemie H Geeraerd, Arnout R Standaert, Frank Devlieghere, Johan Debevere, and Jan F Van Impe. Concepts and tools for predictive modeling of microbial dynamics. *Journal of Food Protection*, 67(9):2041–2052, September 2004.

106 F Poschet, K M Vereecken, A H Geeraerd, B M Nicolaï, and J F Van Impe. Analysis of a novel class of predictive microbial growth models and application to coculture growth. *International Journal of Food Microbiology*, 100(1):107–124, 2005.

107 Ramaiah Sachidanandham and Karina Yew-Hoong Gin. A dormancy state in nonspore-forming bacteria. *Applied Microbiology and Biotechnology*, 81(5):927–941, 2009.

108 Bevelynn Williams, Jamie Paterson, Helena J Rawsthorne-Manning, Polly-Anne Jeffrey, Joseph J Gillard, Grant Lythe, Thomas R Laws, and Martín López-García. Quantifying *in vitro B. anthracis* growth and pa production and decay: A mathematical modelling approach. *npj Systems Biology and Applications*, 10(1):33, 2024.

109 Peter Armitage, G G Meynell, and Trevor Williams. Birth–death and other models for microbial infection. *Nature*, 207(4997):570–572, August 1965.

110 Trevor Williams. The distribution of response times in a birth-death process. *Biometrika*, 52(3/4):581, December 1965.

111 Randal Douc, Eric Moulines, Pierre Priouret, and Philippe Soulier. *Markov Chains: Basic Definitions*, pages 3–25. Springer International Publishing, Cham, 2018.

112 Thomas E Gorochowski, Antoni Matyjaszkiewicz, Thomas Todd, Neeraj Oak, Kira Kowalska, Stephen Reid, Krasimira T Tsaneva-Atanasova, Nigel J Savery, Claire S Grierson, and Mario Di Bernardo. BSim: An agent-based tool for modeling bacterial populations in systems and synthetic biology. *PLoS ONE*, 7(8):e42790, August 2012.

113 Ferdi L Hellweger, Robert J Clegg, James R Clark, Caroline M Plugge, and Jan-Ulrich Kreft. Advancing microbial sciences by individual-based modelling. *Nature Reviews Microbiology*, 14(7):461–471, July 2016.

114 Jack A Gilbert and Erica M Hartmann. The indoors microbiome and human health. *Nature Reviews Microbiology*, 22(12):742–755, December 2024.

115 Matthew F Blair, Emily Garner, Pan Ji, and Amy Pruden. What is the difference between conventional drinking water, potable reuse water, and nonpotable reuse water? A microbiome perspective. *Environmental Science & Technology*, 58(38):16877–16890, September 2024.

116 Yvan Moënne-Loccoz, Patrick Mavingui, Claude Combes, Philippe Normand, and Christian Steinberg. Microorganisms and biotic interactions. In Jean-Claude Bertrand, Pierre Caumette, Philippe Lebaron, Robert Matheron, Philippe Normand, and Télesphore Sime-Ngando, editors, *Environmental Microbiology: Fundamentals and Applications: Microbial Ecology*, pages 395–444. Springer Netherlands, Dordrecht, 2015.

117 N J Ashbolt. Environmental (saprozoic) pathogens of engineered water systems: Understanding their ecology for risk assessment and management. *Pathogens*, 4(2):390–405, June 2015.

118 Nicholas John Ashbolt. Conceptual model to inform Legionella–amoebae control, including the roles of extracellular vesicles in engineered water system infections. *Frontiers in Cellular and Infection Microbiology*, 18(13):1200478, May 2023.

119 Diplina Paul, Praveen Kolar, and Steven G Hall. A review of the impact of environmental factors on the fate and transport of coronaviruses in aqueous environments. *npj Clean Water*, 4(1):7, 2021.

120 Edouard Jurkevitch. The ecology of *Bdellovibrio* and like organisms in wastewater treatment plants. In Edouard Jurkevitch and Robert J Mitchell, editors, *The Ecology of Predation at the Microscale*, pages 37–64. Springer International Publishing, Cham, 2020.

121 Neda Jafarian, Abbas Akhavan Sepahi, Nafiseh Sadat Naghavi, Farzaneh Hosseini, and Jamileh Nowroozi. Using autochthonous Bdellovibrio as a predatory bacterium for reduction of Gram-negative pathogenic bacteria in urban wastewater and reuse it. *Iranian Journal of Microbiology*, 12(6): 556–564, Dec 2020.

122 Charles N Haas. Application of predator–prey models to disinfection. *Journal of the Water Pollution Control Federation*, 53(3):378–386, 1981.

123 Alfred J Lotka. Analytical note on certain rhythmic relations in organic systems. *Proceedings of the National Academy of Sciences*, 6(7):410–415, 1920.

124 Vito Volterra. Variations and fluctuations of the number of individuals in animal species living together. *ICES Journal of Marine Science*, 3(1):3–51, 1928.

125 Jacob D Davis, Daniel V Olivença, Sam P Brown, and Eberhard O Voit. Methods of quantifying interactions among populations using Lotka-Volterra models. *Frontiers in Systems Biology*, 2:1021897, 2022.

126 Lars R Bakken and Rolf A Olsen. Buoyant densities and dry-matter contents of microorganisms: Conversion of a measured biovolume into biomass. *Applied and Environmental Microbiology*, 45(4):1188–1195, 1983.

127 G Siegl and G G Frösner. Characterization and classification of virus particles associated with hepatitis ai size, density, and sedimentation. *Journal of Virology*, 26(1):40–47, 1978.

128 Felicitas Pfeifer. Distribution, formation and regulation of gas vesicles. *Nature Reviews Microbiology*, 10(10):705–715, 2012.

129 Walter R A Goossens. Review of the empirical correlations for the drag coefficient of rigid spheres. *Powder Technology*, 352:350–359, June 2019.

130 Efstathios E Michaelides and Zhigang Feng. Review—Drag coefficients of non-spherical and irregularly shaped particles. *Journal of Fluids Engineering*, 145(6):060801, June 2023.

131 David Leith. Drag on nonspherical objects. *Aerosol Science and Technology*, 6(2):153–161, January 1987.

132 Claure N Lunardi, Anderson J Gomes, Fellipy S Rocha, Jacopo De Tommaso, and Gregory S Patience. Experimental methods in chemical engineering: Zeta potential. *The Canadian Journal of Chemical Engineering*, 99(3):627–639, March 2021.

133 H C van der Mei and H J Busscher. Electrophoretic mobility distributions of single-strain microbial populations. *Applied and Environmental Microbiology*, 67(2):491–494, Feb 2001.

134 Gediminas Mainelis, Klaus Willeke, Paul Baron, Sergey A Grinshpun, and Tiina Reponen. Induction charging and electrostatic classification of micrometer-size particles for investigating the electrobiological properties of airborne microorganisms. *Aerosol Science and Technology*, 36(4):479–491, 2002.

135 Jonathan Popovici, Colin P White, Jill Hoelle, Brian K Kinkle, and Darren A Lytle. Characterization of the cell surface properties of drinking water pathogens by microbial adhesion to hydrocarbon and electrophoretic mobility measurements. *Colloids and Surfaces B: Biointerfaces*, 118(1 June):126–132, 2014.

136 James W Swan and Eric M Furst. A simpler expression for Henry's function describing the electrophoretic mobility of spherical colloids. *Journal of Colloid and Interface Science*, 388(1):92–94, 2012.

137 Virginia P Harden and John O Harris. The isoelectric point of bacterial cells. *Journal of bacteriology*, 65(2):198–202, 1953.

138 B Michen and T Graule. Isoelectric points of viruses. *Journal of Applied Microbiology*, 109(2):388–397, August 2010.

139 Horiba Scientific. Isoelectric point determination an195. web, ND.

140 Chip Appel, Lena Q Ma, R Dean Rhue, and Elizabeth Kennelley. Point of zero charge determination in soils and minerals via traditional methods and detection of electroacoustic mobility. *Geoderma*, 113(1):77–93, 2003.

141 Caroline L Monteil and Christopher T Lefevre. Magnetoreception in microorganisms. *Trends in Microbiology*, 28(4):266–275, 2020.

142 George Antonoff. What is a phase? *Journal of Chemical Education*, 21(4):195, 1944.

143 W W Nazaroff. Indoor bioaerosol dynamics. *Indoor Air*, 26(1):61–78, 2014.

144 Eric Loth and Jeffrey S Marshall. Restitution coefficient models for collisions of airborne particles and drops. *Journal of Aerosol Science*, 173:106186, 2023.

145 Christopher P Weis, Anthony J Intrepido, Aubrey K Miller, Patricia G Cowin, Mark A Durno, Joan S Gebhardt, and Robert Bull. Secondary aerosolization of viable *Bacillus anthracis* spores in a contaminated us senate office. *JAMA*, 288(22):2853–2858, December 2002.

146 Jing Qian, Jordan Peccia, and Andrea R Ferro. Walking-induced particle resuspension in indoor environments. *Atmospheric Environment*, 89:464–481, 2014.

147 Tianren Wu, Manjie Fu, Maria Valkonen, Martin Täubel, Ying Xu, and Brandon E Boor. Particle resuspension dynamics in the infant near-floor microenvironment. *Environmental Science & Technology*, 55(3):1864–1875, February 2021.

148 Heidi K Hyytiäinen, Balamuralikrishna Jayaprakash, Pirkka V Kirjavainen, Sampo E Saari, Rauno Holopainen, Jorma Keskinen, Kaarle Hämeri, Anne Hyvärinen, Brandon E Boor, and Martin Täubel. Crawling-induced floor dust resuspension affects the microbiota of the infant breathing zone. *Microbiome*, 6(25):1–12, 2018.

149 D Licina and W W Nazaroff. Clothing as a transport vector for airborne particles: Chamber study. *Indoor Air*, 28(3):404–414, May 2018.

150 Jacob Kvasnicka, Elaine A Cohen Hubal, Jeffrey A Siegel, James A Scott, and Miriam L Diamond. Modeling clothing as a vector for transporting airborne particles and pathogens across indoor microenvironments. *Environmental Science & Technology*, 56(9):5641–5652, 2022.

151 Julie A Layshock, Brooke Pearson, Kathryn Crockett, Michael J Brown, Sheila Van Cuyk, W Brent Daniel, and Kristin M Omberg. Reaerosolization of *Bacillus* spp. in outdoor environments: A review of the experimental literature. *Biosecurity and Bioterrorism: Biodefense Strategy, Practice, and Science*, 10(3): 299–303, 2012.

152 Jason P Field, Jayne Belnap, David D Breshears, Jason C Neff, Gregory S Okin, Jeffrey J Whicker, Thomas H Painter, Sujith Ravi, Marith C Reheis, and Richard L Reynolds. The ecology of dust. *Frontiers in Ecology and the Environment*, 8(8):423–430, 2010.

153 H Hollis Wickman. Deposition, adhesion, and release of bioaerosols. In *Atmospheric Microbial Aerosols: Theory and Applications*, pages 99–165. Springer, 1994.

154 Marek Korcz, Janina Fudała, and Czesław Kliś. Estimation of wind blown dust emissions in Europe and its vicinity. *Atmospheric Environment*, 43(7):1410–1420, 2009.

155 Michael A Jahne, Shane W Rogers, Thomas M Holsen, Stefan J Grimberg, and Ivan P Ramler. Emission and dispersion of bioaerosols from dairy manure application sites: Human health risk assessment. *Environmental Science & Technology*, 49(16):9842–9849, 2015.

156 Rob C Jamieson, Douglas M Joy, H Lee, R Kostaschuk, and Robert J Gordon. Resuspension of sediment-associated *Escherichia coli* in a natural stream. *Journal of Environmental Quality*, 34(2):581–589, 2005.

157 G Ziskind, M Fichman, and C Gutfinger. Resuspension of particulates from surfaces to turbulent flows—review and analysis. *Journal of Aerosol Science*, 26(4):613–644, 1995.

158 R J Lai and O H Shemdin. Laboratory study of the generation of spray over water. *Journal of Geophysical Research (1896-1977)*, 79(21):3055–3063, 1974.

159 Michael Cunliffe, Anja Engel, Sanja Frka, Blaženka Gašparović, Carlos Guitart, J Colin Murrell, Matthew Salter, Christian Stolle, Robert Upstill-Goddard, and Oliver Wurl. Sea surface microlayers: A unified physicochemical and biological perspective of the air–ocean interface. *Progress in Oceanography*, 109(2):104–116, February 2013.

160 Janina Rahlff. The virioneuston: A review on viral–bacterial associations at air–water interfaces. *Viruses*, 11(2):191, February 2019.

161 Guillaume Bastin, Christophe Gantzer, and Guillaume Sautrey. New method to quantify hydrophobicity of non-enveloped virions in aqueous media by capillary zone electrophoresis. *Virology*, 568:23–30, 2022.

162 Anke Schafer, Hauke Harms, and Alexander J B Zehnder. Bacterial accumulation at the air water interface. *Environmental Science & Technology*, 32(23):3704–3712, 1998.

163 Matthew A Pendergraft, Derek J Grimes, Sarah N Giddings, Falk Feddersen, Charlotte M Beall, Christopher Lee, Mitchell V Santander, and Kimberly A Prather. Airborne transmission pathway for coastal water pollution. *PeerJ*, 9:e11358, June 2021.

164 Matthew A Pendergraft, Pedro Belda-Ferre, Daniel Petras, Clare K Morris, Brock A Mitts, Allegra T Aron, MacKenzie Bryant, Tara Schwartz, Gail Ackermann, Greg Humphrey, et al. Bacterial and chemical evidence of coastal water pollution from the Tijuana River in sea spray aerosol. *Environmental Science & Technology*, 57(10):4071–4081, March 2023.

165 Afeefa A Abdool-Ghany, James S Klaus, Luis E Sosa Villegas, Trent D'Alessio, Maribeth L Gidley, Christopher D Sinigalliano, Cassandra Gaston, and Helena M Solo-Gabriele. Microbial communities in the water surface microlayer and associations with microbes in aerosols, beach sand, and bulk water. *FEMS Microbiology Ecology*, 99(5):fiad039, April 2023.

166 Joby Jacob, Ingrid Veras, Olga Calderon, Holly A Porter-Morgan, Joshua Tan, Harry E Aguilar, Willis T Elkins, Veronica P Martinez Castro, Vania Fulton, and Wesam K Yousri. Possibly pathogenic bacteria in aerosols and foams as a result of aeration remediation in a polluted urban waterway. *Folia Microbiologica*, 69(1):235–246, February 2024.

167 N S Holmes and L Morawska. A review of dispersion modelling and its application to the dispersion of particles: An overview of different dispersion models available. *Atmospheric Environment*, 40(30):5902–5928, September 2006.

168 Irene Korsakissok and Vivien Mallet. Comparative study of Gaussian dispersion formulas within the Polyphemus platform: Evaluation with Prairie Grass and Kincaid experiments. *Journal of Applied Meteorology and Climatology*, 48(12):2459–2473, 2009.

169 F Pasquill. The estimation of the dispersion of windborne material. *Meteorological Magazine*, 90:33–49, 1961.

170 J H Seinfeld and S N Pandis. *Atmospheric Chemistry and Physics: From Air Pollution to Climate Change*. John Wiley & Sons, 2006.

171 Matthew Meselson, Jeanne Guillemin, Martin Hugh-Jones, Alexander Langmuir, Ilona Popova, Alexis Shelokov, and Olga Yampolskaya. The Sverdlovsk anthrax outbreak of 1979. *Science*, 266(5188):1202–1208, 1994.

172 United States Environmental Protection Agency. Air quality dispersion modeling. https://www.epa.gov/scram/air-quality-dispersion-modeling, November 2024.

173 J P G. Van Leuken, A N Swart, A H Havelaar, A Van Pul, W Van der Hoek, and D Heederik. Atmospheric dispersion modelling of bioaerosols that are pathogenic to humans and livestock – A review to inform risk assessment studies. *Microbial Risk Analysis*, 1:19–39, January 2016.

174 Hani M Mohamed, Lindsay Zielinski, Abdoulaye Diedhiou, Nakia Clemmons, Jessica C Smith, Jessica L Rinsky, Troy Ritter, Melisa Willby, Nancy Burton, Karl Feldmann, et al. Cluster of Legionellosis cases associated with manufacturing process, South Carolina, USA, 2022. *Emerging Infectious Diseases*, 31(1):41–49, January 2025.

175 Kerry A Hamilton, Aditya Kuppravalli, Ashley Heida, Sayalee Joshi, Charles N Haas, Marc Verhougstraete, and Daniel Gerrity. Legionnaires' disease in dental offices: Quantifying aerosol risks to dental workers and patients. *Journal of Occupational and Environmental Hygiene*, 18(8):1–16, August 2021.

176 John P Crimaldi, Aaron C True, Karl G Linden, Mark T Hernandez, Lars T Larson, and Anna K Pauls. Commercial toilets emit energetic and rapidly spreading aerosol plumes. *Scientific Reports*, 12(1):20493, 2022.

177 Ciara A Higham, Martín López-García, Catherine J Noakes, Emma Tidswell, and Louise Fletcher. A quantitative microbial risk assessment (QMRA) framework for exposure from toilet flushing using experimental aerosol concentration measurements. *Indoor Environments*, 2(1):100069, 2025.

178 Kerry A Hamilton, Mark T Hamilton, William Johnson, Patrick Jjemba, Zia Bukhari, Mark LeChevallier, Charles N Haas, and P L Gurian. Risk-based critical concentrations of legionella pneumophila for indoor residential water uses. *Environmental Science & Technology*, 53(8):4528–4541, 2019.

179 Lisa Ng, Dustin Poppendieck, Brian Polidoro, W Stuart Dols, Steven Emmerich, and Andrew Persily. *Single-Zone Simulations Using FaTIMA for Reducing Aerosol Exposure in Educational Spaces*. US Department of Commerce, National Institute of Standards and Technology, 2021.

180 Shujie Yan, Liangzhu (Leon) Wang, Michael J Birnkrant, Zhiqiang (John) Zhai, and Shelly L Miller. Multizone modeling of airborne SARS-CoV-2 quanta transmission and infection mitigation strategies in office, hotel, retail, and school buildings. *Buildings*, 13(1):102, 2022.

181 Matthew R Hipsey and Justin D Brookes. Pathogen management in surface waters: Practical considerations for reducing public health risk. In Alfonso J Rodriguez-Morales, editor, *Current Topics in Public Health*, chapter 21. IntechOpen, Rijeka, 2013.

182 Edward A Birge. The annual address of the president: The thermocline and its biological significance. *Transactions of the American Microscopical Society*, 25:5–33, 1904.

183 Thaddeus K Graczyk, Anna C Majewska, and Kellogg J Schwab. The role of birds in dissemination of human waterborne enteropathogens. *Trends in Parasitology*, 24(2):55–59, 2008.

184 Andrea B Hoyer, S Geoffrey Schladow, and Francisco J Rueda. A hydrodynamics-based approach to evaluating the risk of waterborne pathogens entering drinking water intakes in a large, stratified lake. *Water Research*, 83:227–236, October 2015.

185 Robert Coats, Joaquim Perez-Losada, Geoffrey Schladow, Robert Richards, and Charles Goldman. The warming of Lake Tahoe. *Climatic Change*, 76(1):121–148, 2006.

186 Samira Tolouei, Ray Dewey, William J Snodgrass, Thomas A Edge, Robert C Andrews, Milad Taghipour, Michèle Prévost, and Sarah Dorner. Assessing microbial risk through event-based pathogen loading and hydrodynamic modelling. *Science of The Total Environment*, 693:133567, 2019.

187 Paul Dylan Hynds, M Kate Thomas, and Katarina Dorothy Milena Pintar. Contamination of groundwater systems in the US and Canada by enteric pathogens, 1990-2013: A review and pooled-analysis. *PLoS One*, 9(5): e93301, 2014.

188 Jasen M Kunz, Hannah Lawinger, Shanna Miko, Megan Gerdes, Muhammad Thuneibat, Elizabeth Hannapel, and Virginia A Roberts. Surveillance of waterborne disease outbreaks associated with drinking water — United States, 2015–2020. *Morbidity and Mortality Weekly Reports Surveillance Summaries*, 73(1):1–24, 2024.

189 Randall J Hunt and William P Johnson. Pathogen transport in groundwater systems: Contrasts with traditional solute transport. *Hydrogeology Journal*, 25(4):921–930, 2017.

190 Scott A Bradford, Yusong Wang, Hyunjung Kim, Saeed Torkzaban, and Jiri Šimůnek. Modeling microorganism transport and survival in the subsurface. *Journal of Environmental Quality*, 43(2):421–440, March 2014.

191 Daniel Matthews, Simon Bottrell, Landis Jared West, Louise Maurice, Andrew Farrant, Sarah Purnell, and Danny Coffey. Comparing the fate and transport of MS2 bacteriophage and sodium fluorescein in a karstic chalk aquifer. *Pathogens*, 13(2):168, February 2024.

192 Gopinathan R Abhijith and Avi Ostfeld. Contaminant fate and transport modeling in distribution systems: EPANET-C. *Water*, 14(10):1665, May 2022.

193 P Legnani, E Leoni, S Rapuano, D Turin, and C Valenti. Survival and growth of Pseudomonas aeruginosa in natural mineral water: A 5-year study. *International Journal of Food Microbiology*, 53(2-3):153–158, December 1999.

Part III

Mitigating Exposure

9

Characterizing Interventions

It may be that the source(s) of pathogens to the environment and their transport to receptor(s) result in exposure and an undesirably high resulting risk. The formal quantification of exposure will be discussed in Chapter 13 and the resultant risk will be discussed in Chapter 14. To reduce exposure and risk, interventions are necessary.

9.1 Multiple Barriers

As noted in Figure 3.4, multiple types of interventions exist. The focus here is on source control, environmental interventions, and waste management. Note also the potential importance of personal protective equipment and (not shown) medical prophylactic and post-exposure interventions.

In this chapter, a broad characterization of intervention types will be given, and in Chapters 10–12, details of processes acting primarily by physical removal and by inactivation, respectively, will be discussed.

In general, having multiple barriers or intervention types is beneficial rather than a single strong barrier. This is shown descriptively in Figure 9.1, where each barrier or intervention is depicted as a slice of cheese. While each barrier may be imperfect (the holes in the cheese), as long as they do not all line up (i.e., the performance of the barriers is independent), then there is a reduced chance that threats (pathogens) can pass through to cause a loss (infection or illness in a host receptor). Additionally, given the biological variability between the types of pathogens, having multiple barriers that act by different mechanisms reduces the chance that some yet unrecognized pathogen may pose a significant threat.

If only a single barrier can remove a fraction E_1 of a contaminant, *and that fraction is independent of the concentration of the contaminant*, then the overall proportion remaining is $1 - E_1$. However, if this barrier is effective only f_{on} of the time, and at $1 - f_{\text{on}}$ is completely ineffective, then the effective average proportion remaining is given by Equation 9.1.

$$\bar{E} = (1 - E_1)f_{\text{on}} + (1 - f_{\text{on}}) \tag{9.1}$$

If multiple ("*j*") barriers can each respectively remove E_j fraction of a contaminant, and *if the performance of each is independent of the others*, and each barrier has a probability of being functional $f_{\text{on},j}$, then the overall removal must be a weighted sum considering the likelihood that 1, 2, ... *j* processes are functional. This is best illustrated numerically, as shown in Example 9.1.

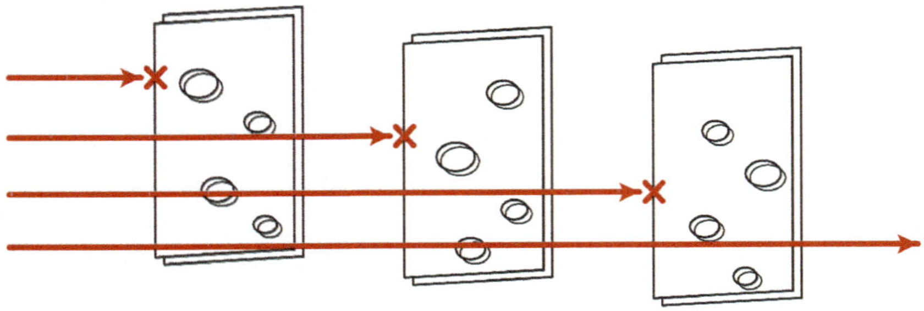

Figure 9.1 Swiss cheese model of multiple barriers.

It should be noted that multiple barriers may be present in a single facility or in a small contiguous space, for example, as in the following cases:

- A ventilation duct contains high-performance filters (e.g., HEPA) followed by in-duct ultraviolet disinfection.
- A water treatment plant may contain pretreatment, sand filtration, and disinfection.
- A facility to process animal manure prior to land disposal may use anaerobic digestion followed by composting.

Example 9.1 Comparison of Single Versus Multiple Barriers

Consider using a single process rather than two processes for pathogen removal. The following table presents the performance in terms of removal and the probability of downtime. Considering downtime, calculate the average removal rate for the single-process system versus the two-process system.

	Removal (E)	p(offline) (f)
Single process	0.99	0.01
Two processes		
Process A	0.92	0.01
Process B	0.95	0.03

Solution:

For the single-process system, Equation 9.1 can be applied directly. Substituting the numbers from the table yields:

$$\bar{E}_{single} = (1 - 0.99)(0.99) + (1 - 0.99)$$

which yields $\bar{E}_{single} = 0.0199$. In other words, any incoming pathogen loading is, on average, reduced to about 2% by the single process.

For the two-process system, the removals and proportions of time the system is in one of four states (both processes functional, both processes not functional, only process A functional, or only process B functional) need to be considered. The calculations are laid out in the following table:

State	Removal (E)	Probability (f)	$(1 - E) \times f$
A and B working	0.996	0.9603	0.0038412
A not B	0.92	0.0297	0.002376
B not A	0.95	0.0097	0.000485
Neither A nor B	0	0.0003	0.0003

In this table, the removal fraction is calculated using Equation 9.1. The probability is computed assuming independence of being offline, i.e.,

$$f_{\text{overall}} = 1 - (1 - f_A)(1 - f_B)$$

The last column is the product of the remaining proportion and the probability, and the sum of this column, which is equal to 0.0070022. Note that this is about one-third of the value for the single-process system—in other words, use of the two-process train, despite the lesser performance of each of the processes, and the poorer (for process B) reliability of one of the components, produces superior overall results. Of course, from a practical point of view, the cost of building and operating the two-process system must also be considered.

One way to compensate for the offline performance of a single process in a multiple-process system is to ensure that sufficient capability to achieve a control objective is available even if the most capable process goes offline.[1] This approach has been assessed in treating wastewater for reuse as drinking water [1].

This analysis describes individual processes as either performing or not performing. This is a conservative assumption, since processes may exhibit variable rather than all-or-none performance. This will be discussed in greater detail in Sections 9.3 and 9.4.

The types of interventions can be categorized in various ways. One important overarching distinction is the use of the terms sterilization, disinfection, and cleaning. The definitions provided by the Centers for Disease Control (CDC) (focused on healthcare applications) are shown as follows. In brief, sterilization is the elimination of all organisms; disinfection is the reduction of pathogens; and cleaning is the removal of debris and contamination that can contain organisms or hinder the action of disinfectants or sterilization processes.

Disinfection, Sterilization, and Cleaning

"**Sterilization** describes a process that destroys or eliminates all forms of microbial life and is carried out in health-care facilities by physical or chemical methods....

(Continued)

1 This approach also assumes that both performance and failure of individual processes are statistically independent. Modification of the analysis for more general assumptions is discussed later. There may be common failure modes, albeit rare; for example, if multiple processes require electric power and there is only one source of electricity, then a loss of electricity is a common failure mode for those processes. Another common failure mode to assess is susceptibility to natural hazards such as floods, tornadoes, hurricanes, or earthquakes.

Disinfection, Sterilization, and Cleaning (Continued)

Disinfection describes a process that eliminates many or all pathogenic microorganisms, except bacterial spores,....

Cleaning is the removal of visible soil (e.g., organic and inorganic material) from objects and surfaces and normally is accomplished manually or mechanically using water with detergents or enzymatic products."

Source: CDC, https://www.cdc.gov/infection-control/hcp/disinfection-sterilization/introduction-methods-definition-of-terms.html.

It is virtually impossible to demonstrate complete sterilization if post-treatment sampling is to be used as a criterion because it is impossible to prove an absolute zero, but only to demonstrate that an intervention may have reduced microorganism levels below a level of detection or quantification, or to have achieved greater than some percentage removal (e.g., 99.99%).

9.2 Typology of Interventions

Broadly speaking, interventions can act primarily by physically removing pathogens (or particles containing pathogens) from a system and/or by inactivating them. The former is discussed in Chapter 10, while the latter is discussed in Chapters 11 and 12. Some processes act by a combination of physical removal and inactivation; for example, using some surface-active cleaning agents on surfaces can result in both the physical removal of pathogens and pathogen-containing material and the inactivation of some organisms.

One issue with physical removal systems is that, inevitably, they will produce a waste stream containing the removed microorganisms, some of which may retain infectivity and therefore may require further treatment before disposal to the environment. This waste stream will depend on the removal process used; for example:

- Cleaning towels and wipes for decontaminating surfaces
- Backwash water from filtering aqueous systems
- Scrubber water from decontaminating air streams

9.3 Quantifying Performance

The examples presented in Section 9.1 introduced the idea of quantifying individual process performance and the performance of systems of processes. One assumption behind that analysis is that the efficiency of each process (E_j) is independent of the microorganism concentration being removed. While this is true for a first-order system, it will not, in general, be true for systems where removal is governed by other kinetics. This is shown in Table 9.1, where both zero-order and second-order removal efficiencies are inversely related to the incoming microorganism concentration.

It is common when referring to the reduction of microorganisms by treatment systems to report them as "logs reduced" or "log reduction." The log reduction value (LRV) of a process or series of processes can be defined by:

$$\mathrm{LRV} = -\log_{10}\left(\frac{N_{\mathrm{out}}}{N_{\mathrm{in}}}\right) \tag{9.2}$$

Table 9.1 Removal expressions ($E = N/N_{in}$) in plug flow reactors (PFRs) and completely stirred tank reactors (CSTRs) for different orders of removal.

Order	Rate expression	PFR performance	CSTR performance
0	$-k_0$	$1 - \frac{k_0\theta}{N_{in}}$	$1 - \frac{k_0\theta}{N_{in}}$
1	$-k_1 N$	$\exp(-k_1\theta)$	$\frac{1}{1 + k_1\theta}$
2	$-k_2 N^2$	$\frac{1}{1 + 2k_2\theta N_{in}}$	$\frac{-1 + \sqrt{8k_2\theta N_{in} + 1}}{4k_2 N_{in}}$

For example, 99% reduction would correspond to two logs reduction, 99.9% to three logs reduction, etc. The word "reduction" is preferred rather than "removal," since the former can encompass both inactivation, as by disinfection processes, and physical removal.

As with microbial concentrations, which are frequently reported as log concentrations, reductions are commonly reported as LRV. This not only provides simpler reporting by avoiding the use of exponential notation but, more importantly, the distribution of reduction values is often skewed, and the log transformation to LRV often reduces this asymmetry.

Using Table 9.1, the performance of a process or set of processes can be assessed for zero-, first-, and second-order systems. This is illustrated in Example 9.2.

Example 9.2 Removal in a PFR with Second-order Kinetics

Consider a process designed as a plug flow system, with microorganism removal described by second-order kinetics. The incoming microorganism concentration is $10\,m^{-3}$. If it is desired to reduce the concentration to $0.1\,m^{-3}$, and the rate constant is $2\,m^3/s$, what should the residence time be?

Solution:

The desired efficiency is given by:

$$E = \frac{0.1}{10} = 10^{-3}$$

This is substituted into the equation for the second-order PFR in Table 9.1 to yield:

$$10^{-3} = \frac{1}{1 + 2(2\,\mathrm{m}^3/\mathrm{s})\theta(10\,\mathrm{m}^{-3})}$$

This can be solved to yield:

$$\theta = 24.975\,\mathrm{s}$$

If there are multiple processes in series, at least one of which is not first-order, then a simple product of efficiencies cannot be used, since at least one process depends on the actual concentration of organisms in its influent. In such a case, the specific concentrations on a process-by-process basis need to be considered.

Example 9.3 A First-order Process Followed by a Second-order Process

Consider two processes in series for pathogen removal. The first process is a PFR characterized by first-order kinetics with a residence time of 120 seconds and a rate constant of $0.01\ s^{-1}$. The second process is a CSTR operating with second-order kinetics, with a residence time of 200 seconds and a rate constant of $0.1\ m^3/s$. If the incoming pathogen contraction to the cascade is 20 organisms per cubic meter, what is the final concentration exiting?

Solution:

Applying the first-order PFR equation, the effluent from the first stage can be computed as follows:

$$\frac{N_1}{20\,\text{m}^{-3}} = \exp(-0.01\,\text{s}^{-1} \times 120\,\text{s})$$

$$N_1 = 6.024\,\text{m}^{-3}$$

N_1 is now used as the influent to the performance equation for the second-order CSTR to yield the final effluent.

$$\frac{N_2}{6.024\,\text{m}^{-3}} = \frac{-1 + \sqrt{8(0.1\,\text{m}^3/\text{s})(120\,\text{s})(6.024\,\text{m}^{-3}) + 1}}{4 \cdot (0.1\,\text{m}^3/\text{s}) \cdot (6.024\,\text{m}^{-3})}$$

$$N_2 = 3.4742\,\text{m}^{-3}$$

Of course, processes may have other performance characteristics than those given by zero-, first-, or second-order kinetics, and also they may be in systems other than an ideal PFR or CSTR. Given the concepts discussed in Chapter 8, the overall mass balances on an individual process can be combined to determine, perhaps numerically, the performance of that process.

As an example, suppose a CSTR is designed to afford time for a Weibull decay process (Table 8.5) to occur. The rate function for the Weibull decay process was derived in Equation 8.61. This can be substituted into a CSTR mass balance to yield the following:

$$N_{\text{in}} - N - kmN\theta\left[-\ln\left(\frac{N}{N_{\text{in}}}\right)\right]^{m-1} = 0 \tag{9.3}$$

Given a fixed value of m, this results in a curve relating $\ln(N/N_{\text{in}})$ to $k\theta$. Since this function is explicit in $k\theta$, the curves are constructed by computing the values of $k\theta$ corresponding to a range of values of N/N_{in} and plotting the locus of points. The results are presented as the curves in Figure 9.2.

9.4 Incorporating Variability and Uncertainty

In Section 3.3, the concepts of uncertainty and variability were defined. In the context of quantifying interventions, both uncertainty and variability can arise. The mechanisms can include the following:

- The removals may be a function of the absolute levels of microorganisms entering the system. This is particularly the case when non-first-order mechanisms for removal occur.
- The other constituents in a material to be treated may alter the performance of a particular process in removing pathogens.

- The external environmental conditions, such as temperature, humidity, wind, etc., may influence the performance of a process or system.
- There may be sampling or laboratory assay issues that add heterogeneity to the measurement of pathogens.
- There may be intrinsic fluctuations in the performance or operation of individual processes.
- For large-scale systems, there may be spatial differences in concentration.
- Mechanical, electrical, or computer control equipment may malfunction, resulting in partial or complete failure of a process or system for some period of time.
- External catastrophic events such as storms or earthquakes may cause temporary or longer-term deviations from performance.

Even from a single process, estimating reductions is more complicated than it may seem. Influents may vary with time; the process may introduce fluctuations and time shifting and smoothing (associated with non-plug flow residence time distributions) as well as reductions as shown schematically in Figure 9.3.

Given this, several approaches characterize removal in a single process. Each has its advantages and disadvantages and depends on the objective of the measurement program.

If the primary objective is to assess average or typical removals, then physical or virtual composites can be used. In a physical composite, actual samples are collected over time into a single pooled sample, which is then analyzed. With virtual composites, a series of discrete samples is taken over time, e.g., perhaps once every 30 minutes during a 24-hour period, and analyzed separately. The

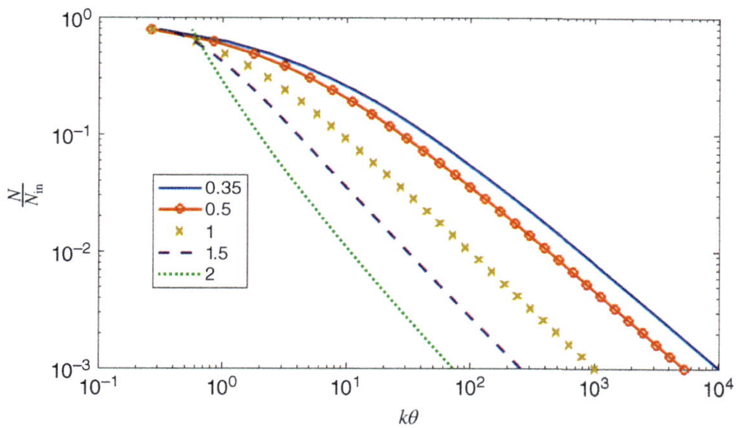

Figure 9.2 Relationship between $k\theta$ and survival ratio for Weibull decay in a CSTR for different values of the exponent "*m*."

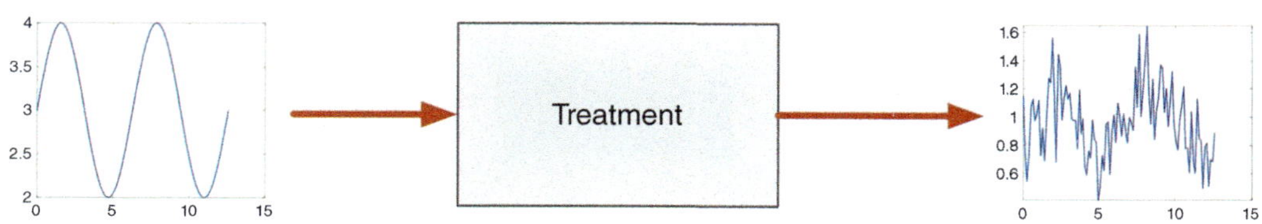

Figure 9.3 Schematic of input and output time series from a process.

results of the individual analyses are then averaged, possibly weighted by a flow rate or other quantitative measure. The advantage of the virtual composite is that individual laboratory analyses of a sample can be completed rapidly, minimizing the opportunity for alteration (decay or growth) of the target population. To obtain the distribution of removals through a process, other sampling techniques must be used.

A second approach is to take paired samples simultaneously (or nearly so). This would be appropriate if the temporal variabilities in the influent and of the process are small; recognizing that there is a time lag inherent to any process. By taking multiple paired samples over time, the system uncertainties can be gauged. This is perhaps the most frequently used strategy. Note, however, that if there are fluctuations in the influent and/or variability in the process, these fluctuations may be underestimated unless the frequency of grab samples was large with respect to the period of system fluctuations.

A third approach is to take paired samples, with the effluent sample from a process time-shifted from the influent sample by a time equal to the mean residence time. This approach, which would be substantially correct for plug flow systems and can be thought of as a set of batch systems (Figure 8.4), would be inexact for systems with more complex and dispersive flow patterns.

A final approach is to take a large number of samples of both incoming and exiting material, ideally randomized over time, and then assess the relationship between the two distributions for a shift associated with reduction.

When removal through a series of processes occurs, there are similar issues with how sampling occurs to obtain measures of typical reductions and their fluctuations.

The output of any measurement program, whether of a single process or through a series of processes, will result in an assessment of LRVs through the system. Assessing the appropriate statistical distributions that characterize such LRVs may be desired.

The various approaches will be illustrated using an example data set. A pilot wastewater treatment plant was studied for almost a year by collecting approximately weekly paired samples of the incoming raw wastewater, secondary effluent, and tertiary effluent, and measuring cysts of *Giardia*. Data are shown in Table 9.2.

Table 9.2 *Giardia* concentrations in a pilot wastewater treatment system (#/L) in raw wastewater, secondary effluent, and tertiary effluent.

Date	Raw wastewater	Secondary effluent	Tertiary effluent
April 9, 2019	6039	19.3	1.66
April 16, 2019	8638	17.2	0.51
April 29, 2019	18,680	39.3	3.28
May 6, 2019	4385	27.7	0.87
May 20, 2019	15,560	10.2	0.83
May 28, 2019	15,280	25.0	1.92
June 10, 2019	2760	20.6	0.80
June 18, 2019	12,611	21.6	2.26
June 24, 2019	11,732	18.2	0.65
July 1, 2019	11,483	19.6	3.51
July 9, 2019	4935	51.4	5.62

Date	Raw wastewater	Secondary effluent	Tertiary effluent
July 23, 2019	6372	85.5	4.77
July 30, 2019	4518	66.8	5.58
August 5, 2019	11,886	23.2	0.84
August 19, 2019	13,523	28.0	0.12
September 23, 2019	8905	8.8	0.73
September 30, 2019	10,679	13.5	0.22
October 7, 2019	3771	12.0	4.26
December 10, 2019	4340	19.4	4.97
January 27, 2020	13,853	55.3	15.22
February 26, 2020	3807	36.3	1.45
March 9, 2020	13,480	22.5	10.96

Table 9.3 Mean and median concentrations of the *Giardia* data from Table 9.2.

		Ln(#/L)	
Statistic	Raw	Secondary	Tertiary
Median	9.185	3.094	0.5808
Mean	9.009	3.196	0.5536

A first approach to analyzing this data is to compute central tendencies directly from Table 9.2. This is shown in Table 9.3. Note that (natural) logarithmic transformations of the observed data are taken, which is a common practice for microbial concentration data since the distribution is expected to be skewed.

From this analysis, several inferences can be made. First, the similarity of mean and median indicates that the logarithmic transformations perform well at making the distribution of raw, secondary, and tertiary wastewater symmetric.[2]

Second, by taking the difference in the mean natural logs and dividing by 2.303, i.e., transforming to $\log_{10}$ values, the LRVs can be determined. From this, the LRV from raw to secondary effluent is 2.524 (99.7% reduction) and from secondary to tertiary effluent is 1.147 (92.8% reduction).

A second, semi-quantitative approach is to plot the cumulative distribution function (CDF) for each data set (columns in Table 9.2). Such plots are shown in Figure 9.4.

Visual inspection of this figure shows that the distribution of secondary effluent is shifted (on the log concentration axis) by a relatively constant amount at all percentiles. However, note that the tertiary effluent distribution is not only shifted but also becomes less steep—corresponding to an increase in the standard deviation. In effect, tertiary treatment superimposes additional variability on the secondary effluent.

The previous two methods ignored the *date* on which samples were taken and also the potential presence of correlations. In addition, without further assumptions, neither of the methods facilitates the computation of confidence limits nor the distributional characteristics of LRVs.

2 If there was significant right-hand skewness, which is not uncommon for *untransformed* microbial data, the mean would be expected to be substantially greater than the median. This is, in fact, the case for the untransformed data in Table 9.2.

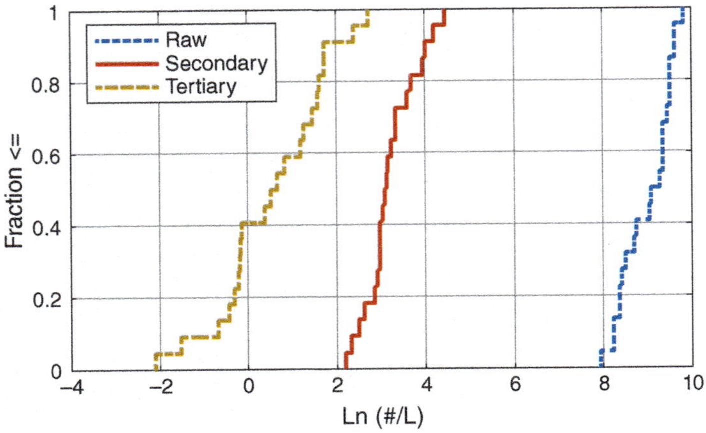

Figure 9.4 CDF for *Giardia* data in Table 9.2.

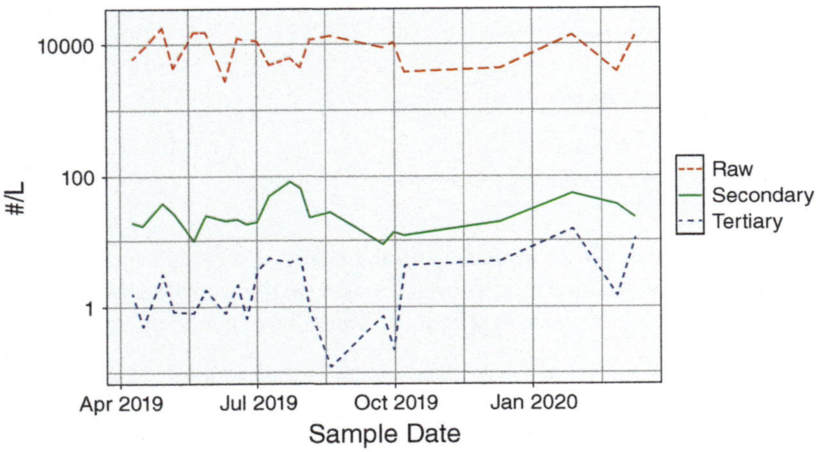

Figure 9.5 Time series plot for *Giardia* data from Table 9.2.

A third approach is to examine the distribution *of changes* between processes on a day-by-day basis with the paired samples. As an initial screening for this, the time series of concentrations is examined. Such a plot is shown in Figure 9.5. Visually, there appears to be a somewhat stronger correlation between the secondary and tertiary samples than between the raw and secondary samples. The calculated Pearson correlation coefficients between log concentrations show a correlation between raw and secondary log *Giardia* concentrations of −0.1314 (which is not statistically significant) and between secondary and tertiary of 0.4813.

This correlation is also seen in the windowpane plot of Figure 9.6. Also note that the raw wastewater appears to be bimodal, and the tertiary wastewater platykurtic, i.e., less peaked than a normal distribution. As a result, there is justification for examining paired differences to assess the LRVs.

Paired LRVs are computed and plotted as the empirical histograms in Figure 9.7. The right-hand panel of that figure shows a scatter plot of the secondary versus tertiary LRVs. There is little evidence of any correlation between the LRVs, and the numerical (Pearson) correlation coefficient is very

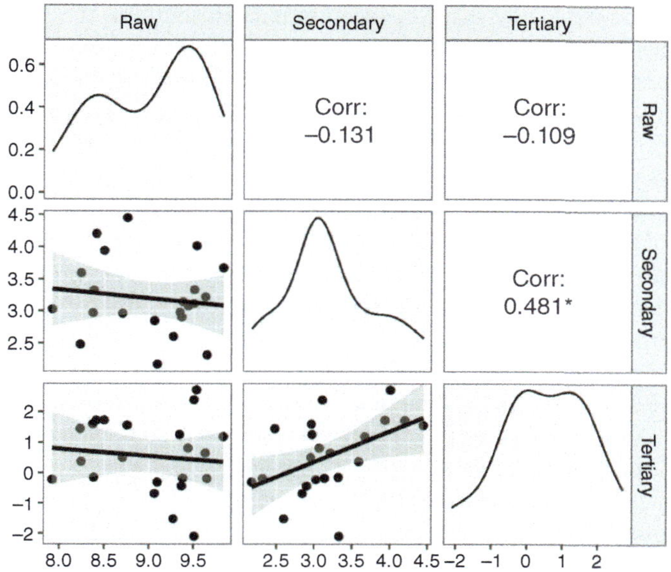

Figure 9.6 Correlation plots and histograms of natural log transformed *Giardia* concentrations for data in Table 9.2.

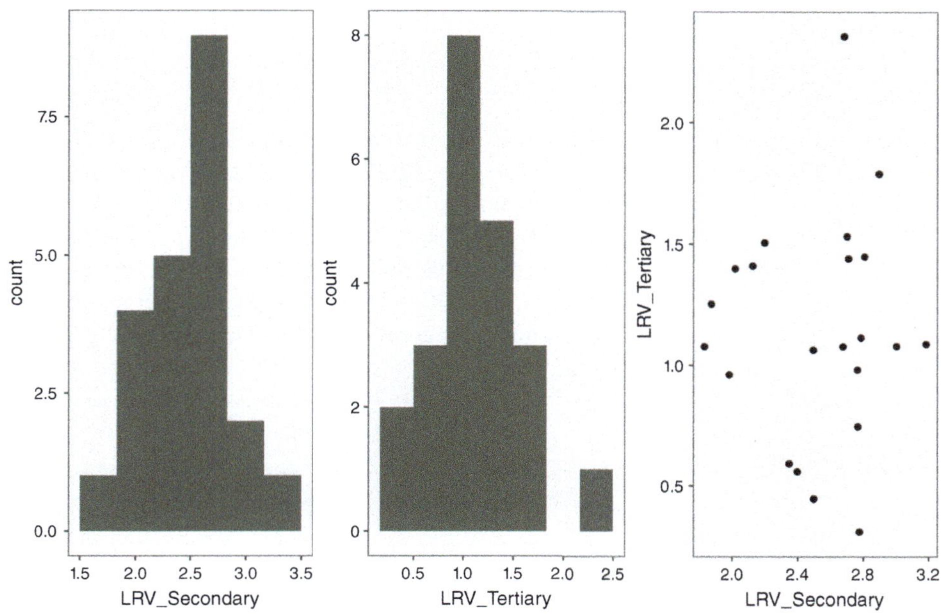

Figure 9.7 Histogram of $\log_{10}$ LRVs for secondary and tertiary *Giardia* removal and correlations. Data from Table 9.2.

small (0.034). Also note that despite the unusually shaped concentration distributions, the LRV distributions appear relatively symmetrical.

While the empirical distributions of LRVs are useful, they are limited in their ability to assess potential extremes due to the small sample size. If it is desired to infer the likelihood of extreme events (e.g., the probability of smaller log reductions), then a fit to a distribution is desirable.

A first step is the test of the adequacy of a normal distribution to fit the LRVs. The Shapiro–Wilk test [2] is widely recognized as the most powerful test for normality. It is implemented in R using the command `shapiro.test`, and in Python as the function `shapiro` in the `scipy.stats` module. Applying this test to the secondary and tertiary LRVs, it is found that the null hypothesis of a normal distribution cannot be neglected (Table 9.4). However, even though the normal distribution cannot be rejected, is it, in fact, the best alternative to use?

Various distributions can be fit to observed data using the method of maximum likelihood. The basic approach described in Section 6.2 is applicable. For the type of data under consideration, the likelihood function of Equation 6.1 can be used. Once the optimum likelihood is computed for a particular distribution, the Bayesian information criterion (BIC) (Equation 6.9) and Akaike information criterion (AIC) (Equation 6.10) can be computed, and the distribution with the smallest value of these criteria is selected as preferable. In R, these computations can be performed expeditiously using the function `fitdist` in the library `fitdistrplus`[3] [3].

A comparison of the fit of a number of alternative distributions to the LRV data is shown in Table 9.5. For the secondary LRV, clearly the Weibull distribution is preferred, while for the tertiary LRV, the Weibull is preferred (although the fits of the normal, gamma, and logistic are quite close). The density function of the Weibull distribution is given in Equation 9.4, where σ is the scale parameter and a is the shape parameter.

$$f(x) = \frac{a}{\sigma}\left(\frac{x}{\sigma}\right)^{a-1} \exp\left[-\left(\frac{x}{\sigma}\right)^{a}\right] \tag{9.4}$$

Although the Weibull appears to be the best fit to the data for both secondary and tertiary LRVs, a separate test should be done to assess whether the fit is sufficient or not—i.e., a test of goodness of fit. This is analogous to the use of the Shapiro–Wilk test for normality used in Table 9.4.

There are three commonly used goodness-of-fit statistics applicable to these types of data, all of which examine the weighted distance between the empirical CDF and the fitted CDF. They differ in the relative weights given to differences in the tails versus the center of the distribution. These statistics are the Anderson–Darling [4], the Kolmogorov–Smirnov [5], and the Cramer–von Mises [6]. All three of these statistics are computed in the `fitdistrplus` library using the function `gofstat`. For the example data, the best-fit parameters of the Weibull distribution and the

Table 9.4 Shapiro–Wilk test of normality of *Giardia* LRVs.

	W statistic	p-value
Secondary	0.939	0.186
Tertiary	0.958	0.458

3 Note that this distribution is one in which all observations are measured. It may be that in some cases, only a less-than or greater-than value may exist, for example, if non-detects after a process occurs. This leads to a censored measurement. The approach used here can be extended to data containing censored distributions, and this R package can analyze such data.

Table 9.5 Summary statistics for fit of $\log_{10}$ *Giardia* reduction values to alternative distributions.

	Secondary			Tertiary		
Distribution	**LogL**	**AIC**	**BIC**	**LogL**	**AIC**	**BIC**
Cauchy	−13.577	31.154	33.33609	−16.39288	36.78575	38.96784
Gamma	−9.850403	23.70081	25.88289	−14.03984	32.07968	34.26176
Log-normal	−10.17399	24.34799	26.53007	−15.15276	34.30551	36.4876
Logistic	−10.03799	24.07599	26.25807	−13.72401	31.44802	33.6301
Normal	−9.36732	22.73464	24.91673	−13.93861	31.87721	34.0593
Weibull	−8.690952	21.3819	23.56399	−13.63164	31.26329	33.44537

Table 9.6 Goodness-of-fit statistics for fit of LRVs from data in Table 9.5 and best-fit parameters of the Weibull distribution.

	Secondary LRV	**Tertiary LRV**
Goodness of fit of Weibull		
Kolmogorov–Smirnov statistic	0.173	0.137
Cramer–von Mises statistic	0.081	0.072
Anderson–Darling statistic	0.462	0.405
Best-fit parameters		
Shape (σ)	8.241	2.691
Scale (a)	2.680	1.290

Note: None of the goodness-of-fit tests are statistically significant.

goodness-of-fit statistics are shown in Table 9.6. In neither case can the null hypothesis of adequate fit be rejected. Therefore, these distributions can be used for further analysis.

9.5 Estimating Performance of Multiple Barriers with Variability and Uncertainty

As noted in Section 9.1, if there are multiple barriers in series, and if they each function with a constant reduction fraction, possibly with an "off-line" fraction of time, then it is straightforward to estimate the overall performance of the cascade. However, if the performance of each process is probabilistic, reflecting variability and uncertainty, a more complex approach is needed. In a general sense, this can be described as in Figure 9.8, where there are distributions, e.g., $f_A(N_A|N_0)$, that describe the effluent probability as a function of the influent. These are then propagated in series to obtain the overall distribution of the product (effluent).

For first-order removals, the individual process distributions become distributions of the ratio of product to input of each process, i.e., $f_A(N_A/N_0)$. This can be related to the ($\log_{10}$) LRV by $f_A = 10^{-\text{LRV}_A}$. Also note that the removal fraction of the process, E_A, can be given by $E_A = 1 - f_A$. Hence, the distribution of LRVs can lead directly to the distribution of f_A or E_A.

With this approach, the distribution of the product/effluent from the cascade can be computed in several ways. Given current computational power, the most straightforward approach to this problem is Monte Carlo. A Monte Carlo method is a method that uses random numbers to solve complex numerical problems [7] and was initially developed as an efficient means to solve (what were then intractable) complex multivariate integrals.

In the context of the problem at hand, and more generically in other problems discussed in this book, the method can be described schematically by Figure 9.9.

The logic of this method proceeds as follows:

1. Each of the inputs is described by a probability distribution, which may be the same or different. There may also be ancillary parameters treated as constants that characterize performance (e.g., reaction rate constants, chemical doses).
2. For each "trial" of the Monte Carlo, a random variate from each of the input distributions is selected.[4]
3. Using those random variates, the effluent/output distribution is calculated using the applicable performance relationship(s). For example, for first-order removal, the overall efficiency can be computed by:

$$E_{\text{total}} = 1 - (1 - E_A) \cdot (1 - E_B) \ldots \tag{9.5}$$

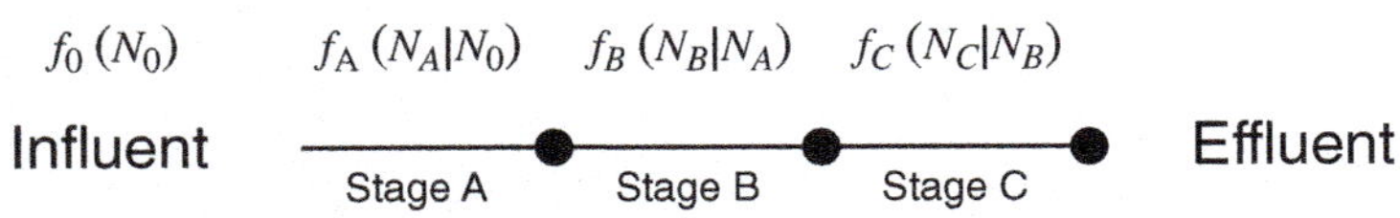

Figure 9.8 Schematic of three-process cascade with probabilistic approach.

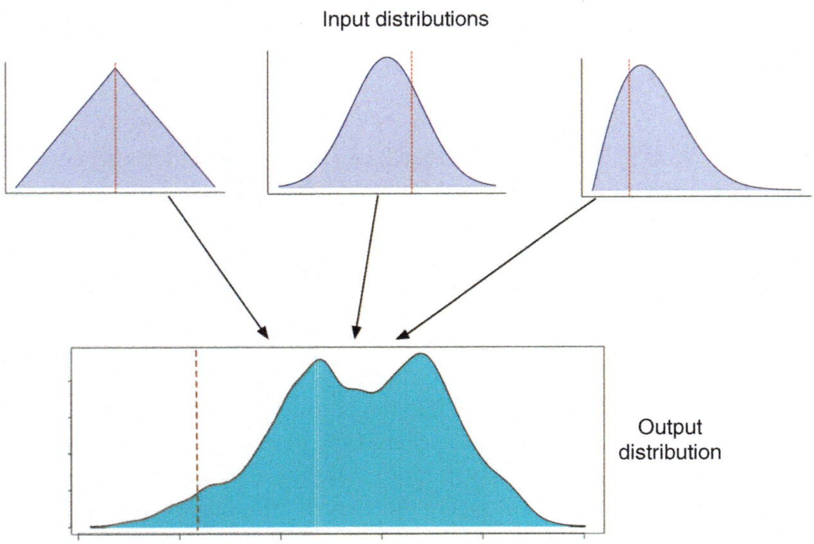

Figure 9.9 Logic flow for Monte Carlo analysis.

4 This approach assumes that each of the inputs is statistically independent of the other inputs. If there are correlations between inputs, then the set of random numbers that involve those correlated inputs needs to be generated jointly. This problem is treated in more detail in Section 13.3.4.

Having defined the concept of LRV for an individual process, with substitutions, Equation 9.5 can be transformed into an additive equation of LRVs, i.e., Equation 9.6:

$$\mathrm{LRV}_{\mathrm{total}} = \mathrm{LRV}_A + \mathrm{LRV}_B \ldots \tag{9.6}$$

4. The calculation is repeated a sufficient number of times to build up the form of the output distribution.

Several implementation details of this approach need to be considered. First, how are the random numbers generated? Second, how many Monte Carlo trials suffice for the problem at hand? These two issues will be considered sequentially.

The selection of points to be sampled for the Monte Carlo trials has most frequently been done using random (actually pseudorandom[5]) numbers. Fundamentally, a univariate distribution can be regarded as a mathematical mapping between <0,1> (the cumulative frequency) and a set of numbers on the real interval. A common, though perhaps not always efficient, way to generate random numbers from a distribution is to draw a random number on the interval <0,1> and compute the corresponding value from the CDF. If two input distributions exist, then a pair of numbers over that interval is necessary. If there are three input distributions, a triplet is required, etc. It is desired that the random points, of whatever the applicable dimensionality, evenly cover the interval, space, volume, or hypervolume.

An alternative approach to the use of pseudorandom numbers is to use algorithms to generate what has been termed "quasi-random" numbers [10]. These, which can also be multidimensional, are designed to cover the entire interval, space, volume, or hypervolume evenly. There have been multiple algorithms designed to achieve this, with the most common being Halton, Sobol, or Hammersley sequences [11]—and these are implemented in R, Python, MATLAB, and other packages.

The difference in pseudorandom versus quasi-random points is shown graphically in an example in Figure 9.10. The left panel, pseudorandom sample, shows "void" areas of coverage, whereas the quasi-random sample shows smaller void spaces. This will manifest itself in the number of samples required for convergence.

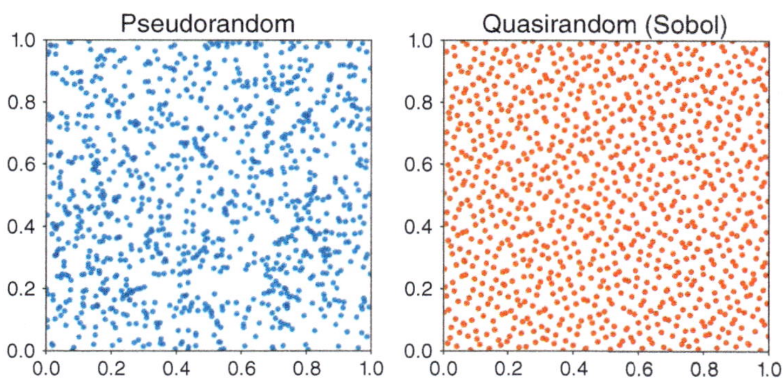

Figure 9.10 Scatter plot of points in two-dimensional space generated by pseudorandom versus Sobol quasi-random algorithms.

---

5 A true random number is one "generated via physical mechanisms such as the timing between successive events in atomic decay, thermal noise in semiconductors, and the like" [8]. In computer simulation, since sequences are defined by (possibly complex) numerically deterministic algorithms given perhaps an initial true random "seed," they are termed pseudorandom numbers. There are many potential algorithms that can be used, and they are evaluated using a complex battery of tests on a variety of properties [9].

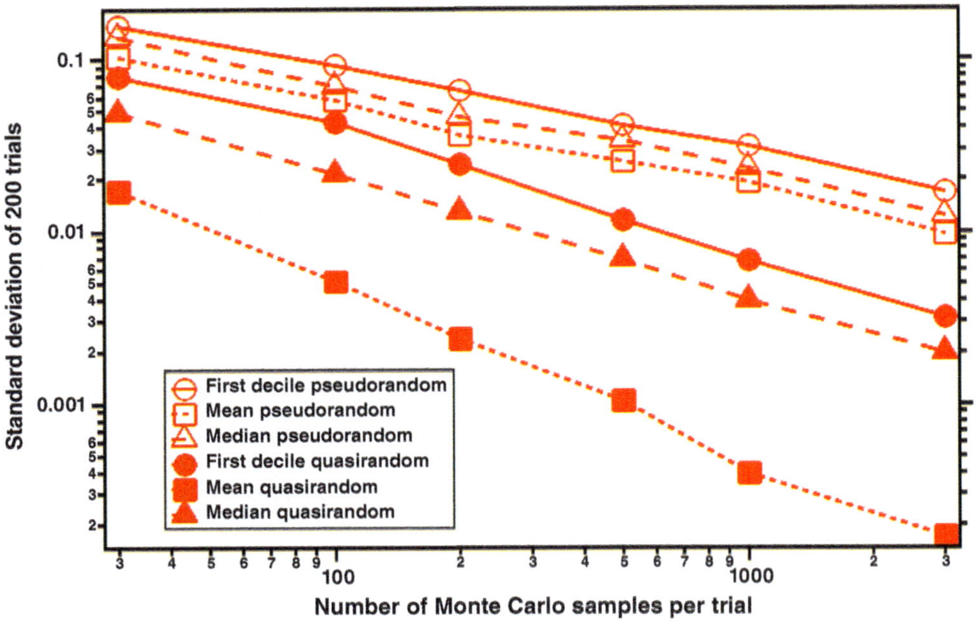

Figure 9.11 Comparison of simulation standard deviations (200 trials) for mean, median, and first decile of the *Giardia* example as a function of number of Monte Carlo replications.

Quasi-random sampling is generally expected to be more efficient and yield smaller standard deviations for estimates than pseudorandom sampling for a given number of trials. Also, in either case, the between-run variability diminishes as the number of samples taken in a Monte Carlo trial increases. Details will vary with the particular problem at hand; however, the concept can be illustrated by use of the *Giardia* example heretofore discussed.

The distribution for LRVs from secondary and tertiary treatment was each shown to be best fit by Weibull distributions (Table 9.5). Using these, with the approach of Figure 9.9, the distribution of the overall LRV of the secondary and tertiary cascade can be found. By repeating this, the distribution (uncertainty) of the LRV for the cascade is determined. If this Monte Carlo process is repeated again and again, the mean, standard deviation, and various percentiles of this LRV distribution are determined. Even with a small number of iterations, the mean converges in a few trials per run.

However, there is substantial run-to-run variability, which decreases systematically with the size of the Monte Carlo sample. This is shown in Figure 9.11. Also shown is that the variability between Monte Carlo runs is significantly lower when quasi-random samples are used than when pseudorandom samples are used. For this analysis, the Sobol sequence approach is used in conjunction with a permutation algorithm, with the calculations implemented in Python.

The numerical code is shown in Listing 9.1.

```
from scipy.stats import qmc
import numpy as np
import pandas as pd

rng = np.random.default_rng()
trials_per_iterate=200  # Number of trials at that MC Sample Size
MCsamplesize=30    # number in Monte Carlo Sample

def weibull_value(p,shape,scale):
    A=-np.log(1-p)
```

```
    return scale*(A**(1/shape))

def output_vector(Ntrials):
    # --- SOBOL ------
    sampler = qmc.Sobol(d=2, scramble=True)
    sample = sampler.random(n=Ntrials)
    p1=sample[:,0]
    p2=sample[:,1]
    w1=weibull_value(p1, 8.241, 2.68)  # Secondary LRV
    w2=weibull_value(p2,2.691,1.290)    # Tertiary LRV
    LRVtotal=w1+w2
    qrmeanLRV=np.mean(LRVtotal)
    qrmedianLRV=np.median(LRVtotal)
    qrLowerDecileLRV=np.percentile(LRVtotal,10)

    # --- Pseudorandom ----
    p1=rng.random(Ntrials)
    p2=rng.random(Ntrials)
    w1=weibull_value(p1, 8.241, 2.68)  # Secondary LRV
    w2=weibull_value(p2,2.691,1.290)    # Tertiary LRV
    LRVtotal=w1+w2
    prmeanLRV=np.mean(LRVtotal)
    prmedianLRV=np.median(LRVtotal)
    prLowerDecileLRV=np.percentile(LRVtotal,10)
    return[ qrmeanLRV,qrmedianLRV,qrLowerDecileLRV,prmeanLRV,prmedianLRV,
    prLowerDecileLRV]

summarytableau=np.empty((13,6))
MonteCarloTrials=[30,100,200,500,1000,3000]
for k in range (0,6):
    MCsamplesize=MonteCarloTrials[k]
    tableau=np.empty((trials_per_iterate,6))
    for i in range(0, trials_per_iterate):
       tableau[i,:]=output_vector(MCsamplesize)
    mean=np.mean(tableau,0)
    sd=np.std(tableau,0)
    A=np.concatenate((mean,sd))
    A=np.insert(A,0,MCsamplesize)
    summarytableau[:,k]=A

Final=np.transpose(summarytableau)   # Transpose Results Matrix
Finaldf=pd.DataFrame(Final,columns=['MC Size','mean qr mean','mean qr median','mean qr
     first decile','mean pr mean','mean pr median','mean pr first decile','sd qr mean'
    ,'sd qr median','sd qr first decile',
'sd pr mean','sd pr median','sd pr first decile'])
```

Listing 9.1 Comparison of pseudorandom versus quasi-random for *Giardia* example.

The example presented here assumes first-order behavior for each process, i.e., the LRV for each process is independent of microorganism concentration. In addition, this analysis assumes that the inputs are uncorrelated: the LRV for process A at a given instant is not correlated with the LRV for process B at that instant. There are several circumstances that can result in correlations between LRV values of individual processes in a cascade, including the following:

- Changes in the characteristics of the stream entering the system may alter the performance of multiple processes.
- The effluent quality from the first process (perhaps for microorganisms or other components in the system) may vary, influencing the second process's performance.

- Factors external to both processes, such as meteorology, power fluctuations or outages, or other unforeseen disasters, may cause coincident fluctuations in the performance of each process.

However, numerical studies have indicated that unless there is a strong correlation (absolute value of the correlation greater than about 0.8) and/or inferences on the extreme tails are needed, the presence of correlations can be neglected [12]. This topic will be discussed further in Section 14.2.

It should be noted that an (older) alternative to characterizing the distribution of a function of several random variables uses the analytical framework of propagation of errors. In this method, each of the inputs is characterized in terms of first moments (i.e., means), second moments (variances and covariances), and higher moments around the mean, and then these moments in the resultant output distribution are calculated using a Taylor series expansion of the functions relating input and output. The detailed theory was described by Ku [13], and specific examples have previously been presented [14]. For example, if w is a function of random variables x and y, then the mean of that function can be given (to the second term of a Taylor series) as:

$$\overline{w(x,y)} \cong w(\overline{x},\overline{y}) + \frac{1}{2}\left\{\frac{\partial^2 w}{\partial x^2}\text{var}(x) + 2\frac{\partial^2 w}{\partial x \partial y}\text{cov}(x,y) + \frac{\partial^2 w}{\partial y^2}\text{var}(y)\right\}\Bigg|_{\overline{x},\overline{y}} \tag{9.7}$$

For the variance of this same derived function, the following is the second-term Taylor expansion:

$$\text{var}(w) = \left(\frac{\partial w}{\partial x}\right)^2 \text{var}(x) + \left(\frac{\partial w}{\partial y}\right)^2 \text{var}(y) + 2\left(\frac{\partial w}{\partial x}\right)\left(\frac{\partial w}{\partial y}\right)\text{cov}(x,y) \tag{9.8}$$

In the absence of correlation in the inputs, then the terms involving covariance in Equations 9.7 and 9.8 vanish. However, this approach has limitations in that only (perhaps higher) moments can be calculated, rather than a full output distribution, and also that the algebraic calculations can be tedious.

9.6 Discussion Questions and Problems

1 Consider the two-process problem in Example 9.1. To relax the process specifications for process A by allowing for a greater probability of out-of-performance time, what would be this value to provide equivalent removal to the single-process system?

2 If three independent and identical processes (in terms of removal fraction and probability of being online) are to be used to provide an overall removal of 99.9%, what should the design performance and off-spec time be (note that multiple solutions are possible)? Assume that having an offline proportion below 1% for any single process is not feasible.

3 Consider the process with kinetics described in Example 8.2. If this is to be conducted in a CSTR, what should the mean residence time be to achieve the same performance objective? This will need to be solved either numerically or graphically.

4 Using Figure 9.2, if Weibull removal in a CSTR characterizes a process, and if a system has a Weibull exponent of 0.5, what value of $k\theta$ is needed to achieve 99% removal?

5 Consider the *Giardia* case discussed in this chapter. From the mean and variances of the LRVs for secondary and tertiary removal, use the Ku approach to compute the mean and variance of the overall LRV for the cascade.

6 Consider the scenario in Example 8.2. Assume that the rate constant mean is as in the example but is distributed as a gamma random variable with a standard deviation of $0.5\ m^3/s$, and that the influent concentration is distributed as a Weibull random variable with a standard deviation of $5\ m^{-3}$. The residence time is as computed in the example; compute the mean and variance of the effluent concentration using the propagation-of-error approach.

7 Repeat the calculation in the previous problem using a Monte Carlo approach. Compare the mean and standard deviation from this calculation to that using the Ku approach, and also compute the upper 95th percentile of the effluent concentration.

References

1 Charles N Haas and R Rhodes Trussell. Frameworks for assessing reliability of multiple, independent barriers in potable water reuse. *Water Science and Technology*, 38(6):1–8, 1998.
2 Nornadiah Mohd Razali and Yap Bee Wah. Power comparisons of Shapiro-Wilk, Kolmogorov-Smirnov, Lilliefors and Anderson-Darling tests. *Journal of Statistical Modeling and Analytics*, 2(1):21–33, 2011.
3 Marie Laure Delignette-Muller and Christophe Dutang. fitdistrplus: An R package for fitting distributions. *Journal of Statistical Software*, 64(4):1–34, 2015.
4 Theodore W Anderson. Anderson-Darling tests of goodness-of-fit. In Midorag Lovric, editor, *International Encyclopedia of Statistical Science*, volume 1, pages 52–54. Springer, Berlin Heidelberg, 2011.
5 Frank J Massey Jr. The Kolmogorov-Smirnov test for goodness of fit. *Journal of the American Statistical Association*, 46(253):68–78, March 1951.
6 J Durbin and M Knott. Components of Cramér–von Mises statistics. I. *Journal of the Royal Statistical Society: Series B (Methodological)*, 34(2):290–307, 1972.
7 W F Bauer. The Monte Carlo method. *Journal of the Society for Industrial and Applied Mathematics*, 6(4):438–451, 1958.
8 Pierre L'Ecuyer. Random number generation, *Working Paper* 2004,21, 2004.
9 Kamalika Bhattacharjee and Sukanta Das. A search for good pseudo-random number generators: Survey and empirical studies. *Computer Science Review*, 45:100471, August 2022.
10 William H Press and Saul A Teukolsky. Quasi-(that is, sub-) random numbers. *Computers in Physics*, 3(6):76–79, 1989.
11 Sushant S Garud, Iftekhar A Karimi, and Markus Kraft. Design of computer experiments: A review. *Computers & Chemical Engineering*, 106:71–95, November 2017.
12 John Bukowski, Leo Korn, and Daniel Wartenberg. Correlated inputs in quantitative risk assessment: The effects of distributional shape. *Risk Analysis*, 15(2):215–219, 1995.
13 H H Ku. Notes on the use of propagation of error formulas. *Journal of Research of the National Bureau of Standards - C: Engineering and Instrumentation*, 70C(4):263–273, 1966.
14 Charles N Haas. Doing engineering calculations. *Electronic*, May 2023.

10

Physical Removal and Reduction

Microorganism occurrence can be reduced in a system through a variety of approaches. One class of approaches involves removing organisms from systems via physical processes. As such, it is not necessarily the properties of microorganisms *per se* that affect removal, but their characteristics as or of the particles they are associated with. Hence, most of the analysis in this chapter recapitulates the approaches used elsewhere in environmental engineering for particle removal.

There may also be coincidental inactivation. The physical removal transfers and potentially concentrates the microorganisms in another medium or venue where they can more readily be managed or disposed of. In this chapter, the physical removal processes will be discussed. In Chapters 11 and 12, processes that principally operate by inactivating microorganisms will be discussed.

As was true in the discussion of fate and transport of organisms (Chapter 8), organisms may be in environmental media alone (or in clumps) or associated with other particles. If they remain associated with such particles, then the adherent microorganisms will also be removed by physically removing those particles (which may be larger, more massive, and/or have other different properties).

10.1 Removal by Action of Gravitational Forces

Microorganisms, or the particles with which they are associated, have a net settling velocity relative to the fluid in which they are suspended if there is a differential density. This velocity is given from the force balances between gravitational, buoyant, and drag forces, as outlined in Section 8.5.1.

Depending on particle density, the sedimentation process may need to be analyzed and designed differently. At relatively low particle densities, individual particles settle independently, without interacting with one another; this is termed discrete or Type 1 sedimentation. At higher particle densities, particle–particle collisions are more frequent, leading to some agglomeration over time. This is termed flocculant or Type 2 sedimentation and occurs via the processes discussed in Section 8.4.1. With Type 2 sedimentation, the effective particle size increases with time, along with the settling velocity. In Type 3 sedimentation, the concentration of solids has increased further to the point where the limitation on additional concentration becomes the ability of the water (or other carrier fluid) to travel against the gravitational force through the pores of the compacting solids. In Type 4, the densest situation, further particle concentration is limited by the mechanical compression of the accumulated solids. The overall scenario is illustrated in Figure 10.1.

10.1.1 Water Applications

In water applications, several common designs are used. Broadly speaking, they can be classified as circular upflow and rectangular horizontal flow. A schematic of a rectangular system is shown in Figure 10.2.

For either an upflow or horizontal sedimentation tank, under ideal conditions with Type 1 settling, the key engineering design variable is the overflow velocity v_o defined by:

$$v_o = \frac{Q}{A_\perp} \tag{10.1}$$

Q is the volumetric flow and $A_\perp$ is the cross-sectional area of the settling tank normal to the gravitational vector.

For both upflow and horizontal-flow ideal systems, particle removal and the microorganisms associated with them are functions of the design variable v_o and the particle settling velocity, v_p.

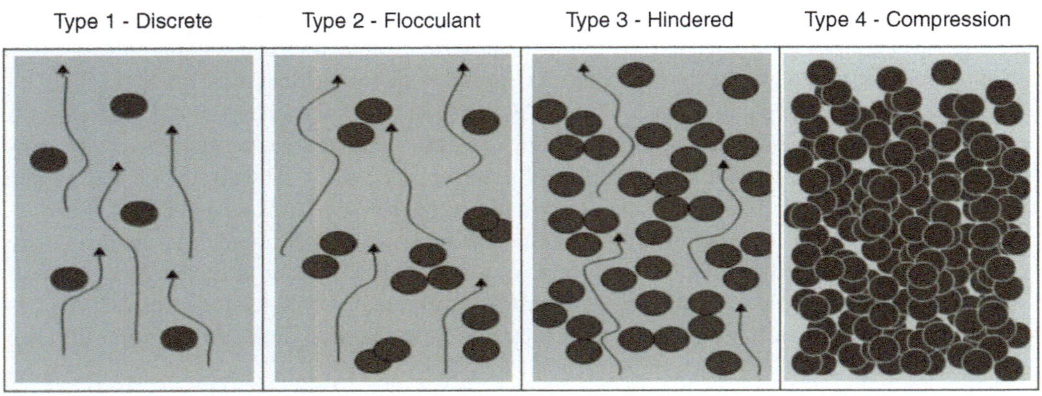

Figure 10.1 Types of settling. *Source:* Shubham Sakhareliya. https://www.slideshare.net/slideshow/sedimentation-238373917/238373917 (last accessed February 14, 2026).

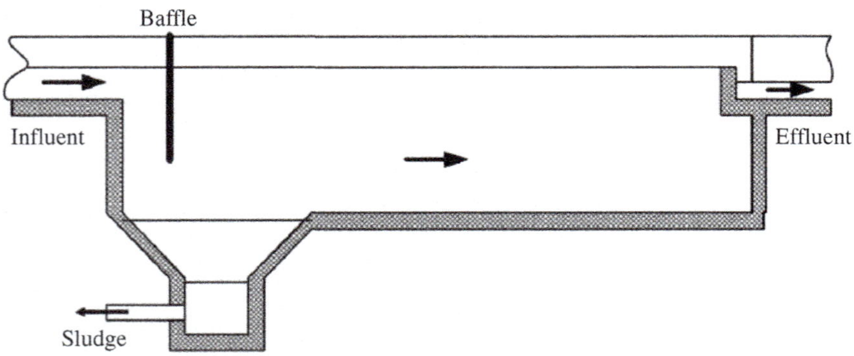

Figure 10.2 Rectangular sedimentation tank. *Source:* Kamal Abdulraheem (2021). Research Square Platform LLC / CC BY 4.0. https://www.researchgate.net/publication/352843587_Numerical_Simulation_of_a_Rectangular_Clarifier_Using_Drift-Flux_Model_in_OpenFOAM (last accessed February 12, 2026).

This latter quantity can be calculated from the particle diameter and density, and the medium density and viscosity using Equation 8.100 at low Reynolds number. For a single particle type (based on density and diameter), termed monodisperse, v_p is a single quantity. If there is a distribution of particle types, there will be a distribution of v_p.

For upflow ideal Type 1 clarifiers, all particles with $v_p > v_o$ are completely removed, and other particles are not removed. This leads to a straightforward design calculation as outlined in Example 10.1.

Example 10.1 Design of Upflow Settler for Monodisperse Particle

What is the minimum area of an upflow clarifier operating under ideal conditions to achieve complete removal of particles with a density of 1.2 g/cm^3 and a diameter (assume perfect sphere) of 100 μm, if the temperature is 25 °C. The flow of liquid to be treated is 100 m^3/d. Type 1 conditions prevail.

Solution:

First, the settling velocity is computed using Stokes law (Equation 8.100). At the stated temperature, the water viscosity is 0.0089 poise $\left(\frac{g}{cms}\right)$. With this information, using Stokes law, the settling velocity of particles is:

$$v_p = 0.00122 \, \frac{\text{m}}{\text{s}}$$

A check that Stokes law is applicable is made by computing the Reynolds number from Equation 8.95:

$$Re = 0.137$$

Since $0.137 < 1.0$, the Stokes law assumption is appropriate.

Now, since Q is given, the minimum area is given by equating v_p to v_o in Equation 10.1 and solving for area:

$$A = \frac{100 \, \text{m}^3/\text{d}}{0.00122 \, \text{m/s}} \times 1.15741 \cdot 10^{-5} \, \frac{\text{d}}{\text{s}}$$

$$A = 0.949 \, \text{m}^2$$

For rectangular settlers, some particles with a sedimentation velocity less than the overflow velocity will still be removed in an ideal settler since a fraction of those particles (assuming uniform entry in the vertical direction) will start closer to the bottom of the tank and thereby be removed. Given this, the fractional removal η can be given as:

$$\eta = \max\left(1.0, \frac{v_p}{v_o}\right) \tag{10.2}$$

In most systems, there is a distribution of particles with different sizes and densities, and as a result, with different settling velocities. There are several means to ascertain this distribution:

- Using calibrated sieves or screens, particles of different effective sizes can be separated, and the mass fraction in each size class determined [1]. In this method, if sampling and separation is done aseptically, the concentration, e.g., number per gram of solids, in each fraction can also be determined.

- A second approach is to place the suspension in a batch settling column, and to withdraw samples over time (the solids remaining at a given time at fixed depth represent those with a lesser settling velocity) to directly determine the complementary cumulative distribution of settling velocities [2]. This can then be used to discretize the solids into bins of settling-time distributions (or, alternatively, to fit the distribution to analytical forms). As in the previous method, if the solids are also analyzed for microorganism concentration, then the distribution of microorganisms with size can be ascertained.

Regardless of the method used to determine the distribution of settling velocities, the result can be used to assess removal in an ideal upflow or horizontal settler. An example for a horizontal settler is given in Example 10.2.

Example 10.2 Horizontal Settler Analysis of a Heterogeneous Suspension

River water is first processed using sedimentation without chemical addition (typically called "plain sedimentation") in a horizontal flow tank. A sample of the water was taken for analysis of the settling velocity distribution and the target pathogen concentration per unit mass of solids, with results shown below. The data result from sampling in different bins, and the average (settling velocity and microorganism concentration) for each bin is tabulated. Use this data to estimate the overflow velocity required to achieve 90% removal of microorganisms, and calculate the corresponding removal of solids.

$\overline{v_{s,i}}\left(\frac{m}{s}\right)$	$m_i\left(\frac{mg}{L}\right)$	$C_i\left(\frac{\#}{mg}\right)$
0.1	1	10
0.3	4	7
1	6	8
3	8	4
10	10	2

Solution:

For each row in the table mentioned earlier, at a given overflow rate, the fractional removal is calculated from Equation 10.2. This is then multiplied by the mass concentration of solids (mass/volume) in each row, or the microbial concentration (#/volume, obtained by multiplying the #/mass solids by the mass solids/volume, then summed up to get the total concentration of solids or the total microorganism concentration remaining. By repeating this calculation for different values of v_o, the graph below shows the fraction of the initial load remaining. From this graph, the overflow velocity that results in a 0.1 fraction of organisms remaining is 0.4 m/s. At that overflow velocity, the fraction of solids remaining is about 0.05, or 95% solids are removed.

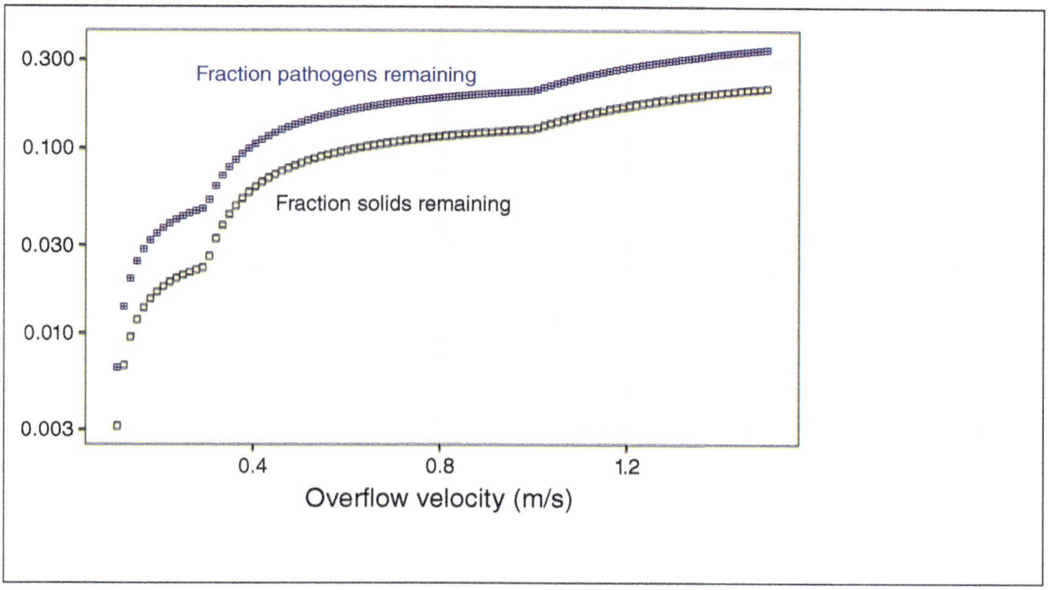

In sedimentation tanks, several factors can lead to nonideal behavior. In water systems, these can include:

- Density currents (from high solids, temperature differentials) can lead to nonideal flow patterns and short-circuiting.
- Wind (especially in rectangular clarifiers with a long "fetch") can cause differential flows at the surface, and flow recirculation.
- Weir hydraulics, particularly in rectangular clarifiers, can result in higher velocities near the outlet of the tank that alter flow patterns.

To more accurately incorporate deviations such as those above, CFD models can be used successfully [3].

10.1.2 Air Applications

Sedimentation of particles in air can be a removal process, as in deposition in rooms. However, as noted in Section 8.4.1, biological containing particles in air may readily shrink in size due to evaporative processes, and hence their effective sedimentation rates will change dynamically. In addition, in the air, the presence of thermal currents, including those from heating and cooling systems, heat-emitting appliances and devices, and people, may influence particle trajectories. Therefore, sedimentation-related removal of particles in air is best handled by CFD [4].

10.1.3 Cyclonic Separators

Closely related to the removal of particles by gravitational sedimentation is the use of devices that remove particles by centrifugal forces, essentially replacing the "mg" force terms in settling with a centrifugal force term, i.e., $m\omega^2 r$.

With air systems, a principal embodiment of centrifugal removal is the use of cyclone separators. These were discussed in the context of air sampling for microorganisms in Section 5.2.4, and a schematic was shown earlier (Figure 5.11).

Such systems can be used to deliberately treat air flows containing suspended particles of concern. A schematic of one type of such systems from [5] is shown in Figure 10.3. These systems are primarily used for removal of coarser (> 10 μm) particles from air.

There are several geometric regions of a cyclone, and the relative proportions affect efficiency. These include:

- Size of the inlet and outlet sections, which will influence the fluid velocity in the system. In addition, the inlet diameter influences the velocity, which in turn affects the number of turns the fluid takes in the system.
- The radii of the top cylindrical section and the conical bottom section, which influence rotational velocity and centrifugal force.
- The relative height of the top cylindrical and the bottom conical sections.

Particle removal occurs by contact with the walls of the device, and the removed material falls to the bottom hopper from which it is periodically removed.

The fluid dynamics within a cyclone separator are complex due to the rotational flow and complex particle trajectories. Often, the design of these systems has proceeded on a semi-empirical basis. A common figure of merit for design is the "cut size," defined as the diameter of particles that are removed by 50% in a device [6].

Cut size can then be used to determine the removal of other particle sizes using an approach as outlined by Rhodes [7]. The starting point of this analysis is the specification of geometric ratios

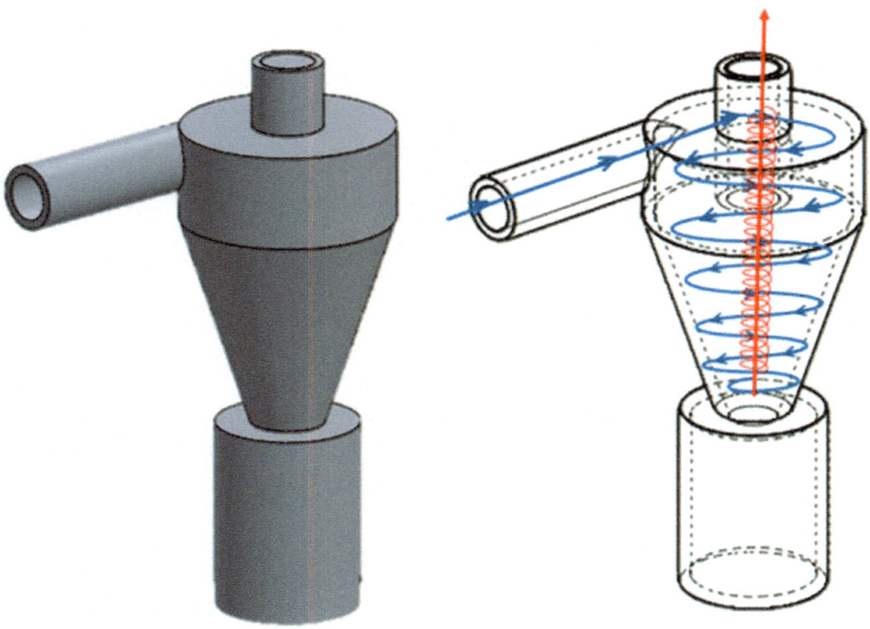

Figure 10.3 Cyclone separator for air treatment. *Source:* [5]/MDPI/CC BY 4.0.

for the cyclone. Let D be the diameter of the upper cylindrical throat of the cyclone. Then, the following ratios to D can be defined:

A Total height of cylindrical and conical sections (A = B + C)
B Height of the conical section
C Height of the cylindrical section
E Diameter of the cone at its lowest narrow point
J Length of the projection of the outlet tube into the body of the cylinder
L Length of the projection of the inlet tube into the body of the cylinder
K Diameter of the inlet tube
N Diameter of the outlet tube

Two important dimensionless numbers have been used to characterize cyclone performance. The Euler number *Eu* relates pressure drop to nominal gas velocity in the cyclone according to Equations 10.3 and 10.4.

$$v_{cyc} = \frac{4Q}{\pi D^2} \tag{10.3}$$

$$Eu = \frac{2\Delta p}{\rho_{air} v_{cyc}^2} \tag{10.4}$$

The Stokes number for the cut diameter x_{50} particle Stk_{50} arises from consideration of centrifugal forces on the particle and can be defined by Equation 10.5:

$$Stk_{50} = \frac{x_{50}^2 \rho_s v_{cyc}}{18\mu D} \tag{10.5}$$

Different values of the Stokes and Euler numbers will occur for different cyclone designs. A tabulation of the geometric factors and these dimensionless numbers is presented in Table 10.1.

Zhao and Su [8] have attempted to produce a generalized model for cut size for cyclones of different geometric characteristics by developing a correlation for Stk_{50} in terms of internal Reynolds number and geometric ratios. This allows generalization to cyclones of different characteristics.

Table 10.1 Geometric ratios and Euler and Stokes numbers for two common cyclone designs.

	Stairmand High efficiency	Stairmand high rate
A	4	4
B	2.5	2.5
C	1.5	1.5
E	0.375	0.575
J	0.5	0.875
L	0.2	0.375
K	0.5	0.75
N	0.5	0.75
Eu	320	46
Stk_{50}	0.00014	0.006

Source: Adapted from [7].

A semi-empirical relationship, termed the grade efficiency curve, can be computed from cut size. If E is the fractional removal of particles of diameter x, this can be expressed via Equation 10.6, which is generally termed the Lapple model.

$$E = \frac{(d/d_{50})^2}{1 + (d/d_{50})^2} \tag{10.6}$$

However, the Lapple model is not the only model for grade-efficiency curves as a function of cut size. At least seven such models have been presented [9], which yield different predictions for size-specific removals. As noted by Rhodes [7]: "Very little is known about how the shape of the grade efficiency curve is affected by operating pressure drop, cyclone size or design, and feed solids concentration." Furthermore, with particular relevance to particles containing microorganisms, issues such as particle density, changes in diameter due to evaporation, ambient temperature and humidity, and possible flocculation have not been incorporated into such models. Due to these limitations, there is a growing body of literature on the application of computational fluid dynamics to the analysis of particle removal and pressure drop in cyclones [10].

Cyclones are used in the water sector [11], where they are often referred to as swirl concentrators. They can be used in drinking water treatment for initial removal of coarse sand or sediment prior to further treatment, or in wastewater or stormwater treatment for removal of similar particles (in this latter context, they are often called grit chambers). These systems are also used in industrial applications such as mining. The design basis for water is less developed than for air cyclones. Pretorius presents a review of the design bases for these systems, with a principal focus on total solids removal [12]. There is an increasing application of CFD to these water systems [13], although it has not been as well developed as in the case of air cyclones.

10.2 Removal by Virtue of Size

Sedimentation and cyclone processes remove particles primarily based on their mass, as indicated by their settling velocity. Other processes operate primarily based on particle size, though there is a relationship between size and mass. These systems operate using concepts of filtration. They can be classified as filtration via passage through a depth of material, or at the surface of a membrane of defined pore characteristics. These will be discussed in turn.

10.2.1 Depth Filters

Depth filtration can occur in natural settings, for example, as mentioned in Section 8.6 in the context of groundwater passage. In engineered systems, depth filtration can be designed for many different configurations in air or water purification. Fluid containing microorganisms (either free or associated with other particles) passes through a layer of material. The particles containing microorganisms and the microorganisms themselves are removed upon collision and remain adhered to the filter material. After substantial particle removal, the filter has reached capacity, determined by its inability to effectively remove further material and/or the pressure drop buildup to unacceptably large values. The filter may then be discarded or cleaned and reused. Hence, filtration is intrinsically an unsteady-state (cyclical) process.

The classification of depth filters is multidimensional (in addition to whether they are used for air or water purification):

Media Type. Arguably, this is the most significant variable, since it will drive the dynamics of particle removal and pressure drop.

Driving Force. Gravity, pressure, vacuum.

Physical Configuration. Column, module (air, water), in personal protective equipment (PPE).

In general, for all types of filters, three key parameters may be interrelated that govern performance and cost. The first is the pressure drop, or head loss, through the filter, which will influence any required pumping. The second is the efficiency of removing particles of different sizes. The third is the cycle time or time in which a particular filter module or filter bed can be in service before either particle removal or pressure drop degrades to unacceptable levels.[1]

10.2.1.1 Air Filtration

The concept of purifying air by passing it through fixed material dates back over 100 years, to the use of activated carbon gas masks in the First World War [14]. It evolved to room, building, and commercial scale for the removal of particles in circulating air, and expanded dramatically with mechanical air conditioning (as well as heating) in the 1920s.

Particulate air filters generally consist of prefabricated modules of woven or non-woven fibrous materials. They may be in flat plate format as in Figure 10.4 or in cylindrical or cartridge format. Different types of materials can be used, yielding different performance in particle size removal. Also note, as shown in the figure, the filter media may be pleated to provide more surface area to the oncoming flow, resulting in a lower surface velocity, i.e., flow per unit area.

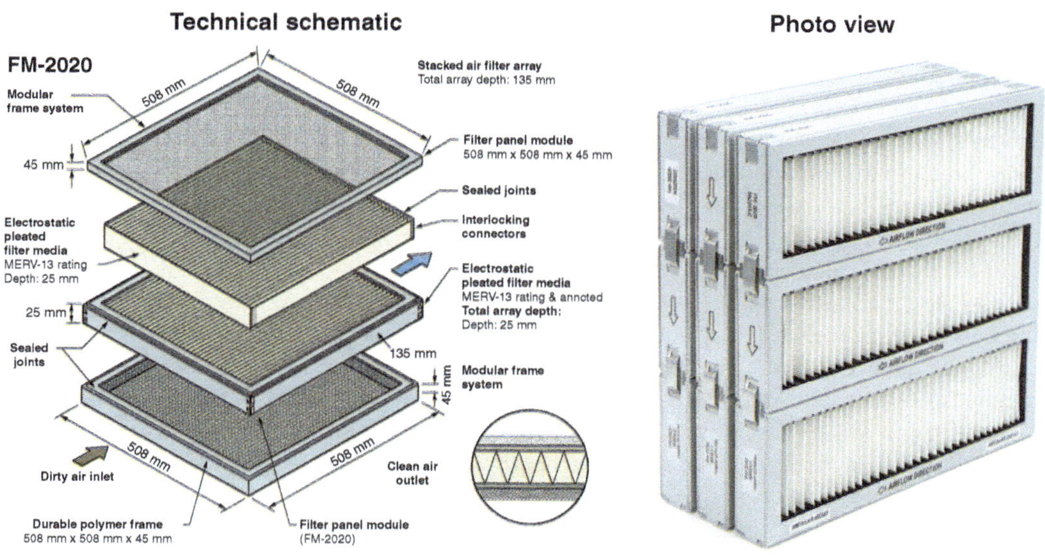

Figure 10.4 Flat air filter. Left panel shows internal construction. Right panel shows modules.

1 While some types of filters are continuous in nature, for example, rotary vacuum filters, in the contexts of this book they are not principally employed for the removal of microorganisms or particles containing microorganisms.

Table 10.2 Particle removal efficiency by MERV category.

MERV rating	0.3–1 μm	1–3 μm	3–10 μm
1–4			<20%
5			>20%
6			>35%
7			>50%
8		>20%	>70%
9		>35%	>75%
10		>50%	>80%
11	>20%	>65%	>85%
12	>35%	>80%	>90%
13	>50%	>85%	>90%
14	>75%	>90%	>95%
15	>85%	>90%	>95%
16	>95%	>95%	>95%

Source: Data from [15].

Air filters can be classified by their ability to remove particles of different sizes. A common classification is the MERV ("minimum efficiency reporting value") rating, originally developed by ASHRAE. This classifies filters as MERV 1 through 16 based on the removal of different sizes of particles as summarized in Table 10.2.

A higher-efficiency than MERV 16 filter is the high-efficiency particulate air (HEPA) filter, which requires 99.97% removal of particles of 0.3 μm, with greater removal of larger or smaller particles. It would be anticipated that as MERV rating increases (and even more so with HEPA), the pressure drop through the filter would increase, the relationships can be complex, and are also influenced (with respect to both particle removal and pressure drop) by imperfect sealing of the edges of the filter in their holders [16].

Air filters can be deployed in different configurations. At the building scale, filters are frequently used to reduce particle deposition on condenser surfaces and to reduce dust circulation. Prior to the COVID-19 pandemic, many buildings had MERV-4 (or less) filtration [17]. It was recognized during the pandemic that upgrading to MERV-13 was often feasible and desirable to improve the removal of bioaerosol-containing particles [18].

Higher-rated MERV and HEPA filters can also be deployed in free-standing devices in rooms where air is forced through one or more filters, with or without additional non-filtration (e.g., UV, oxidative) treatment. Many such devices are in commerce.

In mid-2020, Richard Corsi and Jim Rosenthal developed a do-it-yourself (DIY) air purifier design that has since been eponymously termed the Corsi-Rosenthal Box [19]. This was developed using readily available parts, enabling construction at a lower cost than many commercial purifiers. A photograph of an assembled Corsi-Rosenthal Box is shown in Figure 10.5. It is constructed as follows.[2] An outline of construction is:

2 Dimensions are given in US customary units; however, construction directions and dimensions are also available for UK and EU countries at https://corsirosenthalfoundation.org/instructions/.

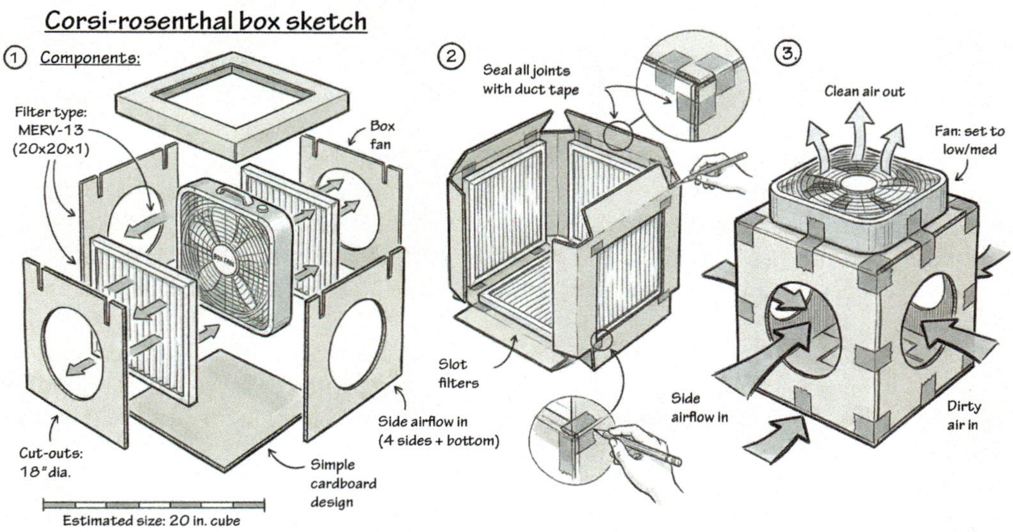

Figure 10.5 Schematic and construction of a Corsi-Rosenthal Box.

- Materials required are:
  - A 20 inch box fan
  - Four filters (20 × 20 or 20 × 25 inch, 1 inch or preferably 2 in thickness) rated as MERV 13 or 3M MPR 1900[3]
  - Scissors, utility knife, and duct tape
- Arrange the four filters to form a hollow cube and tape the edges. The outside of the cube is to be the air intake. Use a piece of cardboard to form the bottom of the cube and tape that to the filters.
- Place the fan on top of the cube such that the exhaust side of the fan is on the top outside.
- Tape the fan to the top of the cube, sealing all edges.
- Take another cardboard sheet to fit the entire box of the fan, and cut a circular hole in the diameter of the fan blades. Tape this upper "shroud" to the fan. Use of this shroud has been found to reduce noise and improve flow through the device.

In addition to bioaerosols, this box has been found effective at removing other indoor air particles, such as dust or particles from fire smoke.

A frequently used metric for the performance characterization of air filters and air purifiers in general has been the clean air delivery rate (CADR). CADR is a metric developed by the Association of Home Appliance Manufacturers (AHAM) to measure the effectiveness of air purifiers. It indicates the volume of air, if completely clean, that would yield the same removal rate in a room expressed in cubic feet per minute (CFM), for three specific pollutants: smoke (0.09–1.0 μm), dust

3 This is a proprietary rating for removal of fine particles.

(0.5–3 μm), and pollen (5–11 μm). The higher the CADR rating, the more efficiently the purifier removes those particles from the air.

In 2006, Shaughnessy and Sextro measured the CADR of commercially available indoor air-cleaning devices. Most devices were rated between 60 and 250 ft^3/min (1.7–7.1 m^3/min for all classes of pollutants), with some devices rating above 350 [20].

Cheek et al. [21] performed a systematic literature review on the impact of portable air purifiers on indoor small particulate matter ($PM_{2.5}$). They concluded that: "All the included studies conducted across all populations in homes and schools found that PAPs reduced PM2.5 concentrations by between 22.6 and 92.0% in homes and 49% in the schools. This variability can be attributed to a variety of factors, including study design, duration of the intervention, internal airflow, and user compliance."

Dal Porto et al. tested the performance of MERV-13-based Corsi-Rosenthal boxes and measured the CADR to be 600–850 ft^3/min (17–24 m^3/min), increasing with fan speed [22]. This indicates that inexpensive, do-it-yourself interventions can be as effective as, or more effective than, many commercial devices.

Given a clean air delivery rate for a device, along with other room ventilation flows, the concentration of a contaminant can be modeled using box-model approaches in subsection 8.3.1. This is illustrated by an example.

Example 10.3 Required CADR for Room Air Purification.

Consider a room of dimensions 20 m^2 with a floor area and ceiling height of 2.5 m. Existing room ventilation provides an ACH of 2.5 h^{-1}. It is desired to increase the effective ACH to 6 h^{-1} using an indoor air purifier. What should the CADR of the device or devices be?

Solution:

1. Given the room dimensions and the ACH, the existing air flow is:

$$20\,m^2 \times 2.5\,m \times 2\,h^{-1} = 1000\,m^3/h$$

2. The air flow at the desired ACH would be:

$$20\,m^2 \times 2.5\,m \times 6\,h^{-1} = 3000\,m^3/h$$

3. Therefore, a CADR of 3000 − 1000 = 2000 m^3/h or 33.3 m^3/min from purifiers is needed. If Corsi-Rosenthal boxes are to be used, this could readily be satisfied by two such units. Or a single commercial purifier of greater capacity could be used.

It should be noted that the use of an air purifier (or, for that matter, ordinary ventilation flow) is designed to reduce the *average* room concentration. If there are localized sources, for example, an infected individual (as in the Guangzhou cluster depicted in Figure 4.6), there may still be focal areas of higher concentration and thus potentially higher risk than described by an average.

Depth filtration is also an important operative mechanism in masks used for personal protective equipment (PPE). There are two principal types of certified masks in wide use [23]:

Medical/Surgical Masks. These may also be called procedure masks. These are typically made of a melt-blown polypropylene or other polymer mesh between internal and external moisture-absorbing layers. These are typically 60–80% efficient at particle removal.

Respirator Masks. These are frequently designated as N95 in the United States, or KN95 or KF94 in other countries. The designation denotes the removal of at least 94 or 95% of particles of all sizes. These are typically composed of an inner layer of filtering material (which might be electrostatically treated) between inner and outer layers that serve as barriers to fluids and moisture.

The nature of filtering action in both types of masks is akin to the mechanisms operative in depth filters.

10.2.1.2 Water Filtration

Depth filtration, a physical mechanism for particle removal in water treatment, is widely used. Symons [24] has traced its first installation in North America to 1882. In original form, these filters were a bed of sand (0.5–1.5 m depth being typical) above a support bed of coarse gravel or porous constructed underdrains. The water to be treated is fed at the top of the bed, forming a pool of some depth that provides the driving force for flow, and the treated water is collected at the bottom.

Current implementations use only the single bed of sand or may have additional layers of different types of media, such as anthracite (less dense than sand) and/or garnet (more dense than sand), such that after the filter is backwashed, the differential media (with different diameters) can resettle. This allows the provision of a gradient of filter media particle sizes from coarse at the top to fine at the bottom.

While textiles and fibers can be used in commercial water filters, they are primarily confined to filters used for sample collection, industrial applications, and purposes such as dewatering and thickening solids collected from water by other processes [25]. An exception worth noting is a DIY filter developed for use in water treatment in developing country settings.

Colwell et al. found that in trials in Bangladesh, filtration of polluted source water through a homemade filter comprising eight layers of indigenous sari cloth could provide substantial microbial reduction and a reduction in risk from cholerae [26]. This has been found to be sustainable over multiple years based on repeated visits to villages [27]. The advantage of sari cloth in this setting is its low cost and widespread availability. Others have shown that multiple layers of deliberately engineered textiles may have equivalent or superior performance [28], though likely of higher cost.

10.2.1.3 Depth Filter Removal Efficiency

There are three principal design considerations for depth filtration: the removal of target particles, the pressure drop (and its evolution with cumulative loading), and, if possible, the efficiency of regenerating and reusing the filter. These will be discussed in turn.

Depth filters, whether in air or water and whether granular or fibrous material is used, operate by removing particles throughout their depth. This is in contrast to membrane-type filters that have their primary removal of particles on the surface (Figure 10.6).

Perhaps the earliest model for depth filtration performance was that of Iwasaki [29], who described removal of particles through a filter bed as first order with length (Equation 10.7), where z is the dimension of length through the bed, C is the concentration of particles to be removed, and λ is phenomenological filtration coefficient of dimension $\frac{1}{\text{Length}}$.

$$\frac{\partial C}{\partial z} = -\lambda C \tag{10.7}$$

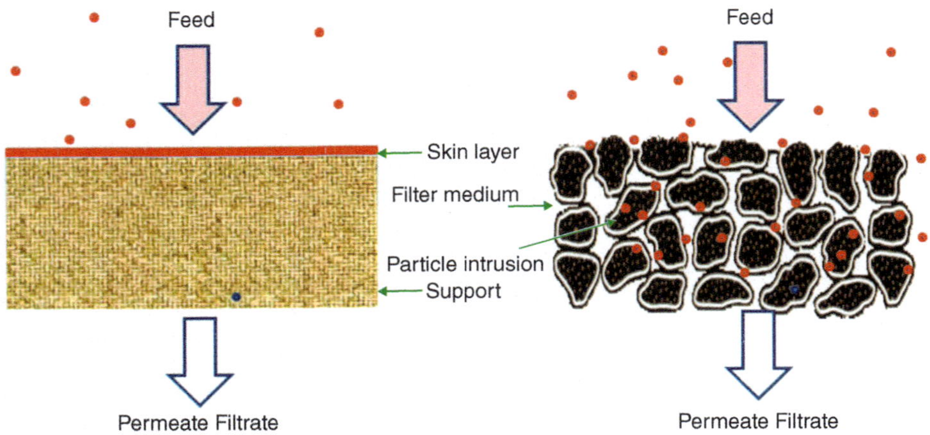

Figure 10.6 Depth versus surface filtration. *Source:* [18]/IntechOpen/CC BY 3.0.

This is written first as a partial derivative since filter performance can vary with time. At (quasi-) steady state, this can be written as an ordinary first-order derivative, which implies (assuming that λ is constant with depth)[4] an exponential decay of particles with depth.

Further developments have led to a theory that describes the filtration coefficient as a function of the filter media and operational characteristics. Development of this theory first requires an understanding of the potential mechanisms of particle removal by filter media — these fundamental mechanisms are relevant both in air and water.

Schematically, removal mechanisms are depicted in Figure 10.7. The curves denote fluid flow around the individual filter element (either a grain or a fiber), and the small circles denote particles to be removed. The mechanisms can be described by:

Impaction occurs when particles, typically of larger mass, deviate from the fluid path as the fluid travels around the filter media, and come sufficiently close to the media to collide and be retained. This is in part driven by the sedimentation velocity that the free particles would have in the fluid.

Interception occurs when particles, traveling along with the fluid, come sufficiently close to the filter media surface to collide. As shown schematically in Figure 10.7, larger particles would present a greater potential area for collision with filter media, and so increase in importance for larger particles.

Diffusion or Brownian removal occurs when small particles, which are more influenced by random processes, deviate randomly from the flow path, and when these excursions result in collision with the media surface. Since the extent of Brownian motion increases as particle size decreases, this mechanism is more important for small particles.

4 This may not be a true assumption, particularly in water filters that are backwashed and returned to service. In such systems, if there is only one type of filter medium, the finer filter grains will be at the top of the filter, and the grain size will become coarser towards the bottom. The grain size distribution with depth in multimedia filters will be more complex. With stratification, either in single or multimedia filters, this equation can be applied to describe individual filter layers.

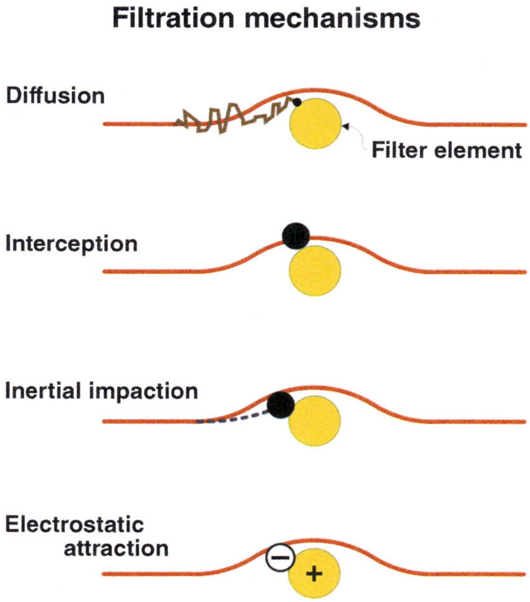

Figure 10.7 Mechanisms of filtration.

Electrostatic attraction is more important in the air than in water due to the greater dielectric constant of water. If the particles and the filter media have opposite surface charges, then a Coulombic force of attraction will occur, deflecting the charged particle away from a fluid streamline toward a surface. This is analogous to the influence of electrostatics on transport discussed in Section 8.5.1.

A first step in modeling is to consider particle removal in a single "layer" of filter media. A parallel approach to development has occurred in both air [30] and water [31] filtration, with minor variation. In air, where filter media are primarily (approximately) cylindrical fibers, Equation 10.8 describes this coefficient, while in water, where filter media are primarily (approximately) spherical, Equation 10.9 is applicable.

$$\lambda = \frac{4}{\pi} \frac{\eta(1-\epsilon)}{\epsilon \, d_f} \tag{10.8}$$

$$\lambda = \frac{3}{2} \frac{\eta(1-\epsilon)}{d_c} \alpha \tag{10.9}$$

The terms in these equations are as follows:

- α is termed the collision efficiency factor and is the fraction of collisions between particles and filter media that result in the particles adhering to the filter. In the air, this is implicitly assumed to be unity. In water, as noted in Section 8.4.1, chemical interventions can increase α.
- ϵ is termed the porosity and is the fraction of the filter's total volume occupied by the fluid.
- η is termed the single-layer collection efficiency and is the probability that a particle passing through a layer of filter media will collide with the media by one of the mechanisms noted above.

d_c is the diameter of a filter grain in granular filtration.
d_f is the diameter of a cylindrical fiber in fibrous depth filtration.

The single-layer collection efficiency η can in turn be decomposed into terms comprising the different mechanisms of removal. The development of this approach is slightly different for granular media filters used for water treatment and for fibrous media filters generally used in air.

Single-layer Removal Efficiency in Granular Water Filters

The single-layer removal efficiency in granular filters used for water can be computed via an approach developed by Tufenkji and Elimelech [32]. This decomposes the single-layer removal efficiency into terms for removal by diffusion (η_D), interception (η_I), and inertial impaction/gravitational sedimentation (η_G). These are, in turn, computed as separate terms.

Diffusion

$$\eta_D = 2.4A_S^{1/3}N_R^{-0.081}N_V^{0.052}Pe^{-0.715} \tag{10.10}$$

where

A_S = porosity parameter = $\dfrac{2(1-\gamma^5)}{2-3\gamma+3\gamma^5-2\gamma^6}$

$\gamma = \sqrt[3]{1-\epsilon}$

N_R = relative size number = $\dfrac{d_p}{d_c}$

N_V = van der Waals number = $\dfrac{H_a}{k_B T}$

H_a = Hamaker constant, which is a function of the surface property [33] of the material, and is typically 10^{-19} - 10^{-20} J

Pe = Peclet number = $\dfrac{3\pi\mu d_p d_c v_f}{k_B T}$ with v_f = filtration rate (flow rate divided by area)

Impaction

$$\eta_G = 0.22N_R^{-0.24}N_V^{0.053}N_G^{1.11} \tag{10.11}$$

where

N_G = Gravity number= $\dfrac{g(\rho_p-\rho_w)d_p^2}{18\mu v_f}$

Interception

$$\eta_I = 0.55A_S N_A^{1/8}N_R^{1.675} \tag{10.12}$$

where:

N_A = Attraction number = $\dfrac{H_a}{3\pi\mu d_p^2 v_f}$

Typical design parameters for water filtration are (summarized from [34, 35]):

- Medium bed depth is from 0.5–2 m.
- Filtration rates of 5–15 m/h (under special circumstances, up to 33 m/h) are used. This is conventionally called the “superficial velocity” of filtration, and it is the rate at which water approaches the upstream end of a filter column.
- The filter bed may consist of only sand or layers of sand and one or more of several additional minerals. The properties of these materials are:

Media	Effective size (mm)[a]	Density $\left(\frac{g}{cm^3}\right)$
Anthracite	0.7–1.7	1.4
Sand	0.3–0.7	2.6
Garnet	0.4–0.6	3.8
Magnetite	0.3–0.5	4.9

[a] Effective size is the lower tenth percentile of the grain size distribution.

- Length of filter run before backwashing is needed — 1–4 days, depending on influent quality.
- Backwashing needed is typically 2–5% of the amount of water produced.

Since the diffusion removal mechanism increases at small particle sizes, while the impaction and interception mechanisms increase at large particle sizes, a particle size exists for which removal is at a minimum. For water filtration, this typically falls within the 1–2 μm size range, as confirmed experimentally [31].

The use of these equations is illustrated by example.

Example 10.4 Computation of Filter Bed Length for Pathogen Removal.

It is desired to remove 90% of pathogens from the water. Assume they are 2 μm in diameter, and compute the depth of the filter required given the design parameters: $\rho_p = 1.05$, $d_c = 0.5$ mm, v_f = 10 m/h, ϵ-0.45, $H_a = 10^{-20}$J, T = 15 °C, α = 1.0).

Solution:

First, the single-layer removal efficiencies for all three mechanisms are computed and summed.

- Compute terms required for η_D and the diffusion single-layer removal.

$$\gamma = \sqrt[3]{1 - 0.45} = 0.819$$

$$A_S = \frac{2(1 - 0.819^5)}{2 - 3 \cdot 0.819 + 3 \cdot 0.819^5 - 2 \cdot 0.819^6} = 28.145$$

$$N_R = \frac{2 \cdot 10^{-6}\,\text{m}}{5 \cdot 10^{-4}\,\text{m}} = 0.004$$

$$N_V = \frac{10^{-20} J}{1.381 \cdot 10^{-23} \frac{J}{K} \times 288K} = 2.514$$

$$Pe = \frac{3\pi \times 0.0011375 Pa \cdot s \times 5 \cdot 10^{-4} m \times 2 \cdot 10^{-6} m \cdot 10 \frac{m}{hr} \cdot 0.000278 \frac{hr}{s}}{1.381 \cdot 10^{-23} \frac{J}{K} \times 288K} = 6.59 \cdot 10^9$$

$$\therefore \eta_D = 2.4(28.145)^{1/3}(0.004)^{-0.081}(6.59 \cdot 10^9)^{-0.715} = 1.089 \cdot 10^{-6}$$

(Continued)

Example 10.4 (Continued)

- Compute terms required for η_G and the impaction single-layer removal.

$$N_G = \frac{9.8\frac{m}{s^2}(1050 - 1000)\frac{kg}{m^3} \times (2 \cdot 10^{-6}m)^2}{18 \cdot (0.0011375Pa \cdot s)(10\frac{m}{hr} \cdot 0.000278\frac{hr}{s})} = 3.44 \cdot 10^{-5}$$

$$\therefore \eta_G = 0.22(0.004)^{-0.24}(2.514)^{0.053}(3.44 \cdot 10^{-5})^{1.11} = 9.654 \cdot 10^{-6}$$

- Compute terms required for η_I and the interception single-layer removal.

$$N_A = \frac{10^{-20}J}{3\pi \cdot (0.0011375Pa \cdot s)(2 \cdot 10^{-6}m)^2(10\frac{m}{hr} \cdot 0.000278\frac{hr}{s})} = 8.388 \cdot 10^{-5}$$

$$\therefore \eta_I = 0.55(28.145)(8.388 \cdot 10^{-5})^{1/8}(0.004)^{1.675} = 4.610 \cdot 10^{-4}$$

- Summing up, the total single-layer removal efficiency is
  $\eta = 1.089 \cdot 10^{-6} + 9.654 \cdot 10^{-6} + 4.610 \cdot 10^{-4} = 4.72 \cdot 10^{-4}$
  and it is clear that the interception mechanism is the dominant one.

Now using the single-layer removal efficiency, the filtration coefficient can be computed from Equation 10.9.

$$\lambda = \frac{3}{2}\frac{(4.72 \cdot 10^{-4})(1 - 0.45)}{5 \cdot 10^{-4}m} \cdot 1 = 0.779\ m^{-1}$$

With a constant filtration coefficient (and removal efficiency) with depth, Equation 10.7 integrates to:

$$C = C_0 \exp(-\lambda L)$$

Since λ is known, and the problem specifies the required removal, the necessary bed length is computed as:

$$L = -\frac{1}{\lambda}ln\left(\frac{C}{C_0}\right) = -\frac{1}{0.779\ m^{-1}}ln(0.1) = 2.9\ m$$

In many applications, this would be considered undesirably high, and so alternative filter media characteristics, sizes, or filtration velocities might be explored.

An analogous approach has been used to describe particle removal by fibrous filters used in air. The porosity of these filters is typically much greater (>0.9) than that of filters used for water applications. In these systems, the dominant mechanisms are diffusion and interception. The theory for a filter composed of randomly packed fibers has been presented by Shou et al. [36] and is described below:

Single-layer Removal Efficiency in Randomly Packed Fibrous Filters

The single-layer removal efficiency (η) is modeled as the sum of removal from diffusion (η_D) and interception (η_I). These terms are calculated as follows using the approach of Shou et al. [36].

The Kuwabara factor corrects for the flow field around the packing and is given for random fiber packing by:

$$Ku = -0.64ln(\epsilon) - 0.737 + \epsilon - 0.05\epsilon^2$$

N_R is the geometric factor, defined as the ratio of particle diameter to filter fiber diameter.
Pe is a Peclet number defined by:

$$Pe = \frac{d_f v_f}{D}$$

with D, the diffusion coefficient for the particles estimated by the Stokes–Einstein equation incorporates a Cunningham-like correction:

$$D = k_B T \frac{1 + \frac{l_f}{d_p}\left[2.33 + 0.966exp\left(-0.4985\frac{d_p}{l_f}\right)\right]}{3\pi\mu d_p}$$

with l_f the mean free path of air molecules (65 nm under standard conditions).
Then the single-layer removal from diffusion can be computed from Equation 10.13.

$$\eta_D = 2.6\sqrt[3]{\frac{\epsilon}{Ku}}Pe^{-2/3} \tag{10.13}$$

The single-layer removal from interception can be computed from Equation 10.14.

$$\eta_I = \frac{\epsilon}{Ku} \cdot \frac{N_R^2}{1 + N_R} \tag{10.14}$$

Given the overall single-layer removal ($\eta = \eta_D + \eta_I$), the filtration coefficient can be computed using Equation 10.8, and in turn, the removal through a defined thickness of filter medium is determined.

10.2.1.4 Depth Filter Pressure Drop

To maintain fluid flow through a filter, a positive pressure on the feed side relative to the product side is required. This headloss may require pumping upstream or reducing the pressure head through the filter. Although, in principle, filtration could also be performed by maintaining a vacuum on the product side, this is rare.

Therefore, it is necessary to understand the initial pressure drop through the filter and how the accumulation of removed material may influence the pressure drop and the further removal of particles.

The underlying approach to estimating the initial pressure drop for granular water filters and fibrous air filters is similar, using the Ergun equation, which includes losses that are linear (dominant at low Reynolds numbers) and those that are quadratic (dominant at higher Reynolds numbers) with respect to velocity through the filter itself [37, 38]. This can be written as:

$$\frac{\Delta P}{\Delta L} = A\frac{\mu v_f}{\mathfrak{d}^2}\frac{(1-\epsilon)^2}{\epsilon^3} + B\frac{\rho v_f^2}{\mathfrak{d}}\frac{(1-\epsilon)}{\epsilon^3} \tag{10.15}$$

where
$\mathfrak{d} = d_c$ for granular media and 1.5 times the diameter of the fiber for fibrous media.
A and B are empirical coefficients for filter media determined experimentally.

Table 10.3 Porosity and Ergun equation parameters for some water and air filter media.

Application	Medium	ϵ	A	B	Reference
Granular water	Crushed anthracite	0.47–0.52	210–245	3.5–5.3	[37]
Granular water	Crushed sand	0.4–0.43	110–115	2–2.5	[37]
Fibrous air	PET nonwoven	0.95–0.98	150–220	0.82–1.09	[38]
Fibrous air	PP nonwoven	0.85–0.91	153–242	1.24–1.97	[38]

Notes: PET, polyethylene terephthalate; PP, polypropylene

The Ergun equation parameters incorporate aspects of filter media shape (angularity, surface-to-volume ratio, and potentially deformability). These have been determined experimentally for various media, including air and water; data for some materials are summarized in Table 10.3. Use of the Ergun equation is shown by Example 10.5.

Example 10.5 Use of the Ergun Equation.

A filter is to be designed for 25 °C air using a 5 cm deep bed of nonwoven PET fibers of diameter 0.05 mm. If the acceptable pressure drop is 0.8 cm of water, what is the corresponding velocity?

Solution:

1. Midpoint values from Table 10.3 will be assumed: $\epsilon = 0.965$, A = 185, B = 0.95
2. It will be easiest to work in a consistent set of units; in this case, the SI system is chosen. With this, the inputs to the Ergun equation are:
   - $\mu = 1.85 \cdot 10^{-5} \dfrac{k}{m \cdot s}$
   - $\rho = 1.185 \dfrac{kg}{m^3}$
   - For this fiber, $\mathfrak{d} = 1.5 \times 0.05\, mm = 7.5 \cdot 10^{-2}\, m.$
3. The target pressure drop of 0.8 cm of water is equivalent to 78.5 Pa. Therefore, the left-hand side of the Ergun equation is $\dfrac{78.4\, Pa}{0.05\, m} = 1566 \dfrac{Pa}{m}$
4. Substitution of these terms into Equation 10.15 and evaluating, produces a quadratic equation in the unknown velocity:

$$18315\, v_f^2 + 8.29\, v_f - 1566 = 0$$

5. Application of the quadratic formula and taking the positive root gives the acceptable velocity to be:

$$v_f = 0.29 \frac{m}{s}$$

As the removal of particles in filters increases, several phenomena occur:

- The particles collect on the filter media surface, and thus change the geometry of the filter itself.
- The porosity of the filter decreases due to the accumulation of solids.

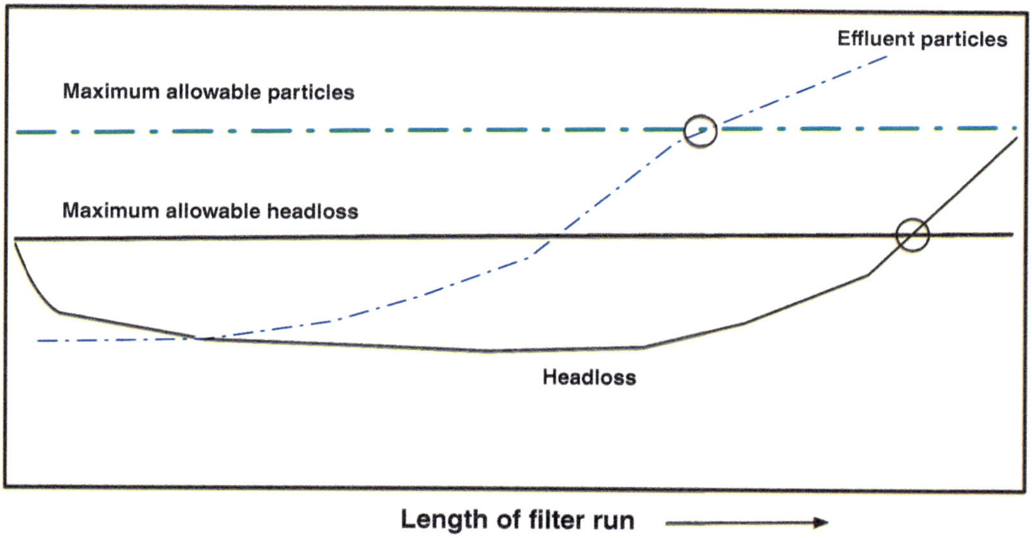

Figure 10.8 Evolution of head loss and effluent quality during filtration.

- As the aforementioned occurs, there may be changes both in the single-layer collection efficiency (typically degrading removal[5]) and as a result, the effective velocity relative to the filter media within the filter increases.

Conceptually, the time course of head loss and effluent quality in a filter is shown in Figure 10.8. The horizontal lines indicate the maximum allowable head loss and effluent particle concentration. At the point where they intersect with the actual performance curves, the filter must be taken offline and either replaced or regenerated. The earliest point at which either the head loss curve or the product quality curve is crossed indicates the end of the filter cycle. In the example shown, since the breakthrough in effluent quality occurs first, it governs the duration of the cycle. Whether head loss or quality breakthrough occurs first is a function of filter design and operation, as well as feed quality. Ideally, both breakthroughs would occur simultaneously, but this is not common.

A theory for the interaction among particle removal, head loss, and deposition within the filter was first elucidated by Ives for granular water filtration [40] in the form of three coupled partial differential equations (PDEs), as described below. In a subsequent paper, Ives [41] developed numerical simplifications, with increasing capabilities for the direct solution of PDEs, e.g., by computational fluid dynamics, such approximations have decreasing utility.

10.2.1.4.1 Mass Balance on Particles

With C as the local instantaneous particle concentration anywhere in the filter, and σ defined as the volume of local instantaneous removed particle volume per volume of the filter, then Equation 10.16 is a mass balance (equation of continuity) assuming that the particles are incompressible.

$$v_f \frac{\partial C}{\partial x} + \frac{\partial \sigma}{\partial t} = 0 \tag{10.16}$$

5 An exception may occur in water filtration where, in the very initial phases of filtration, there can be a "ripening" phase which actually results in a short-term improvement in performance, potentially due to flocculation of particles within the filter and conditioning of the filter media surface[39].

10.2.1.4.2 Particle Removal

The removal of particles is described by the Iwasaki equation (Equation 10.7), recognizing that the filter coefficient (λ) will vary as particle deposition occurs. There is an initial improvement in removal (the "ripening effect" at small times as depicted in Figure 10.8) followed by degradation due to decreasing porosity, straightening of velocity trajectories, and alteration of filter media surface area. buildup and porosity decrease. This is described as a linear/quadratic form where λ in Equation 10.7 is replaced by:

$$\lambda = \lambda_0 + c \cdot \sigma - \frac{\phi\sigma^2}{\epsilon_0 - \sigma} \tag{10.17}$$

where ϵ_0 is the initial porosity, λ_0 is the filter coefficient at the run initiation, and ϕ and c are rate parameters, which need to be determined experimentally.

10.2.1.4.3 Evolution of Head Loss

To describe head loss during filter runs, Ives used a variant of the Kozeny equation,[6] which incorporates the dynamic change in effective porosity (including the accumulation of solids) as shown in Equation 10.18.

$$\frac{\partial H}{\partial x} = K_k \frac{\mu}{\rho g}\left(\frac{v_f}{\epsilon_0 - \sigma}\right)\left(\frac{1 - \epsilon_0 + \sigma}{\epsilon_0 - \sigma}\right)^2 \tag{10.18}$$

K_k is an empirical coefficient, and [40] suggests it may be a function of the value of the bed load (σ) itself.

Ives's approach has been updated to a somewhat more mechanistic one in a review article by Tien [42]. However, ultimately, with modern computational resources, a full CFD model is appropriate, with additional assumptions as to the effect of bed deposition of removed solids on the geometry of the collector grains themselves, more experience with these models is needed [43].

For fibrous air filtration, an analogous theoretical framework has been developed, as outlined by Berry [44]. However, with the greater variability in geometry and materials used in air filtration and the recency of this approach, more experience is needed to validate these models under loading.

10.2.1.5 Depth Filter Regeneration

After the end of a filter run, whether by reaching terminal headloss or unacceptable pressure drop, filter replacement or regeneration is needed.

In granular water filtration, due to the process's capital intensity, filter regeneration is required. This generally occurs by reversing the flow and partially expanding the bed. Air may be introduced, or a surface-scouring flow of water may be used. The resultant agitation causes abrasion and loosening of adhered solids into the backwash water, which must be treated or disposed of. This results in a waste product of typically 1–5% of the water treated. After backwashing, the filter is drained, the media is settled, and service is resumed. The hydraulics of backwashing are relatively well understood [45], but the duration of backwashing required for effective cleaning remains largely empirical.

In the case of fibrous filters used for air treatment, these are principally single use and then disposed.

---

6 The Kozeny equation is an earlier version of the Ergun equation, Equation 10.15, using only the linear term in velocity, which dominates under low Reynolds number conditions. In principle, a modification of the full Ergun relationship could be used, but it is not clear that this has been explored in a dynamic context.

Note that, both in the case of backwash water and used air filters, the accumulation of any microorganisms, as well as other contaminants, may require either further treatment for inactivation of pathogens and/or management as a biohazardous or mixed bio and chemical hazardous waste.

10.2.2 Membranes

In contrast to depth filtration, membrane filtration operates by virtue of a material with defined porosity to remove material by virtue of size, typically at or near the filter surface (Figure 10.6). Particle removal efficiencies may vary with pore size and material properties.

Membrane filters can be classified in several ways. The first way is by pore size, which is shown schematically in Figure 10.9. In both water and air applications, the flux through the membranes is driven by a pressure gradient through the membrane, and in general, membranes that can exclude smaller particles have greater pressure drop than membranes that have larger size cutoffs. This is summarized in Table 10.4.

Note that based only on the size cutoffs, bacteria and eukaryotic microorganisms, even if present as free organisms, would be expected to be removed by UF, NF, and RO, and also by most MF systems. Viruses would be removed by NF and RO, with some removal in UF. For all organisms,

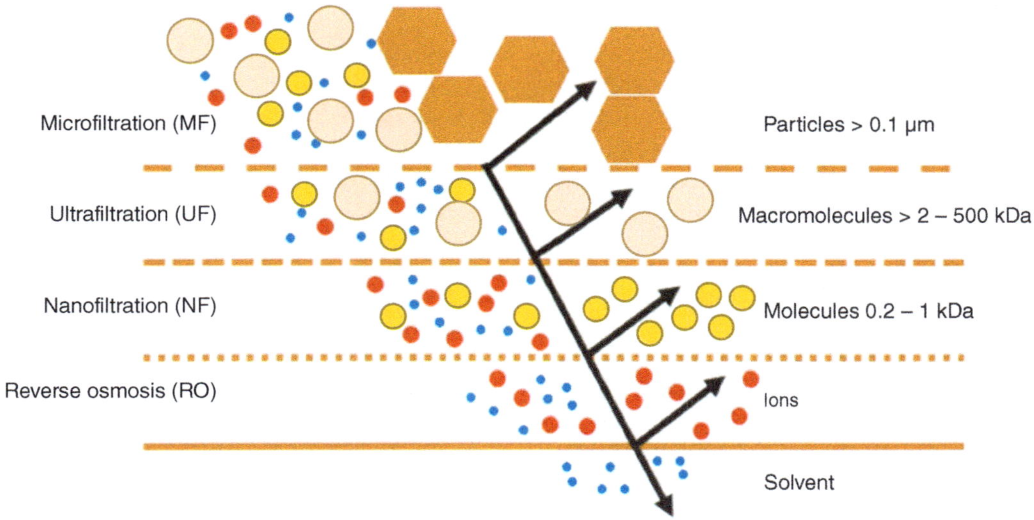

Figure 10.9 Categorization of membrane types. *Source:* [46]/MDPI/CC BY 4.0.

Table 10.4 Size cutoffs, pressure drop, and permeability of various membrane types used in water systems.

Process	Size cutoff μm	Typical pressure difference (bar)	Average clean membrane permeability $\frac{L}{m^2 hr\ bar}$
MF	0.1–10	1–3	500
UF	10^{-3}–1	2–5	150
NF	10^{-3}–10^{-2}	5–15	10–20
RO	10^{-4}–10^{-3}	15–75	5–10

Source: Adapted from [47].

association with solids would facilitate their removal by membranes with larger size cutoffs. In practice, complete removal does not always occur, as membrane aging and deterioration can introduce imperfections that permit some passage of contaminants that would otherwise be rejected [48]. In addition, failure or improper sealing of the seals between membranes and the module housing may provide passage for some materials [49].

As material is filtered through a membrane, a buildup of rejected material occurs on the feed side. This results in an increase in pressure drop due to the buildup, necessitating cleaning. Furthermore, especially with UF and RO systems, this fouling layer will also contain chemical contaminants that are rejected from passage through the membrane, which may encourage biological fouling on the feed side of the membrane to occur [50].

Membranes may be composed of different materials, depending on the manufacturer and type of process. These are typically organic, although inorganic membrane materials are infrequently used [51]. There is also intense development of novel materials, principally for RO and UF to improve selectivity and permeability [52].

During operation, the pressure drop across membranes increases as solids are retained, and in RO and UF, a chemical gel-like fouling layer forms. To counteract this performance degradation, membrane operations often use a two-phase approach:

- On a frequent, short-term basis (perhaps every 15 minutes to an hour), the flow is briefly (less than a few minutes) reversed (termed backflushing) to dislodge and recover removed solids. This will partially restore permeability.
- After more prolonged operation, when the buildup of more tightly bound materials causes an unacceptable headloss even after the periodic backflushing, the flow is reversed while also feeding a chemical cleaning solution to remove chemical and biological material. This cleaning cycle might occur every week for an hour. Depending on the membrane material, its chemical compatibility, and the predominant nature of fouling, chemicals used might be caustics (sodium hydroxide), oxidants (sodium hypochlorite, hydrogen peroxide, peroxyacetic acid), acids (citric acid, nitric acid), chelating agents[7] (citric acid, EDTA), or surfactants.

As with granular filtration, there is waste produced in the form of the flow from backflushing and the spent cleaning solution. Both of these will contain concentrated levels of any contaminants removed, possibly requiring additional treatment before disposal.

Membrane filters have rarely been used for air treatment applications. However, increasing interest in removing ultrafine particles has led to greater advances in knowledge of these systems [53, 54]. Membranes (such as nanofiltration membranes) can be used in such applications. In addition to removal by surface sieving of larger particles, there is some potential for removal of particles smaller than the pore openings due to the mechanisms discussed above for granular and fibrous filters. The scalability of such systems for large scale application merits future study.

10.3 Removal by Virtue of Charge

As noted in Section 8.5.1, microorganisms, and the particles with which they may be associated, can carry net electrical charges. In the presence of an imposed electric field, this will result in motion. The extent of the motion depends on the nature of the fluid and the opposing forces on the particles.

---

7 A chelating agent is an organic chemical that forms complexes with multivalent cations, including heavy metals, calcium, and magnesium. In the context of fouling control and cleaning, they can be very useful at solubilizing calcium and magnesium that may be part of the fouling layer.

In aqueous systems, the separation of charged particles, whether microorganisms, particles, or ions, by electrical fields occurs in two processes. The first, which is primarily used for analytical separations, is electrophoresis. The second embodiment, which has found more utility in treatment applications, is electrodialysis.

In water applications, electrodialysis is primarily used for desalinating saline and brackish waters [55]. It functions by imposing a direct current through a stack of alternating, selectively permeable membranes to anions or cations. As with other membrane systems, this produces a treated stream and a higher-concentration wastewater stream. However, there has been little study of these systems for pathogen removal.

In the air, electrostatic precipitators (ESPs) have been widely used to remove fine and ultrafine particles from waste gas streams. However, these are not typically designed for pathogen removal *per se*, although removal will occur if microorganisms are associated with removable particles. ESPs function by allowing gas to pass through one or more chambers across which a DC field is imposed. Particles move to the electrode of their opposite charge, where they are collected for disposal. There is a strong interaction among the particle dynamics imposed by the electric field, the turbulent flow field, and the gravitational field. In addition, depending on the field strength, ionization effects may occur. Analysis of these systems shows typical minimum removals in the 0.1–1 μm particle size range [56].

Removal by charge can also occur in air filtration. Filter materials can be manufactured that have defined surface charges [14]. These materials, often labeled as "electrets", have enhanced particle removal in fibrous filters by providing an additional flux of particles to the media surface. Electret materials have also found wide application in personal protective equipment, notably in N95 and related face masks [57].

While there are a number of consumer-grade air purifiers that use electrostatic removal, these are generally only effective for coarser particles and achieve relatively low levels of removal. So these will not be discussed further.

10.4 Surface Cleaning

Physical removal of microorganisms can occur through surface cleaning. The target of such processes is the reduction of potential disease transmission via a fomite route (Section 2.2).

Surface cleaning may be assisted by chemical agents. Such agents may have primarily detergent or surface-active properties where the objective is to reduce forces of adhesion between microorganisms or particles containing them and the surface(s) on which they occur. Chemical agents may also have antimicrobial, i.e., disinfectant, attributes — discussion of such inactivation characteristics is reserved to Chapter 11.

Surface cleaning, *per se*, is not an absolute barrier to fomite exposure, even if it were completely effective. Surfaces, even after cleaning, may be contaminated by individuals touching them and transferring contamination (if they have not thoroughly washed their hands) [58]. Respiratory particles that deposit on surfaces can also result in recontamination. Over time, therefore, even with cleaning, a quasi-steady-state surface contamination is attained unless sources to the surface are eliminated.

The efficiency of mechanical removal of pathogens and solids from surfaces is a complex function of a number of potential factors, which could include:

- The nature of the surface itself — including porosity, hydrophobicity, and presence of electrostatic or other forces that could alter particle adhesion.

- The nature of the microorganism, or the particles on which microorganisms are found, including any coating that could alter particle properties (e.g., coating with mucus or other biological fluids, attachment to room dust particles). Different microorganisms and particles may exhibit different hydrophobicities or surface interaction potentials.
- Humidity, which might influence the chemical or physical aspects of both the surfaces and the particles to be removed.
- The material being used to mechanically remove the particles or microorganisms from the surface and its chemical and physical properties.
- Whether there are chemical additives applied to the surface or the cleaning material to enhance dislodgment from the surface.
- The force applied for removal and the duration of the removal process.

As a result of the numerous factors that can influence performance, there has not been a systematic development of quantitative models for physical removal processes. Therefore, a series of examples will be presented for entry into the literature.

Barker et al. [59] used controlled contamination of nonporous (melamine) surfaces with fecal samples from Norovirus-infected patients to examine the effects of cleaning. The surface was allowed to dry and cleaned with a cloth soaked with detergent. It was found that this provided no discernible reduction in the number of PCR-positive samples from the surface. Only when the surfaces were treated with a hypochlorite/detergent mixture could a reduction in contamination be observed.

The mechanical cleaning of surfaces has a similarity to the sampling of surfaces for microorganisms, as discussed in Section 5.2.5. In cleaning, the goal is to maximize removal of infectious material from a surface, whereas in sampling, the goal is to provide a quantitative, reproducible method to measure the current contamination level. Hence, experimental studies on sampling can inform approaches to cleaning and *vice versa*.

Using the ms2 bacteriophage, Julian et al. [60], investigated recovery by swabbing onto various nonporous surfaces to which broth prepared viruses were applied and dried before sampling. Even the most effective swab and soaking combination only recovered 40% (median) of the infectivity and 7% of the RNA (determined by qPCR).

Gibson et al. [61] tested the removal of four viruses from either stainless steel or nonporous acrylic surfaces by wiping with commercial cloths used in the food service industry. The viruses were prepared in pure culture and dried. Wiping resulted in (generally) 3–4 logs removal from the surface; there was evidence that cotton/cellulose and microfiber woven cloths performed significantly poorer than either non-woven cloths or cotton/terry bar towels.

Tuladhar et al. [62] found that wiping stainless steel surfaces with water-dampened cloths resulted in 2–4 log reduction. Experiments were performed with six animal viruses, and also *Staphylococcus aureus* and *Salmonella enteriditis*.

10.5 Discussion Questions and Problems

1 A Stairmand high-efficiency cyclone is to be used to treat an air stream. The diameter of the upper cylindrical throat is to be 1 m. For pressure drops between 5 cm and 20 cm of water, plot the curves of the corresponding air flows and cut diameters versus pressure drop.

2 Repeat the previous computation for a High Rate Cyclone.

3 For either cyclone in the two previous problems, if the design basis is that there be 90% removal of particles of 5 μm diameter, what is the design airflow and pressure drop that should be used?

4 For the filter of media type specified in Example 10.4, how deep of a bed is required to achieve 95% removal of particles that are 0.5 μm in diameter?

5 For a 2-m deep filter of characteristics given in Example 10.4, estimate the clean bed head loss using the Ergun equation.

6 A fibrous filter composed of polyethylene (PE) fibers is used to remove 1-micron (1 μm) particles from an air stream. The filter operates under conditions where Brownian diffusion and interception are the dominant capture mechanisms. The target efficiency is **95% removal** of these particles.

Data:

- Airflow velocity through the filter: $U = 0.2$ m/s
- Fiber diameter: $d_f = 5$ μm
- Packing density of the filter (solidity): $\alpha = 0.05$
- Air viscosity: $\mu = 1.8 \times 10^{-5}$ Pa·s
- Air density: $\rho = 1.2$ kg/m^3
- Particle density: $\rho_p = 1000$ kg/m^3

Using this information,

a **Determine the collection efficiency per unit length of fiber,** η_f: Consider single-fiber efficiency contributions from **interception** and **diffusion**.

b **Calculate the required filter depth** L to achieve the desired overall efficiency of **95% removal**.

c **Discuss the implications** of increasing the fiber diameter or changing the packing density on the required depth.

References

1 Harold P Guy. Chapter C1: Laboratory theory and methods for sediment analysis. *Techniques of Water-Resources Investigations of the United States Geological Survey*. US Geological Survey, Washington, D.C., 1969.

2 Thomas R Camp. Sedimentation and the design of settling tanks. *Transactions of the American Society of Civil Engineers*, 111(1):895–936, January 1946.

3 Haiwen Gao and Michael K Stenstrom. Development and applications in computational fluid dynamics modeling for secondary settling tanks over the last three decades: A review. *Water Environment Research*, 92(6):796–820, 2020.

4 Marko Korhonen, Alpo Laitinen, Gizem Ersavas Isitman, Jose L Jimenez, and Ville Vuorinen. A GPU-accelerated computational fluid dynamics solver for assessing shear-driven indoor airflow and virus transmission by scale-resolved simulations. *Journal of Computational Science*, 78: 102265, 2024.

5 M O Rivera-García, M A Reyna, M A Camarillo-Ramos, M A Reyna-Vargas, Roberto L Avitia, Daniel Cuevas-González, and A R Osornio Vargas. Cyclone separator for air particulate matter personal monitoring: A patent review. *Atmosphere*, 14(4):624, 2023.

6 David Leith and Dilip Mehta. Cyclone performance and design. *Atmospheric Environment (1967)*, 7(5):527–549, 1973.

7 M Rhodes. Chapter 9: Separation of particles from a gas: Gas cyclones. *Introduction to Particle Technology*. John Wiley & Sons Ltd, 2nd edition, 2008.

8 Bingtao Zhao and Yaxin Su. Particle size cut performance of aerodynamic cyclone separators: Generalized modeling and characterization by correlating global cyclone dimensions. *Journal of Aerosol Science*, 120(June):1–11, 2018.

9 Constantine Marinos, Alexia A Economopoulou, and Alexander P Economopoulos. Comparative fractional efficiency predictions by selected cyclone simulation models. *Journal of Environmental Engineering*, 133(5):556–567, May 2007.

10 Sakura Ganegama Bogodage and Andrew Y T Leung. CFD simulation of cyclone separators to reduce air pollution. *Powder Technology*, 286:488–506, December 2015.

11 Richard Field. The dual functioning swirl combined sewer overflow regulator/concentrator. *Water Research*, 9(5-6):507–512, 1975.

12 Coenraad F Pretorius. A review of vortex grit basin design. In *WEFTEC 2012-85th Annual Technical Exhibition and Conference*, volume 9, pages 5715–5734, 2012.

13 Prudhvinath Reddy Ankireddy, Sudikondala Purushotham, and Mangadoddy Narasimha. Fluid flow modeling and analysis of low- and high-gravity spiral concentrators: Experimental and analytical approaches. *Chemical Engineering & Technology*, 46(8):1619–1629, 2023.

14 H E Barney Burroughs, Brian Krafthefer, and Marilyn Listvan. History of air filtration and air cleaning: A 100 year review of significant advances and related influences in the advancement of the art and science of filtration and air cleaning. *ASHRAE Transactions*, 126:601–615, 2020.

15 US Environmental Protection Agency. What is a MERV rating? https://www.epa.gov/indoor-air-quality-iaq/what-merv-rating, March 2024.

16 Matthew Ward and J A Siegel. Modeling filter bypass: Impact on filter efficiency. *ASHRAE Transactions*, 111(1):1091–1100, 2005.

17 D H Bennett, W Fisk, M G Apte, X Wu, A Trout, D Faulkner, and D Sullivan. Ventilation, temperature, and HVAC characteristics in small and medium commercial buildings in California. *Indoor Air*, 22(4):309–320, 2012.

18 Meghan K McNulty, Jamie Kono, and Barry Abramson. From guidance to implementation: Applying ASHRAE epidemic task force building readiness strategies in 95 commercial office buildings. *ASHRAE Transactions*, 128(1):393, 2022.

19 Wikipedia Contributors. Corsi–Rosenthal box — Wikipedia, the free encyclopedia. https://en.wikipedia.org/w/index.php?title=CorsiFebruary 2025. [Online; accessed 17 February 2025].

20 R J Shaughnessy and R G Sextro. What is an effective portable air cleaning device? A review. *Journal of Occupational and Environmental Hygiene*, 3(4):169–181, April 2006.

21 Emily Cheek, Valentina Guercio, Clive Shrubsole, and Sani Dimitroulopoulou. Portable air purification: Review of impacts on indoor air quality and health. *Science of The Total Environment*, 766:142585, April 2021.

22 Rachael Dal Porto, Monet N. Kunz, Theresa Pistochini, Richard L Corsi, and Christopher D Cappa. Characterizing the performance of a do-it-yourself (DIY) box fan air filter. *Aerosol Science and Technology*, 56(6):564–572, March 2022.

23 Sonali Das, Sagarika Sarkar, Anusree Das, Shreyosree Das, Pallab Chakraborty, and Joy Sarkar. A comprehensive review of various categories of face masks resistant to Covid-19. *Clinical Epidemiology and Global Health*, 12:100835, 2021.

24 George E Symons. Water treatment through the ages. *Journal AWWA*, 98(3):87–98, 2006.

25 Murat Eyvaz, Serkan Arslan, Ercan Gürbulak, and Ebubekir Yüksel. Textile materials in liquid filtration practices: Current status and perspectives in water and wastewater treatment. In Bipin Kumar and Suman Thakur, editors, *Textiles for Advanced Applications*, chapter 11. IntechOpen, Rijeka, 2017.

26 Rita R. Colwell, Anwar Huq, M. Sirajul Islam, K. M. A. Aziz, M. Yunus, N. Huda Khan, A. Mahmud, R. Bradley Sack, G. B. Nair, J. Chakraborty, et al. Reduction of cholera in Bangladeshi villages by simple filtration. *Proceedings of the National Academy of Sciences*, 100(3):1051–1055, 2003.

27 Anwar Huq, Mohammed Yunus, Syed Sohel Salahuddin, Abbas Bhuiya, Michael Emch, Stephen P Luby, Estelle Russek-Cohen, G Balakrish Nair, R Bradley Sack, and Rita R Colwell. Simple sari cloth filtration of water is sustainable and continues to protect villagers from cholera in Matlab, Bangladesh. *mBio*, 1(1):00034-10, 2010.

28 Stephen Siwila and Isobel C Brink. Low cost drinking water treatment using nonwoven engineered and woven cloth fabrics. *Journal of Water and Health*, 17(1):98–112, 2018.

29 Tomihisa Iwasaki, J J Slade, and W E Stanley. Some notes on sand filtration [with discussion]. *Journal (American Water Works Association)*, 29(10):1591–1602, 1937.

30 Hsu-Chi Benjaminy and Benjamin Y H Liu. Aerosol filtration by fibrous filters—I. Theoretical. *Aerosol Science*, 5(2):191–204, 1974.

31 Kuan-Mu Yao, Mohammad T Habibian, and Charles R O'Melia. Water and waste water filtration. Concepts and applications. *Environmental Science & Technology*, 5(11):1105–1112, November 1971.

32 Nathalie Tufenkji and Menachem Elimelech. Correlation equation for predicting single-collector efficiency in physicochemical filtration in saturated porous media. *Environmental Science & Technology*, 38(2):529–536, January 2004.

33 Hideyuki Takagishi, Takashi Masuda, Tatsuya Shimoda, Ryo Maezono, and Kenta Hongo. Method for the calculation of the Hamaker constants of organic materials by the Lifshitz macroscopic approach with density functional theory. *The Journal of Physical Chemistry A*, 123(40):8726–8733, October 2019.

34 Veolia Water Technologies & Solutions. Handbook of industrial water treatment. https://www.watertechnologies.com/handbook/, 2025.

35 Kerry J Howe, David W Hand, John C Crittenden, R Rhodes Trussell, and George Tchobanoglous. *Principles of Water Treatment*. John Wiley & Sons, 2012.

36 Dahua Shou, Jintu Fan, Heng Zhang, Xiaoming Qian, and Lin Ye. Filtration efficiency of non-uniform fibrous filters. *Aerosol Science and Technology*, 49(10):912–919, 2015.

37 R Rhodes Trussell and Melissa Chang. Review of flow through porous media as applied to head loss in water filters. *Journal of Environmental Engineering*, 125(11):998–1006, November 1999.

38 Tao Yang, Lizhu Hu, Deyou Yu, Xiaoman Xiong, Jiri Chvojka, Mohanapriya Venkataraman, Michal Petru, Blanka Tomkova, Hideaki Morikawa, and Jirf Militký. Simple determination of key structural parameters for fibrous materials enabled by Ergun-type and Kozeny-type equations. *Polymer Testing*, 108:107514, 2022.

39 Jeannie L Darby and Desmond F Lawler. Ripening in depth filtration: Effect of particle size on removal and head loss. *Environmental Science & Technology*, 24(7):1069–1079, 1990.

40 K J Ives. Rational design of filters. *Proceedings of the Institution of Civil Engineers*, 16(2):189–193, June 1960.

41 K J Ives. Simplified rational analysis of filter behaviour. *Proceedings of the Institution of Civil Engineers*, 25(3):345–364, July 1963.

42 Chi Tien and Alkiviades C Payatakes. Advances in deep bed filtration. *AIChE Journal*, 25(5):737–759, 1979.

43 David Fernandes Del Pozo, Arslan Ahmad, Usman Rehman, Arne Verliefde, and Ingmar Nopens. A novel CFD model to predict effluent solids concentration and pressure drop in deep bed granular filters for water treatment. *Separation and Purification Technology*, 295:121232, 2022.

44 Gentry Nathaniel Berry. Advancing air filtration analysis: A comprehensive approach to particle loading models. *PhD Thesis*, Mississippi State University, 2023.

45 Onno J I Kramer, Peter J de Moel, Johan T Padding, Eric T Baars, Sam B Rutten, Awad H E Elarbab, Jos F M Hooft, Edo S Boek, and Jan Peter van der Hoek. New hydraulic insights into rapid sand filter bed backwashing using the Carman–Kozeny model. *Water Research*, 197:117085, 2021.

46 Aldo Saavedra, Hugo Valdés, Juan Velásquez, and Sebastián Hernández. Comparative analysis of Donnan steric partitioning pore model and dielectric exclusion applied to the fractionation of aqueous saline solutions through nanofiltration. *ChemEngineering*, 8(2):0039, 2024.

47 Elorm Obotey Ezugbe and Sudesh Rathilal. Membrane technologies in wastewater treatment: A review. *Membranes*, 10(5):89, 2020.

48 Marie-Laure Pype, Bogdan C Donose, Llucia Martí, Dominique Patureau, Nathalie Wery, and Wolfgang Gernjak. Virus removal and integrity in aged RO membranes. *Water Research*, 90:167–175, 2016.

49 Dwani Venkataswamy Gowda, Ariane Berthet, Danny Harmsen, Arnout D'Haese, and Emile R Cornelissen. Impact of membrane surface and module damage on virus removal and integrity in RO membranes. *Desalination*, 593:118244, 2025.

50 Lauren Weinrich, Mark LeChevallier, and Charles N Haas. Contribution of assimilable organic carbon to biological fouling in seawater reverse osmosis membrane treatment. *Water Research*, 101:203–213, 2016.

51 Anna Lee, Jeffrey W Elam, and Seth B Darling. Membrane materials for water purification: Design, development, and application. *Environmental Science: Water Research & Technology*, 2(1):17–42, 2016.

52 Kuichang Zuo, Kunpeng Wang, Ryan M. DuChanois, Qiyi Fang, Eva M. Deemer, Xiaochuan Huang, Ruikun Xin, Ibrahim A. Said, Ze He, Yuren Feng, et al. Selective membranes in water and wastewater treatment: Role of advanced materials. *Materials Today*, 50:516–532, November 2021.

53 Pavel Bulejko, Mirko Dohnal, Jiří Pospíšil, and Tomáš Svěrák. Air filtration performance of symmetric polypropylene hollow-fibre membranes for nanoparticle removal. *Separation and Purification Technology*, 197:122–128, 2018.

54 Pavel Bulejko, Ondřej Krištof, Mirko Dohnal, and Tomáš Svěrák. Fine/ultrafine particle air filtration and aerosol loading of hollow-fiber membranes: A comparison of mathematical models for the most penetrating particle size and dimensionless permeability with experimental data. *Journal of Membrane Science*, 592:117393, December 2019.

55 Sajjad Al-Amshawee, Mohd Yusri Bin Mohd Yunus, Abdul Aziz Mohd Azoddein, David Geraint Hassell, Ihsan Habib Dakhil, and Hassimi Abu Hasan. Electrodialysis desalination for water and wastewater: A review. *Chemical Engineering Journal*, 380:122231, 2020.

56 Hsunling Bai, Chungsying Lu, and Chung Liang Chang. A model to predict the system performance of an electrostatic precipitator for collecting polydisperse particles. *Journal of the Air & Waste Management Association*, 45(11):908–916, November 1995.

57 Na Wang, Ming Cai, Xue Yang, and Yuyan Yang. Electret nanofibrous membrane with enhanced filtration performance and wearing comfortability for face mask. *Journal of Colloid and Interface Science*, 530:695–703, 2018.

58 Hao Lei, Shenglan Xiao, Benjamin J Cowling, and Yuguo Li. Hand hygiene and surface cleaning should be paired for prevention of fomite transmission. *Indoor Air*, 30(1):49–59, January 2020.

59 J Barker, I B Vipond, and Sally F Bloomfield. Effects of cleaning and disinfection in reducing the spread of Norovirus contamination via environmental surfaces. *Journal of Hospital Infection*, 58(1):42–49, 2004.

60 Timothy R Julian, Francisco J Tamayo, James O Leckie, and Alexandria B Boehm. Comparison of surface sampling methods for virus recovery from fomites. *Applied and Environmental Microbiology*, 77(19):6918–6925, 2011.

61 K E Gibson, P G Crandall, and S C Ricke. Removal and transfer of viruses on food contact surfaces by cleaning cloths. *Applied and Environmental Microbiology*, 78(9):3037–3044, 2012.

62 Era Tuladhar, Wilma C Hazeleger, Marion Koopmans, Marcel H Zwietering, Rijkelt R Beumer, and Erwin Duizer. Residual viral and bacterial contamination of surfaces after cleaning and disinfection. *Applied and Environmental Microbiology*, 78(21):7769–7775, November 2012.

11

Inactivation Fundamentals

This and the subsequent chapter will cover systems and processes designed for the deliberate inactivation of microorganisms. This may occur regardless of the physical removal of the organisms or particles with which they are associated. As discussed in Chapter 6, there are different methods to assess the occurrence of microorganisms, from culture techniques to the measurement of nucleic acid to microscopy. In particular, with inactivation processes, different methods will yield differential responses across assays depending on the mode(s) of action of the inactivating agent.

In this chapter, the agents used for disinfection will be discussed. This will follow with a discussion of the kinetic models used to describe inactivation processes. The subsequent chapter will discuss fitting such kinetic models to data, by-products of disinfection, and applications to different environmental venues.

Generally, inactivation processes are nonspecific and act against a broad spectrum of microorganisms. This is distinct, for example, from pharmaceuticals and antibiotics that may be targeted to different microorganisms, e.g., Gram-negative bacteria, DNA viruses, etc.

Also, the distinction between disinfection and sterilization noted in Section 9.4 is essential. Generally, only disinfection is required rather than, as with sterilization, removal of all biological activity from a system.

Many different physical and chemical systems can be used to achieve inactivation. First, the agents' production and interaction with other constituents in the environment will be discussed. Then, the kinetics of inactivation, common to inactivation processes, will be addressed. Finally, the application to inactivating pathogens in liquids, air, and surfaces will be addressed.

11.1 Disinfecting Agents

There are many disinfecting agents available. As a first classification, chemical disinfectants and physical disinfectants are used. Chemical disinfectants are applied either from externally supplied sources or from on-site generation. Physical disinfection systems achieve inactivation by imposing a physical input, typically thermal or radiation (ionizing or nonionizing), although they can also result in chemical transformations.

The key properties of each agent and system will be summarized, with consideration of their generation, chemical and physical interactions, and interactions with other environmental components.

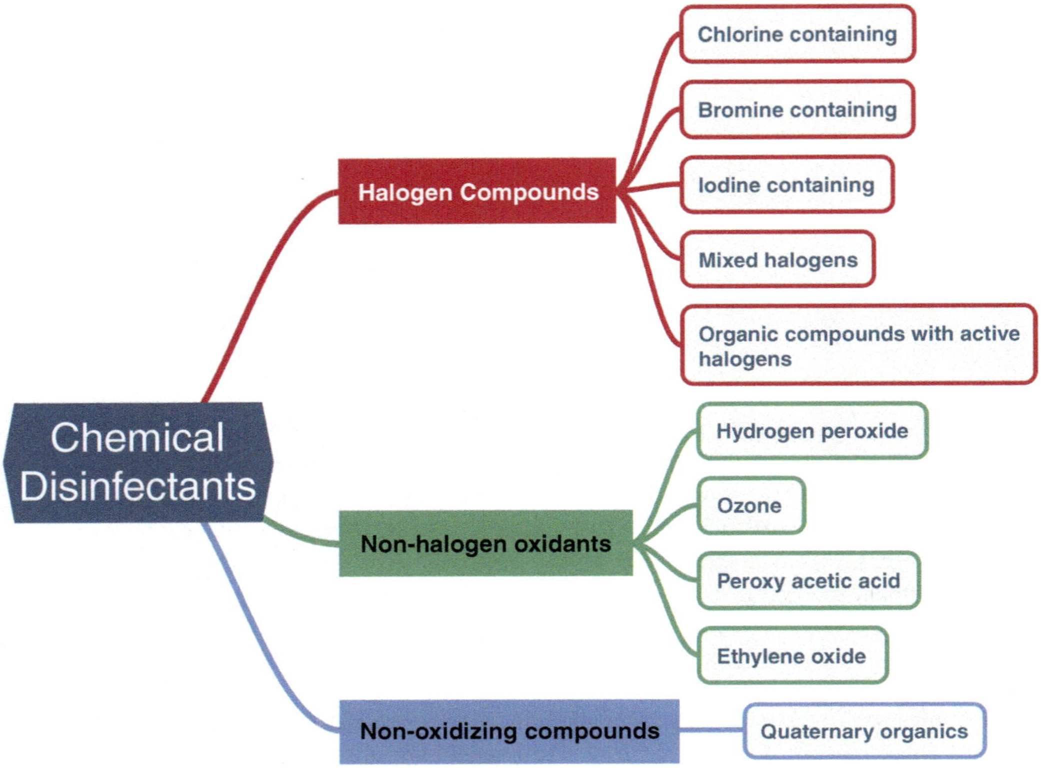

Figure 11.1 Classification of chemical disinfectants.

11.1.1 Chemical Disinfectants

There are many types of chemical disinfectants, as classified in Figure 11.1. The broad classes are halogen-containing compounds, non-halogen oxidants, and nonoxidizing compounds. Chemical disinfectants are classified as pesticides under US law and are subject to regulation under the Federal Insecticide, Fungicide, and Rodenticide Act (FIFRA). Following is a summary of FIFRA provisions pertinent to these chemicals.

Regulatory Requirements for Chemical Disinfectants Under FIFRA

The Federal Insecticide, Fungicide, and Rodenticide Act (FIFRA), administered by the US Environmental Protection Agency (EPA), provides the main regulatory framework for antimicrobial products, including disinfectants used on surfaces, in air, and in water treatment. Under FIFRA, disinfectants are treated as pesticides and must be registered with the EPA before they can be sold or distributed in the United States. Registration requires evidence that the product works as claimed, is properly formulated, and can be used safely for both people and the environment.

For surface disinfectants, manufacturers must show effectiveness against specific microorganisms using standardized test methods, such as those developed by the Association of Official Analytical Collaboration (AOAC) [1]. Claims for air disinfection are evaluated more cautiously, reflecting the added technical challenges of controlling airborne pathogens; as a result, relatively few products qualify for approved air-disinfection claims.

FIFRA also applies to disinfectants used in non-potable water applications, including cooling towers, recreational waters, and wastewater systems. In these settings, products must demonstrate antimicrobial performance and safe use under realistic operating conditions.

Disinfectants used for potable water may fall under additional oversight through the Safe Drinking Water Act (SDWA), administered by the EPA's Office of Water. FIFRA labeling requirements further ensure safe use by specifying directions, precautions, and first-aid information. The EPA also maintains lists such as List N, which identifies disinfectants effective against emerging pathogens, including SARS-CoV-2, based on FIFRA efficacy criteria [1]. Taken together, FIFRA helps ensure that disinfectants used across different media meet consistent federal expectations for performance and safety.

11.1.1.1 Halogens

Halogens are among the most widely used chemical disinfectants, particularly chlorine and related chlorine-containing compounds. Chlorine has been used as a disinfectant for nearly 200 years [2].

Chlorine, Cl_2 at room temperature, is a gas and is transported as a liquefied gas under pressure. As such, it can be dissolved in water to form aqueous dissolved chlorine ($Cl_{2(aq)}$), with the equilibrium described by Henry's law.

The aqueous chlorine gas will hydrolyze to form hypochlorous acid, HOCl, which is a weak acid and can dissociate to hypochlorite, OCl^- according to the following equilibria:

$$Cl_{2(aq)} + H_2O \leftrightarrow HOCl + H^+ + Cl^- \tag{11.1}$$

$$HOCl \leftrightarrow OCl^- + H^+ \tag{11.2}$$

One molecule of Cl_2 can be reduced to two chloride ions (Cl^-) by accepting two electrons. Similarly, one hypochlorous acid or hypochlorite molecule can be reduced to one chloride ion by accepting two electrons. Hence, on an electron equivalent basis, Cl_2, HOCl, and OCl^- have the same oxidizing power, and it is conventional to refer to the concentrations on the basis of "as chlorine." Furthermore, these three species are also referred to as "free chlorine" in contrast to other oxidized forms bound to ammonia, as described subsequently.

Henry's law can describe the solubility of chlorine gas, with an equilibrium constant,[1] K_H in units of $\frac{L-atm}{mol}$ given as a function of temperature [3] by:

$$K_H = (6.553 \cdot 10^5)\exp\left(-\frac{3163.7}{T}\right) \tag{11.3}$$

This equation describes the relationship between the partial pressure of Cl_2 gas and the dissolved species $Cl_{2(aq)}$. Temperature (T) is in absolute degrees Kelvin.

1 In this discussion, equilibrium constants are written based on liquid concentrations and atmospheric partial pressures, assuming ideal conditions. Under nonideal conditions, such as at high ionic strength (e.g., >0.1 M), equilibria should be written using activities, incorporating nonideality corrections.

The hydrolysis or disproportionation reaction (Equation 11.1) has an equilibrium constant and value given by Equation 11.4 [3].

$$K_d = \frac{[HOCl][H^+][Cl^-]}{[Cl_{2(aq)}]}$$
$$log_{10}(K_d) = -\frac{982,798}{T^2} + \frac{5485.7}{T} - 10.7484 \tag{11.4}$$

In Figure 11.2, combinations of chloride and hydrogen ion concentration at which hypochlorous acid and aqueous chlorine concentrations are equal are plotted. As noted, except at highly acidic conditions *and high chloride concentrations*, the aqueous dissolved chlorine concentration is expected to be much lower than the hypochlorous acid concentration.

Hypochlorous acid behaves as a weak acid with a dissociation constant that can be expressed as a pK_a ($pK_a = -log_{10}(K_a)$). It has a temperature dependence correlated by the following relationship [3]:

$$pK_a = 0.0253\ T - 10.0686 + \frac{3000}{T} \tag{11.5}$$

Over the range of 0–35 °C, the pK_a values are 7.83 and 7.46, respectively. Hence, undissociated hypochlorous acid predominates under acidic conditions, while hypochlorite ion predominates under alkaline conditions. This is significant since it is well established that biocidal efficiency is strongly dependent on pH, with HOCl being a more potent disinfectant than hypochlorite [4].

In passing, it is noted that the trichloride species, Cl_3^- can also exist. However, except in very high-chloride systems such as seawater, its presence is negligible [3].

Since there is a rapid equilibration among the forms of aqueous free chlorine, this oxidant can be supplied in multiple ways. While gaseous chlorine is used in high-volume applications, such as large water or wastewater treatment plants, its shipping and storage pose several safety and security hazards. Alternative sources of free chlorine include sodium hypochlorite (e.g., commercial liquid bleach) and calcium hypochlorite. It is also possible to continuously generate hypochlorite from the electrolysis of brines; depending on the cost of power, this may be cost-competitive.

When free chlorine is added to water, it can react with numerous constituents that may be present. From the standpoint of affecting the chemistry of the disinfectant itself, the most important reactions are those with inorganic ammonium/ammonia and organic ammonium-containing compounds.

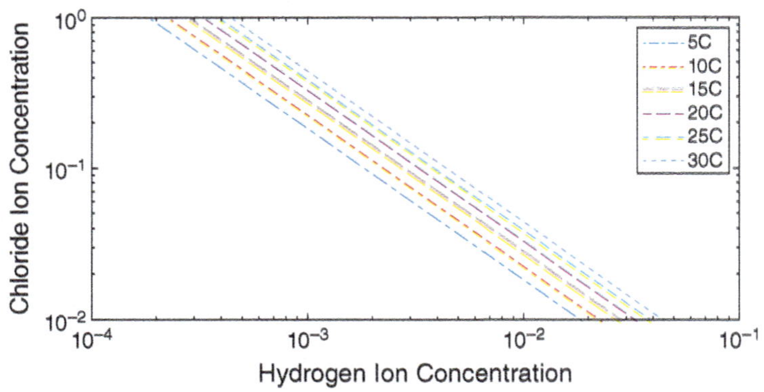

Figure 11.2 Combinations of hydrogen ion and chloride at which $[Cl_{2(aq)}]$=[HOCl].

It has long been known that as the dose of chlorine applied to a water (or wastewater) increases, there is a complex dependence of the residual remaining [5]. Phenomenologically, this is shown in Figure 11.3. At low chlorine doses, there may be an immediate demand not shown in the curve. Then, up to the "hump," the residual increases in a roughly proportional manner with dose (Zone 1). However, beyond the "hump," in Zone 2, as the dose increases, there is oxidation of ammonium until the "breakpoint" is reached, wherein nearly all of the ammonium is oxidized. After that, in Zone 3, the residual increases again. Qualitatively, this phenomenon is governed by a complex reaction scheme involving the successively chlorinated ammonia derivatives, NH_2Cl (monochloramine), $NHCl_2$ and NCl_3 (trichloramine). The formation of these chloramine species can be initially considered as the following processes, which are generally not at equilibrium:

$$HOCl + NH_3 \leftrightarrow NH_2Cl + H_2O \tag{11.6}$$

$$HOCl + NH_2Cl \leftrightarrow NH_2Cl + H_2O \tag{11.7}$$

$$HOCl + NHCl_2 \rightarrow NCl_3 + H_2O \tag{11.8}$$

The reactions in Equations 11.6–11.8 should not be considered elementary. Jafvert and Valentine [6] and Deborde et al. [7] have reviewed the kinetics and mechanisms of the chlorine ammonia system, which also involve transient, and incompletely known intermediates. The underlying finding is that once substantial dichloramine is formed—as in Zone 2 in Figure 11.3, the kinetics for oxidation of the nitrogen in ammonia become favorable relative to the kinetics of formation of trichloramine, and so the latter species is rarely present at substantial concentrations.

Collectively, monochloramine, dichloramine, and trichloramine are denoted as "combined chlorine." The sum of combined chlorine and free chlorine is denoted as "total chlorine." Although Figure 11.3 suggests that the hump and dip of the breakpoint curves occur at a 5:1 and 7:1 weight ratio of chlorine to ammonia nitrogen, the precise ratios will depend on multiple water quality

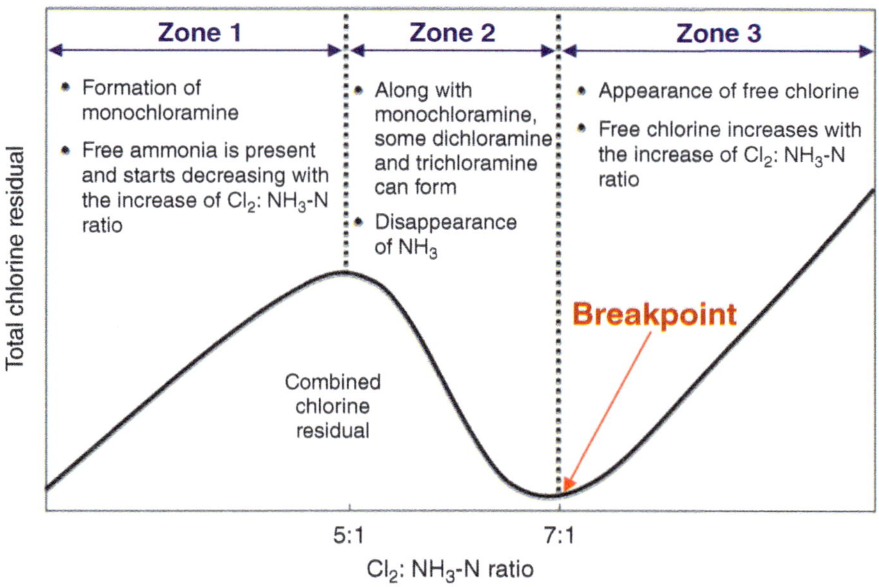

Figure 11.3 Idealized chlorine breakpoint curve. *Source:* [8]/MDPI/CC BY 4.0.

factors, including pH and presence of organic amine compounds, as well as the contact time between chlorine and ammonia nitrogen.

The chemical form in which chlorine is present is important in understanding its potency as a disinfectant. Free chlorine is more effective than combined chlorine for microorganisms.[2] Furthermore, free chlorine at low pH (where HOCl predominates) tends to be more potent than at high pH (where OCl^- predominates).

Chlorine can react with a broad suite of organic compounds by either oxidation (and hence reduction of the amount of active chlorine present) or by substitution in the case of amines to form (at least transiently) organo-amine structures [7]. In general, however, although some organic chloramine compounds may be measurable as total chlorine, their efficacy as disinfectants is small [10].

A significant exception is the mono-, di-, and trichlorinated forms of isocyanuric acid. The structure of trichloroisocyanuric acid is shown in Figure 11.4. These are approved as disinfectants in drinking water, and also used in disinfection of pools. These compounds are widely available commercially and are relatively easy to transport and ship. The amine chlorine groups are in equilibrium with free chlorine in solution and can act as a constant reservoir of free chlorine in a system [11].

Stable by-products of chlorination can be significant sources of human health risk. These will be discussed in more detail in Section 12.2.

Another common chlorine-containing disinfectant used in water (and wastewater), surface, and air applications is chlorine dioxide (ClO_2). The chlorine atom in this species is at the +IV oxidation state; hence, this is a more oxidized form.

For larger-scale disinfection applications, chlorine dioxide must be generated at or near the point of use. This has generally been done using a reaction between either hydrochloric acid or chlorine and chlorite (ClO_2^-) [12], although it should be noted that in very high quantity use cases (tons per day), such as in industrial scale pulp and paper manufacture, chlorine dioxide can efficiently be generated by reduction of chlorate (ClO_3^-).

Chlorine dioxide can be generated to produce solutions for disinfecting liquids or as a surface disinfectant. It is also possible to generate gaseous chlorine dioxide for fumigation of contaminated rooms; this was one of the cleanup methods used following the 2001 anthrax contamination episodes [13].

The generation of chlorine dioxide by the various routes may not produce a 100% pure stream of the disinfectant, but there may also be chlorine, hypochlorous acid, or hypochlorite ion present. In

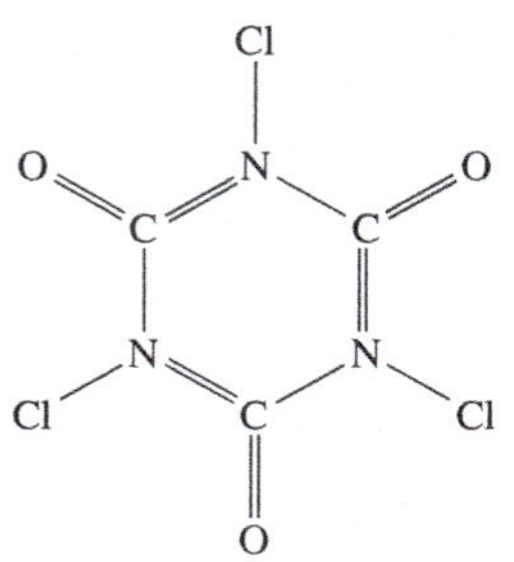

Figure 11.4 Structure of trichloroisocyanuric acid.

2 With an exception being that if the target microorganisms are embedded in biofilms, chloramines may be more effective since they can penetrate into biofilms to a greater extent than free chlorine [9].

such mixed solutions, the non-chlorine dioxide components will react with other constituents in the environment in a similar manner as if there was no ClO_2 present. In solution, there is no effect of pH on chlorine dioxide speciation. ClO_2 can react with constituents in water to form either the chlorite ion $(Cl)_2^-$) by a one-electron reduction, or more infrequently to the chloride ion.

Bromine and iodine compounds can be used as disinfectants, although their primary uses are in medical applications.

11.1.1.2 Hydrogen Peroxide

Hydrogen peroxide (H_2O_2) is a widely used disinfectant known for its strong oxidizing properties and broad-spectrum antimicrobial activity. As a clear, colorless liquid, it is commercially available at concentrations ranging from 3% for household use to over 30% for industrial and medical applications. Its effectiveness lies in its ability to produce reactive oxygen species (ROS), particularly hydroxyl radicals ($^{\bullet}OH$), which can disrupt the cellular membranes, proteins, and DNA of bacteria, viruses, fungi, and spores. Because it decomposes into water and oxygen, hydrogen peroxide is considered an environmentally friendly disinfectant with minimal toxic residue.

Hydrogen peroxide can also be used for gaseous decontamination as vaporized hydrogen peroxide (VHP) [14].

The disinfecting power of hydrogen peroxide stems from its chemical instability and tendency to break down into water (H_2O) and oxygen (O_2), by oxidizing other substances — note that the formal oxidation state of the oxygen atom in peroxide is -I, compared to 0 in elemental O_2 and -II in H_2O. Peroxide reactions can be caused by light, heat, or various chemical or biological materials. This decomposition releases nascent oxygen and, more importantly, can lead to the formation of highly reactive free radicals, such as hydroxyl radicals, through processes like the Fenton reaction [15]:

$$Fe^{2+} + H_2O_2 \rightarrow Fe^{3+} + \bullet OH + OH^- \tag{11.9}$$

These radicals are highly effective in damaging lipids, nucleic acids, and proteins, thereby enhancing the antimicrobial action of hydrogen peroxide [16]. Unlike chlorine-based disinfectants, hydrogen peroxide does not leave harmful by-products or persistent residues. Its efficacy can be further enhanced when combined with other agents (e.g., silver ions or surfactants) or applied in vaporized or plasma states, offering versatile solutions across different disinfection needs.

11.1.1.3 Ozone

O_3 is a colorless gas that is highly reactive and, as such, needs to be continually produced at the point of use. There are several commercial technologies for ozone generation: photochemical using very short wavelength UV, and corona discharge. The latter is by far the most common [17].

In corona discharge O_3 generation, a source gas is passed through an electrode stack across which a significant electric potential is maintained, and the capacitive discharge generates a plasma in which ozone is formed. This is an energy-intensive process. The theoretical minimum energy required is the free energy of formation of ozone from oxygen, which is 163.3 $\frac{kJ}{mol}$ under standard conditions [18]. The practical energy requirements are generally (substantially) greater than this due to heating associated with the dielectric system, back reactions that cause decomposition, and the formation of other products, depending on the presence of other constituents in the gas.

If air is used as the feed to ozone generation, it must be dried to a very low relative humidity — ordinarily expressed as a dew point below −54°C [17]. Dry air is needed for two reasons. First, moisture can reduce ozone generation efficiency. Second, the electric field will also oxidize the nitrogen gas present to produce various oxides of nitrogen, some of which, in turn, if significant moisture is present, would produce nitric acid vapor that can corrode materials. Using dry air as the feed gas, the typical resulting ozone concentration is 2–4% [3].

Pure oxygen can be used as a feed gas for ozone production. The feed can be from commercially supplied liquified oxygen (LOX) or by continuous production on-site of oxygen. The latter is often preferred for larger installations, particularly where other onsite uses for oxygen exist.[3] Intrinsically, pure oxygen sources are very dry and have low nitrogen gas content. With O_2 as a feed gas, it is possible to achieve 6–12% [3] ozone concentration in the product.

While there are some applications of gaseous ozone, as produced for air disinfection (Section 12.4.2), most applications of ozone rely on either dissolution into water or the production of a concentrated ozone solution for application. Ozone is a soluble gas; therefore, its equilibrium can be described by Henry's law.

Levanov et al. [20] reviewed data on the solubility of ozone in water. By theory, this can be modeled by a Henry's law constant that is dependent on the ionic composition of the solution (a so-called "salting out" effect). They report a function, with H defined in units of $\frac{mol}{L \cdot bar}$ in Equation 11.10 for equilibrium in water containing NaCl at different concentrations.

$$\log_{10}(\mathrm{H}) = -5.76 - 0.234[\mathrm{Cl}^-] + \frac{1120}{T} \tag{11.10}$$

In the presence of other ions than NaCl, the coefficient describing the effect of ionic concentration will differ, and must be determined empirically [21].

Example 11.1 Calculation of Ozone Solubility.

The product of an ozone generator has 4% gaseous O_3 concentration. If this is bubbled into a solution, what is the equilibrium liquid concentration? Assume the temperature is 20 °C and the solution can be described as NaCl at 0.01 M concentration.

Solution:

1. The partial pressure of ozone is 0.04 atm. This is equal to 0.04 times 1.01325 = 0.0405 bar.
2. Employing Equation 11.10, the Henry's law constant can be calculated:

$$log_{10}(H) = -5.76 - 0.234[Cl^-] + \frac{1120}{T}$$

$$= -5.76 - 0.234(0.01) + \frac{1120}{293}$$

$$= -1.9398$$

$$\therefore H = 0.0115 \frac{mol}{L \cdot bar}$$

3. Using the value of H, the solubility can be calculated as (also noting the ozone molecular weight of 48 g/mol):

$$[\mathrm{O_3}] = 0.0115 \cdot 0.04 = 4.59 \times 10^{-4} \frac{\mathrm{mol}}{L} = 22.05 \frac{\mathrm{mg}}{L}$$

The solubility of ozone is a driving force for ozone transfer from the gas phase, although it is rarely achieved due to ozone's reactivity.

3 For example, pure oxygen can be used in high-rate activated sludge systems for wastewater treatment [19]. The incremental cost of additional oxygen generation for supply to an ozone generator for disinfection would generally be relatively minor on a unit cost basis.

When added to water, O_3 can enter into a complex series of reactions. In particular, the hydroxyl ($OH^{\bullet}$) and superoxide ($HO_2^{\bullet}$) radicals can be formed [22]. The hydroxyl radical is particularly important in oxidizing organic trace contaminants. Although it has frequently been regarded that molecular ozone is the more important biocide, in at least some circumstances, the hydroxyl radical can function as a disinfectant [23]. The deliberate formation of hydroxyl radical via various routes, including the use of ozone, is a feature of advanced oxidation processes, which are discussed in Section 11.1.3.

The formation of radicals from ozone, and hence the decomposition of ozone residual, occurs through a number of pathways, including radical formation promotion and inhibition, and is therefore a function of the detailed chemistry of a system. The rate of ozone decay is often modeled as first order in ozone concentration. Like most rates, this is temperature dependent, although, as reviewed by [22], there is no consensus value for the activation energy of the rate constant.

There is a water quality dependence of the ozone decay rate. Gardoni et al. [22] restate the correlation of Laplanche, which gives the first-order decay rate constant, k_d in units of s^{-1} as a function of pH, alkalinity (ALK in mg/L as $CaCO_3$) dissolved organic carbon (DOC, in units of mg/L), and UV absorbance at 254 nm[4] (UVA, in units of cm^{-1} shown in Equation 11.11

$$\begin{aligned}\log_{10}(k_d) = &-3.93 + 0.24\ \text{pH} + 0.7537\ \log_{10}(\text{UVA}) \\ &+ 1.08\ \log_{10}(\text{DOC}) - 0.19\ \log_{10}(\text{ALK})\end{aligned} \tag{11.11}$$

Neither [22] nor the original paper gives the temperature associated with this decay rate, which was obtained by analyzing results on numerous natural waters, so this should be interpreted as being useful over a range of typical natural waters, perhaps 20-25 Celsius.

Example 11.2 Calculation of O_3 Decay Rate and Half-life in Water.

Using the correlation from Equation 11.11, calculate the ozone decay rate for a water with the following characteristics:

pH	7.5
UVA	0.03 cm^{-1}
DOC	1.5 mg/L
ALK	50 mg/L

Solution:
Substituting the water values into the equation yields:

$$\begin{aligned}log_{10}(k_d) = &-3.93 + 0.24(7.5) + 0.7537\ log_{10}(0.03) \\ &+1.08\ log_{10}(1.5) - 0.19\ log_{10}(50)\end{aligned}$$

Evaluating this and taking the antilog:

$$k_d = 0.00172\ s^{-1}$$

(Continued)

4 UV absorbance at 254 nm is a frequently used measure of the quality of organic material in a water which is particularly sensitive to the aromatic organic compounds in water.

Example 11.2 (Continued)

The half-life can be determined by:

$$t_{\frac{1}{2}} = \frac{ln(2)}{0.00172} = 403.3\ s$$

Therefore, half the ozone residual in this water (in a batch system) is expected to decay in 6.7 minutes.

In very pure (low DOC) cold water, ozone stock solutions can be prepared and remain stable for extended periods,[5] but for larger applications, continuous generation is needed. The generated ozone can be applied as a gas in air disinfection. In water disinfection, the gaseous ozone must be contacted with water and be transferred into the solution phase. There are different types of contacting, and these systems will be discussed in Section 12.4.1.

11.1.1.4 Peracetic and Other Peroxy Acids

There has been increasing interest in the use of peracids for disinfection. In general, these are compounds that have a peroxide functional group (–O–O–) as part of their structure. The most widely used member of this class of compounds is peracetic acid (PAA), which is also known as peroxyacetic acid. PAA is a potent oxidizing agent widely utilized for its broad-spectrum antimicrobial properties.

PAA is increasingly used in environmental applications, notably in wastewater and drinking water treatment. It is advantageous for drinking water treatment because it does not generate significant organic disinfection by-products. In wastewater management, it serves as an alternative to chlorine-based disinfectants because it is effective at reducing microbial loads without generating harmful, stable ecotoxic residuals and by-products. This characteristic makes PAA particularly suitable for environmentally sensitive areas, where compliance with stringent discharge regulations is essential [25]. Furthermore, its rapid decomposition into harmless by-products — acetic acid, oxygen, and water — minimizes ecological impact.

Peracetic acid is an organic peroxide with the formula CH_3COOOH. Its structure is shown in Figure 11.5.

PAA is produced from the equilibrium reaction of acetic acid and hydrogen peroxide in the presence of an acid catalyst and is typically supplied in 15–50% stabilized aqueous solutions containing varying concentrations of acetic acid, hydrogen peroxide, and water [26]. The equilibrium reaction can be represented by:

$$CH_3COOH + H_2O_2 \rightleftharpoons CH_3CO_3H + H_2O \tag{11.12}$$

Peracetic acid is also a weak acid with a pKa of 8.2 under standard conditions [26].

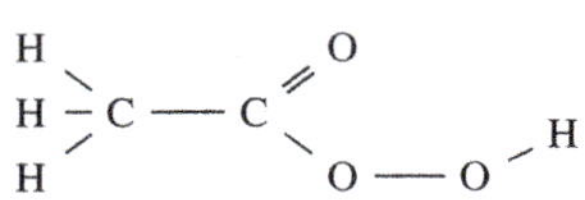

Figure 11.5 Structure of peracetic acid.

5 In laboratory disinfection studies, it is entirely possible to prepare stock solutions of ozone in cold demand-free water, and under dark refrigerated conditions, the residual can be stable for hours [24].

Peracetic acid can decompose by two routes: spontaneously (Equation 11.13) and by base-catalyzed hydrolysis (Equation 11.14).

$$2\,CH_3C(O)OOH \rightarrow 2\,CH_3COOH + O_2 \tag{11.13}$$

$$CH_3C(O)OOH + H_2O \rightarrow CH_3COOH + H_2O_2 \tag{11.14}$$

The overall decomposition rate tends to be greatest near pH 8.2. PAA solutions in excess of 15% can be explosive, and this common solutions in use are in the neighborhood of 10%. While dilute, 1%, PAA solutions are relatively unstable, with a half-life of six days, common 10% solutions exhibit greater stability [25]. Zhang et al. have comprehensively reviewed the thermodynamic properties of PAA [27].

A more recent emerging peracid is peroxymonosulfuric acid, H_2SO_5. Although this shows promise both for disinfection, particularly on surfaces [28], and as an organic oxidant [29], its commercial use has been limited.

11.1.1.5 Ethylene Dioxide

Ethylene oxide (C_2H_4O) is a colorless gas used as a sterilant. Its structural diagram is shown in Figure 11.6.

Ethylene oxide has been used as a medical sterilizing agent, particularly for heat-sensitive instruments and equipment, although careful cleaning before treatment is essential for effective inactivation of organisms [30]. However, the International Agency for Research on Cancer's classification of this agent as a group 1 carcinogen has limited its broader application, although systematic literature reviews may have cast some uncertainty on this designation [31].

11.1.1.6 Organic Disinfecting Compounds

A final class of chemical agents to be considered are organic compounds that can be supplied commercially. Many of these serve as both cleaning agents and inactivating agents. Three broad classes will be considered — quaternary ammonium compounds, or quats, alcohols, and phenolic compounds.

Quaternary ammonium compounds are based on ammonium, but with each of the hydrogen atoms replaced by an organic moiety. A general schematic is shown in Figure 11.7. Depending on the specific groups, R_1, R_2, R_3 and R_4, a quat will have different physical, chemical and antimicrobial properties. As ionic compounds, quats are frequency distributed as salts, neutralized with an anion, such as chloride.

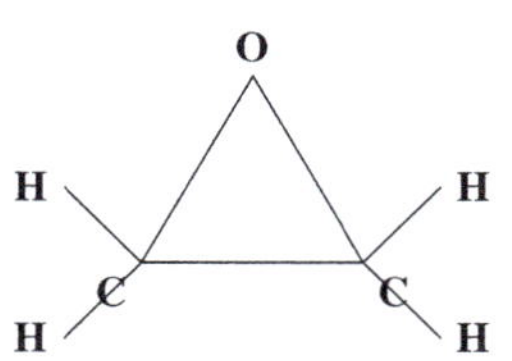

Figure 11.6 Structure of ethylene oxide.

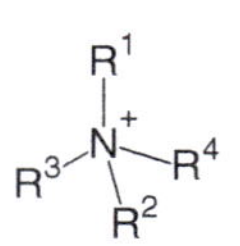

Figure 11.7 Structure of quaternary ammonium ion.

Because of the four, potentially distinct, organic groups, the quats are a very broad set of compounds. They have different uses, potencies, ecological effects upon release, and environmental fate and transport properties [32]. The earliest classes used as disinfectants were the alkyl dimethyl benzyl ammonium chlorides (ADBAC) and the dialkyl dimethyl ammonium chloride (DDAC) compounds [33].

The surface-active properties of the quats — particularly ones where one or more of the functional groups are hydrophilic and one or more are hydrophobic — give them a broad applicability both as a cleaning agent and also as an antimicrobial. As antimicrobials, they act by disrupting membranes and cell walls of vegetative organisms [34]; they can also be virucidal against enveloped viruses, although they are less effective against nonenveloped viruses [35].

In use, quats *per se* are generally stable; however, their efficacy as disinfectants may be diminished in the presence of other organic material, such as soils, biological matter, or humic material [36].

A second broad class of organic disinfectants are the alcohols. Broadly speaking, these are compounds that have a –C–OH bond, other than a carboxyl group, where the C is not on an aromatic ring. The most commonly used are ethyl alcohol and isopropyl alcohol, which are frequently used in disinfection of surfaces and some devices in medical settings [37].

A final class of organic agents is phenolic compounds. Phenolic compounds are a broad class of aromatic organic molecules characterized by one or more hydroxyl groups (–OH) directly attached to a benzene ring. They are structurally derived from phenol (C_6H_5OH), the simplest member of the group. Phenolics exhibit weak acidity, hydrophobicity, and the ability to denature proteins and disrupt cell membranes, which underpin their antimicrobial action [38].

Phenolic compounds have been utilized since the 19th century, beginning with phenol (carbolic acid) in antiseptic surgery. Their mechanism of action involves disruption of microbial cell walls and precipitation of proteins, making them effective against a broad range of bacteria and some viruses, though generally less effective against spores [39].

The principal phenolic disinfectants include:

- **Phenol (carbolic acid)**: Historically used in surgical antisepsis but largely phased out due to corrosiveness and toxicity.
- **Cresols**: Including ortho-, meta-, and para-cresol, used in formulations like early Lysol® products. Effective against Gram-positive and Gram-negative bacteria.
- **Chlorinated phenols**: Notably ortho-phenylphenol (OPP) and para-chloro-meta-xylenol (PCMX or chloroxylenol), which were common in household and healthcare disinfectants, but whose use is being diminished due to environmental and health concerns.[6]

Despite their efficacy, concerns about phenolics include toxicity, environmental persistence, and the potential to promote antimicrobial resistance. These concerns have led to a reevaluation of their use and substitution with other disinfectants in some applications.

11.1.2 Physical Disinfectants

In contrast to chemical agents, physical agents can be used for inactivating microorganisms. These can be broadly divided into heat and three forms of electromagnetic radiation: solar and near UV, UV, and ionizing radiation.

6 Though their use remains in such controlled applications as surgical site preparation and wound care under medical supervision.

11.1.2.1 Heat

The use of heat as an antimicrobial agent was known in antiquity for food preservation. Even prior to the acceptance of the germ theory of disease, the autoclave, as a device for providing steam at increased temperature and pressure, was employed as a substitute for quarantine. The work of Pasteur and Koch provided a basis for the use of either dry or wet heat for sterilization [40].

The principal variable characterizing heat-induced inactivation is temperature. Although at low and moderate temperatures, depending on the particular microorganism,[7] vegetative organisms exhibit increasing growth rates. However, as temperature increases beyond the thermal optimum, inactivation occurs.

Temperature can be increased using various methods, including external heating with gas-fired or electric devices, microwave heating, or infrared radiation. Inactivation occurs through the denaturation of critical biological molecules, such as proteins and nucleic acids, leading to death. The rate of inactivation may be altered by water activity, or equivalently in air, relative humidity, and the presence of chemical or biological substances that may enhance or protect against such adverse effects. The inactivation is a rate process, and thus the extent of inactivation is a function of time and temperature. A discussion of kinetics and these alterations is in Section 11.2.3.

11.1.2.2 Light

Light, like all forms of electromagnetic (EM) radiation, is classified by wavelength—the distance between the peaks of successive waves. UV has a wavelength less than visible light, ranging from the size of microorganisms to the size of large molecules (Figure 11.8). As a practical matter, water (either in vapor or as liquid) has a high absorbance of light with wavelengths less than about 200 nm (termed "vacuum UV") and above 700 nm (the infrared region and above). In addition, some containers, such as ordinary glass, absorb UV radiation, thereby limiting the practical wavelength range for light inactivation.

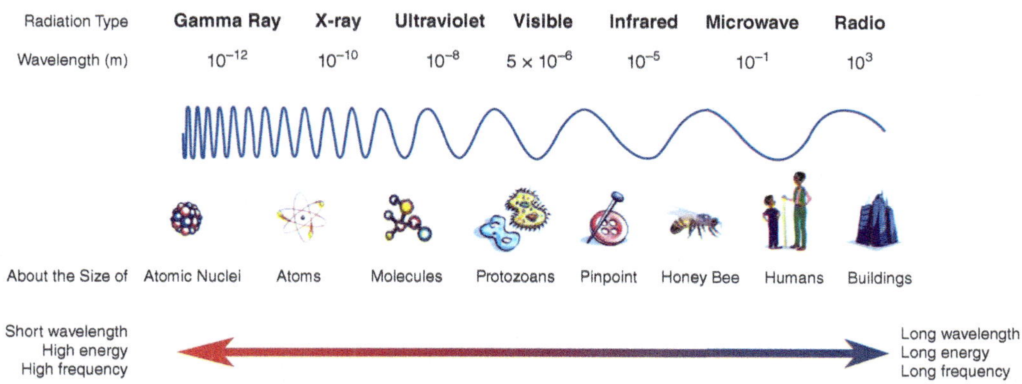

Figure 11.8 The electromagnetic spectrum. *Source:* Modified from NASA. https://mynasadata.larc.nasa.gov/basic-page/electromagnetic-spectrum-diagram (last accessed February 11, 2026).

---

7 For example, psychrophiles have optimum growth rates at 15 °C or lower, mesophiles between 20 and 45 °C, and thermophiles typically with an optimum above 45 °C. There are some extreme thermophiles with optima above 70 °C; however, none of these are known to be pathogens [41].

The most common source of ultraviolet (UV) radiation is sunlight, which emits three primary types of UV rays:

- **UVA** (320–400 nm)
- **UVB** (280–320 nm)
- **UVC** (100–280 nm)

Among these, UVA rays have the longest wavelengths, followed by UVB, while UVC rays have the shortest. The Earth's atmosphere, particularly the ozone layer, absorbs all UVC radiation and most UVB rays. Consequently, the majority of UV radiation that reaches the Earth's surface consists of UVA rays, with a smaller portion of UVB.

- **UVB rays**, with wavelengths ranging from 280 to 320 nm, primarily affect the *epidermis*, the outermost layer of the skin.
- **UVA rays**, with longer wavelengths of 320–400 nm, penetrate more deeply into the *dermis*, the middle layer of the skin.

UV disinfection systems used for air, water, and surface treatment are regulated in the United States primarily based on their intended use. For water treatment applications, particularly in potable water systems, UV systems fall under the purview of the Safe Drinking Water Act (SDWA) and are subject to guidance from the US Environmental Protection Agency (EPA). For wastewater and non-potable applications, UV systems must comply with relevant state regulations and performance standards, often involving bioassay-based validation to demonstrate efficacy. When UV systems are marketed for antimicrobial use in air and on surfaces, they may fall under the jurisdiction of the Federal Insecticide, Fungicide, and Rodenticide Act (FIFRA) if pesticide claims are made — i.e., if the device is claimed to kill, inactivate, or control microorganisms. However, unlike chemical disinfectants, UV devices are often exempt from EPA registration as pesticides under FIFRA, provided they do not involve a chemical substance and are not marketed with pesticide claims beyond physical or mechanical means [42]. Nonetheless, such devices remain subject to enforcement by the Federal Trade Commission (FTC) for truthful advertising, and the EPA may take action against misleading or unsupported efficacy claims.

When UV disinfection systems are used in healthcare settings — such as hospitals, clinics, or long-term care facilities — the US Food and Drug Administration (FDA) may have a regulatory role, particularly if the device is intended for use on medical devices, in surgical environments, or in applications directly related to patient care.

The FDA requires manufacturers of such UV systems to provide evidence of safety and effectiveness, including data on microbial inactivation, compatibility with medical equipment or materials, and labeling that includes clear instructions for use and warnings. If the device emits UV radiation, it is also subject to performance standards for electronic products under Federal regulations that address exposure limits, labeling, and safety mechanisms. In addition to FDA oversight, healthcare facilities using UV disinfection technologies must also comply with infection control guidelines from the Centers for Disease Control and Prevention (CDC) and recommendations from professional bodies. Therefore, the use of UV systems in healthcare is governed by a complex interplay of FDA device regulation, EPA/FIFRA rules (if pesticidal claims are involved), and facility-level infection prevention protocols.

11.1.2.3 Ionizing Radiation

Ionizing radiation has sufficient energy such that it is capable of altering the chemical composition of substances by the addition or deletion of electrons [43]. They can be produced by emissions of radioisotopes (alpha particles and beta and gamma rays) or by electronic sources, such as X-ray emissions, electron-beam sources, and proton-beam sources. Their ability to produce chemical alterations includes the radiolysis of water and the generation of free radicals, hydrogen peroxide, and other oxidizing substances [44]. Hence, the effect of ionizing radiation may arise from both the direct action of the radiation itself and from by-products produced by reactions with water and other materials.

11.1.3 Advanced Oxidation – Combinations of Processes

Advanced oxidation processes refer to a group of chemical treatment procedures designed to remove organic and inorganic contaminants from water and air by oxidation through reactions with hydroxyl ($OH^\bullet$) and other radicals. While these radicals are extremely reactive and nonselective, capable of decomposing a wide range of persistent pollutants, including pharmaceuticals, pesticides, dyes, and endocrine-disrupting compounds, they may not be as potent as the disinfectants used to generate them.

These radicals can be generated, albeit at low concentrations (e.g., $10^{-11} - 10^{-9}\frac{\text{mol}}{\text{L}}$) by combinations of oxidants including hydrogen peroxide, ozone, persulfate, and UV light [45]. Due to their relative lack of efficacy as disinfectants, they will not be discussed further. A good review is provided by Feijoo et al. [46].

11.2 Kinetics of Disinfection

Disinfection is intrinsically a rate process between the microorganisms to be inactivated and the agent(s) producing disinfection. As with any other rate process, other reactions may interfere with or potentiate inactivation. There are thus two fundamental variables that will describe the process, potentially with others:

- The instantaneous *concentration* of viable microorganisms in a particular region (possibly infinitesimal), denoted by υ, at any location in our system,[8] typically in units of $\frac{\#}{\text{Volume}}$. If surfaces are disinfected, the areal concentration $\frac{\#}{\text{Area}}$ might be more useful. If a solid or semi-solid material is being disinfected, $\frac{\#}{\text{Mass}}$ might be more useful.
- The instantaneous amount of disinfectant in a particular region, denoted generically by $\mathcal{C}$. The specific units will depend on the particular disinfectant. With chemical agents, this will generally be $\frac{\text{Mass}}{\text{Volume}}$ denoted by the symbol C. With light or ionizing radiation, the analogous particularization of $\mathcal{C}$ is as $\frac{\text{Power}}{\text{Volume}}$, where power is the particular instantaneous amount delivered to that specific region in space. With thermal disinfection, an appropriate dose variable is simply the instantaneous temperature, T, at any location.

Using these definitions, some general concepts of inactivation kinetics can be outlined.

8 This symbol (Greek lowercase upsilon) is adopted here for microbial concentration. Many references use μ for concentration; however, this can be confused with growth rate. Note in particular that count data (generally given the symbol N) is an integer, while υ is a continuous variable.

11.2.1 General Inactivation Models

Amongst the first to apply kinetic concepts to disinfection was Chick [47], who described the rate of disinfection (using modern nomenclature) as first order, i.e., Equation 11.15, where r_d is the rate of inactivation by a disinfectant. Note that this is identical to the first-order decay models discussed in Section 8.4.2.

$$r_d = -kv \tag{11.15}$$

Shortly thereafter, Watson proposed that the inactivation rate constant itself would be proportional to the disinfection concentration raised to a particular exponent, "n," frequently termed the coefficient of dilution. Conceptually, if $n > 1$, a marginal increase in disinfectant produces a greater increase in inactivation than that same marginal increase in time, while if $n < 1$, the opposite is true. The resulting Chick–Watson law is shown in Equation 11.16.

$$r_d = -k'C^n v \tag{11.16}$$

In a completely mixed batch system, if C is constant with time, and the parameters remain invariant, then Equation 11.16 can be integrated to yield:

$$\frac{v}{v_0} = \exp(-k'C^n t) \tag{11.17}$$

where v_0 is the initial concentration of viable microorganisms. The ratio $\frac{v}{v_0}$ is denoted as the survival ratio, with symbol S.

Plots of the Chick–Watson relationship are shown in Figure 11.9 for a particular value of the kinetic coefficients in a batch system with no disinfectant decay. Note that the plots of survival ratio versus time (left-hand panel) are linear on a semi-log plot, and when the loci of combinations yielding a fixed survival ratio are plotted (right-hand plot), they are linear on a log-log plot.

Chick's law describes an exact exponential decay of organisms over time. As noted in the discussion of decay models, survival curves in the presence of inactivating agents often deviate from such (log-)linear behavior, exhibiting "shoulders," "tailing off" and combinations of such forms (Section 8.4.2). It has long been recognized that the response of microorganisms to inactivating agents may also deviate from such log-linear relationships [48]. Therefore, several other models have been developed to describe these characteristics.

In considering alternative models, it is preferable that the rates be expressible only as functions of "state variables," rather than of descriptors of the prior history of the element under consideration.[9] A state variable in thermodynamic terms, which is generalized here, is a property of a system that defines its current condition or state, independent of the system's history [49]. For example, this might include the instantaneous local disinfectant concentration, temperature, light intensity, pH, viable microorganism concentration, and total microorganism concentration.

Models expressible as rates involving only state variables can be used to describe behavior in batch systems and also open systems with different mixing characteristics. If the rate expression(s) truly depict the process(s), then there should be equal fidelity in modeling in all types of systems.

Gyurek and Finch [50] have comprehensively reviewed many elaborations on Chick–Watson kinetics. The first set of elaborations can be expressed generically as deriving from the following modification of Equation 11.16:

9 This is analogous to the distinction between autonomous and nonautonomous models for microbial growth presented in Section 8.4.3. The use of nonautonomous disinfection models in complex reactor flows would require particle-tracking models for analysis.

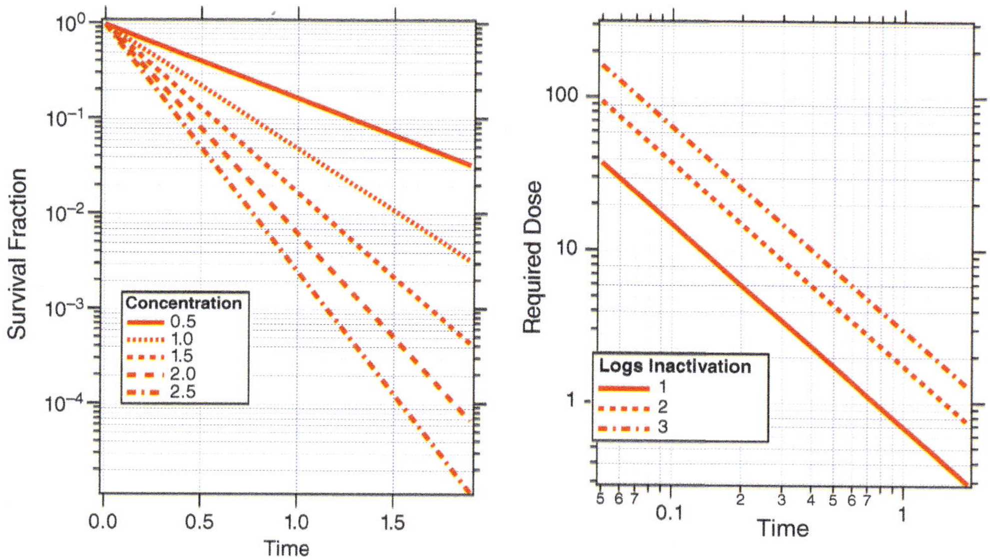

Figure 11.9 Plots of Chick–Watson kinetics in a batch system with no decay. Chick's law in left panel and Watson's law in right panel. Drawn for $k' = 3$, $n = 0.75$.

$$r_d = -k' m C^n v^x t^{m-1} \tag{11.18}$$

In batch systems, for particular cases, as well as the general case, under demand-free conditions, the differential mass balance can be solved. A discussion of the particular forms follows.

The Chick–Watson law is obtained when $x = 1$ and $m = 1$. The model where $x = 1$ and $m \neq 1$ was first developed and used by Hom [51] and is referred to as the Hom model. In batch systems with constant disinfectant concentration, this can be integrated to yield[10] :

$$\ln\left(\frac{v}{v_0}\right) = -k' C^n t^m \tag{11.19}$$

If $m > 1$, this model will produce a shoulder on the plot of log survival versus t, while it $m < 1$, it will exhibit tailing off. Also note that the power-law dependence on time is reminiscent of the Weibull distribution for survival times discussed in Section 8.4.2, so the Hom model can be interpreted in that context as well. The Hom model can also be derived when an intermediate microorganism-disinfectant complex must be formed prior to inactivation, yielding a Hom m value of 2 [52].

In its original form, the Hom law in the special case of Equation 11.18, since by containing "t," the rate is not a function only of state variables. This can be transformed into a relationship containing only state variables, as shown by Anotai [53], by using the integrated equation to back-substitute for time. This yields the following rate expression for the Hom rate model:

$$r_d = -m(k' C^n)^{\left(\frac{1}{m}\right)} v \left[-ln\left(\frac{v}{v_0}\right)\right]^{\left(\frac{m-1}{m}\right)} \tag{11.20}$$

10 The symbol k' will be retained as the rate constant for all models; however, note that for the different rate laws, it will have different dimensions depending on the particular rate law, and perhaps order exponents.

Note that in this equation, unlike in the Chick–Watson rate, there is an explicit dependency of the rate on the total (initial) microorganism concentration, υ_0.

A kinetic relationship that is power law in microorganisms appears to have first been proposed by Majumdar et al. [54]. This has been referred to either as power law or rational kinetics. The rate equation is derived from Equation 11.18 with $m = 1$ and $x \neq 1$, and expressed as Equation 11.21, which when integrated with constant C in a batch system becomes Equation 11.22.

$$r_d = -k'C^n\upsilon^x \tag{11.21}$$

$$ln\left(\frac{\upsilon}{\upsilon_0}\right) = -\frac{1}{x-1} ln\left[1 + \upsilon_0^{x-1}(x-1)k'C^n t\right] \tag{11.22}$$

Note that, as with the Hom model, the power law model depends on the total microorganism concentration. As with the Hom model, the power law model can describe shoulders or tailing off depending on the value of x.

For the Hom power law (HPL) model, the rate, when back-substituting for time, and the batch performance equation in the absence of disinfectant change become Equations 11.23 and 11.24.

$$r_d = -m(k'C)^{\frac{1}{m}}\ \upsilon^x \left[\frac{\upsilon_0^{1-x} - \upsilon^{1-x}}{1-x}\right]^{\frac{m-1}{m}} \tag{11.23}$$

$$ln\left(\frac{\upsilon}{\upsilon_0}\right) = -\frac{1}{x-1}\ ln\left[1 + (x-1)\upsilon_0^{x-1}k'C^n t^m\right] \tag{11.24}$$

Note that the Chick–Watson law has two parameters, the Hom and Power law models have three parameters, and the HPL model has four parameters. Fitting data, determining the optimal parameter values, and selecting the best model, balancing fit and parsimony, become statistical exercises analogous to those discussed in Section 8.4.2. The specific fitting of disinfection models is covered in Section 12.1.

With four adjustable parameters, the HPL model provides tremendous flexibility in the shape of the survival-versus-time function. This is shown in Figure 11.10 with examples of curves with different parameter values. Shoulders, tailing off, and a precipitous decay resembling a threshold-type phenomenon can be described with varying parameter choices.

With both the power law and HPL model, unlike the Hom model (Equation 11.19), the batch survival curves have an explicit dependence upon the initial microorganism concentration (υ_0). This is shown graphically by an example in Figure 11.11, where the effect of initial concentration depends on the value of the parameter x.

The mechanism for the dependence of survival on the initial concentration of microorganisms remains unclear. Deliberate investigation of this effect is uncommon; however, several studies demonstrating this effect with chemical disinfectants in water [55, 56] are consistent with power law model kinetics.

The previous models that generalized from the Chick–Watson relationship were phenomenological, in that they were not based on specific mechanistic hypotheses. The next class of models proceeds from mechanistic hypotheses to derive a model from those hypotheses. These can be regarded as structured models.

However, it must be stressed that finding that an inactivation data set is consistent with a kinetic model does not *ipso facto* demonstrate the truth of the hypotheses from which that model was derived. The rationale for this statement is that functional forms consistent with a set of data can be derived from multiple approaches, and therefore, further information is generally needed to infer the underlying mechanisms. This information could include detailed microbiological characterization and/or demonstration that the kinetics obtained in one type of system, such as a batch reactor,

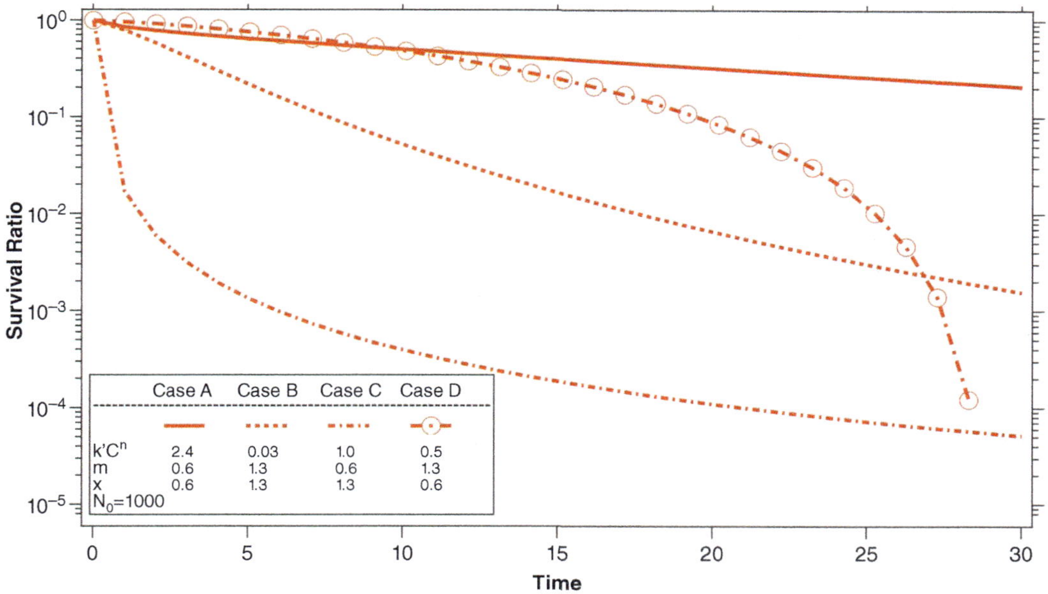

Figure 11.10 Examples of the HPL model with different parameter values.

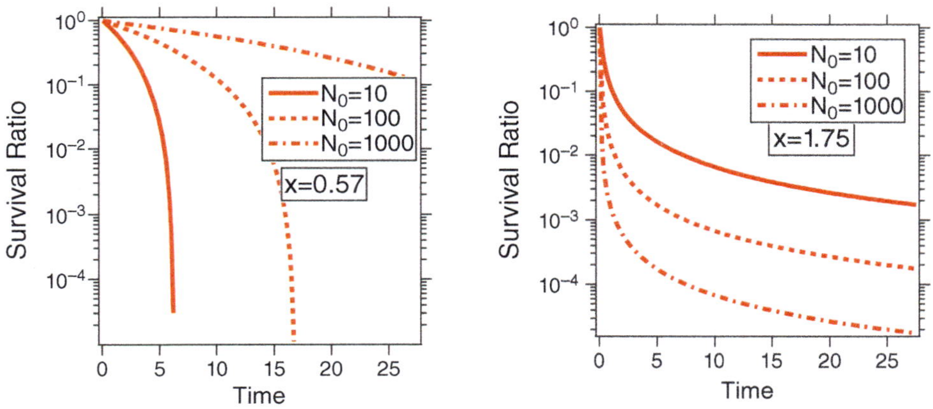

Figure 11.11 Effect of υ_0 on power law kinetics ($k'C^n = 1$).

provides good prediction of performance in other kinds of systems, such as completely mixed or partially mixed flow-through systems.

Two such structured models are the series-event (also called the multistage) models, which have a long history of analysis for describing survival curves [57]. Severin et al. [58] employed the series-event model for water disinfection. In this model, an individual microorganism is conceptualized to require a certain number of "hits" before being inactivated, with the rate of a hit being proportional to the disinfectant level, i.e., $k'C$.

With this model, the survival ratio is given by Equation 11.25.

$$\ln\left(\frac{v}{v_0}\right) = -k'Ct + \ln\left[\sum_{i=0}^{n-1} \frac{(k'Ct)^i}{i!}\right] \tag{11.25}$$

An alternative equivalent form to this relationship can be written in terms of the lower incomplete gamma function γ as Equation 11.26.

$$\ln\left(\frac{v}{v_0}\right) = \ln\left[1 - \frac{\gamma(n, k'Ct)}{(n-1)!}\right] \tag{11.26}$$

The series event model produces inactivation curves with shoulders.

The multi-target model is based on the assumption that a microbial cell contains multiple individual critical sites or targets that must *all* be inactivated for the organism to lose viability. In contrast to the multi-hit model, which assumes cumulative damage from identical hits, the multi-target model postulates that all targets must be independently damaged before cell death occurs. In the simple multi-target model, a single "hit" is required to inactivate a target. This model also describes survival curves with shoulders. The survival curve can be expressed as Equation 11.27.

$$\frac{v(t)}{v_0} = \left[1 - \left(1 - e^{-k'Ct}\right)^n\right] \tag{11.27}$$

where n is the number of critical targets that must *all* be hit for inactivation to occur. It is also assumed that hitting each target is independent and identical to hitting any other target. There are various biological processes that can lead to a multi-target model, for example:

- If individual organisms to be inactivated are "clumped" such that all organisms in a clump must be inactivated before the clump is completely killed. This can occur when the enumeration process does not dissociate all microorganisms in a clump into individual cells.
- If multiple copies of a key target exist in an organism (such as multiple genomes) and retention of any copy is sufficient to maintain viability. There is evidence that many microorganisms can exhibit polyploidy, in which they maintain multiple copies of their chromosomes [59].

While the series-event and multi-target models are derived from different underlying assumptions and, like the Hom model with $m > 1$, will produce shoulders, as a practical matter their fits to data may yield very similar results. This is illustrated in Figure 11.12. Unless a large number of very precise data points were available, use of the survival data *per se* to differentiate between models of this type would be difficult. The process of fitting inactivation models to data is discussed in Section 12.1.

Another class of survival models is derived from quasimechanistic assumptions, in which it is assumed that, during exposure to an inactivating agent, there is a change in the microorganism's intrinsic sensitivity to inactivation. Perhaps the first example of this was the work of Gard [60], motivated by difficulties in assuring high-level inactivation in the production of the Salk polio vaccine, developed a "tailing off" model to describe such a process:

$$\log\left(\frac{v}{v_0}\right) = -\log(1 + Ct) \tag{11.28}$$

In a more general sense, however, this approach can be generalized. If the Chick–Watson model is used as the basis and the inactivation rate constant depends on contact with the disinfectant, then two rate processes are coupled. As an example, for a batch system, if the rate is declining via a first-order process to a lower asymptotic value, the coupled system of Equation 11.29.

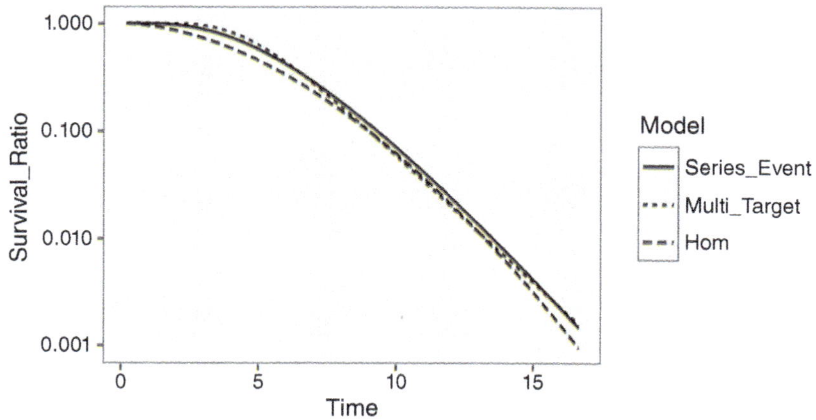

Figure 11.12 Comparison of series event, multitarget and Hom models. In all cases, $C = 1$. For series event, $k = 1$ and $n = 5$. For multi-target, $k = 0.55$ and $n = 15$. For Hom, $k = 0.044$ and $m = 1.8$.

$$\begin{aligned} \frac{dv}{dt} &= -k'C^n \\ \frac{dk'}{dt} &= -k_h \cdot (k' - k'_\infty) \end{aligned} \tag{11.29}$$

The solution of this, if k'_0 is the initial rate is

$$\left(\frac{v}{v_0}\right) = -C^n \left\{ k'_\infty t + \frac{(k'_0 - k'_\infty)[1 - \exp(-k_h t)]}{k_h} \right\} \tag{11.30}$$

This would produce tailing in a quasi-mechanistic formulation.

It is also possible to generalize the Chick–Watson, or indeed any of the other models described previously, by positing intrinsic heterogeneity among the population of microorganisms being inactivated. Mechanisms for such heterogeneity may include:

- Presence of microorganisms in multiple types of clumps, perhaps also including unassociated or "free" organisms.
- Presence of organisms both in a "free" state, and also enmeshed within extraneous matter, of biotic and/or abiotic origin, that can be partially protective against the effect of inactivating agents.
- Different subpopulations of organisms resulting from the enumeration method being nonspecific. For example, the coliform assays (either fecal or total) will enumerate multiple genera and species that could have different sensitivities.
- Genetically different subpopulations, even of single species, that manifest themselves in a phenotypic difference in sensitivity to disinfection.
- For vegetative organisms, existence in different physiological states results in different sensitivities.

Regardless of the mechanism, heterogeneous populations can be described (assuming they do not interact with each other during the inactivation process) by a set of parallel rate processes. For example, if there are "q" subpopulations, then there could be a separate rate for each subpopulation, i.e., $r_{d,1}, r_{d,2}, \ldots, r_{d,q}$, and therefore a separate mass balance for each subpopulation. In the case of a batch system, this would result in the system of Equation 11.31.

$$\frac{dv_1}{v_{0,1}} = -r_{d,1}$$
$$\frac{dv_2}{v_{0,2}} = -r_{d,2}$$
$$\cdots \tag{11.31}$$
$$\frac{dv_q}{v_{0,q}} = -r_{d,q}$$

In this approach, the decay of each subpopulation can be computed independently using a particular kinetic model, and the total remaining viable organism concentration is the sum of the concentrations of each subpopulation.

Examples are shown in Figure 11.13 for the case of two Chick–Watson subpopulations, and the case of one Hom subpopulation (with m>1) and one Chick–Watson subpopulation. Note that in the former case, the overall curve is of a tailing-off form, whereas in the latter case, there is a shoulder (from the Hom model) followed by tailing off. Clearly, this approach allows a quite flexible framework for fitting data. More than two subpopulations can be used. A related approach is to use a probability distribution for the inactivation parameter(s) and then formally convolute the rate law, e.g., Chick–Watson, with that mixing distribution.

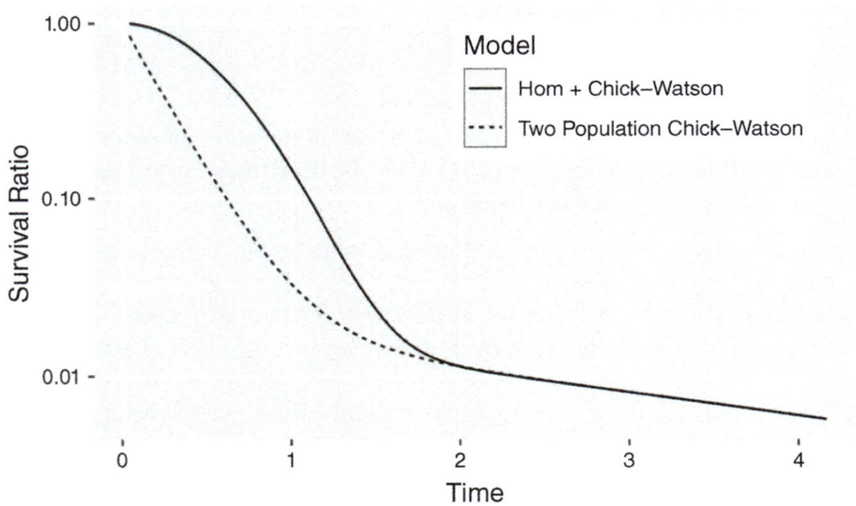

Figure 11.13 Inactivation curves for two population mixtures of Chick–Watson and Hom–Chick–Watson. In both cases, there are two populations at proportions of 0.98 and 0.02. For the two Chick–Watson models, these have inactivation rates ($k'C^n$) of 4.0 and 0.3. The Hom–Chick–Watson model has the majority subpopulation decaying according to the Hom model with $m = 2$ and $k'C^n$ of 0.90, and the minority subpopulation decaying according to a Chick–Watson relationship with an inactivation rate ($k'C^n$) of 0.02.

This is illustrated by example, for Chick–Watson kinetics (Equation 11.17) in which the inactivation rate (k') has a distribution given by a gamma probability distribution, but n is constant.[11]

Under these assumptions, the probability density function for the inactivation parameter is given by Equation 11.32.

$$g(k'|\alpha,\beta) = \frac{1}{\Gamma(\alpha)\qquad\beta^{\alpha}} k'^{(\alpha-1)} \exp\left(-\frac{k'}{\beta}\right) \tag{11.32}$$

Then, the mixture distribution predicting $\frac{v}{v_0}$ is the integral of Equation 11.33.

$$\frac{v}{v_0} = \int_0^{\infty} g(k'|\alpha,\beta) \cdot \exp(-k'C^n t)dk' \tag{11.33}$$

This has a simple analytical solution given in Equation 11.34.

$$\frac{v}{v_0} = (1 + \beta C^n t)^{-\alpha} \tag{11.34}$$

The gamma mixture of exponentials gives a tailing off inactivation, which further reinforces the idea that similar shapes of inactivation curves can be derived from multiple sets of (quasi-)mechanistic assumptions.

This subsection is concluded with the observation that all kinetic models discussed herein treat the disinfecting agent, C as being constant in space and time. If there is nonuniformity or non-constant levels, then modifications in the predicted survival ratio functions are required. The approaches differ by the type of disinfectant (chemical, thermal, UV, and radiation) and will be discussed subsequently for each specific disinfectant type.

11.2.2 Chemical Disinfection

The kinetics of chemical disinfection need to account for the form(s) of the disinfectant that may be present and its spatial and temporal variation in the system being disinfected. In Section 11.1.1, the effect of chemical forms, including the impact of pH on speciation, was discussed. Some disinfectants, such as quaternary ammonium compounds, may be less affected by chemical speciation than others, such as free chlorine.

For chemical disinfection, the disinfectant level, C, is specified by the local concentration of active agent, denoted by the variable C. C may vary with space and time, and there may be different chemical forms that could be denoted by superscripts, e.g., for chlorine one might denote free and combined chlorine by C^{free} and C^{combined}. To fully understand inactivation kinetics, it is necessary to determine how these concentrations vary in space and time.

When a reactive disinfectant is added to a batch system, there is frequently a rapid decline in concentration, followed by a slow decline. This can be modeled as a process consisting of different subcomponents of residual reacting at different rates, analogously to the multi-population microbial decay in Equation 11.31. For a well-mixed batch system with two subcomponents reacting at different first-order rates, this can be described by:

$$C = x \cdot e^{-k_1 t} + (1 - x) \cdot e^{-k_2 t} \tag{11.35}$$

---

11 This could be generalized even further by assuming that k and n were jointly distributed by a bivariate distribution. However, the complexities of this formulation, which would require a double integration over the bivariate distribution, although computationally interesting, are beyond the scope of this treatment. In particular, such bivariate distribution would require consideration of the marginals, as well as of the correlation structure.

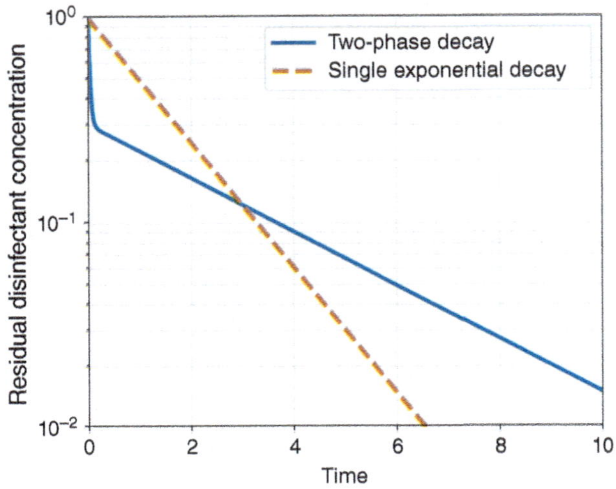

Figure 11.14 Uniform decay vs. biphasic decay of a chemical disinfectant in a batch well mixed system.

In Equation 11.35, $k_1 > k_2$ and x is the fraction of residual decaying at the rapid rate. This produces a curve such as that in Figure 11.14, where it is compared with a single exponential decay process.

As an approximation, if the initial rapid decay is very fast, then little error is made by replacing the actual initial concentration, C_0 by the concentration at the end of the rapid decay period, denoted as C_i, with the difference, $C_0 - C_i$ regarded as instantaneous demand.

A simple example of combining inactivation kinetics with decay kinetics will be illustrated, using the Chick–Watson relationship for inactivation (Equation 11.16) with biphasic first order reaction decay (Equation 11.35).

The differential equation for a batch mass balance with Chick–Watson inactivation is given in Equation 11.36. The time-varying disinfectant residual with the two-phase model is given by Equation 11.37.

$$\frac{dv}{dt} = -k'C^n v \tag{11.36}$$

$$C = C_0\left[x\, e^{-k_1 t} + (1 - x)\, e^{-k_2 t}\right] \tag{11.37}$$

The latter equation is substituted into the former equation, and analytical integration is possible, producing the solution shown in Equation 11.38. When plotted, this will have "tailing off" behavior.

$$log\left(\frac{v}{v_0}\right) = -k'C_0^n\left[\left(\frac{x^n}{n\,k_1}\right)\left(1 - e^{-nk_1 t}\right) + \left(\frac{(1-x)^n}{n\,k_2}\right)\left(1 - e^{-nk_2 t}\right)\right] \tag{11.38}$$

Conducting a similar derivation in the case of the Hom, power law and HPL models with a single-phase first-order decay, with decay coefficient defined as k_d, the relationships with decay for these three laws can be developed to yield the functions in Table 11.1.

If the expressions for disinfectant decay rate and the inactivation rate cannot be solved analytically, it is always possible to solve the system numerically as a set of ordinary differential equations. The following Python code shows this in the case of a Hom inactivation model with a second-order decay rate for disinfectant residual. This example is constructed using the following parameters:

$k' = 0.05,\ n = 0.8,\ m = 1.9$
$k_d = 2,\ C_0 = 2$

Table 11.1 Inactivation expressions for batch systems with first-order demand.

Relationship	$\log\left(\frac{v}{v_0}\right) =$	Reference
Hom	$-\frac{k'mC_0^n}{(n\ k_d)^m}\gamma(m, nk't)$	[61]
Power law	$-\frac{1}{x-1}log\left[1+(x-1)\frac{k'C_0^n}{k_d n}\left(1-e^{-k_d t}\right)^n v_0^{x-1}\right]$	[53]
HPL	$-\frac{1}{x-1}log\left[1+(x-1)\frac{k'mC_0^n}{(k_d n)^m}\gamma(m, nk't)v_0^{x-1}\right]$	[53]

```
import numpy as np
import scipy as sp
import matplotlib.pyplot as plt

kprime=0.05 ; n=0.8 ; m=1.9 ; kd=2 ; C0=2
lasttime=90.0

def decay_rate(Conc):
    rate=-kd*(Conc**2)
    return rate
def hom_rate(Conc,time):
    #note - written as dlnN/dt
    rate=-kprime*n*(Conc**n)*(time**(m-1))
    return rate
def model(time,y):
    Conc=y[0]
    lnS=y[1]
    dy0=decay_rate(Conc)
    dy1=hom_rate(Conc, time)
    dy=np.array([dy0,dy1])
    return dy
A=sp.integrate.solve_ivp(model,(0.001,lasttime),np.array([C0,0]),max_step=0.1)
Conc=A.y[0,:] ; S=np.exp(A.y[1,:])  ; t=A.t
plt.semilogy(t,S,label='Survival')
plt.semilogy(t,Conc,label='Disinfectant Concentration',linestyle="dashed")
plt.legend()
plt.xlabel("Time")
plt.grid()
plt.show()
```

The output of this computation is shown in Figure 11.15.

In flow-through systems, such as water disinfection tanks or rooms where chemical disinfectant is used, rather than a batch mass balance, other mass balances pertinent to the physical situation must be used. For a flowing system with plug-flow characteristics, as noted in Section 8.2.1, steady-state mass balances are identical to batch mass balances, with physical time t replaced by hydraulic residence time θ.

If a system has characteristics of a completely mixed volume, then performance can be predicted by applying the relevant rates to the complete-mix mass balance, which, in general terms, is given in Equation 8.4. For chemical disinfectants, a mass balance must be written for each viable microorganism and the disinfectant. This will be illustrated by an example for a completely mixed volume

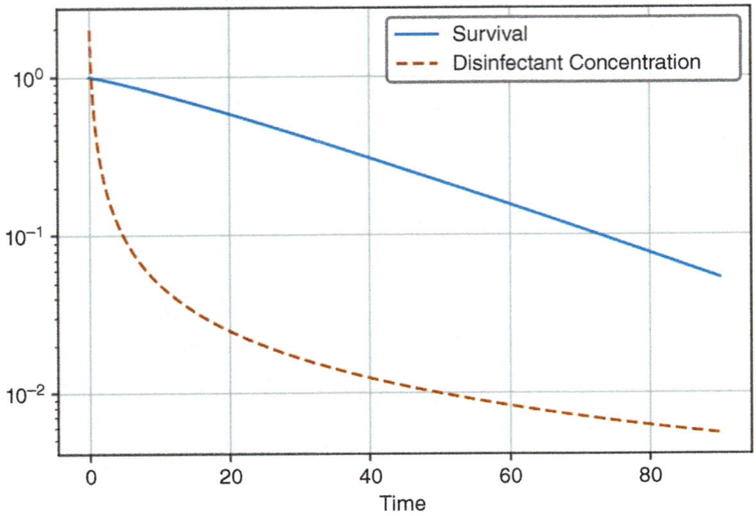

Figure 11.15 Time course of disinfectant residual and survival for the example of Hom inactivation with second-order decay kinetics.

with an internal source of microorganisms and an internal flow of disinfectant, with inactivation given by the Chick–Watson relationship and disinfectant decay as a first-order process. For this situation, the mass balances for microorganisms and the disinfectant are given by Equations 11.39 and 11.40, respectively.

$$\frac{dv}{dt} = E_N - Qv - k'C^n vV \tag{11.39}$$

$$\frac{dC}{dt} = E_C - QC - k_d CV \tag{11.40}$$

In these equations, E_N and E_C are the internal source terms for microorganisms (in # per time) and disinfectant (in amount per time), respectively.

At steady state, the derivatives are zero, and two algebraic equations for C and N result:

$$v = \frac{E_N}{Q + k'C^n V} \tag{11.41}$$

$$C = \frac{E_C}{Q + k_d V} \tag{11.42}$$

The application of these relations is illustrated by Example 11.3.

Example 11.3 Calculation for Disinfectant Addition to a Completely mixed System at Steady State.

Consider a room 10 m by 10 m with 2.5 m ceilings and a ventilation rate of 6 air changes per hour. Assume it is completely mixed and has an internal source of infectious pathogens at 10,000 organisms per hour, such as from one or more infectious individuals. This is to be managed with the continuous addition of a chemical disinfectant, and the target acceptable viable microorganism concentration is $1\frac{\#}{m^3}$. Inactivation follows a Chick–Watson relationship, and disinfectant decay follows a first-order relationship. Given the following kinetic parameters, what should be the disinfectant dosing rate (E_C) to achieve the target result:

- $k_d = 0.05 s^{-1}$
- $k' = 0.05 \frac{L}{mg - s}$
- $n = 1$

Solution:

1. From the data, the room volume and the flow rate can be computed:

$$V = (10\,\text{m})(10\,\text{m})(2.5\,\text{m}) = 250\,\text{m}^3$$

$$.Q = (250\,\text{m}^3)(6\,\text{h}^{-1}) = 1500\,\frac{\text{m}^3}{\text{h}} = 0.417\,\frac{\text{m}^3}{\text{s}}$$

2. Equation 11.41 can be rearranged to solve for the concentration of disinfectant that needs to be maintained in the room:

$$C = \left(\frac{E_N - NQ}{\upsilon k' V}\right)^{1/n}$$

Substituting in the information given:

$$C = \left(\frac{\frac{10,000}{3600} - 1 \cdot 0.417}{\frac{10,000}{3600} \cdot 0.05 \cdot 250}\right)^{1/1} = 0.068 \frac{mg}{L}$$

3. Equation 11.42 can be rearranged to solve for E_C:

$$E_C = C(Q + k_d V)$$
$$= 0.068 \frac{mg}{L}\left(0.417 \frac{m^3}{s} + 0.05 s^{-1} \cdot 250 m^3\right)\left(1000 \frac{L}{m^3}\right)$$
$$= 878 \frac{mg}{s} = 3.16 \frac{kg}{hr}$$

In other cases, where the system has intermediate mixing, more complex numerical approaches may be needed. There have been several approaches to estimating inactivation in such systems.

The most straightforward such approach has been to regard inactivation as a function of the concentration-time product, which would be the case in an ideal plug flow system with Chick–Watson kinetics with $n = 1$; recall the depiction of a plug flow system as a conveyor belt of batch reactors (Figure 8.4). For real systems, there are deviations from these idealized assumptions in several respects:

1. Not all fluid spends an equal and identical time θ in the system.
2. The concentration of disinfectant may change over time due to demand and decay.
3. Inactivation kinetics may not be in accord with Chick–Watson, $n = 1$.

An approach to overcoming the first two assumptions and partially overcoming the third has been embodied in regulations for disinfecting surface water used for drinking water under the surface water treatment rule (SWTR) [62]. In this approach, the following modifications to calculate "ct" are used, and the level of inactivation corresponding to the "ct" is tabulated based on experimental data[12]:

- To account for the shorter contact time exhibited by some proportion of material traveling through the system, the time experienced by the most rapid ten percent of material, termed t_{10} is used instead of θ. It is possible to use the geometric properties of a contactor to estimate the ratio of $\frac{t_{10}}{\theta}$ [63]. For an ideal plug flow system, this ratio would be 1.0. For an ideal CSTR, this ratio is 0.105.
- To account for disinfectant decay in a system, three approaches have been employed. The first, and most conservative, is to use the final (effluent) disinfectant concentration, termed C_f, as a predictor. A second approach is to use the average concentration between initial concentration and final concentration, C_{avg} in the system as a metric.[13] A third approach is to measure the concentration at various intermediate points within the system, and numerically integrate the product of concentration and time increment to get an integrated "ct" directly.
- It is recognized that even when Chick–Watson is adequate to describe the system, but when n is not equal to unity, the value of "ct" needed to attain a given degree of inactivation may be dependent on concentration itself. Therefore, tabulations at a given concentration may be made to predict inactivation, which may also incorporate other characteristics, such as pH or temperature, that can influence performance.

A second approach is to combine information on inactivation (and perhaps disinfectant decay) kinetics, possibly obtained in batch systems, with the residence time distribution of the flow through system to estimate performance in the latter. However, even this approach does not yield a unique estimate, since the degree of segregation or micromixing must be specified.

The residence time distribution (RTD) characterizes the distribution of latency between entry into and exit from a system. Material (in nonideal cases) will then exit *along with* material that had entered at a different time. However, the RTD itself does not define *how soon* after entering a system, a quantity of material mixes with other material that will exit along with it. Zweitering [64] defines two extreme cases:

Complete Segregation Material entering a system remains associated with other material entering at the same time until the last possible moment prior to exit, consistent with the RTD.

Maximum Micromixing Material entering a system immediately mixes with material with which it will exit at the first possible moment consistent with the RTD.

The RTD itself can be determined using experimental tracer techniques on a full-scale system [62] or on a scale model system.[14]

These two extremes can be illustrated by the example of a CSTR connected to a plug-flow reactor, as shown in Figure 11.16. Whether flow goes first into the CSTR, thence into the plug flow reactor, or vice versa, the residence time distribution is the same. Furthermore, for a unimolecular first

12 In addition, to provide an implicit factor of safety, the lower confidence limits to inactivation from experimental results are used to calculate the inactivation based on corrected "ct."

13 This would be exactly true in the case where the Chick–Watson with $n = 1$ rate was exact, and there was a straight line decay of concentration.

14 An example of a physical scale model test for assessing hydraulics of contactors is work by Kothandaraman and Evans [65]. They suggest that keeping the geometric ratios proportionate and scaling the velocity using Froude number similitude is the preferred approach. However, with the rapid development of mathematical simulation, this approach has not received significant subsequent evaluation.

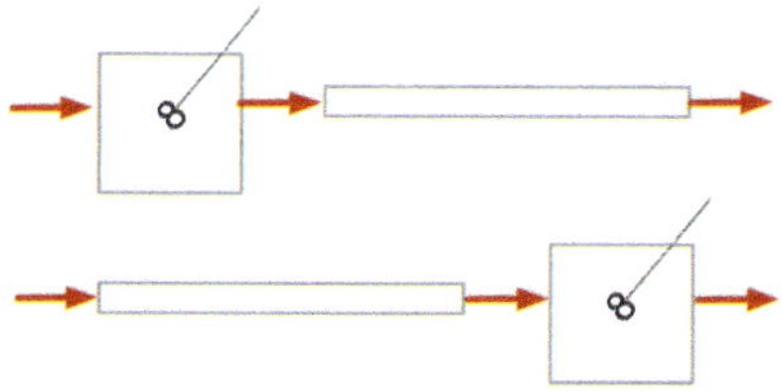

Figure 11.16 Illustration of a CSTR and a PFR connected in series in two different manners.

(or zero) order reaction, the overall extent of conversion/removal is the same. However, for other reactions, the order, i.e., how soon after entry a mass of material interacts with previously entering material, affects the extent.

If maximum segregated flow is assumed, then the performance of a flow through a system can be determined by convoluting the residence time distribution with the batch performance function [64]. In the context of water disinfection, this approach was first outlined by Trussell and Chao [66] and has more recently been termed the integrated disinfection design framework (IDDF) [67]. It involves integrating the batch performance equation with the residence time distribution function, the E curve defined in Section 8.3, to evaluate performance.

The applicable performance, using the segregated flow assumption, if $f(\tau)$ is the function describing survival, i.e., $\frac{\upsilon}{\upsilon_0}$ in a batch system at time τ (considering disinfection demand and decay), and $E(\tau)$ is the residence time distribution curve, then the predicted survival at the effluent of a continuous system is given by Equation 11.43.

$$\frac{\upsilon}{\upsilon_0} = \int_0^{\infty} f(\tau)E(\tau)d\tau \tag{11.43}$$

The analysis of the maximum micromixing problem was developed by Zweitering [64] who presents the solution as a set of ordinary differential equations in X_i, where these represent (i.e., $X_1, X_2 \ldots$) represent the set of concentrations of reactive species, such as disinfectant residual, viable microorganisms, etc. in the form of Equation 11.44.

$$\begin{aligned} \frac{dX_i}{d\lambda} &= r_{X_i} + (X_i - X_{i,0})\frac{E(\lambda)}{1 - \int_0^{\lambda} E(z)dz} \\ \text{at } \lambda = \infty, &\ \frac{dX_i}{d\lambda} = 0 \end{aligned} \tag{11.44}$$

where

$X_{i,0}$ is the influent concentration of species "i."
λ is the residual time a parcel of fluid has *before exit*/.
r_{X_i} is the reaction rate of constituent "i" (as a function of the local state of the system).

This system can be solved as an initial value problem starting at a sufficiently large value of λ and a small value of $\frac{dX_i}{d\lambda}$ with a negative time step [68]. The numerical computation is somewhat more difficult than the IDDF, and so this approach has not been widely used. Note that $E(\lambda)$ has a functional form of a probability density function, and if it is analytical,[15] often the integral in this equation can be expressed as the cumulative distribution function (cdf).

The complete segregation and maximum micromixed solutions provide extremes in performance, given an assumed set of reaction kinetics and residence time distribution. An intermediate approach can be obtained by approximating a residence time distribution by an integral number of CSTRs in series, and then by computing the effluent from that cascade the result can be computed. This CSTR approximation typically lies intermediate between the segregated and maximum micromixed solutions [68, 69].

Use of these three solutions to the second approach to estimating the performance of a flow through system will be shown by example. It will be assumed that the Chick–Watson relationship is applicable with $k' = 0.1$ and $n = 0.8$, and that disinfectant residual decay is given by a first-order process with $k_d = 0.02$, with an initial disinfectant concentration of $2\frac{mg}{L}$. The time units are minutes, concentration is in $\frac{mg}{L}$ and the mean residence time (θ) is 30 minutes. The residence time distribution is an Erlang (Equation 8.13) with N = 4, which would result from four CSTRs in series each with a mean residence time of $\frac{\theta}{4}$.

The complete segregation solution (IDDF) to this example problem can be done using numerical quadrature of Equation 11.43. Since the system is first order in microorganism concentration, there is no loss in generality by solving the performance equation in terms of $\frac{v}{v_0}$ and the result is given by Equation 11.38 with $x = 0$ (i.e., single-phase decay of residual). The following Python code shows the result from the IDDF solution to this problem. The built-in quadrature algorithm is used to convolute the RTD (the function `Erlang_RTD`) with the batch kinetic survival function (`CW_Demand`).

```
# Example of IDDF Segregated Flow
import numpy as np
import scipy as sp
kprime=0.1; n=0.8; kd=0.01; NCSTRs=4;theta=30; C0=2
def Erlang_RTD(t): # Erlang RTD E curve
    A=(t**(NCSTRs-1))/sp.special.factorial(NCSTRs-1)
    B=(NCSTRs/theta)**NCSTRs
    C=np.exp(-NCSTRs*t/theta)
    E=A*B*C
    return(E)
def CW_Demand(t):  # Chick-Watson S with first order demand
    A=-kprime*(C0**n)
    B=1.0-np.exp(-n*kd*t)
    C=n*kd
    S=np.exp(A*B/C)
    return(S)
def IDDF_argument(t): # Argument to be integrated for IDDF
    y=Erlang_RTD(t)*CW_Demand(t)
    return(y)
S,err=sp.integrate.quad(IDDF_argument,0.0,np.inf)
```

The output of this code gives a survival ratio, $\frac{N}{N_0} = 0.0417$ with an estimated error of $3.8 \cdot 10^{-10}$.

---

15 For example, the Erlang distribution, which is applicable for multiple CSTRs in series, has an analytical form of the cdf expressed in terms of the incomplete gamma function. Furthermore, for many common distributions that might be used to fit the residence time distribution of a system, many software platforms have efficient approximations, if not exact numerical solutions, for their cdf.

The maximum micromixedness equations can be solved numerically, although they require some atypical approaches since the boundary conditions are given at $\lambda = \infty$. Craik [69] and Pfeiffer and Barbeau [70] approach the numerical solution in a disinfection context by transforming the equation into an initial value problem with a negative timestep, starting at λ equal to a large multiple of the mean residence time where the conditions are set equal to the system influent, e.g., survival ratio (S) of 1.0 and disinfectant concentration equal to C_0. For the conditions in this example, this approach is used starting at $\lambda = 10 \cdot \theta$. For the Erlang distribution, both the pdf and the cdf are built-in functions in Python, and this is used in the following code.

```
# Maximum Micromixed Example
import scipy as sp
import matplotlib.pyplot as plt
kprime=0.1; n=0.8; kd=0.01; NCSTRs=4;theta=30; C0=2

def S_rate(C,S): #Chick Watson Inactivation Rate
    r_S=-kprime*(C**n)*S
    return(r_S)
def C_rate(C): #Disinfectant Decay Rate
    r_C=-kd*C
    return(r_C)
def ODEs(llambda,X): #Zweitering Maximum Micromixed for C,S
    C,S=X
    H1=sp.stats.erlang.pdf(llambda,NCSTRs,scale=theta/NCSTRs)
    H2=sp.stats.erlang.cdf(llambda,NCSTRs,scale=theta/NCSTRs)
    H=H1/(1-H2)
    dCdlambda=-C_rate(C)+(C-C0)*H
    dSdlambda=-S_rate(C, S)+(S-1)*H
    dXdlambda=(dCdlambda,dSdlambda)
    return(dXdlambda)
t_span=(10*theta,0.00)
y0=(1e-12,1e-12)
OUT=sp.integrate.solve_ivp(ODEs, t_span, y0, method='BDF')
plt.plot(OUT.t,OUT.y[0])
plt.plot(OUT.t,OUT.y[1])
Ceff=OUT.y[0][-1]
Seff=OUT.y[1][-1]
print(Ceff, Seff)
```

The result of this calculation predicts an effluent S value of 0.05170, which is somewhat greater than the result from complete segregation, so in this case, the effect of micromixing is small.

The intermediate approach is to regard the system, which has a residence time distribution that would result from a number (in this case, four) of CSTRs in series, as *actually consisting of* four CSTRs in series. Therefore, for each CSTR for each constituent (disinfectant and microorganisms), there is an algebraic equation resulting from the steady-state mass balance. In the case of disinfectant and microorganism reacting, this yields two equations for each CSTR that need to be solved simultaneously; in each of the form of the Equation 11.45 (where the subscripts "in" and "out" refer to that particular reactor), NCSTR is the number of CSTRs in the cascade, and θ is the overall mean residence time of the cascade. If the reaction rates, particularly that for microorganisms (r_S), are nonlinear in disinfectant and/or microorganisms, the algebraic solution may not be possible.

$$\begin{aligned} C_{\text{in}} - C_{\text{out}} + r_C * \frac{\theta}{\text{NCSTR}} &= 0 \\ S_{\text{in}} - S_{\text{out}} + r_S * \frac{\theta}{\text{NCSTR}} &= 0 \end{aligned} \tag{11.45}$$

However, such a series of simultaneous equations with nonlinearities can be solved numerically. This can be done in Python using the package `scipy.optimize.fsolve` as shown as follows.

```
import numpy as np
from scipy.optimize import fsolve

kprime=0.1; n=0.8; kd=0.01; NCSTRs=4;theta=30; C0=2
def S_rate(C,S): #Chick Watson Inactivation Rate
    r_S=-kprime*(C**n)*S
    return(r_S)
def C_rate(C): #Disinfectant Decay Rate
    r_C=-kd*C
    return(r_C)
def func(X):
    # X(0-3)  #Disinfectant Residual for CSTR 1-4
    # X(4-7)  #Survival for CSTR 1-4
    Y=np.zeros(2*NCSTRs)  #output values of equations
    Y[0]=C0-X[0]+C_rate(X[0]*theta/NCSTRs)
    Y[4]=1-X[4]+S_rate(X[0],X[4]*theta/NCSTRs)
    for i in (1,2,3):
        Y[i]=X[i-1]-X[i]+C_rate(X[i]*theta/NCSTRs)
        Y[i+4]=X[i+3]-X[i+4]+S_rate(X[i],X[i+4]*theta/NCSTRs)
    return(Y)

initial_guesses=[C0,.9*C0,.8*C0,.7*C0,.8,.6,.5,.2]
roots=fsolve(func,initial_guesses,full_output=False)
print(roots)
print(func(roots))
```

The output is that the estimated survival ratio at the exit of the final CSTR is 0.0485, which lies between the IDDF and the maximum micromixedness solution.

It is clear that different assumptions about micromixing lead to different estimates of the process efficiency of chemical disinfection. What guidance can be given with respect to the use of the alternative assumptions?

Nauman and Buffham [71] defined a dimensionless segregation number (Sg) given by:

$$Sg = \frac{1}{4\pi^2 \mathcal{D}\theta}\sqrt{\frac{\mu^3 Q}{\rho^3 P}} \tag{11.46}$$

where P is the power imparted to the fluid in the reactor, θ is the mean residence time, Q is the flow rate, and $\mathcal{D}$ is the molecular diffusivity of the reactant. It was proposed that if $Sg < 0.1$, segregation was unimportant, while if $Sg > 1$ it could be important unless reactions were rapid on the time scale of the residence time.

Evaluation of the segregation number for chlorine contactors [68] and ozone contactors [69] in water indicates that typical conditions yield $Sg < 0.1$, indicating that segregated flow is usually a less applicable assumption than maximum micromixedness. However, as noted in the example calculations comparing the IDDF and maximum micromixed solutions, the differences, while real, can potentially be small, especially considering the experimental uncertainties in measuring disinfection efficiency performance.

Pfeiffer and Barbeau [72] compared the maximum micromixing, the IDDF, and the N CSTR method to observed performance at several full-scale ozone and chlorine disinfection systems. They concluded that the segregated-flow approach (IDDF) tended to over-predict performance, and that while the maximum micromixed formulation performed better, the N CSTR approach provided a practical and more tractable result and also provided reasonable agreement with experimental results.

The IDDF, maximum micromixing approach, and the N CSTR approach share a common disadvantage: they require the residence time distribution (RTD) as an input to predict performance. If a disinfection system has not yet been built, or if modifications to a system are being considered, the RTD may not be known *a priori*. If the system already exists, a full-scale tracer study can be conducted to determine t_{10} or RTD [62], but this is experimentally difficult, expensive, and potentially disruptive to an ongoing operating system. In addition, with ozonation, there is a multiphase flow, with coupled gas-liquid mass transfer, which complicates the analysis. With the increasing availability of advanced computational methods, a third approach, directly employing computational fluid dynamics (CFD), has been developed.

In Section 8.3.3, a brief outline of the fundamental transport equations for material and energy (momentum) governing fluid flow was presented. To describe disinfection processes, separate material transport equations, incorporating reaction rates, need to be written for disinfectant (or each separate disinfectant species to be considered) and for viable microorganisms. In general, this set of partial differential equations must be solved numerically using computational fluid dynamics. Application software packages for conducting these calculations are available, and also these calculations can be done in a general-purpose computational environment, such as MATLAB™ or Python. The following box lists some commercial and open-source CFD packages.

A Partial List of Computational Fluid Dynamic Software

The following is a partial list of CFD software available. No representation is made with respect to the adequacy of any of these packages, nor should an inference be drawn about the packages not on this list.

- **ANSYS Fluent** (Commercial): Finite-volume solver with built-in species transport, reaction kinetics, and UV radiation modules; widely used for chlorine contact tanks and UV reactors. https://www.ansys.com/products/fluids/ansys-fluent
- **ANSYS CFX** (Commercial): Robust multiphase and turbulent reacting-flow solver; applied to CT tank hydrodynamics and airborne disinfection studies. https://www.ansys.com/products/fluids/ansys-cfx
- **COMSOL Multiphysics** (Commercial): Finite-element framework with tightly coupled fluid flow and chemical reaction modules; ideal for detailed UV disinfection simulations. https://www.comsol.com/comsol-multiphysics
- **STAR-CCM+** (Commercial): Advanced multiphysics solver supporting discrete particles, aerosol dynamics, and species transport – perfect for HVAC-based air disinfection modeling. https://totalsim.us/software/star-ccm/
- **Autodesk CFD** (Commercial): Integrates turbulent-mixing and reactive transport workflows; useful for rapid prototyping of disinfection reactor geometries. https://www.autodesk.com/products/cfd/overview
- **SimScale** (Commercial, Cloud): Cloud-based CFD platform with species transport and reacting-flow capabilities, eliminating the need for local HPC resources. https://www.simscale.com
- **OpenFOAM** (Open Source): C++ toolbox for custom solver development; extensive libraries for species transport and reaction kinetics; widely used in environmental engineering contexts. https://www.openfoam.com
- **MOOSE** (Open Source): Multiphysics finite-element framework supporting fully coupled flow–reaction systems; suitable for advanced disinfection kinetics modeling. https://www.mooseframework.org

(Continued)

A Partial List of Computational Fluid Dynamic Software (Continued)

- **Elmer** (Open Source): Finite-element suite configurable for reactive transport and fluid flow; often used in academic disinfection research. https://www.elmerfem.org/blog/
- **FLOW-3D HYDRO** (Specialized/Open): Specializes in free-surface and shallow-water modeling with scalar transport; ideal for mixing-driven disinfection analyses in water treatment. https://www.flow3d.com/products/flow-3d-hydro/

The past several decades have seen a progressive increase in the sophistication of CFD models used for inactivation processes, and further development is anticipated. The ability to model increasingly detailed aspects has been enabled by improvements in computer hardware and software and a dramatic reduction in computational time for more sophisticated calculations. Table 11.2 categorizes the levels of CFD models for inactivation processes in terms of outputs and needed inputs.

The earliest uses of CFD for disinfection modeling were to predict the residence time distribution of a system, as noted in Level 1 of Table 11.2. Hannoun et al. [73] developed CFD models for the flow pattern in several full-scale chlorine contactors and used the models as tools to predict how engineering modifications, such as insertion or alteration of baffles, could favorably affect the flow pattern, e.g., increase the $\frac{t_{10}}{\theta}$ ratio. The full RTD from a CFD computational model could also be used as input to either the IDDF or complete micromixed analysis, or it could be approximated by an equivalent number of CSTRs and used in the N CSTR analysis.

Peplinski and Ducoste [74] modeled the RTD of a full-scale water chlorination contactor and compared the computed results to the experimental findings. Furthermore, they considered a number of the model inputs to be uncertain, and characterized these as probability distributions, and thence used a Monte Carlo approach to propagate the results into uncertainty model predictions.[16] Sources of uncertainty included:

- Kinetic energy dissipation constant
- Inlet turbulence length scale
- Turbulence model used
- Wall roughness length
- Turbulent Schmidt number

Table 11.2 Levels of elaboration of CFD models for inactivation processes.

	Desired output(s)	Needed input(s)
Level 1	Residence time distribution	Geometry, flow rate
Level 2	Integrated disinfection intensity (e.g., "ct")	Above plus disinfectant dose, decay rate(s)
Level 3	Estimated inactivation	Above plus influent microorganism load, inactivation rate(s)
Level 4	Above plus by-product concentration	Above plus rate model(s) for by-product production

16 A more detailed discussion of this approach to quantify uncertainty is in Chapter 13.

They showed that, considering these sources of uncertainty, the computational model outputs tightly bracketed the experimental results.

The first Level 2/3 CFD analysis of chemical disinfection was apparently that of Greene et al. [75] who compared an Eulerian–Eulerian CFD model to experimental results in a pilot-scale water disinfectant contactor. The system was operated on either Willamette or Bull Run, Oregon water, with either free chlorine or preformed monochloramine disinfectants, and the inactivation of *E. coli*, *Giardia lamblia* cysts, or bacteriophage ms2 were studied. Inactivation kinetics were determined by independent laboratory experiments and were found to be consistent with the Hom model. Key findings of this study were:

- The predicted RTDs were in concordance with the RTDs determined by tracer experiments.
- The CFD-based predictions of survival were superior in terms of bias and variance compared to the IDDF-derived predictions, and this was especially true when inactivation was high.

Wols et al. [76] modeled the performance of a full-scale multi-chamber baffled ozone contactor using CFD and reported Level 1, 2, and 3 analyses. In their system, the first chamber mixed ozone bubbles with incoming water to transfer dissolved O_3 into the system, and subsequent chambers were where the bulk of the disinfection process occurred. Key findings of this study were:

- A Level 1 analysis showed good correspondence between the results of an experimental tracer study and the CFD output for a simulated experiment.
- A Level 2/3 analysis was conducted in which the concentration field was modeled using an Eulerian (continuum) CFD model, while the inactivation process was modeled either by the Eulerian method or by a particle tracking (Lagrangian) method. In either case, inactivation kinetics were Chick–Watson with $n = 1$. The mean of the Lagrangian results and the results from the Eulerian method were in agreement. The computational time for the Eulerian–Eulerian model was less than that of the Eulerian–Lagrangian model. No comparison with experimental results was provided.

A Level 3 model for inactivation of microorganisms in an indoor air environment during chemical decontamination was developed by Hoque et al. [77]. They used an Eulerian–Lagrangian approach in which they tracked the deactivation of the microorganisms as they flowed through the space. They also found that the solutions could be approximated by a simplified artificial neural network [77]. Zhang et al. [78] comprehensively reviewed the application of CFD methods in the field of water disinfection by chemical agents as well as UV. While there have been numerous Level 3 applications, as of that review, there did not appear to be any full Level 4 applications in which a comprehensive suite of disinfection by-products was considered.

There have not been many formal investigations of the kinetics of inactivation of microorganisms on surfaces (fomites). It would be expected that in a surface in contact with disinfectant in the gas phase, the kinetics would obey one of the relationships discussed in Section 11.2.1, e.g., Chick–Watson, with modifications for the effect of water activity/relative humidity [79]. For microorganisms embedded in large amounts of soil or debris, there may be additional resistance due to the protection afforded by such material.

11.2.3 Thermal Inactivation

As noted in Section 8.4.2, temperature is an important master variable in affecting the decay (and growth) of microorganisms in the environment. By deliberately increasing the temperature, the decay can be accelerated — this has been known since (at least) the early 19th century [40].

Thermal inactivation has been used in a variety of applications, such as:

- Food preservation and processing (which is beyond the direct scope of this book; however, much of the conceptual and theoretical underpinning of thermal inactivation arises from research in that venue)
- Hot water heaters and use of hot water for decontaminating the plumbing system
- Hot air/steam decontamination of indoor environments
- Disinfection and sterilization of equipment and devices
- Sterilization of laboratory media and equipment

The fundamental analysis of thermal inactivation starts from the models in Section 11.2.1. The analogy to Chick's law with thermal inactivation in rate form is:

$$r_d = -k_d(T) \cdot N \tag{11.47}$$

where the rate $k_d(T)$ is a function of temperature (and perhaps other variables). This can be integrated into a simple exponential decay if the temperature is constant. It is common to model this temperature dependence as an exponential as in

$$k_d(T) = k_d(T_{\text{ref}}) \cdot \exp\left(\frac{T - T_{\text{ref}}}{z}\right) \tag{11.48}$$

with T_{ref} being a reference temperature and z being a coefficient of sensitivity of the decay rate to temperature. For a number of bacterial pathogens in various food matrices, the parameters have been tabulated,[17] e.g., in Asselt and Zwietering [80]. Note that this approach, which has frequently been used, assumes that the *instantaneous* inactivation rate is only a function of the *instantaneous* temperature, and not, for example, a function of the thermal *history*. If the time history of temperature, e.g., the *rate of heating* influenced subsequent sensitivity, then a more complex mathematical formulation would be required.

As with chemical disinfection, it is common that in a batch system, even under isothermal conditions, the survival of microorganisms versus time is non-log-linear (i.e., Chick's law may not be followed). Instead, there could be "shoulders," tailing off or more complex behaviors. As a result, many have proposed alternative forms such as a Weibull [81] (which is functionally identical to the Hom model), a biphasic [82], and others. Indeed, any of the general inactivation models in Section 11.2.1 can be modified by replacing the term $k'C^n$ by a function of temperature, $k'_d(T)$.

It is frequently the case that systems may have temperatures that vary in time and/or space, and therefore, the performance of the process will be influenced by thermal transport. Even in the simplest case of a well-stirred homogeneous volume with a constant heat source, there will be an exponential temperature profile.[18] Even if Chick's law were followed, this temporal variation in T will manifest itself as a deviation from log-linearity of the survival function, as illustrated in Example 11.4.

17 Conventionally, in food microbiology, rather than writing Equations 11.47 and 11.48 in terms of a rate constant, k_d, it is written in terms of the decimal reduction time (D) — time required for a tenfold decrease in microorganisms. The relationship is that $D = \frac{2.303}{k_d}$.

18 This will be a function of the heat input, the heat transfer rates through the boundaries of the system, the properties of the bulk fluid, the mixing rate, and any other heat losses. For example, see [83].

Example 11.4 Thermal Inactivation in a Batch System with Chick's Law and Time-varying Temperature.

Consider a well-mixed batch system that receives a continuous, constant input of heat such that the increase in temperature towards an equilibrium state has a rate constant, $k_T = 0.3\ min^{-1}$. Inactivation is given by Chick's law (Equation 11.47) with a temperature dependent k_d given by Equation 11.48. The system starts at a temperature of 25 °C and reaches an equilibrium temperature of 90 °C upon heating. Given the following parameters, compute the time required to achieve three-log inactivation (99.9%), and the temperature when it is attained.

$$k_d(T_{\text{ref}}) = 0.1\ \text{min}^{-1}$$

$$z = 10\,^{\circ}\text{C}$$

$$T_{\text{ref}} = 20\,^{\circ}\text{C}$$

Solution:

The problem can be defined by two differential equations and an algebraic equation:

$$\frac{dS}{dt} = -k_d(T) \cdot S$$

$$\frac{dT}{dt} = k_T(T_{\text{eq}} - T)$$

$$k_d(T) = k_d(T_{\text{ref}}) \cdot \exp\left(\frac{T - T_{\text{ref}}}{z}\right)$$

This system can be solved numerically or analytically. The Python code for the numerical solution is shown as follows. The advantage of the numerical solution is that it can be edited to account for different initial conditions, inactivation conditions, etc.

```
import numpy as np
import scipy as sp
import matplotlib.pyplot as plt
Tref=20; z=10; kd_Tref=0.1
kT=0.3; Teq=90
def temperature_rate(T):
    rate=kT*(Teq-T)
    return rate
def CW_rate(S,T):
    #note - written as dS/dt
    kprime=kprime_T(T)
    rate=-kprime*S
    return rate
def kprime_T(T):
    k=kd_Tref*np.exp((T-Tref)/z)
    return k
def model(time,y):
    T=y[0]
    S=y[1]
    dy0=temperature_rate(T)
    dy1=CW_rate(S, T)
    dy=np.array([dy0,dy1])
    return dy
A=sp.integrate.solve_ivp(model,(0,4),np.array([25,1]),method='LSODA')
T=A.y[0,:] ; S=A.y[1,:]  ; t=A.t
```

(Continued)

Example 11.4 (Continued)

```
plt.subplot(2, 1, 1)
plt.semilogy(t,S,label='Survival')
plt.legend()
plt.grid()
plt.subplot(2, 1, 2)
plt.plot(t,T,label='Temperature',linestyle="dashed")
plt.legend()
plt.xlabel("Time (min)")
plt.grid()
plt.show()
```

By inspection of these graphs, three-log inactivation occurs at approximately 2.75 minutes, corresponding to approximately 63 °C.

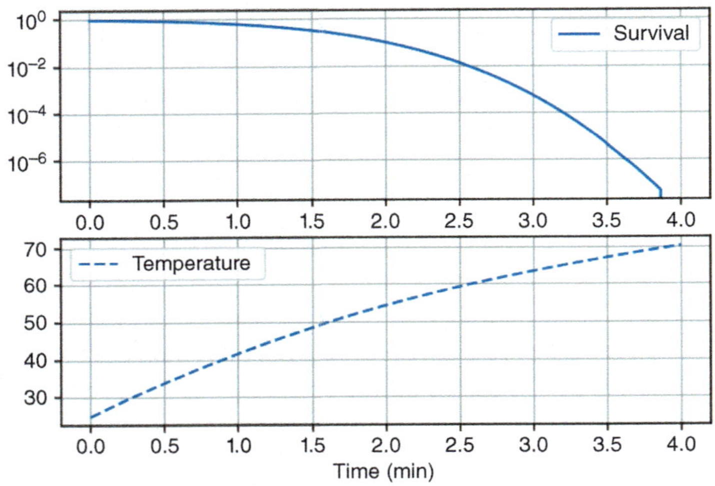

Clearly, a key aspect of describing the inactivation of microorganisms by heat is predicting the inactivation rate constant(s). It has long been known that these rates are influenced by a variety of parameters, including water activity (a_w), specific ionic composition of the environment, and the impact or organic constituents. For example, it is known that the maximum rates of inactivation of bacterial spores occur at a_w values in the range of 0.2–0.4 [84].

Espinosa et al. [85] conducted a systematic review for thermal inactivation of pathogens of particular concern in water, sanitation, and food protection in under-resourced locations. From their compiled data, they provided guidance for time-temperature combinations[19] yielding one, two, and three log inactivation.

John et al. [86] systematically reviewed the literature on heat inactivation of norovirus, hepatitis A virus, and hepatitis E virus, and based on their information also identified time-temperature combinations corresponding to different degrees of inactivation in particular matrices.

The US Department of Agriculture has developed an open online database of pathogen survival (motivated by food safety), including the rates of inactivation by elevated temperature, COMBASE, which is available at https://combase.errc.ars.usda.gov/.

19 Note that the concept of a time-temperature product as a key variable in predicting inactivation is quite analogous to the concept of "ct" in chemical disinfection designed previously.

Ordinary pure water at 1 atm pressure (1.01325 bar) has a boiling point of 100 °C. By pressurizing water, such as in a pressure cooker or steam autoclave, the boiling point can be raised. This is often used to sterilize liquids and equipment. The relationship between pressure and temperature at boiling can be described by the Antoine equation (Equation 11.49).

$$\log_{10}(P) = A - \frac{B}{T + C} \tag{11.49}$$

where P is pressure in bar, T is absolute temperature (K), and A, B, and C are empirical parameters. For water, these are given by values in Table 11.3.

It is common to sterilize microbiology media and equipment in an autoclave, where steam is added to achieve and maintain a temperature of 121 °C (374.15 K). The Antoine equation can be used to estimate the corresponding pressure. Note that this temperature lies between the limits specified for the coefficients in Table 11.3, so both values will be calculated.

$$\begin{aligned} \log_{10}(P) &= 4.6543 - \frac{1435.264}{374.15 - 64.848} = -0.0010834 \\ \log_{10}(P) &= 3.55959 - \frac{643.748}{374.15 - 198.043} = -0.1167222 \end{aligned} \tag{11.50}$$

Since the desired temperature is between the range of applicability of these two estimates, the arithmetic mean is taken and then the antilog used to estimate the pressure, which is 0.873 bar, or 0.862 atm.

In large, and/or flowing systems, there can be spatial variation of temperature, in addition to velocity and concentration. This is schematically illustrated in a conceptual diagram of a domestic hot water heater (Figure 11.17), where there is flow in from the bottom along with heating from an external source. This sets up a thermal field in the system as show by the isocontour lines.

Furthermore, thermal variation will result in local differences in a variety of properties, such as in fluids, including density, viscosity, and molecular diffusivity of constituents. As noted previously, the local temperature will result in locally variable inactivation rates. If there are any chemical reactions, they will also be affected by the thermal field.

As a result of this complexity, flow systems with heating require additional relationships to solve using computational fluid dynamics. The transport of heat through a medium is a function of its thermal conductivity (which in itself may be a function of temperature), and heat transfer through the system boundaries is a function of the temperature difference between the system and its surroundings (ΔT), and the effective heat transfer coefficient. An example approach is given in Sedeh and Khodadadi [87] for optimizing the internal structures of a domestic water heater. This is equivalent to modeling the concentration field within a chemical disinfection system. By inclusion of an advection-reaction-transport model for viable microorganisms into a CFD simulation with inactivation rates as a function of the local temperature, the inactivation performance in such flowing systems can be computed [88].

Table 11.3 Antoine equation parameters for water vapor pressure.

Temperature range (K)	A	B	C
255.9–373	4.6543	1435.264	−64.848
397–573	3.55959	643.748	−198.043

Source: Data from NIST Standard Reference Database 69 (https://doi.org/10.18434/T4D303).

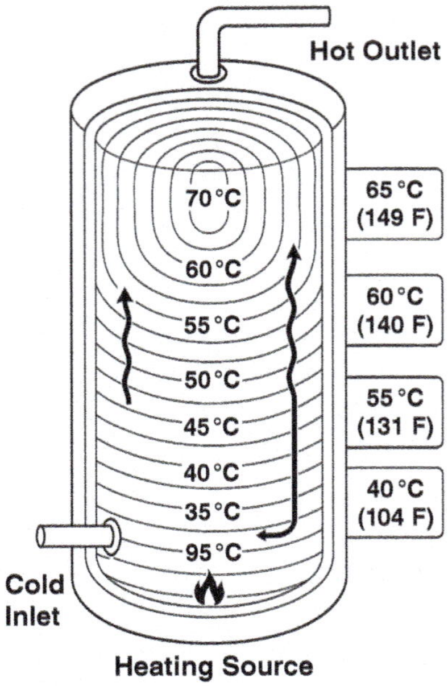

Figure 11.17 Idealized conceptual diagram of a water heater.

11.2.4 UV and Radiation Disinfection

UV light is a portion of the electromagnetic spectrum (Figure 11.8), and in particular, UVB and UVC range wavelengths can be good disinfectants. Shorter wavelengths include X-rays and gamma rays, which will be discussed subsequently to UV.

The wavelength of electromagnetic radiation is of significance due to the duality principle in which such radiation can also be described as a stream of photons of particular energy, which is given by Planck's law (Equation 11.51):

$$E = \frac{h \cdot c}{\lambda} \tag{11.51}$$

where E is the energy per photon,[20] λ is the wavelength, h is Planck's constant ($6.663 \cdot 10^{-34} J - sec$), and c is the speed of light in a vacuum ($3 \cdot 10^8 \frac{m}{s}$). By applying this relationship, UV at 200 nm (considered far UVC) has photons with energies of 6.2 eV.

Historically, the term "ionizing radiation" was used for radiation of sufficient energy to dislodge valence electrons from atoms or molecules, which is about 10 eV [43]. Therefore, UV is considered non-ionizing radiation, although it can induce chemical changes (including within microorganisms). Since chemical changes require the absorption of a photon to proceed, if the energy of the reaction is greater than the energy of a photon, such a reaction is not possible [89].

UV is emitted into a medium from a source, which may be a lamp, a laser, or a light-emitting diode. It is conventional to talk about fluence rate and fluence. The fluence rate is the energy of light passing through a unit area per unit time, i.e., in power per unit area. The fluence is the energy delivered to such a unit area, which is the product of fluence rate and time under illumination, and has units of energy per unit area.. It has been argued by some [89] that it is preferable to discuss

20 Note that 1 electron-volt (ev) equals $1.602 \cdot 10^{-19}$ J.

UV inactivation in terms of photon fluence rate, or moles of photons per unit area per unit time, however, this has not been widely accepted.[21]

Note also that the fluence rate is based on the energy of the emitted light. Physical sources of UV will draw more electric power than the light they emit, and thus, the overall fluence rate over an entire area emitted by a source is less than the electric power used by the source. This has been termed the wall plug efficiency, and ranges between 0.0065–0.35 depending on the nature of the source [90]. The higher values of wall plug efficiency pertain to very commonly used low-pressure mercury vapor lamps, which have peak emission at 254.7 nm.

Microorganisms exhibit different sensitivities across wavelengths. A typical pattern is higher sensitivity below 240 nm, perhaps corresponding to greater formation of internal reactive oxygen species within the organisms [90], and then a lower peak sensitivity at approximately 260 nm, which corresponds to the maximum absorbance of nucleic acids [91].

The inactivation kinetics by UV in a system will be a function of the fluence intensity, e.g., in units of power per cross-sectional area, given the symbol I (which may be spatially or temporally variant) effectively with C in Equation 11.16 and subsequent equations in Section 11.2.1 being replaced by the fluence intensity, given a symbol I.[22]

When light is emitted from an open point source into a medium such as water or air, its propagation is governed not only by geometric spreading and wavelength-dependent dispersion but also by spectrally selective absorption and angular redistribution via scattering from suspended particles. These processes jointly determine the spatial, spectral, and directional distribution of radiance.

In the absence of absorption and scattering, there would be an inverse-square dependence on the fluence rate. With effective losses due to absorption and scattering (neglecting the redistribution from the scattering), a Beer–Lambert law describes the fluence versus radius from the source (Equation 11.52):

$$I_\lambda(r) = \frac{P_\lambda}{4\pi r^2} e^{-\kappa_\lambda r} \tag{11.52}$$

where κ_λ is the *spectral absorption coefficient* [m^{-1}]. This parameter reflects absorption due to both molecular and particulate constituents, and P_λ denotes the specific wavelength light-emission rate.

In real UV systems, this simple analysis is complicated by additional factors, including the following:

- The source is generally not a single point, but cylindrical, circular, or other areal form.
- The source may be in a reflecting fixture, resulting in emission from a complex geometry.
- Scattering from particles suspended in the medium may result in re-emission (either at the original emitted wavelength or less energetic, longer-wavelength photons).
- Particularly in water systems, the source may be in an enclosed jacket, such as quartz.
- In water systems, the lamp-water or jacket-water interface will exhibit refraction due to the differential refractive indices.
- There may be reflection or backscattering from light impinging on walls or surfaces within the space, which can result in indirect fluence within particular zones of a system.

21 Fluence rate and fluence can be inter-converted using Planck's law (Equation 11.51) if the wavelength is known. Bolton et al. express fluence rate in units of Einsteins (Ei) per unit area per unit time, where an Einstein is considered to be a mole of photons. However, Ei has been deprecated from the SI as a derived unit and will not be used here.

22 Strictly speaking, this should be the fluence at a specific wavelength, and then the rate constants also need to be wavelength specific, with the overall rate being integrated over the wavelength distribution at a point. This will be neglected in the subsequent discussion, and the intensities and rates should be regarded as being conditional on the light source used. However, this dependency becomes significant when polychromatic sources of light are used.

Ignoring the reflection, refraction, and scattering complexities, except to the degree that particle scattering might be regarded as light absorption and lumped into a phenomenological absorption coefficient, there are multiple approaches to modeling the fluence distribution within a space. These have been summarized by Kowalski et al. [92] and Liu et al. [93] in the air and water contexts, respectively. A high-level sketch of several approaches used is:

Point source summation model: This approach treats the UV source as a set of discrete, approximately point sources. At any location in a system, the total fluence is the sum of the fluence from all of these points, each described by a relationship such as Equation 11.52.

Multiple segment source summation model: This approach, developed for cylindrical sources, also accounts for reflection and refraction at the emitter surface and incorporates the cylindrical geometry into the emission calculation.

View factor method: This approach considers the "line of sight" between a particular location and the (areal) source, and bases a local fluence intensity on that concept. This generally neglects absorption or scattering. A good introduction to this approach is given by Kowalski et al. [92] in the context of air disinfection by UV.

The detailed modeling of the fluence intensity distribution within a volume is beyond the scope of this book. However, using approaches such as the above, the fluence intensity I_λ at any spatial location and time can be predicted. Then, using this field, it can be incorporated into a model for microbial survival using either an Eulerian or Lagrangian (particle-tracking) approach to estimate inactivation, once an inactivation rate function applicable to all locations and times in the system is available [94].

There are several aspects of formulating a rate function for UV inactivation that parallel the considerations for chemical inactivation. In a homogeneous batch system receiving a constant fluence rate, the survival-versus-time curve may not be strictly exponential and may exhibit "shoulders," "tails," or a combination of the two. The causal mechanisms discussed in Section 11.2.1 could explain such deviations. In addition, particular features of UV may introduce other causal mechanisms for deviation from Chick's law:

- At low doses of UV, some bacteria and other organisms can repair the damage from the light [95], either in the absence of further light (dark repair) or in the presence of further light energy. This could result in both an apparent shoulder at low doses and a reduction of the effective inactivation rate.
- At high doses of UV, especially at shorter wavelengths (< 220 nm) free radicals such as $OH^\bullet$ and other reactive oxidant species can be produced via the photochemical reactions of UV with water and other molecules present. These radicals can also serve as potent disinfectants [90]; however, they are short-lived, and their disappearance could result in a tailing-off.
- At extremely high doses of UV, likely well above those used in either air or water inactivation processes, the energy transferred could result in some local heating, resulting in inactivation by temperature rise.

It must be re-emphasized that the form of the inactivation curve in itself does not confirm a particular causal mechanism since multiple mechanisms can lead to the same phenomenological behavior.

A second consideration is whether the inactivation rate depends linearly or nonlinearly on dose rate, I_λ. In 1862, Bunsen and Roscoe postulated that the rates of photochemical reactions were proportional to the fluence (E_λ) delivered [96], which can be viewed as the time integral of the fluence rate, i.e.:

$$E_\lambda = \int_0^t I_\lambda(t)dt \quad (11.53)$$

If the fluence rate is constant, then this integral is simply the product, $I_\lambda \cdot t$; however, in some systems where UV pulses are used, this integral needs to be evaluated based on the pulse frequency and shape. Note then that this concept is equivalent to the use of ct product based on the Chick–Watson relationship with n = 1 for chemical disinfectants. With photochemical reactions, Schwarzschild [97] noted deviations from this Bunsen–Roscoe law, also termed time dose-rate reciprocity, analogous to the Chick–Watson relationship (Equation 11.16); in the context of UV, the "n" value is termed the Schwarzschild exponent.

In the context of UV disinfection processes, there have been few tests of the Bunsen–Roscoe law, since exact dosimetry is required. Sommer et al. [98] found that the Bunsen–Roscoe law held for the inactivation of liquid suspensions of *Escherichia coli*, *Bacillus subtilis* spores, and *Staphylococcus aureus* phage by 254 nm UV. However, for vegetative cells (but not spores) of the yeast *Saccharomyces cerevisiae* , lower dose rates led to greater inactivation (consistent with a Schwarzschild exponent less than 1). In a subsequent study, this group also found that some, but not all, strains of *E. coli* differed from the Bunsen–Roscoe law, with higher dose rates leading to greater inactivation [99]. There is also evidence that inactivation of spores of the fungus *Aspergillus brasiliensis* (formerly *Aspergillus niger*) did not follow the Bunsen–Roscoe law either for 254 nm light or sunlight, with larger fluence rates leading to greater inactivation [100].

Martin et al. [101] comprehensively reviewed the evidence for reciprocity in UV disinfection (as well as photochemical reactions). They concluded that only in a small number of cases was there deviation from the Bunsen–Roscoe law.

Historically, much UV disinfection was conducted using monochromatic sources, particularly low-pressure mercury vapor lamps with a narrow peak emission at 254 nm. More recent work has used polychromatic sources, either from medium-pressure lamps or from light-emitting diode sources. These are described in more detail in Section 12.4.1. There is evidence that such polychromatic sources result in increased inactivation rates per unit fluence due to the greater sensitivity of many organisms at wavelengths between 260 and 270 nm [91].

The greater efficiency of polychromatic UV, or UV with peak emissions in the 260–270 nm range may also be due to the formation of reactive oxygen species that can provide a second mechanism for inactivation [90, 102]. The occurrence of this second pathway has been found to be consistent with a kinetic model for the inactivation of *E. coli* by LED UV radiation at peak emissions from 280 to 308 nm [103].

Hijnen et al. performed a review of inactivation data for numerous organisms in water under both monochromatic 253.7 nm UV and polychromatic UV [104]. They used the Bunsen–Roscoe law, modified in some cases for "shoulders" in the following form for a batch system:

$$\log_{10} \frac{v}{v_0} = -k_{UV} \cdot I_\lambda \cdot t + b \quad (11.54)$$

The intercept, b, was only found to be discernible for bacterial spores, and k_{UV} has units of reciprocal fluence, e.g., $\frac{cm^2}{mJ}$: selected values of this rate parameter are given in Table 11.4. Note the substantial range in inactivation rates among the different organisms.

To obtain kinetic data, such as those in Table 11.4, experiments must be conducted in well-controlled batch systems in which the fluence rate is carefully maintained and calibrated. This is

Table 11.4 Rates for UV inactivation of selected microorganisms in water for monochromatic 254 nm light ($\log_{10}$ basis).

Organism	$k_{UV}\ \frac{cm^2}{mJ}$	Maximum $\log_{10}$
Adenovirus 40	0.018	3.0
Hepatitis A	0.181	5.4
Campylobacter jejuni	0.880	5.3
Escherichia coli O157:H7	0.642	5.4
Cryptosporidium parvum oocysts	0.225	3.0

Source: Data from [104].

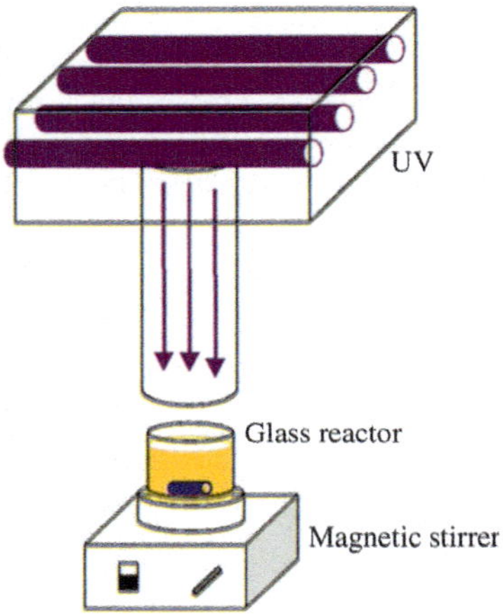

Figure 11.18 Schematic of batch UV collimated beam apparatus. *Source:* [105]/The Royal Society/CC BY 4.0.

typically done using a collimated beam system, such as in Figure 11.18. The optical path of this system is designed to produce a constant fluence rate across the sample area, with relatively small divergence. Sample mixing is included to achieve homogeneous exposure to the fluence. However, if absorbing materials are present in the sample, the delivered dose must be corrected accordingly.

UV inactivation data for microorganisms in air have generally been obtained from exposure of inocula on flat surfaces, although it is understood that an effect of relative humidity on UV inactivation may exist [92], akin to the effect of RH on the decay of airborne organisms as noted in Section 8.4.2. However, the absence of data obviates further consideration of this effect.

Kowalski et al. have analyzed multiple datasets for microbial inactivation on surfaces by UV, all monochromatic at 254.7 nm, and found that these often exhibit discernible shoulders that can be described by the multi-hit model Equation 11.27. Recall, however, from Figure 11.12, that multiple models, such as the series event and the Hom model can yield virtually indistinguishable shaped survival (versus time or fluence) curves as the multi-hit model.

Example 11.5 illustrates the use of these models.

Example 11.5 Dose Required for Batch Water UV Disinfection.

A volume of water is to be treated by monochromatic UV light at 254.7 nm in a batch system. It is desired to inactivate 99% (2 logs) of Cryptosporidium parvum. What should the UV dose (fluence) be?

Solution:
From Table 11.4, the inactivation rate is 0.225 $\frac{cm^2}{mJ}$. The Chick–Watson relationship with exponent equal to one can be written (using log_{10} survival as a basis for the rate constant):

$$log_{10}(S) = -k_{UV} I_\lambda t$$

Substituting known information yields:

$$log_{10}(0.01) = -0.225 \cdot I_\lambda t$$

Inverting this expression solves for the required value:

$$I_\lambda t = 8.89 \, \frac{mJ}{cm^2}$$

A similar approach can be used to determine the fluence required for UV disinfection of air or surfaces, as in Example 11.6.

Example 11.6 Dose Required for Surface Inactivation of Rotavirus.

It is desired to inactivate Rotavirus Type I on a potentially contaminated surface with UV. A four-log reduction (99.99%) is desired. Calculate the required fluence.

Solution:
From Table 11.5, $k_{UV} = 0.0574 \frac{cm^2}{mJ}$ and n (the shoulder parameter) is 1.29.

The multi-hit equation (Equation 11.27) can be used with these parameters, starting by substituting the known values. To use this equation, the rate parameter needs to be expressed in natural log units, i.e., $k'_{UV} = 2.3 \cdot 0.0574 = 0.13202$. With this, the following can be written:

$$0.0001 = 1 - \left[1 - e^{(-0.13202 \cdot I_{UV} \cdot t)}\right]^{1.29}$$

This can be rearranged to solve for the fluence:

$$I_{UV} \cdot t = -\frac{1}{0.13202} ln\left[1 - (1 - 0.0001)^{\frac{1}{1.29}}\right]$$

Finally, evaluating the required fluence:

$$I_{UV} \cdot t = 71.6935 \, \frac{mJ}{cm^2}$$

Table 11.5 Rates for UV inactivation of selected microorganisms on surfaces (log_{10} basis). Includes the best estimate for the number of hits in the multi-hit model.

Organism	$k_{UV}\,\frac{cm^2}{mJ}$	n
Rotavirus Type I	0.0574	1.29
Mycobacterium avium	0.1742	5.97
Pseudomonas aeruginosa	0.2783	1.77
Legionella pneumophila	1.0883	1.67
Bacillus anthracis spores	0.0135	2.6

Source: Adapted from [92].

From an engineering design point of view, to deliver the total light energy required based on fluence, the light absorbance characteristics of the **medium** to be treated must be characterized. If there is no absorbance, then the entire volume exposed receives the same fluence[23] and therefore knowing only the ratio between the surface area receiving illumination and the volume, in other words, what the depth of the volume is, allows the characterization of the total illumination required.

As an example, consider a volume of 10 L (10,000 cm^3) of a water that is to be disinfected by UV with the design objectives and parameters of Example 11.5. If the volume is contained in cube, the depth is $\sqrt[3]{10,000} = 21.5\ cm$, and the cross-sectional area is 464 cm^2. Therefore, the total irradiance required to be delivered is:

$$8.89\,\frac{mJ}{cm^2} \times 464\,cm^2 = 4126\,mJ \tag{11.55}$$

If the volume to be treated has light absorbance, then the incident fluence at the surface (written as $\overline{I_{UV}}$) is not received at all depths in the system.[24] There are two possible extreme states to be considered [106] in describing how to correct for absorbance.[25] In one extreme, the entire volume is regarded as well mixed with respect to distance from the UV source, so that microorganisms effectively see a depth-averaged fluence.[26] This analysis also assumes that the volume being exposed to UV is sufficiently close to the source(s) so that divergence of the illumination can be neglected. Applying Beer's law, where A is the optical absorbance (which conventionally is given on a $\log_{10}$ basis), then the depth-averaged fluence, where d is the depth, can be given by

$$\overline{I_{UV}} = \frac{1}{d}\int_0^d I_{UV}^0 10^{-zA} dz \tag{11.56}$$

---

23 Ignoring any effects due to shadowing in the volume or reflection from bounding surfaces.

24 It is assumed here and in subsequent considerations of UV that the Bunsen–Roscoe law is followed. Extension of these concepts to situations where this is not the case is straightforward, but leads to more complex mathematical formulations.

25 This is akin to the discussion previously on complete segregation versus maximum micromixing in flowing systems.

26 This assumes uniform cross-section with respect to depth. If there is a nonuniform cross-section — for example, if the volume is in a conical volume — then the time at a distance from the source will be proportional to the cross-sectional area at that distance.

which when evaluated yields Equation 11.57.

$$\overline{I_{UV}} = I_{UV}^{0}\left[\frac{1 - 10^{-A \cdot d}}{A \cdot d \cdot ln(10)}\right] \tag{11.57}$$

In this case of mixing in the direction of light passage, the survival ratio of the system can be written by substitution of $\overline{I_{UV}}$ into the survival equation, e.g., the Chick–Watson with an exponent of 1. This relationship indicates that there is still an exponential relationship between surface incident light intensity and the survival ratio; however, the average effective fluence over the volume needs to be reduced by the factor in square brackets in Equation 11.57 due to absorbance in the medium as shown in Equation 11.58.

$$S = exp\left\{-k \cdot I_{UV}^{0}\left[\frac{1 - 10^{-A \cdot d}}{A \cdot d \cdot ln(10)}\right]\right\} \tag{11.58}$$

When there is complete mixing, i.e., complete stratification, in the direction of light passage, any single organism experiences the fluence at a single depth. Hence, the actual inactivation for each layer needs to be computed and averaged over the system. To do this, the integral in Equation 11.59 needs to be evaluated.

$$\overline{S} = \frac{1}{d}\int_{0}^{d} exp(-k_{UV} \cdot I_{UV}^{0} \cdot 10^{-Az})dz \tag{11.59}$$

This can be solved analytically in terms of the exponential integral function, $Ei_1(x) = \int_x^{\infty} \frac{e^{-x}}{x}dx$ to yield Equation 11.60.

$$\overline{S} = \frac{Ei_1(k \cdot I_{UV}^{0} \cdot 10^{-Ad}) - Ei_1(k \cdot I_{UV}^{0})}{Ad \cdot ln(10)} \tag{11.60}$$

In the presence of stratification, there is no longer a strict exponential dependence of survival ratio on intensity, but rather a "tailing off." This is shown graphically in Figure 11.19.

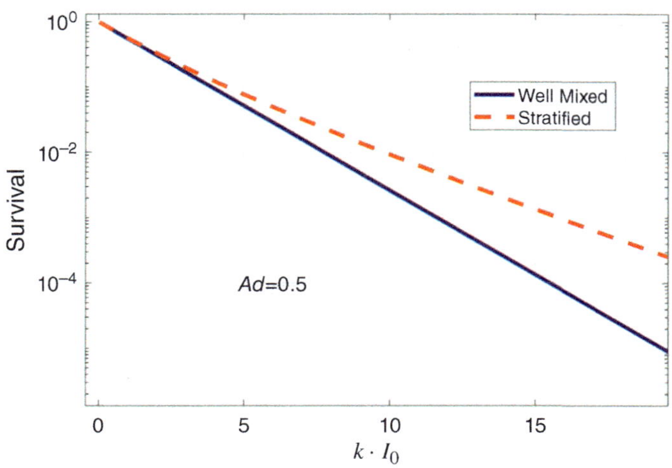

Figure 11.19 Comparison of batch UV systems for the mixed versus stratified models at $Ad = 0.5$.

In flowing systems, whether in water or air, the rate of inactivation can be expressed in terms of the local fluence rate (E_{UV}). The simplest case is when the conditions of a completely mixed box model (Section 8.3.1) hold. In that case, Equation 8.5 for viable microorganism concentration (N) can be written by expressing the reaction rate in terms of the average delivered fluence rate:

$$\frac{dv}{dt} = -k_{UV}\overline{E_{UV}}v + \frac{v_{in} - v}{\theta} + \frac{E}{V} \tag{11.61}$$

If the Bunsen–Roscoe law does not hold, then the first term after the equal sign must be modified with the exponent characterizing the lack of dose-time reciprocity.

Due to the inter-relationship between variation in fluence rate in space (and potentially in time for pulsed systems), hydrodynamic fluctuations associated with fluid flow, and potential absorbance in the volume treatment, more complex approaches are needed to assess performance in flowing systems. Three approaches have been used — computational fluid dynamics (CFD), biodosimetry, and chemical actinometry. Much of the work on flowing systems has been in the context of water; however, several air treatment performance studies have been reported.

UV flow through reactors can be modeled using either a Lagrangian (particle-tracking) or an Eulerian approach, as defined in Section 8.3. In the UV context, the principal difference is that the local fluence rate at each point in the system is defined and used to determine the local microbial inactivation rate. In the Lagrangian approach, based on the trajectory of a "particle" through the system, the probability of its remaining viable is tracked. In the Eulerian approach, the pathogen concentration is treated as a continuous variable with a local reaction rate. The Eulerian method directly provides the average effluent viable microorganism concentration, whereas the Lagrangian method yields a distribution of survival probabilities that must be averaged to obtain the mean performance. These two approaches have been compared in the context of a full-scale UV disinfection contactor for water and found to yield comparable results [94].

CFD has the advantage that it can be used to optimize the various variables affecting design, such as geometry, baffling, residence time, and fluence rate [107]. It can also be applied to test designs that have not yet been constructed. However, the setup and use of CFD models still require advanced skills and specialized software, which may become more accessible and user-friendly over time.

Biodosimetry has been developed as a means to challenge existing physical systems under a variety of conditions and stems from the original proposals of Qualls and Johnson [108]. If a reproducibly consistent microorganism that can be produced in sufficiently large quantities is used, and if that organism has a (batch) UV inactivation that is exponentially dependent on fluence, then the measurement of the system effluent survival ratio can be used to calculate a "reduction equivalent fluence." However, the approach introduces statistical biases, for which the US EPA [109] has partially introduced correction factors [110]. Additionally, if the inactivation of the bioassay organism exhibits nonlinearities, such as shoulders or tailing off, the validity of the back calculation may be inappropriate outside the linear range.

Actinometry is conceptually related to biodosimetry. In this approach, reagents that react quantitatively and specifically with UV light are added to the system, and the extent of the reaction is assessed in the system to compute the (average) delivered fluence. There are a number of such calibrated systems including [111]:

- The photolysis of ferrioxalate proceeds by the following process:

$$2\,Fe(C_2O_4)_3^{-3} \longrightarrow 2\,Fe(C_2O_4) + 3\,C_2O_4^{-2} + 2\,CO_2 \tag{11.62}$$

- The photocatalysis of the reaction between iodide and iodate

$$8I^- + IO_3^- + 3\,H_2O \longrightarrow 3\,I_3^- + 6\,OH^- \quad (11.63)$$

- The oxidation of uridine to 6-hydroxy-5,6-dihydrouridine.

Other chemical systems can be used if their reactions are quantitatively related to fluence and if they can be measured with precision. Comparison of actinometric, CFD, and biodosimetric measurements in UV disinfection systems has shown that the methods yield comparable fluence determinations [112].

There are downsides to directly adding actinometric reagents to a flowing reactor. These include costs, potential interferences with substances in the water, and the greater degree of mixing of dissolved reagents vis-a-vis microorganisms. As a result, in the assessment of UV systems used for water, an alternative has developed in which the reagents are bound to, or contained within, microparticles, and these spheres are dosed to the inlet flow. By measuring the conversion *distribution* amongst the effluent microparticles, such as by flow cytometry, the actual fluence distribution can be attained. Because this method is similar to particle-tracking methods used in simulation, it has been termed Lagrangian actinometry [113].

Blatchley et al. used the photoreactive dye, (E)-5-2-methoxycarbonyl ethenyl cytidine bound to microspheres to challenge a lab-scale UV flow-through reactor [113]. The results were compared with biodosimetry using *Bacillus* spores and with Lagrangian CFD simulations and found to be comparable across the approaches. A similar comparison was made by Wols et al. who found good comparison between Lagrangian CFD and biodosimetry; however, the Lagrangian actinometric method yielded lower estimates of fluence, which were explained as due to fluorescence decay within the UV reactor [114].

Actinometry has been used in air applications by placing the actinometric reagents in closed but transparent containers and determining the fluence at discrete locations. Examples of this approach are:

- In one study, 0.8–1.25 cm diameter quartz spheres were filled with an iodide-iodate actinometric solution and placed at multiple locations in a 90 m^3 room containing five UV lamps [115]. Five spheres were placed at various locations within the room, and from their response, the fluence rate was determined to be 65–85 $\frac{\mu W}{cm^2}$, with a mean of 42–44 $\frac{\mu W}{cm^2}$ (depending on exposure time).
- In follow-up work, it was found that the actinometric solution contained in cylinders was more suited to measuring the output close to UV lamps mounted in louvered fixtures (to protect occupants from direct exposure to the illuminance [116].

Ionizing radiation can be used for disinfection. These radiations are at shorter wavelengths, higher frequencies, and higher energies per quanta than UV (Figure 11.8). It is common to express the dose of such radiation in the derived SI unit of Grays, which is the *absorbed dose* of ionizing radiation. It is measured by the *energy deposited by radiation per unit mass* of the absorbing material, i.e. $Gy = 1\,\frac{\text{joule}}{\text{kilogram}}$. An alternative unit for absorbed radiation is the rad, where 1 rad = 0.01 Gy.

Ionizing radiation is a powerful tool for disinfection, widely applied in healthcare sterilization, water and wastewater treatment, and industrial settings. The principal sources used are **gamma rays**, **electron beams (e-beams)**, and **X-rays**, each with unique characteristics and operational considerations.

Gamma radiation is produced from sealed radioisotope sources, predominantly cobalt-60. It emits high-energy gamma photons (approximately 1.17 and 1.33 MeV) with excellent penetration, delivering uniform doses suitable for sterilizing medical equipment, surgical instruments, pharmaceuticals, and industrial products.

Electron beams are generated by linear accelerators that produce high-energy electrons (typically 4–10 MeV). While their penetration is limited (a few centimeters in dense materials), e-beams deliver high dose rates and are highly efficient for sterilizing surfaces, packaging, disposable medical supplies, and wastewater sludge. E-beam facilities have been piloted and implemented at municipal wastewater treatment plants for reducing pathogens in sludge, offering a chemical-free, on-demand alternative to thermal or chemical methods [117].

Radiation can induce a variety of chemical effects in exposed materials. This can include the formation of free radicals such as $OH^{\bullet}$ and direct radiolysis of organic compounds. At high doses, radiation can be used for decolorization of humic materials from natural waters [118]. As a result, the inactivation of microorganisms by radiation can be caused by direct damage and by indirect effects of the reactive products generated.

As with UV, the inactivation of microorganisms by radiation has often been modeled as a function solely of the total delivered dose, e.g., Grays. However, a recent study on the inactivation of *E. coli* by X-rays has shown a dose rate effect; the nature of this effect depended on whether the treated cells were in exponential or stationary phase growth prior to irradiation [103]. In further discussion, the dose-time reciprocity of the Bunsen–Roscoe relationship will be employed, i.e., inactivation is a function of total delivered dose.

With the reciprocity, the inactivation kinetics by ionizing radiation can be described by the same classes of models as elsewhere in this chapter. In many, but not all, cases, exponential dependency on delivered doses has been noted. Some particular examples are summarized below.

- A dose of 5 kGy of electron beam radiation to wastewater secondary effluent resulted in three logs inactivation of coliphage, total coliform, and total plate count. Gamma radiation of the same dosage yielded four logs of inactivation. In all cases, a tailing-off inactivation curve was observed [119].

  Taghipour studied the inactivation of *E. coli* in primary and secondary wastewater. A dose of 170 and 150 Gy from gamma irradiation, respectively, was required to achieve one log inactivation in the two wastewaters. There was no evidence of a dose-rate effect, and the inactivation was exponential over at least 5 logs [120]. Interestingly, this study also investigated UV inactivation and estimated that the energy cost per unit inactivation when UV was used was lower than that for gamma irradiation.
- The inactivation of naturally occurring spores in soil by gamma radiation was found to exhibit an exponential decay with dose, with a dose of 2.05 kGy required per log of inactivation. There was some evidence of synergistic inactivation as temperature increased [121].
- Inactivation of various organisms in aerobically and anaerobically digested sewage sludges by e-beam irradiation found that inactivation was exponential with dose, and that the doses for one log inactivation were 0.2–0.3 kGy for vegetative bacteria, 3.1–5 kGy for aerobic or anaerobic spores, and 1.5–4 kGy for virus and bacteriophage. There was some evidence of potential matrix effects due to undefined protective matter [122].
- Irradiation has been widely used in the food industry. A study of the inactivation of *E. coli* by electron beam irradiation in a variety of meat and poultry matrices showed exponential decay with some matrix effects, and a dose for one log inactivation of 0.27–0.7 kGy [123].

- Historically, there has been interest in the use of radiation for the disinfection or sterilization of medical devices [124] and also archival (library) material [125]. During the SARS-CoV-2 pandemic, there was concern about contamination of surfaces, including food packaging. Liue et al. [93] studied the inactivation of porcine epidemic diarrhea virus, a SARS-CoV-2 surrogate, by electron beam irradiation on surfaces inoculated with the virus and found that a 10 kGy dose gave an approximate 1.7 log inactivation on a variety of surfaces that could be used in packaging.

11.3 Discussion Questions and Problems

1. A surface is being wiped with a chlorine solution containing 10 mg/L free chlorine at a pH of 8.5. What is the equilibrium chlorine gas concentration above this solution if the solution has $100\frac{mg}{L}$ and is at a temperature of 25 °C?

2. Ozone gas containing 8% O_3 is bubbled through a solution that has 0.03 M of chloride. If the temperature is maintained at 4 °C, what is the solution concentration that can be achieved at equilibrium?

3. A water to be ozonated has the following characteristics:

pH	8.1
UVA	0.02 cm^{-1}
DOC	1.1 mg/L
ALK	100 mg/L

   If a CSTR with a mean residence time of 10 minutes is to be dosed with sufficient ozone to maintain a residual of 1.5 $\frac{mg}{L}$, what is the required ozone dose?

4. A high-pressure steam autoclave is to be designed for a temperature of 145 °C. What is the required steam partial pressure?

5. For the calculation in Example 11.4, what would be the time required if the rate constant for temperature increase, k_T was 0.1 min^{-1} ?

6. A continuous flow 254 nm UV system is to be designed to achieve 99% inactivation of *Cryptosporidium parvum* oocysts in water. If the mean residence time is 10 minutes and there is no light absorption, what fluence is required for a plug-flow system? What would it be for a CSTR?

7. It is desired to use UV for reduction of microorganisms on a surface. The target reduction for all organisms is 99.9%. Assuming no light absorbance or interference from other materials in the fomites, which organism from Table 11.5 would be the most resistant, i.e., require the greatest fluence delivered to the surface?

References

1 US Environmental Protection Agency. Pesticide registration manual. https://www.epa.gov/pesticide-registration/pesticide-registration-manual, February 2025.

2 C Averill. Facts regarding the disinfecting powers of chlorine: Letter to Hon. J I Degraff, *Mayor of the City of Schenectady NY*. SS Riggs Printer, Schenectady, NY, 1832.

3 Black & Veatch Corporation. *White's Handbook of Chlorination and Alternative Disinfectants*. Wiley, Hoboken, NJ, 5th edition, 2010.

4 Gordon M Fair, J Carrell Morris, Shih Lu Chang, Ira Weil, and Robert P Burden. The behavior of chlorine as a water disinfectant. *Journal (American Water Works Association)*, 40(10):1051–1061, 1948.

5 A E Griffin and N S Chamberlin. Some chemical aspects of breakpoint chlorination. *Journal of the New England Water Works Association*, 55:371–374, 1941.

6 Chad T Jafvert and Richard L Valentine. Reaction scheme for the chlorination of ammoniacal water. *Environmental Science and Technology*, 26:577–586, 1992.

7 Marie Deborde and Urs Von Gunten. Reactions of chlorine with inorganic and organic compounds during water treatment—kinetics and mechanisms: A critical review. *Water Research*, 42(1-2):13–51, January 2008.

8 Sharif Hossain, Christopher W K Chow, David Cook, Emma Sawade, and Guna A Hewa. Review of nitrification monitoring and control strategies in drinking water system. *International Journal of Environmental Research and Public Health*, 19(7):4003, 2022.

9 Jacob L Kool, Joseph C Carpenter, and Barry S Fields. Effect of monochloramine disinfection of municipal drinking water on risk of nosocomial legionnaires disease. *Lancet*, 353:272–277, 1999.

10 Fariba Amiri, Maria M F Mesquita, and Susan A Andrews. Disinfection effectiveness of organic chloramines, investigating the effect of pH. *Water Research*, 44:845–853, 2010.

11 David G Wahman. Chlorinated cyanurates: Review of water chemistry and associated drinking water implications. *JournalAmerican Water Works Association*, 110(9):E1–E15, September 2018.

12 Gilbert Gordon and Aaron A Rosenblatt. Chlorine dioxide: The current state of the art. *Ozone: Science & Engineering*, 27(3):203–207, June 2005.

13 Vipin K Rastogi, Shawn P Ryan, Lalena Wallace, Lisa S Smith, Saumil S Shah, and G Blair Martin. Systematic evaluation of the efficacy of chlorine dioxide in decontamination of building interior surfaces contaminated with anthrax spores. *Applied and Environmental Microbiology*, 76(10):3343–3351, 2010.

14 Dustin Poppendieck, Heidi Hubbard, and Richard L Corsi. Hydrogen peroxide vapor as an indoor disinfectant: Removal to indoor materials and associated emissions of organic compounds. *Environmental Science & Technology Letters*, 8(4):320–325, April 2021.

15 Christine C Winterbourn. Toxicity of iron and hydrogen peroxide: The fenton reaction. *Toxicology letters*, 82:969–974, 1995.

16 James A Imlay. Pathways of oxidative damage. *Annual Reviews in Microbiology*, 57(1):395–418, 2003.

17 Cameron Tapp and Rip G Rice. Generation and control of ozone. In Colm O'Donnell, B K Tiwari, P J Cullen, and Rip G Rice, editors, *Ozone in Food Processing*, chapter 4, pages 33–54. Blackwell Publishing Ltd., 2012.

18 Peter J Linstrom and William G Mallard. NIST Chemistry WebBook, NIST Standard Reference Database Number 69. https://webbook.nist.gov/, 2001, accessed 7 April 2025.

19 Abdallah Abdelfattah, Hassan Ramadan, Tamer Elsamahy, Reham Eltawab, Somaia Mostafa, Xiangtong Zhou, and Liang Cheng. Multifaced features and sustainability of using pure oxygen in biological wastewater treatment: A review. *Journal of Water Process Engineering*, 53:103883, 2023.

20 Alexander V Levanov, Oksana Ya Isaikina, Ramiya B Gasanova, and Valery V Lunin. Solubility of ozone and kinetics of its decomposition in aqueous chloride solutions. *Industrial & Engineering Chemistry Research*, 57(43):14355–14364, 2018.

21 Eberhard Rischbieter, Hendrik Stein, and Adrian Schumpe. Ozone solubilities in water and aqueous salt solutions. *Journal of Chemical & Engineering Data*, 45(2):338–340, 2000.

22 Davide Gardoni, A. Vailati, and Roberto Canziani. Decay of ozone in water: A review. *Ozone: Science & Engineering*, 34(4):233–242, July 2012.

23 Min Cho, Hyenmi Chung, and Jeyong Yoon. Effect of pH and importance of ozone initiated radical reactions in inactivating *Bacillus subtilis* spore. *Ozone: Science & Engineering*, 24(2):145–150, January 2002.

24 Charles N Haas and Gordon R Finch. *Methodologies for the Determination of Disinfection Effectiveness*. AWWA Research Foundation, Denver, CO, 2001.

25 Mehmet Kitis. Disinfection of wastewater with peracetic acid: a review. *Environment International*, 30(1):47–55, 2004.

26 Xiu-wei Ao, Jussi Eloranta, Ching-Hua Huang, Domenico Santoro, Wen-jun Sun, Ze-dong Lu, and Chen Li. Peracetic acid-based advanced oxidation processes for decontamination and disinfection of water: A review. *Water Research*, 188: 116479, January 2021.

27 Chiqian Zhang, Pamela J B Brown, and Zhiqiang Hu. Thermodynamic properties of an emerging chemical disinfectant, peracetic acid. *Science of The Total Environment*, 621:948–959, 2018.

28 W Tulalamba, A Assawamakin, A Thayananuphat, and V Viprakasit. Evaluation of potassium peroxymonosulfate (MPS) efficacy against SARS-CoV-2 virus using RT-qPCR-based method. *International Journal of Infectious Disease*, 110:162–164, September 2021.

29 Dariusz Kiejza, Urszula Kotowska, Weronika Polińska, and Joanna Karpińska. Peracids-new oxidants in advanced oxidation processes: The use of peracetic acid, peroxymonosulfate, and persulfate salts in the removal of organic micropollutants of emerging concern a review. *Science of the Total Environment*, 790:148195, 2021.

30 Centers for Disease Control and Prevention. Ethylene oxide "gas" sterilization. https://www.cdc.gov/infection-control/hcp/disinfection-sterilization/ethylene-oxide-sterilization.html, November 2023.

31 Heather N Lynch, Jordan S Kozal, Anthony J Russell, William J Thompson, Haley R Divis, Rachel D Freid, Edward J Calabrese, and Kenneth A Mundt. Systematic review of the scientific evidence on ethylene oxide as a human carcinogen. *Chemico-Biological Interactions*, 364:110031, 2022.

32 William A Arnold, Arlene Blum, Jennifer Branyan, Thomas A Bruton, Courtney C Carignan, Gino Cortopassi, Sandipan Datta, Jamie DeWitt, Anne-Cooper Doherty, Rolf U Halden, et al. Quaternary ammonium compounds: A chemical class of emerging concern. *Environmental Science & Technology*, 57(20):7645–7665, May 2023.

33 Paul C DeLeo, Carolyn Huynh, Mala Pattanayek, Katherine Clark Schmid, and Nathan Pechacek. Assessment of ecological hazards and environmental fate of disinfectant quaternary ammonium compounds. *Ecotoxicology and Environmental Safety*, 206:111116, 2020.

34 Dobrawa Kwaśniewska, Ying-Lien Chen, and Daria Wieczorek. Biological activity of quaternary ammonium salts and their derivatives. *Pathogens*, 9(6):459, 2020.

35 Charles P Gerba, Stephanie Boone, Raymond W Nims, Jean-Yves Maillard, Syed A Sattar, Joseph R Rubino, Julie McKinney, and M Khalid Ijaz. Mechanisms of action of microbicides commonly used in infection prevention and control. *Microbiology and Molecular Biology Reviews*, 88(3):e00205-22, 2024.

36 Paula A Araújo, Madalena Lemos, Filipe Mergulhão, Luís Melo, and Manuel Simões. The influence of interfering substances on the antimicrobial activity of selected quaternary ammonium compounds. *International Journal of Food Science*, 2013(1):237581, 2013.

37 John M Boyce. Alcohols as surface disinfectants in healthcare settings. *Infection Control & Hospital Epidemiology*, 39(3):323–328, 2018.

38 McDonnell Gerald and Russell A Denver. Antiseptics and disinfectants: Activity, action, and resistance. *Clinical Microbiology Reviews*, 12(1):147–179, 1999.

39 William A Rutala, David J Weber, and Healthcare Infection Control Practices Advisory Committee. Guideline for disinfection and sterilization in healthcare facilities, 2008 (updated 2024). *Technical Report*, Centers for Disease Control and Prevention, 2024.

40 W B Hugo. A brief history of heat, chemical and radiation preservation and disinfection. *International Biodeterioration & Biodegradation*, 36(3-4):197–217, October 1995.

41 Benjamin M Zeldes, Matthew W Keller, Andrew J Loder, Christopher T Straub, Michael W W Adams, and Robert M Kelly. Extremely thermophilic microorganisms as metabolic engineering platforms for production of fuels and industrial chemicals. *Frontiers in Microbiology*, 6:1209, 2015.

42 US Environmental Protection Agency. Pesticide devices: A guide for consumers. *Technical Report*, US Environmental Protection Agency, October 2024.

43 Robert L Platzman. What is ionizing radiation? *Scientific American*, 201(3):74–83, 1959.

44 B G Ershov and A V Gordeev. A model for radiolysis of water and aqueous solutions of H_2, H_2O_2 and O_2. *Radiation Physics and Chemistry*, 77(8):928–935, 2008.

45 Kerry J Howe, David W Hand, John C Crittenden, R Rhodes Trussell, and George Tchobanoglous. *Principles of Water Treatment*. John Wiley & Sons, 2012.

46 Sara Feijoo, Xiaobin Yu, Mohammadreza Kamali, Lise Appels, and Raf Dewil. Generation of oxidative radicals by advanced oxidation processes (AOPs) in wastewater treatment: A mechanistic, environmental and economic review. *Reviews in Environmental Science and Bio/Technology*, 22(1):205–248, March 2023.

47 Harriette Chick. An investigation of the laws of disinfection. *The Journal of Hygiene*, 8(1):92–158, January 1908.

48 C W Hiatt. Kinetics of the inactivation of viruses. *Bacteriological Reviews*, 28(2):150–163, 1964.

49 J M Smith, Hendrick C Van Ness, Michael M Abbott, and Mark T Swihart. *Introduction to Chemical Engineering Thermodynamics*. McGraw-Hill Education, New York, 8th edition, 2018.

50 Lyndon L Gyürék and Gordon R Finch. Modeling water treatment chemical disinfection kinetics. *Journal of Environmental Engineering*, 124(9):783–793, 1998.

51 Leonard W Hom. Kinetics of chlorine disinfection in an ecosystem. *Journal of the Sanitary Engineering Division*, 98(1):183–194, 1972.

52 Charles N Haas. A mechanistic kinetic model for chlorine disinfection. *Environmental Science and Technology*, 14:339–340, 1980.

53 Jin Anotai. Effect of calcium ion on chemistry and disinfection efficiency of free chlorine at pH 10. *Ph.D. Thesis*, Drexel University, 1996.

54 Somendu B Majumdar, William H Ceckler, and Otis J Sproul. Inactivation of poliovirus in water by ozonation. *Journal (Water Pollution Control Federation)*, 45(12):2433–2443, 1973.

55 Charles N Haas and Baris Kaymak. Effect of initial microbial density on inactivation of Escherichia coli by monochloramine. *Journal of Environmental Engineering and Science*, 7:237–245, 2008.

56 Charles N Haas and Baris Kaymak. Effect of initial microbial density on inactivation of Giardia muris by ozone. *Water Research*, 37:2980–2988, 2003.

57 J F Fowler. Differences in survival curve shapes for formal multi-target and multi-hit models. *Physics in Medicine & Biology*, 9(2):177, 1964.

58 Blaine F Severin, Makram T Suidan, and Richard S Engelbrecht. Series-event kinetic model for chemical disinfection. *Journal of Environmental Engineering*, 110(2):430–439, 1984.

59 Jörg Soppa. Polyploidy and community structure. *Nature Microbiology*, 2(2):16261, 2017.

60 Sven Gard. Chemical inactivation of viruses. In G E W Wolstenholme and Elaine C P Millar, editors, *Novartis Foundation Symposia*, pages 123–146. John Wiley & Sons, Ltd., Chichester, UK, 1957.

61 Charles N Haas and Josh Joffe. Disinfection under dynamic conditions: Modification of Hom's model for decay. *Environmental Science & Technology*, 28(7):1367–1369, 1994.

62 Susan M Teefy and Philip C Singer. Performance and analysis of tracer tests to determine compliance of a disinfection scheme with the SWTR. *Journal of the American Water Works Association*, 82(12):88–98, 1990.

63 D M Marske and J D Boyle. Chlorine contact tank design—A field evaluation. *Water & Sewage Works*, 120(1):71–77, 1973.

64 T N Zwietering. The degree of mixing in continuous flow systems. *Chemical Engineering Science*, 11:1–15, 1959.

65 V Kothandaraman and Ralph L Evans. Hydraulic model studies of chlorine contact tanks. *Journal of the Water Pollution Control Federation*, 44(4):625–633, 1972.

66 R Rhodes Trussell and Junn-Ling Chao. Rational design of chlorine contact facilities. *Journal of the Water Pollution Control Federation*, 49(4):659–667, 1977.

67 William D Bellamy, Gordon R Finch, and Charles N Haas. *Integrated Disinfection Design Framework*. American Water Works Association, 1998.

68 Charles N Haas. Micromixing and dispersion in chlorine contact chambers. *Environmental Technology Letters*, 9:35–44, 1988.

69 Stephen A Craik. Effect of micro-mixing conditions on predictions of *Cryptosporidium* inactivation in an ozone contactor. *Ozone: Science & Engineering: The Journal of the International Ozone Association*, 27:487–494, 2005.

70 Felicitas Pfeifer. Distribution, formation and regulation of gas vesicles. *Nature Reviews Microbiology*, 10(10):705–715, 2012.

71 E Bruce Nauman and B A Buffham. *Mixing in Continuous Flow Systems*. John Wiley and Sons, New York, 1983.

72 V Pfeiffer and B Barbeau. Validation of a simple method for predicting the disinfection performance in a flow-through contactor. *Water Research*, 49(1):144–156, 2014.

73 I A Hannoun, P F Boulos, and E J List. Using hydraulic modeling to optimize contact time. *Journal of the American Water Works Association*, 90:77–87, 1998.

74 Daniel Kevin Peplinski and Joel J Ducoste. Modeling of disinfection contactor hydraulics under uncertainty. *Journal of Environmental Engineering*, 128(11):1056–1067, 2002.

75 D J Greene, B Farouk, and C N Haas. CFD design approach for chlorine disinfection processes. *Journal of the American Water Works Association*, 96:138–150, 2004.

76 B A Wols, J A M H Hofman, W S J Uijttewaal, L C Rietveld, and J C Van Dijk. Evaluation of different disinfection calculation methods using CFD. *Environmental Modelling & Software*, 25(4):573–582, April 2010.

77 Shamia Hoque, Bakhtier Farouk, and Charles N Haas. Development of artificial neural network based metamodels for inactivation of anthrax spores in ventilated spaces using computational fluid dynamics. *Journal of the Air and Waste Management Association*, 61(9):968–82, 2011.

78 Jie Zhang, Andrés E Tejada-Martínez, and Qiong Zhang. Developments in computational fluid dynamics-based modeling for disinfection technologies over the last two decades: A review. *Environmental Modelling & Software*, 58:71–85, 2014.

79 Saima Farooq and Chedly Tizaoui. A critical review on the inactivation of surface and airborne SARS-CoV-2 virus by ozone gas. *Critical Reviews in Environmental Science and Technology*, 53(1):87–109, January 2023.

80 E van Asselt and M Zwietering. A systematic approach to determine global thermal inactivation parameters for various food pathogens. *International Journal of Food Microbiology*, 107(1):73–82, March 2006.

81 Maria G Corradini and Micha Peleg. A Weibullian model for microbial injury and mortality. *International Journal of Food Microbiology*, 119(3):319–328, November 2007.

82 G Stone, B Chapman, and D Lovell. Development of a log-quadratic model to describe microbial inactivation, illustrated by thermal inactivation of *Clostridium botulinum*. *Applied and Environmental Microbiology*, 75(22):6998–7005, November 2009.

83 Maha Mahir, Anas El Maakoul, Ismail Khay, Said Saadeddine, and Mohamed Bakhouya. An investigation of heat transfer performance in an agitated vessel. *Processes*, 9(3):468, 2021.

84 W G Murrell and W J Scott. The heat resistance of bacterial spores at various water activities. *Journal of General Microbiology*, 43(3):411–425, June 1966.

85 Maria Fernanda Espinosa, A Natanael Sancho, Lorelay M Mendoza, César Rossas Mota, and Matthew E Verbyla. Systematic review and meta-analysis of time-temperature pathogen inactivation. *International Journal of Hygiene and Environmental Health*, 230:113595, September 2020.

86 Reimar Johne, Johannes Scholz, and Alexander Falkenhagen. Heat stability of foodborne viruses – Findings, methodological challenges and current developments. *International Journal of Food Microbiology*, 413:110582, March 2024.

87 Mahmoud Moeini Sedeh and J M Khodadadi. Energy efficiency improvement and fuel savings in water heaters using baffles. *Applied Energy*, 102:520–533, 2013.

88 Manuel Feurhuber, Ralf Neuschwander, Thomas Taupitz, Valentin Schwarz, Carsten Frank, and Christoph Hochenauer. A computational fluid dynamics (CFD) model to simulate the inactivation of geobacillus stearothermophilus spores in different moist heat sterilization environments. *Physics in Medicine*, 12:100039, 2021.

89 James R Bolton, Ian Mayor-Smith, and Karl G Linden. Rethinking the concepts of fluence dose and fluence rate: The importance of photon–based units – A systemic review. *Photochemistry and Photobiology*, 91(6):1252–1262, November 2015.

90 Wenjun Sun, Zibo Jing, Zhinan Zhao, Ran Yin, Domenico Santoro, Ted Mao, and Zedong Lu. Dose–Response behavior of pathogens and surrogate microorganisms across the Ultraviolet-C Spectrum: Inactivation efficiencies, action spectra, and mechanisms. *Environmental Science & Technology*, 57(29):10891–10900, July 2023.

91 Sara E Beck, Harold B Wright, Thomas M Hargy, Thomas C Larason, and Karl G Linden. Action spectra for validation of pathogen disinfection in medium-pressure ultraviolet (UV) systems. *Water Research*, 70:27–37, March 2015.

92 W J Kowalski, W P Bahnfleth, D L Witham, B F Severin, and T S Whittam. Mathematical modeling of ultraviolet germicidal irradiation for air disinfection. *Quantitative Microbiology*, 2:249–270, 2000.

93 Dong Liu, Joel Ducoste, Shanshan Jin, and Karl Linden. Evaluation of alternative fluence rate distribution models. *Journal of Water Supply: Research and Technology-Aqua*, 53(6):391–408, September 2004.

94 Joel J Ducoste, Dong Liu, and Karl Linden. Alternative approaches to modeling fluence distribution and microbial inactivation in ultraviolet reactors: Lagrangian versus Eulerian. *Journal of Environmental Engineering*, 131(10):1393–1403, October 2005.

95 Rajeshwar P Sinha and Donat-P Häder. UV-induced DNA damage and repair: A review. *Photochemical & Photobiological Sciences*, 1(4):225–236, 2002.

96 W T Bovie. A law of the relation between exposure and biological effect of radiation. *The Journal of Cancer Research*, 10(2):161–174, July 1926.

97 K Schwarzschild. On the deviations from the law of reciprocity for bromide of silver gelatine. *Astrophysical Journal*, 11: 89, 1900.

98 Regina Sommer, Thomas Haider, Alexander Cabaj, Erich Heidenreich, and Michael Kundi. Increased inactivation of saccharomyces cerevisiae by protraction of uv irradiation. *Applied and Environmental Microbiology*, 62(6):1977–1983, 1996.

99 R Sommer, T Haider, A Cabaj, W Pribil, and M Lhotsky. Time dose reciprocity in UV disinfection of water. *Water Science & Technology*, 38(12):145–150, 1998.

100 Lizbeth Taylor-Edmonds, Tovit Lichi, Adi Rotstein-Mayer, and Hadas Mamane. The impact of dose, irradiance and growth conditions on *Aspergillus niger* (renamed *A. brasiliensis*) spores low-pressure (LP) UV inactivation. *Journal of Environmental Science and Health, Part A*, 50(4):341–347, March 2015.

101 Jonathan W Martin, Joannie W Chin, and Tinh Nguyen. Reciprocity law experiments in polymeric photodegradation: A critical review. *Progress in Organic Coatings*, 47(3-4):292–311, 2003.

102 Dana Pousty, Ron Hofmann, Yoram Gerchman, and Hadas Mamane. Wavelength-dependent time–dose reciprocity and stress mechanism for UV-LED disinfection of Escherichia coli. *Journal of Photochemistry and Photobiology B: Biology*, 217:112129, April 2021.

103 Takahiro Matsumoto, Ichiro Tatsuno, Yukiya Yoshida, Makoto Tomita, and Tadao Hasegawa. Time-dose reciprocity mechanism for the inactivation of *Escherichia coli* explained by a stochastic process with two inactivation effects. *Scientific Reports*, 12(1):22588, December 2022.

104 W Hijnen, E Beerendonk, and G Medema. Inactivation credit of UV radiation for viruses, bacteria and protozoan (oo)cysts in water: A review. *Water Research*, 40:3–22, 2006.

105 Yuqing Tang, Xueting Shi, Yongze Liu, Li Feng, and Liqiu Zhang. Degradation of clofibric acid in uv/chlorine disinfection process: kinetics, reactive species contribution and pathways. *Royal Society Open Science*, 5(2):171372, 2018.

106 Charles N Haas and G P Sakellaropoulous. Rational analysis of ultra-violet disinfection reactors. In *Proceedings of the Sanitary Engineering Division of ASCE*. ASCE, 1979.

107 Jainil Shah, Audrius Židonis, and George Aggidis. State of the art of UV water treatment technologies and hydraulic design optimisation using computational modelling. *Journal of Water Process Engineering*, 41:102099, June 2021.

108 R G Qualls and J D Johnson. Bioassay and dose measurement in UV disinfection. *Applied and Environmental Microbiology*, 45(3):872–877, 1983.

109 United States Environmental Protection Agency, Office of Water. Ultraviolet disinfection guidance manual for the final long term 2 enhanced surface water treatment rule. *Technical Report* EPA 815-R-06-007, US Environmental Protection Agency, 2006.

110 Joel Ducoste, Karl Linden, David Rokjer, and Dong Liu. Assessment of reduction equivalent fluence bias using computational fluid dynamics. *Environmental Engineering Science*, 22(5):615–628, September 2005.

111 Joseph Rabani, Hadas Mamane, Dana Pousty, and James R Bolton. Practical chemical actinometry—A review. *Photochemistry and Photobiology*, 97(5):873–902, September 2021.

112 Kenichiro Deguchi, Satoshi Yamaguchi, and Hiroshi Ishida. UV-disinfection reactor validation by computational fluid dynamics and relation to biodosimetry and actinometry. *Journal of Water and Environment Technology*, 3(1):77–84, 2005.

113 Ernest R. Blatchley III, Chengyue Shen, Zorana Naunovic, Lian-Shin Lin, Dennis A. Lyn, J. Paul Robinson, Katherine Ragheb, Gérald Grégori, Donald E. Bergstrom, Shiyue Fang, et al. Dyed microspheres for quantification of UV dose distributions: Photochemical reactor characterization by Lagrangian actinometry. *Journal of Environmental Engineering*, 132(11):1390–1403, 2006.

114 Bas A Wols, C H M (Roberta) Hofman-Caris, Danny J H Harmsen, Erwin F Beerendonk, Johannis C Van Dijk, Po-Shun Chan, and Ernest R Blatchley. Comparison of CFD, biodosimetry and Lagrangian actinometry to assess UV reactor performance. *Ozone: Science & Engineering*, 34(2):81–91, March 2012.

115 Ronald O Rahn, Peng Xu, and Shelly L Miller. Dosimetry of room-air germicidal (254 nm) radiation using spherical actinometry. *Photochemistry and Photobiology*, 70(3):314–318, September 1999.

116 Ronald O Rahn. Spatial distribution of upper-room germicidal UV radiation as measured with tubular actinometry as compared with spherical actinometry. *Photochemistry and Photobiology*, 80(2):346–350, September 2004.

117 M Siwek and T Edgecock. Application of electron beam water radiolysis for sewage sludge treatment—A review. *Environmental Science and Pollution Research*, 27(34):42424–42448, 2020.

118 A K Pikaev. Current status of the application of ionizing radiation to environmental protection: I. Ionizing radiation sources, natural and drinking water purification (A Review). *High Energy Chemistry*, 34(1):1–12, January 2000.

119 Shaukat Farooq, C N Kurucz, T D Waite, and W J Cooper. Disinfection of wastewaters: high-energy electron vs. gamma irradiation. *Water Research*, 27:1177–1184, 1993.

120 Fariborz Taghipour. Ultraviolet and ionizing radiation for microorganism inactivation. *Water Research*, 38(18):3940–3948, November 2004.

121 Marcel C Reynolds, Kermit F Lindell, Thomas J David, Martin S Favero, and Walter W Bond. Thermoradiation inactivation of naturally occurring bacterial spores in soil. *Applied Microbiology*, 28(3):406–410, September 1974.

122 Chandni Praveen, Palmy R Jesudhasan, Robert S Reimers, and Suresh D Pillai. Electron beam inactivation of selected microbial pathogens and indicator organisms in aerobically and anaerobically digested sewage sludge. *Bioresource Technology*, 144:652–657, September 2013.

123 E Mayer-Miebach, M R Stahl, U Eschrig, L Deniaud, D A E Ehlermann, and H P Schuchmann. Inactivation of a non-pathogenic strain of *E. coli* by ionising radiation. *Food Control*, 16(8):701–705, October 2005.

124 Donald J Dempsey and Ramachandran R Thirucote. Sterilization of medical devices: A review. *Journal of Biomaterials Applications*, 3(3):454–523, 1988.

125 Giuseppe Magaudda. The recovery of biodeteriorated books and archive documents through gamma radiation: Some considerations on the results achieved. *Journal of Cultural Heritage*, 5(1):113–118, 2004.

12

Inactivation Applications

This chapter continues the discussion of inactivation processes. In the first portion of the chapter, the details of fitting experimental inactivation data to kinetic models will be covered. Next, by-products of applying various inactivation technologies will be discussed. Finally, the details of using the different technologies for specific environmental media will be described.

12.1 Fitting Performance Models to Data

If experiments are conducted to assess the removal of microorganisms by a process, it may be desirable to ascertain the parameters characterizing performance from the data. These parameters can then be used for design, including by the approaches discussed in Chapter 11. In some sense, this is an extension of the basic concepts of analyzing microbial occurrence data discussed in Section 6.2.

The fitting process requires several design questions to be answered, from which the approach to parameter estimation can be determined:

1. What is the nature of the dependent variable—is it quantal, count, or continuous? This depends on the method used for enumerating the remaining microorganisms (Section 6.2). For continuous data, there are some observations below a detection or quantification limit, or above a limit of calibration. For count data, there are some observations too numerous to count (TNTC).
2. What is the error distribution of the dependent variable expected to be? For example, with count data, the counts might be expected to be Poisson distributed with a mean given from the model prediction; for a continuous dependent variable, the values (possibly after transformation) might be expected to be normally distributed with a mean given from the model prediction.
3. What are the independent (descriptor) variables? It is assumed that these are measured with minimal error; if not, more specialized approaches that will not be discussed further here, termed error in variable methods [1] must be used.
4. What is/are the underlying relationship(s) between the descriptor variables and the dependent variable? In other words, if we had the actual value of parameters in a model, how would we use them to calculate the dependent variable? In general, for batch systems, these are the equations for $\frac{v}{v_0}$, i.e., survival ratio, S, given for the models in Section 11.2.1.
5. Of the parameters, which are known with certainty, and which do we need to estimate from the data?

The first two items will define the nature of the likelihood function. The remaining items will define the calculation approach needed. The entire system is connected to an optimizer to find the best estimates for the unknown parameters and possibly an estimate for the uncertainty given the data.

The overall process is described in the flowchart in Figure 12.1. The computational steps are as follows:

1. The predictors are identified, which may include items such as time, chemical dose, temperature, and loading rates, among others.
2. An initial guess to parameters is made, possibly by intuition or information from prior studies.
3. Using the process model, the predicted performance is computed based on the parameters and predictors.
4. The quality of the prediction, in terms of agreement with the observations, is assessed using an objective function, such as deviance or the sum of squares of differences. Note that, for count or quantal data, the volume (or mass) of the sample is generally required. In the for count or continuous data with censoring, the TNTC limit or detection limit is required.
5. If there is a significant change from the previous iteration, or if this is the first iteration, then an updated set of parameter guesses is made, and the process returns to step 3. Otherwise, the most recent set of parameter guesses is accepted as the final values, along with the final value of the objective function.

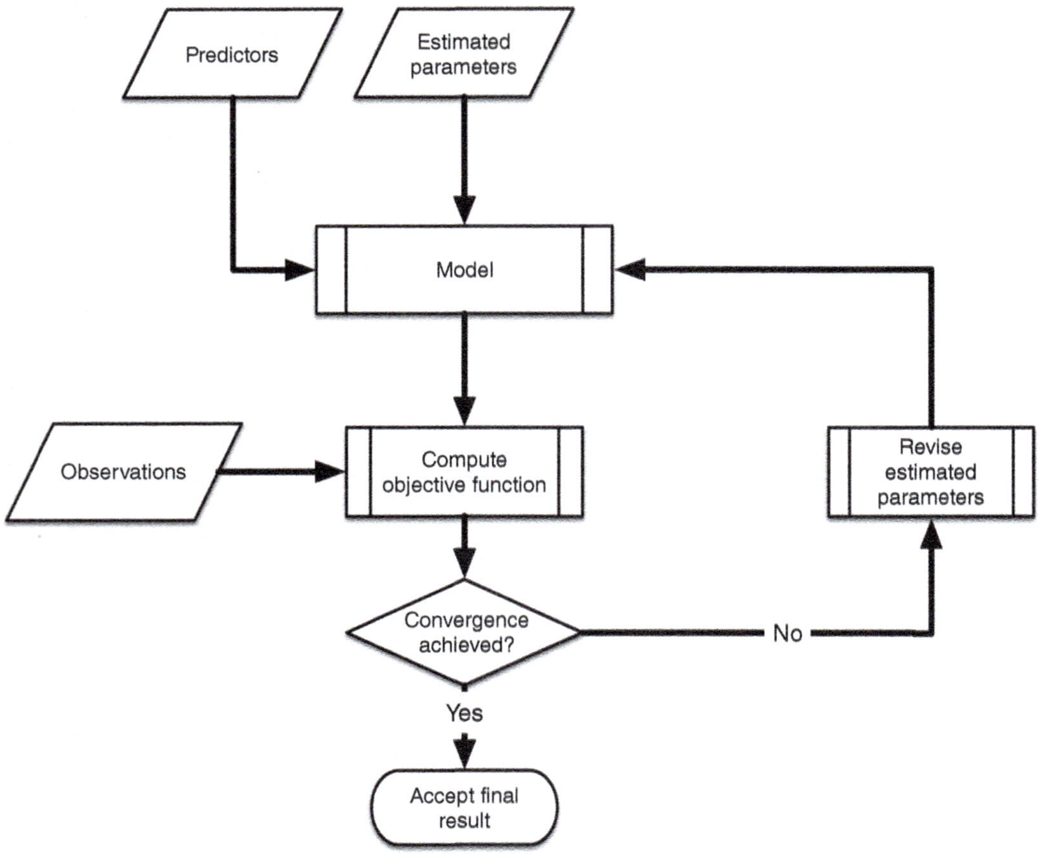

Figure 12.1 Flowchart for parameter estimation of models.

Following the fitting of a model, several post-fit questions can be addressed:

1. Does the model provide a statistically significant improvement in fit (relative to no model)?
2. Is the fit adequate, or is there evidence of a lack of fit that needs to be addressed? This is often termed goodness-of-fit testing [2].
3. When the process can be repeated for multiple models, the performance of each model can be determined. From this comparison, do more complex models with additional parameters provide a statistically significant improvement in fit?

The first two questions will be addressed for each of the data types examined. In all cases, when embarking on such data analysis, it is useful to have and store observations in a "tidy data" format. As defined by Wickham [3], tidy data has[1]

- each variable (type of data) in a column
- each observation forms a row
- each type of observational unit forms a table.

The process is best illustrated by a series of examples.

12.1.1 Basic Count Data

For count data, such as plate count or plaque count data, the fundamental data (columns in a tidy presentation) for the dependent variable would include a column indicating censoring or not,[2] the volume or mass of sample examined,[3] and the actual integer count (or in the case of TNTC data, the maximum number of countable units).

Descriptive data might include codes for experiment number, water examined, etc. Formal predictor variables could include disinfectant dose,[4] contact time, and other attributes of the system that might influence inactivation, possibly including temperature, pH, alkalinity, absorbance, etc.

The analysis of disinfection experiments with count data uses maximum likelihood, modifying Equation 6.17 to incorporate TNTC (censoring). The vector of parameters in this likelihood function, θ, corresponds to the distribution of counts given an expected value, which defines one parameter of the distribution. In the case of the Poisson distribution for counts, the Poisson mean μ_i for observation i is given by:

$$\mu_i = \upsilon_i \cdot V_i \tag{12.1}$$

and with a specification of an inactivation model, for example, Hom, in a batch system:

$$\upsilon_i = \upsilon_{i,0} \exp\left(-k'_d C_i^n t_i^m\right) \tag{12.2}$$

Note that the kinetic parameters, i.e., k'_d, n, and m, are regarded as constant over all observations in an experiment. This method has been shown to yield parameter estimates for an inactivation model

1 This is an extension of the tidy data concept seen in Table 8.8.

2 Since this is a logical flag, it makes sense to code 0 for absence of censoring and 1 for censoring — commonly observations that are TNTC.

3 Best practice would be to calculate the equivalent amount of original sample, after correcting for any dilution that occurs in processing.

4 First, the case of constant disinfectant residual, i.e., when dose=residual and is constant over time, is discussed. Then, the case where there is decay or demand will be treated.

with lower bias and variance than other methods, such as regressing microbial concentrations on predictors [4].

Therefore, given a data set and a candidate inactivation (or for other removal processes, a removal equation), if the Poisson distribution for counts is assumed, the best set of parameters is obtained by maximizing the likelihood function (or equivalently, minimizing the negative log of the likelihood function). This is shown by example.

Anotai [5] performed experiments on the inactivation of *E. coli* by chlorine in a batch mixed system. Results from one set of experiments (at pH 10 and 25 °C) are shown in Table 12.1. In this case, there were no TNTC observations.[5] Experiments were performed on four dates, each with a different chlorine concentration (which was unchanged due to the short duration of the experiments). Note that a standard date code of YYYYMMDD is used as an identifier for the experiment.

At time zero, the initial counts for each experiment were recorded, and based on that, the initial concentrations[6] υ_0 were as noted in Table 12.2.

The code in Listing 12.1, in R, performs the maximum likelihood optimization of this problem. It uses the built-in maximum likelihood optimizer and the Poisson distribution.

Table 12.1 Batch inactivation of *E. coli* by free chlorine at pH 10, 25 °C.

Date	Time	Chlorine	Volume	Counts	Date	Time	Chlorine	Volume	Counts
19940208	0.5	0.095	2.22E-04	160	19941116	0.17	0.285	2.20E-04	181
19940208	1	0.095	2.22E-03	337	19941116	0.33	0.285	2.20E-04	142
19940208	1.5	0.095	2.22E-02	145	19941116	0.5	0.285	2.22E-03	586
19940208	2	0.095	2.22E-01	65	19941116	1	0.285	2.22E-01	225
19940208	2.5	0.095	2.22E+00	34	19941116	1.5	0.285	2.22E-01	48
19940208	3	0.095	2.22E+00	120	19941116	2	0.285	2.22E+01	386
19940208	4	0.095	2.22E+01	118	19941116	3	0.285	2.22E+01	249
19940208	5	0.095	2.22E+01	0	19941116	5	0.285	2.22E+01	39
19940208	6	0.095	2.22E+01	1	19941122	0.17	0.485	2.22E-04	194
19940223	0.5	0.08	2.22E-04	271	19941122	0.33	0.485	2.22E-03	568
19940223	1	0.08	2.22E-03	230	19941122	0.5	0.485	2.22E-03	213
19940223	1.5	0.08	2.22E-02	221	19941122	1	0.485	2.22E-01	14
19940223	2	0.08	2.22E-01	213	19941122	1.5	0.485	2.22E+00	2
19940223	2.5	0.08	2.22E-01	154	19941122	2	0.485	2.22E+01	8
19940223	3	0.08	2.22E+00	102	19941122	3	0.485	2.22E+01	0
19940223	4	0.08	2.22E+01	355	19941122	5	0.485	2.22E+01	0
19940223	5	0.08	2.22E+01	30					
19940223	6	0.08	2.22E+01	2					

Source: Data from [5]. Time in minutes and volume in mL.

5 Had there been, then an additional field (column) for the indicator of censoring would be added.

6 Here we treat υ_0 as a fixed value for each date that the experiment was done. Strictly speaking, this value should also be estimated from the data, since it is most often only known through experimental measurement. For simplicity, this is treated here as a constant per experiment, and in this data set, there is little effect on the derived parameters and fits.

Table 12.2 Initial *E. coli* concentrations in Anotai experiments.

Date	$\upsilon_0 \left(\frac{\#}{mL}\right)$
19940208	9.32E+05
19940223	1.23E+06
19941116	1.01E+06
19941122	9.28E+05

```
library(readxl)
library(maxLik)
library(stats)
# ----- Read in data -----
experiments<-read_excel('experiments.xlsx')
controls<-read_excel('controls.xlsx')
# ----- Add control concentrations -----
experiments$upsilon_0 <- controls$upsilon_0[match(experiments$date, controls$date)]

Hom_Survival<-function(k,n,m,C,t){
  y<-k*(C^n)*(t^m)
  return(exp(-y))
  }

# ---- Note the Optimization is Done for Log Parameters ----
LogLikelihood<-function(parameters){
  y=exp(parameters)
  k=y[1];n=y[2];m=y[3]
  pred_Survival<-Hom_Survival(k,n,m,experiments$chlorine,experiments$time)
  pred_Counts<-pred_Survival*experiments$volume*experiments$upsilon_0
  LogL<- dpois(experiments$counts,pred_Counts,log=TRUE) # returns log L
  return(sum(LogL))
  }
# ----- Built in Function to Maximize Log Likelihood -----
results<-maxLik(LogLikelihood,grad=NULL,hess=NULL,start=c(1,-.1,.2),
     method='BFGS',control=list(printLevel=1))
summary(results)
exp(results$estimate)
```

Listing 12.1 Fitting of disinfection count experiment.

The output of this fitting yields the Hom model parameters of $k_d = 9.267$, $n = 0.402$, and $m = 0.928$, with an optimized negative log likelihood of 4531.342. This could be repeated with alternative inactivation models to ascertain the comparative fitting, and decide which model should be accepted, m, as will be discussed subsequently.

A first statistical test is whether this model represents an improvement over no model (i.e., a null model). From the raw data, generalizing Equation 6.4, a mean concentration over all the observations, $\overline{\upsilon}$, ignoring treatment conditions, can be computed from the individual colony counts N_i and volumes V_i from:

$$\overline{\upsilon} = \sum_i \frac{N_i}{V_i} \tag{12.3}$$

Then the value of this pooled mean can be used in Equation 6.17 to compute the null negative log likelihood as 32367.17—since this is substantially greater than that with the Hom model it is clear that there is a significant improvement from the use of the model.

Test of Significance of Model Versus Null

A formal test of the significance of a given model relative to the null model can be performed using a likelihood ratio approach. If L_0 is the negative log likelihood for the null model, i.e., using the pooled mean, and L_1 is the negative log likelihood of the fitted model, and df_1 is the number of parameters in the fitted model, then twice the difference in likelihood is computed:

$$\Delta = 2(L_0 - L_1)$$

This is compared to the χ^2 distribution at degrees of freedom equal to $df_1 - 1$, and if the value of Δ exceeds the upper critical percentile of the χ^2 distribution, then a null hypothesis of no improvement can be rejected.

In the example presented, $\Delta = 2(32367.17 - 4531.34) = 55671$. For two degrees of freedom, i.e., three Hom parameters minus 1, this is a vanishing small percentile, and thus the fit is highly significant.

A lack-of-fit (LOF) test assesses whether unexplained variance remains after the model is fit. The null hypothesis is that the fit from the model is as good as if each data point were predicted from its own observation. For count data, the observed concentration, i.e., $v_i = \frac{N_i}{V_i}$ is used as the predicted mean in the Poisson likelihood (Equation 6.17), and the negative log likelihood for this "saturated" model[7] is computed (L_s). This is used for a test of significance.

Lack of Fit Test for Count Data

As with the test of significance, twice the difference from the model fit is computed:

$$\Delta_{\text{LOF}} = 2(L_1 - L_s)$$

Δ_{LOF} is compared to the χ^2 distribution at degrees of freedom equal to the number of data points minus the number of parameters.

For the example presented, there are 34 data points, and the degrees of freedom are therefore $34 - 3 = 31$. Evaluating the saturated negative log likelihood yields $L_s = 97.633$. Therefore, $\Delta = 2(4531.34 - 97.633) = 8867$ and the upper percentile is vanishingly small.

It is concluded that the null hypothesis that the Hom model fits the saturated model provides equivalent fits *is rejected.*

As seen in the analysis of the LOF, the rejection of the goodness of fit indicates that there is substantial unexplained variance in the data that is not captured by the model. This may arise from several sources, including the following alone or in combination:

- The model may inadequately capture significant performance details, and a more complex model may perform better—in this case, perhaps a Hom–power law model would perform better.
- The assumption of Poisson variability in counts (from the model predictions) may be incorrect, and a more overdisperse model, such as negative binomial, may be more appropriate.

7 It is called the saturated model since the maximum number of parameters, i.e., one per observation, is used.

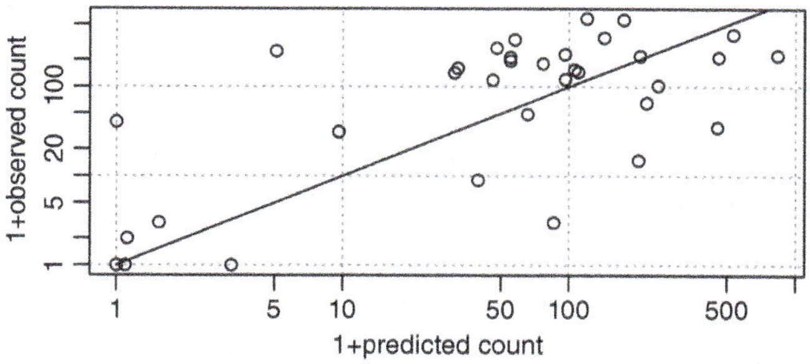

Figure 12.2 Observed versus predicted colonies for Hom fit to Anotai data.

The lack of fit may be further understood by examining a plot of the observed number of colonies versus the predicted number of colonies (Figure 12.2). Several things are clear: first, there are more points below the line of equality, indicating a tendency of the model to underpredict the number of colonies, and also a significant degree of scatter around the fit. The treatment of experiments with count data characterized by overdisperse distributions will not be covered in detail; however, when such lack of fit is noted, an alternative approach is the conversion of the count data to concentrations and analyzing the experiment as if it was continuous — although in this case, with zero count data, consideration of censoring will be needed, which will be discussed subsequently.

As with any estimation, the parameters will have some uncertainty. The quantification of such uncertainties will be discussed in Section 13.3.

If there is disinfectant demand during the experiment, then this approach can be modified in one of two ways:

- If the disinfectant decay process is known and the decay rate(s) have been determined, then an analytical model (for example, one of those in Table 11.1) incorporating decay can be used as a predictor.
- Where an analytical solution is not available, the inactivation process in the presence of changing disinfectant can be described by a set of ordinary differential equations, which would need to be embedded in the maximum likelihood fitting routine.

12.1.2 Basic Quantal Data

Example quantal data (Section 6.2.2) may originate from a variety of experiments in which a positive or negative response is recorded in a sample of a certain amount that has been subjected to specific treatment conditions. The ensemble of such observations, under various treatment conditions, can be used to determine the best set of parameters characterizing the particular process performance.

There are many types of quantal data. These include:

- Use of dilution tube (MPN) methods for coliform or other bacteria.
- Use of similar protocols when viruses or protozoa are assayed by cell culture, with positivity scored by the presence of cytopathogenic effect.
- Use of digital PCR (Section 6.6.2) where the response in a particular trial is the number of positive aliquots and the total number of aliquots examined.

- Use of animal bioassays for assessment, where the endpoint is measured by whether the animal has or does not have the target effect. The effect may be the presence of infection, illness, or death.[8]

For all types of quantal data, the same analysis approach can be used.

For a disinfection experiment, typical data fields might include experiment code, dose, time, initial concentration (υ_0), amount assayed per unit of test, number of units inoculated, and number of units positive (i.e., containing viable organisms). The framework for this analysis is an extension of Equation 6.22, taking into account that at each combination of treatment conditions and volume (given by the subscript k), there is a particular version of υ, so the following can be written:

$$L = \prod_{k=1}^{i} \{[1 - \exp(-\upsilon_k V_k)]^{P_k} [\exp(-\upsilon_k V_k)]^{(T_k - P_k)}\} \tag{12.4}$$

where υ_k is given by the selected performance model, e.g., for Chick–Watson:

$$\upsilon_k = \upsilon_0 \cdot \exp\left(-k_d C_k^n t_k\right) \tag{12.5}$$

And therefore the best values of the inactivation parameters, k_d and n are given by maximizing L (or equivalently minimizing $= \log(L)$) in Equation 12.4.

An example of these data is the study by Hibler [6] of the inactivation of cysts of *Giardia lamblia* by free chlorine. This consisted of over 600 experiments in which samples of 10,000 viable cysts were exposed to chlorine at different concentration/time combinations at three temperatures and three pH values.[9] After each treatment, the sample was inoculated into five gerbils, and the number of infected animals was scored. These data were more fully described earlier [7]. For this example, the experiments at pH 7 and 5 °C are analyzed. The experimental data are shown in Table 12.3.

Table 12.3 Survival of *Giardia muris* after chlorination at pH 7 and 5 °C. Measured by animal infectivity. Time in minutes and chlorine in $\frac{mg}{L}$. Each of five animals was administered 10,000 cysts that had been chlorinated.

Chlorine	Time	Positive	Chlorine	Time	Positive	Chlorine	Time	Positive
0.41	50	5	1.475	107	0	3	54	3
0.49	60	5	1.475	117	0	3	150	0
0.528	326	0	1.475	127	0	3.079	30	5
0.528	336	0	1.475	137	0	3.12	60	5
0.528	346	0	1.5	101	5	3.14	56	1
0.528	356	0	1.5	114	0	3.15	67	1
0.54	150	5	1.57	106	3	3.16	42	3
0.59	254	0	1.9	70	0	3.16	52	2

8 This same type of experiment will be encountered in the determination of dose–response relationships in Section 14.1.

9 Unfortunately, only the disinfectant dose was recorded, and residual disinfectant during or at the end of the experiment was not noted, and so this kinetic analysis is biased by potentially underestimating the potency of disinfectant.

Chlorine	Time	Positive	Chlorine	Time	Positive	Chlorine	Time	Positive
0.597	300	0	1.9	80	0	3.16	62	0
0.614	278	0	1.9	90	0	3.16	72	0
0.614	288	0	1.9	100	0	3.19	300	0
0.614	298	0	1.9	110	0	3.31	67	2
0.614	308	0	2	80	0	3.835	32	5
0.73	227	1	2.02	86	0	3.835	42	0
0.75	213	5	2.02	132	0	3.835	52	0
0.85	176	0	2.02	178	0	3.835	62	0
0.937	300	0	2.02	223	0	4	40	0
1.04	170	0	2.183	67	0	4.01	50	0
1.068	150	0	2.183	77	0	4.06	29	0
1.08	152	2	2.183	87	0	4.06	39	0
1.08	162	0	2.183	97	0	4.06	49	0
1.08	172	0	2.22	66	1	4.06	59	0
1.08	182	0	2.22	76	0	4.07	43	5
1.099	50	5	2.22	86	0	4.08	50	1
1.11	60	5	2.22	96	0	10.065	14	0
1.36	106	1	2.51	67	3	10.065	16	0
			2.54	80	0	10.065	17	0
			2.59	80	0	10.065	19	0
			2.61	60	2	10.065	20	0
			2.69	52	5			
			2.69	63	1			
			2.69	72	0			
			2.69	82	0			

Source: Data from [6].

The analysis of this data can be done with a slight modification of the code for count data by replacing the likelihood function with that of Equation 12.4. The code for fitting the Hom model is given in Listing 12.2.

```
library(readxl)
library(maxLik)

# Hibler Survival Analysis
# Global Constants
initial_cysts=10000
animals_per_assay=5
# ---------------------
experiments<-read_excel('pH7_5C.xlsx')
Hom_Survival<-function(k,n,m,C,t){
  y<-k*(C^n)*(t^m)
  return(exp(-y))
}
LogLikelihood<-function(parameters){
  y=exp(parameters)
  k=y[1];n=y[2];m=y[3]
  pred_Survival<-Hom_Survival(k,n,m,experiments$chlorine,experiments$time)
```

```

  pred_Counts<-pred_Survival*initial_cysts
  y=exp(-pred_Counts)
  L=(1-y)^experiments$positive
  L=L*(y^(animals_per_assay-experiments$positive))
  return(sum(log(L)))
}
# ----- Built in Function to Maximize Log Likelihood -----
results<-maxLik(LogLikelihood,grad=NULL,hess=NULL,start=c(-1,-0.5,-0.1),
                method='BFGS',control=list(printLevel=1,reltol=1e-12))
summary(results)
exp(results$estimate)
```

Listing 12.2 Fitting of quantal experiment.

Fitting this data set with the Hom model results in an optimized log(L) of −147.374, $k_d = 0.983$, $n = 0.457$, and $n = 0.474$. For the complete analysis of this data, the Hom model was found to be superior to Chick–Watson for each pH/temperature combination [7].

The significance of the fit can be tested versus a model in which no disinfection effect is incorporated. In such a case, the expected proportion of positive responses for each experiment is the average of all experiments, i.e., the total positives among all the experiments divided by the total subjects,[10] i.e., for this data set $\frac{87}{440} = 0.1977$. Using this to evaluate the log likelihood, $\log(L) = 591.333$. A test of significance of twice the difference between the Hom likelihood and this null likelihood evaluated at two degrees of freedom (three Hom parameters minus one for the null model) shows a highly significant improvement due to the fit.

The lack-of-fit test is conducted using a saturated model, in which the proportion of positives for each observation equals the observed fraction of tests at that observation that are positive and then evaluates the resulting log likelihood. The saturated log likelihood computed in this manner is −76.790. Twice the difference of the Hom likelihood and the saturated log likelihood is $2(-76.790 + 147.374) = 141.168$. This is evaluated at 85 degrees of freedom (88 observations minus three Hom parameters) and is highly significant ($p \approx 8 \cdot 10^{-6}$), indicating some residual lack of fit.

12.1.3 Basic Continuous Data

Continuous data may arise in several ways. Frequently, the actual raw data from observations (colony counts, positive and negative replicates, etc.) are not available, but they have been used to compute concentration (i.e., v). Alternatively, measurements intrinsically yielding continuous data, such as turbidity, enzyme activity, etc., may have been made.

In this case, assuming no censoring (either below the detection limit or above the quantifiable levels), this problem reduces to least-squares regression. It is generally the case that the regression using the logarithm of the ratio $\frac{v}{v_0}$ is appropriate; however, it is good practice, following least squares regression, to check the validity of the assumptions, as will be demonstrated following the example.

10 The design of this experiment makes the computation of the null model straightforward. If there were multiple dilutions for each observation, then the overall best MPN for the experiment would need to be estimated.

Assumptions of Least Squares Regression

The underlying assumptions for least squares regression that formally allow inferences to be made with respect to statistical significance of the estimated parameters and their fit, and their confidence limits using parametric methods include:

1. That the independent variables (predictors) are measured with little error relative to the dependent variable [8].
2. That the residuals from the fitted model are normally distributed with constant variance.
3. That the residuals are uncorrelated with any of the predictors.

Item 1 must be based on a physical understanding of the measurement methods. Items 2 and 3 may be assessed by a *post hoc* analysis of the fitted model. If these assumptions are not satisfied, then transformation of independent or dependent variables, or the use of an alternative model, is indicated.

A typical tidy data format for this type of experiment would include fields for the experiment identifier, disinfection dose (or residual), exposure time, and either the surviving concentration or the survival ratio. Since it is common to log-transform microbial concentrations and survival ratios, the field may be either transformed or untransformed.

An example of this type of data is work by Wang et al. [9] who exposed paper disks inoculated with spores of *Bacillus subtilis* and assessed survival following exposure to air concentrations of ClO_2 at controlled humidity and temperature in a well-mixed batch air chamber. Table 12.4 presents one set of their experiments.[11]

This can be analyzed in R using the built-in function for nonlinear least squares estimation, `nls` with the code given in Listing 12.3.

```
library(stats)
experiment<-read.csv('fig2_digitized.csv')
Hom<-as.formula(Log10S~log10(exp(-k*(C^n)*(t^m))))
results<-nls(Hom,experiment,start=list(k=0.1,n=1,m=1),trace=TRUE)
summary(results)
```

Listing 12.3 R code for analysis of continuous disinfection model.

Fitting of the data produces a residual sum of squares of 2.406, with parameter estimates $k_d = 0.412$, $n = 0.242$, $m = 0.824$. The value of m below 1 indicates a tailing off inactivation.

A scatter plot of the fitted model versus observations (Figure 12.3) shows generally a good fit, but with deviations at low survival. At these low survivals, there is greater estimation error, so this is not surprising, though perhaps a model with more adjustable parameters, such as the Hom–power law, would yield a better fit with improved statistical significance.

The formal test of significance of the regression (versus a null model) can be performed by conducting a hypothesis test for each parameter versus zero. Running the regression and examining these hypotheses show that each of the coefficients (k, m, and n) is highly significant.

A formal check of residuals shows that the assumption of normality cannot be rejected using the Shapiro–Wilks test. In R, the library `nlstools` is useful for testing assumptions and generating

11 The authors presented the survival ratio data in a figure, which for analysis has been digitized to produce the table.

Table 12.4 Survival of *Bacillus subtilis* spores on paper in presence of gas-phase chlorine dioxide. Temperature = 22–24 °C and relative humidity = 70%.

C $\left(\frac{mg}{L}\right)$	*t* (min)	$\log_{10}\left(\frac{v}{v_0}\right)$
1	5	−0.4
1	15	−1.5
1	30	−3
1	50	−5
1	80	−6.8
2	5	−0.5
2	15	−1.8
2	30	−3.3
2	50	−5.5
2	70	−6.5
3	5	−0.6
3	15	−2
3	30	−4
3	45	−6
3	70	−7
5	5	−0.7
5	10	−1.6
5	20	−3
5	30	−4.8
5	45	−6.5

Source: Data from [9].

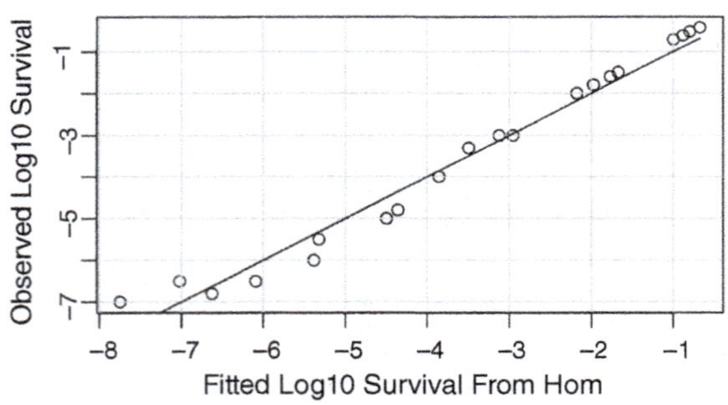

Figure 12.3 Plot of Hom fitted model versus observations for data in Table 12.4.

diagnostic plots. To some extent, this test served the same role as lack-of-fit tests for the other data types.

Another type of lack-of-fit test can be undertaken if there are duplicate experiments. In that case, the residual sum of squares from the regression of a particular model can be compared to the sum of squares of a saturated model in which the value of $\log \frac{v}{v_0}$ is set equal to the mean of the observations under that condition, and the sum of squares from this saturated model is compared to the sum of squares for the candidate model using a classical F test.

12.1.4 Continuous Data with Censoring

The most common situation treated here is data for which it is impossible to quantify observations below a certain value. Consider, for example, an experiment in which the concentration of pathogens in an air stream is 10 organisms per cubic meter, and one cubic meter of air can be examined for the organisms. Hence, if there is removal in excess of 99%, it is unlikely to observe organisms and the concentration is reported as $\leq 1\ m^{-2}$, and therefore the survival fraction is ≤ 2. It would be inappropriate to ignore such data, since they are informative. It would be equally inappropriate to replace this unquantifiable value with a number, such as half the detection limit, since that could introduce bias. However, by using a full maximum likelihood analysis, this censored regression can be performed.[12]

With such data, the analysis begins with considering a likelihood function. Recall that in the consideration of count data, the situation was discussed in which specific observations were censored as TNTC, i.e., censoring to the right, and a modified likelihood involving both a density function and a *complementary cumulative distribution function* was written (Equation 6.17). In the present case, where there is censoring to the left, an analogous equation can be written involving the *cumulative distribution function* (Equation 12.6).

$$L(\overline{v}|\theta) = f(v_1|\theta)^{\delta_1} f(v_2|\theta)^{\delta_2} \ldots f(v_n|\theta)^{\delta_n} \times F(v_1|\theta)^{1-\delta_1} F(v_2|\theta)^{1-\delta_2} \ldots F(v_n|\theta)^{1-\delta_n} \tag{12.6}$$

In the case of regression, f and F are the density function and the cumulative distribution function for the normal distribution. To maximize the likelihood, the parameters of the inactivation relationship, θ, as well as the mean (μ) and standard deviation (σ) of the normal distribution must also be estimated. This censored regression approach has been widely used in a number of applications [10, 11].

Data of this nature in tidy form requires a column, in addition to experiment ID, time, concentration and survival ratio — an indicator for whether the survival ratio is quantified or censored. In Equation 12.6, this would correspond to 0 and 1, respectively. In the R routine used to analyze this, the indicator variable convention is 1 and -1, respectively. The tidy data format for the Anotai experiment is given in Table 12.5. When zero colonies were recorded, the upper limit of the survival ratio is computed on the basis of $v_i = \frac{1}{V_i}$.

This censored regression can be fit using the `gnlr` function within the R `gnlm` library.[13] The code for this computation is shown in Listing 12.4. Using this, the best-fit values of the Hom model are found to be $k = 8.764$, $n = 0.303$, and $m = 0.753$. The residual log-likelihood is -44.38.

12 In some less common circumstances, there may be censoring from above. For example, if plate counts are performed and the results are TNTC, then one only knows that the organisms present are at a level equal to or greater than the maximum countable number.

13 In this example, the `gnlr` function is used assuming a normally distributed error term. However, for more advanced analyses, this function also allows other distributional forms for the error function, and also situations where the variance of error is itself a function of the predicted mean response.

Table 12.5 Anotai data presented as censored concentration data.

Date	Time	Chlorine	Uncensored	S	Date	Time	Chlorine	Uncensored	S
19940208	0.5	0.095	1	7.73E-01	19941116	0.17	0.285	1	8.12E-01
19940208	1	0.095	1	1.63E-01	19941116	0.33	0.285	1	6.37E-01
19940208	1.5	0.095	1	7.00E-03	19941116	0.5	0.285	1	2.60E-01
19940208	2	0.095	1	3.14E-04	19941116	1	0.285	1	1.00E-03
19940208	2.5	0.095	1	1.64E-05	19941116	1.5	0.285	1	2.13E-04
19940208	3	0.095	1	5.80E-05	19941116	2	0.285	1	1.72E-05
19940208	4	0.095	1	5.70E-06	19941116	3	0.285	1	1.11E-05
19940208	5	0.095	−1	4.83E-08	19941116	5	0.285	1	1.73E-06
19940208	6	0.095	1	4.83E-08	19941122	0.17	0.485	1	9.42E-01
19940223	0.5	0.08	1	9.89E-01	19941122	0.33	0.485	1	2.76E-01
19940223	1	0.08	1	8.39E-02	19941122	0.5	0.485	1	1.03E-01
19940223	1.5	0.08	1	8.07E-03	19941122	1	0.485	1	6.80E-05
19940223	2	0.08	1	7.77E-04	19941122	1.5	0.485	1	9.71E-07
19940223	2.5	0.08	1	5.62E-04	19941122	2	0.485	1	3.88E-07
19940223	3	0.08	1	3.72E-05	19941122	3	0.485	−1	4.85E-08
19940223	4	0.08	1	1.30E-05	19941122	5	0.485	−1	4.85E-08
19940223	5	0.08	1	1.09E-06					
19940223	6	0.08	1	7.30E-08					

Note: The value for "uncensored" is 1 for quantified data and −1 for left censored data (this is the convention used in the R software). This would correspond to 0 and 1 in Equation 12.6.

```
library(gnlm)
library(readxl)
library(ggplot2)

experiments<-read_xlsx('AnotaiCensoredData.xlsx')
tt<-experiments$time
C<-experiments$chlorine
censoring_indicator<-experiments$uncensored
log10_survival<-log10(experiments$S)

# Hom Model Prediction
log10_S_pred <- function(p) {
  p=exp(p) # note optimization is on log parameters

  k=p[1];n=p[2];m=p[3]

  ypred=log10(exp(-k*(C^n)*(tt^m)))
  return(ypred)
}
y=cbind(log10_survival,censoring_indicator)
results<-gnlr(y, distribution="normal", mu=log10_S_pred, pmu=c(lnk=-2,lnn=0.1,lnm=0.1)
     ,pshape=0.5)
bestvalues<-exp(results$coefficients[1:3])
experiments$fitted_S<-10^results$fitted.values
experiments$uncensored=factor(experiments$uncensored)
a=ggplot(experiments,aes(x=fitted_S,y=S,shape=uncensored,color=uncensored))
a=a+geom_point(size=2.5)+theme_bw()
a=a+scale_x_log10()+scale_y_log10()
```

```
a=a+geom_abline(aes(intercept=0,slope=1))
a=a+labs(x="Fitted Survival Ratio",y="Observed Survival Ratio")
last_plot()
```

Listing 12.4 R code for analysis of Anotai data as continuous censored survival.

A plot of predicted versus observed survival ratios on a log-log plot is shown in Figure 12.4. The scatter around the line of perfect prediction is relatively homogeneous. However, there is some indication of systematic underprediction of survival at higher survivals and overprediction at lower survival, almost an "S-shaped" pattern. This suggests the possibility that a more parameterized model, such as the Hom–power law, might provide a better fit.

12.1.5 Complex Models

The performance models discussed previously in this section all shared the characteristic that the predicted microorganism response could be written as an explicit function of unknown parameters (to be determined by fit to the data) and other explanatory variables (such as time, concentration, etc.). There are more complex cases in which these characteristics do not hold. The nature of the deviation from these characteristics is problem-specific, so a general classification is not possible. However, three particular classes of examples (that may occur in combination) are the following:

1. The prediction relationship is time-dependent and can only be solved numerically. Therefore, a numerical integrator is embedded within the optimizer. However, given the parameters at a particular iteration, the performance can be computed explicitly.
2. The prediction relationship involves one or more implicit equations for performance, which require an embedded equation solver within the optimizer.

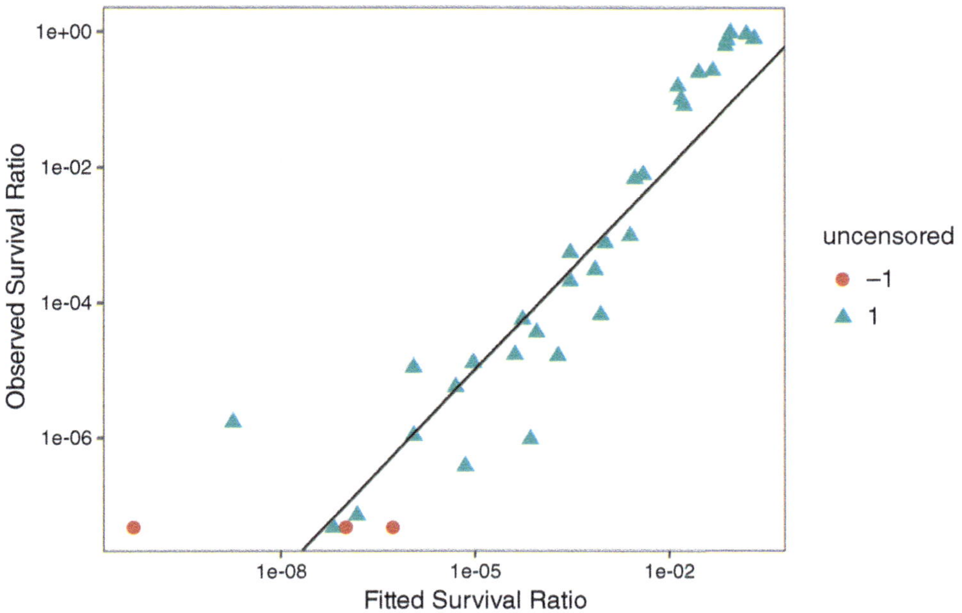

Figure 12.4 Comparison of predicted Hom fit to Anotai data analyzed using the censored regression method.

3. A situation in which the performance with respect to pathogen removal is also a function of another state variable whose presence is determined by unknown parameters that also need to be fitted to the data. In this case, observations of both pathogens and the other state variable(s) are available, and the problem is a multiobjective optimization problem.

These three cases will be considered in sequence.

The first case, where the performance is given by a dynamic model requiring solution of (ordinary or partial) differential equations, has a number of examples presented in this chapter and in Chapter 10. As one example, the evolution of removal of pathogens via particle removal in depth filtration can be given by a solution to the differential equations coupling particle transport, head loss, removal, and mass accumulation (Section 10.2.1).

Another example of this first case occurs when there is disinfectant demand concurrent with inactivation, leading to a model that can readily be stated as a set of ordinary differential equations, in the case of a batch system, but may be difficult or impossible to solve analytically. Consider, for example, inactivation by power law kinetics (Equation 11.21) occurring in a batch system with second-order disinfectant decay. The system can be described by the coupled differential system of Equations 12.7 and 12.8.

$$\frac{d\upsilon}{dt} = -k'C^n\upsilon^x \tag{12.7}$$

$$\frac{dC}{dt} = -k_1C^2 \tag{12.8}$$

If k_1 is known from other experiments, and measurements of υ versus time and initial concentration (C_0) are available, then the kinetic parameters for inactivation (k', n, x) can be determined by fitting the numerical predictions from the ODE system. To implement this, the routine in the curve fitting program, e.g., the function `log10_S_pred` in Listing 12.4, needs to be modified to include a numerical solution of the differential equations. In R, this can be done using the `deSolve` library. A similar approach can be taken in Python.[14]

The second type of complex model is one in which performance is only given by the solution of one or more implicit equations. This can arise when disinfection experiments are conducted in a CSTR, for example. In such a case, generically, for each observation, which is at a given internal reactor concentration of disinfectant and a given mean hydraulic residence time, a mass balance using a reaction rate can be written at steady state for each observation (i) as Equation 12.9.

$$0 = (\upsilon_{0,i} - \upsilon_i) - r_d(\upsilon_i, C_i)\theta_i \tag{12.9}$$

If the inactivation rate function, r_d, is linear, as in the case of Chick–Watson kinetics, this equation is readily rearranged to solve explicitly for υ_i and can be used for fitting as in the previous cases in this section (count, quantal, continuous). However, if r_d is more complex, such as power law or Hom–power law, the solution of Equation 12.9 for υ_i must be obtained by root-finding. This will be shown by example. Table 12.6 has data on survival of spores of *Bacillus subtilis* in a CSTR receiving ozone in water at pH 8 and 15 °C.

14 Although not discussed in this book, platforms are becoming readily available that allow for integrated programming using R, Python, and Julia languages, with the passing of variables between modules. One such platform is `Jupyter` (https://jupyter.org/).

Table 12.6 Survival of spores of *Bacillus subtilis* exposed to ozone in water in a CSTR at 15 °C and pH 8.

Date	Residence time (min)	C (mg/L)	N_o (CFU/mL)	N (CFU/mL)
3/02/2003	2.2	2.106	1605	74
3/02/2003	4.11	2.032	1630	40
3/02/2003	6.2	1.873	1747	33
3/03/2003	2.2	2.19	6940	617
3/03/2003	4.11	2.032	6690	360
3/03/2003	6.2	1.831	6006	290
3/06/2003	2.2	1.513	1648	182
3/06/2003	4.11	1.577	1722	77
3/06/2003	6.2	1.566	1894	53
3/10/2003	2.2	2.085	1732	80
3/10/2003	4.11	2.148	1755	65
3/10/2003	6.2	2.074	1830	38
3/14/2003	2.2	1.492	19186	2169
3/14/2003	4.11	1.683	16850	567
3/14/2003	6.2	1.64	19853	567
3/17/2003	2.2	2.402	1393	49
3/17/2003	4.11	2.339	1346	38
3/17/2003	6.2	2.36	1193	40
3/18/2003	2.2	2.444	15666	941
3/18/2003	4.11	2.413	12896	534
3/18/2003	6.2	2.455	14281	430
3/20/2003	2.2	2.614	99600	5539
3/20/2003	4.11	2.593	98265	2269
3/20/2003	6.2	2.413	105272	2836
3/21/2003	2.2	2.413	14264	621
3/21/2003	4.11	2.466	14264	440
3/21/2003	6.2	2.37	14298	487
3/25/2003	2.2	1.45	97097	12246
3/25/2003	4.11	1.661	100267	4738
3/25/2003	6.2	1.46	90757	3403
3/26/2003	2.2	1.566	116617	11778
3/26/2003	4.11	1.545	108775	4872
3/26/2003	6.2	1.556	116617	5239
3/31/2003	2.2	2.032	85419	5873
3/31/2003	4.11	2.042	118952	3437
3/31/2003	6.2	2	105606	2736

Source: Data from [12].

These data are to be fit to a power law model (Equation 11.21), resulting in a mass balance for each observation (Equation 12.10).

$$0 = (\upsilon_{0,i} - \upsilon_i) - k' C_i^n \upsilon_i^x \theta_i \tag{12.10}$$

Each row in Table 12.6 consists of an observation with predictors C_i, θ_i, and $\upsilon_{0,i}$. The best-fit parameters (k', m, and x) are found by minimizing the sum of squares of predicted and observed log-transformed concentrations (υ_i). The code for this analysis is given in Listing 12.5.

```
library(readxl)
library(rootSolve)
experiment<-read_excel('OzoneBacillusSpores.xlsx')
nrows<-dim(experiment)[1]
#-------------------
pred_upsilon_i<-function(params,row_number){
# given parameters and observation i, compute predicted i
  params<-exp(params)
  kprime<-params[1]; n<-params[2]; x<-params[3]
  Theta<-experiment[row_number,]$Theta
  C<-experiment[row_number,]$C
  upsilon_0<-experiment[row_number,]$upsilon_0

  predictors=c(Theta,C,upsilon_0)
  inputs<-c(Theta,C,upsilon_0,kprime,n,x)

  mass_bal<-function(upsilon_i,inputs){
    y<-inputs[3]-upsilon_i -inputs[4]*(inputs[2]^inputs[5])*
    (upsilon_i^inputs[6])*inputs[1]
    return(y)}
  foundroot<-multiroot(mass_bal,parms=inputs,start=experiment[row_number,]$upsilon,
                          verbose=FALSE,positive=TRUE)
  return(foundroot$root)}
#-------------------
objective <- function(guessedparams) {
  ss<-0
  for(i in 1:nrows){
  pred_y<-  pred_upsilon_i(guessedparams,i)
  ss<-ss+(log10(pred_y) - log10(experiment$upsilon[i]))^2}
  return(ss)}
#-------------------
# Optimize parameters
fit <- optim(par = c(-1,0.1,0.1), fn = objective,control=list(trace=1))
exp(fit$par)
fit$value
```

Listing 12.5 Analysis of CSTR disinfection data by power law model.

Results of this fitting give parameter values of $k' = 3.480$, $n = 1.052$, and $x = 0.971$, with a sum of squares of differences of $\log_{10}$ microorganism concentration equal to 0.4934. Given the closeness of n and x to unity, a test of a basic Chick–Watson relationship would be important to determine the statistical significance of a more complex model.

The third type of complex model arises when data on microbial removal must be fit *alongside* data on another parameter. Consider, for example, that survival and disinfectant concentration are being measured simultaneously. In chemical inactivation kinetics, the inactivation rate depends on the residual disinfectant concentration. Therefore, the inactivation rate constants along with the disinfectant residual decay rates must be estimated from the data.

There are two approaches to handling this type of situation. One approach is to use the chemical decay data separately, estimate the decay rate law and rate coefficients, and incorporate these into an inactivation expression incorporating decay, such as in Table 11.1. This "two-phase" approach is exemplified by the work of Faile et al. who studied inactivation of bacterial spores, *Giardia* and *Cryptosporidium* by ozone in water [13].

A second single-phase approach is possible, in which each observation is considered bivariate, i.e., with respect to disinfectant concentration and survival ratio. Then the decay and inactivation rates can be determined *simultaneously* by optimizing an objective function that combines metrics (e.g., the sum of squares of log ratios) of the two components of the observations, weighted by a factor that accounts for the relative precision of the measurements. This problem is termed a multiresponse nonlinear regression. While this may lead to more efficient, i.e., more precise, estimation of parameters for coupled models [14], there do not appear to have been any datasets on pathogen reduction that have used such an approach or presented data amenable to it.

12.1.6 Comparing Models

The same dataset can be analyzed using different models. As seen previously, there are different potential predictive equations for disinfection (Section 11.2.1). Furthermore, in statistical fitting, the concept of a model encompasses both the constitutive relationship, i.e., the prediction of expected performance given predictor variables, and the distribution assumed for residual errors (normal, log-normal, gamma, inverse Gaussian, Poisson, negative binomial, etc.). The question then is for a given data set with a particular set of predictor variables, what is the best model taking into account the balance of improvement of fit and parsimony?

The question of selection of the best model in this context follows the discussion of selection of the best model for the probability distribution of pathogen occurrence presented in Section 6.2. In that context, BIC (Equation 6.9) and AICc (Equation 6.10) were presented.

The objective of comparing models in the context here is determining which one(s) are most suitable for *prediction*.

AIC and BIC are grounded in different theoretical frameworks and serve different purposes. AIC is derived from information theory and is designed to minimize the expected Kullback–Leibler divergence between the true model and the approximating model, making it suitable for predictive modeling [15]. In contrast, BIC is based on Bayesian principles and aims to identify the true model from a set of candidates, assuming that such a model exists within the set being considered [16]. This fundamental difference leads to AIC being preferred when the goal is **prediction**, as is the case here, while BIC is more appropriate for **model identification**.

Unlike BIC, AIC does not assume that the true model is among the candidate models. Instead, it aims to find the model that best approximates the data-generating process, which is often unknown or complex in real-world scenarios [17].

While BIC is consistent in asymptotically selecting the true model, AIC is **asymptotically efficient** in choosing the model that minimizes prediction error [18]. As noted previously, it is useful to correct AIC for smaller samples, and therefore the corrected AIC in terms of the likelihood function will be used (Equation 6.10).

For count, quantal, and censored data, the likelihood is computed directly from the fitting using the maximum likelihood method. For uncensored continuous data fit by least squares regression, the likelihood can be calculated from the residual sum of squares (RSS) of the regression via Equation 12.11 where n_{obs} is the number of observations.

$$\log(L) = -\frac{n_{obs}}{2}[\log(2\pi) + 1] - \frac{n_{obs}}{2}\log\left(\frac{RSS}{n_{obs}}\right) \tag{12.11}$$

For example, the Hom model fit to the data on spore inactivation by gas-phase chlorine (data in Table 12.4) yielded an RSS of 2.406 with $n_{obs} = 20$. Therefore according to Equation 12.11, the log likelihood can be calculated as:

$$\log(L) = -\frac{20}{2}[\log(2\pi) + 1] - \frac{20}{2}\log\left(\frac{2.406}{20}\right) = -8.7845 \tag{12.12}$$

Using this, the AICc can be computed using Equation 6.10 as:

$$\text{AICc} = -2\log(-8.7845) + 2 \cdot 3\left(\frac{20}{20-3-1}\right) = 24.819 \tag{12.13}$$

By comparing this AICc value with fits of the same data to alternative models, the most suitable model for prediction can be ascertained. Note that other than by comparison of the AICc values between fits of alternative models, no inference can be made based on the specific numerical value itself.

As an example of model comparison, consider the *Giardia muris* data already presented in Table 12.3 and analyzed by the Hom model. That analysis can be repeated using the Chick–Watson, power law and Hom–power law models, and the results are summarized in Table 12.7.

The Chick–Watson model has two parameters, the Hom and power law models have three parameters, and the Hom–power law has four parameters. It would be expected that a model with a greater number of parameters would have a better (or at least no worse) fit—i.e., a greater likelihood, which is the case. By computing the AICc, it is clear that the Hom and PL models are each superior to the Chick–Watson and perform essentially equivalently. The HPL model does not provide any additional improvement. Therefore, based on this analysis, either the HPL or PL model would be accepted.

Another type of comparison that may be done using a similar approach is when an overall dataset has a "natural" split by another category. The following are examples of potential situations that may arise:

- When a series of experiments is performed at different times, and it is desired to determine the temporal consistency of results.
- When tests are done on different environmental samples, and it is desired to determine if there is an effect of the particular environment on performance.
- When experiments are done at different laboratories or test facilities, and it is desired to determine if the performance can be characterized by a single relationship.

Conceptually, these situations can be considered as partitioning an entire data set into multiple subsets, as shown in Figure 12.5.

The formal question then is whether it is sufficient to describe the entire dataset with a single model, or whether, by partitioning and fitting each dataset to an individual model, recognizing the

Table 12.7 Comparison of models fit to Hibler data on *Giardia muris* inactivation.

Model	-ln L	AICc	kd	*n*	m	x
Chick–Watson	−170.0711	344.283	0.068	1.010		
Hom	−147.374	301.034	0.983	0.457	0.474	
Power law	−147.5351	301.356	0.006159	0.959		1.187
Hom–power law	−147.374	303.230	7.80E-08	0.484	0.502	2.754

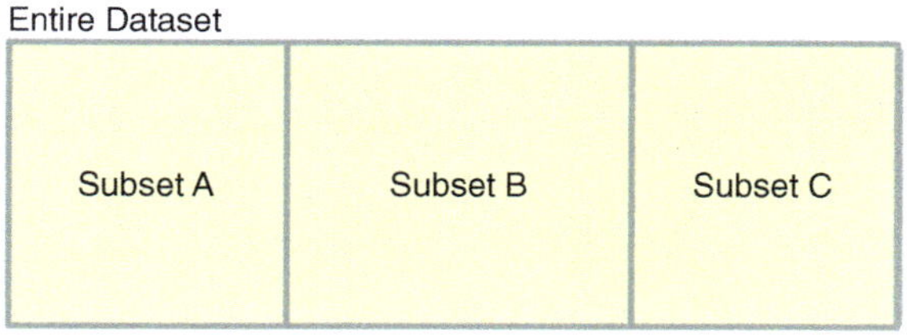

Figure 12.5 Conceptual subsetting of data.

increased number of parameters required, there is a statistically significant improvement in fit. This can be approached in a systematic sense in the following manner:

1. Fit the entire dataset (with n_{obs} observations) to a single model. This model has k_{total} parameters. The fit results in a log likelihood equal to $\log(L_{\text{total}})$. Use these quantities to compute a value for $\text{AICc}_{\text{total}}$ via Equation 6.10.
2. Fit each subset to an individual model. Each subset will have a number of fitted parameters, $k_A, k_B, k_C, \ldots$, and an optimized log likelihood, $\log(L_A), \log(L_B), \log(L_C), \ldots$. Then compute the following:

$$k_{\text{subsetted}} = \sum_{Q=A,B,C,\ldots} k_Q$$

$$\log(L_{\text{subsetted}}) = \sum_{Q=A,B,C,\ldots} \log(L_Q)$$

3. Using $k_{\text{subsetted}}$ and $\log(L_{\text{subsetted}})$, compute an $\text{AICc}_{\text{subsetted}}$ via Equation 6.10.
4. If $\text{AICc}_{\text{subsetted}} < \text{AICc}_{\text{total}}$, then the individual subset models provide a better fit; in other words, the partitioning of data is appropriate. Otherwise, the data should be pooled, and the pooled model should be retained.

Use of this approach will be illustrated by an example using one of the data sets previously discussed.

The data taken by Anotai (Table 12.1) naturally split into two roughly equal subsets — experiments performed in February and experiments performed in November. It is desired to examine whether these two subsets can be described by a pooled model. The entire data set has already been fit (using the Hom model), resulting in a log likelihood of −4531.342. Fitting the two subsets separately yields log-likelihood values of −1511.59 and −2947.68, with n_{obs} equal to 18 and 16, respectively. Therefore, the log likelihood for the subset fits is the sum of −4459.27; note that, by fitting the subsets, there are six parameters, whereas for the pooled model there are three parameters. Applying the *AICc* formula, it is found that $\text{AICc}_{\text{total}} = 9069.084$ and $\text{AICc}_{\text{subsetted}} = 8932.762$. Since the AICc for the subsetted fits is smaller, the two subsets cannot be pooled.

There may be multiple reasons why subsetting improves performance. There could have been unknown, but significant differences between the two sets of data, for example, the microbial stock used to prepare the inoculum could have drifted, or the enumeration media lots may have changed, resulting in differences in quantification efficiency. Another potential reason is that the February subset experiments were conducted at lower disinfectant concentrations than the November subset, and the Watson power law dependency for rate on disinfectant concentration may be insufficient to characterize the influence of this variable on performance.

When multiple models are fit to a data set, they can be ranked by their AICc values. The relative AICc values can be used to assess the weight of evidence that an individual model is the best by comparing their differences. If $AICc_i$ is the value for the ith model, then differences from the best model in the set can be computed via:

$$\Delta_i = AICc_i - \min(\overline{AICc}) \tag{12.14}$$

Then a strength of evidence in support of model i is given by:

$$l_i = \exp\left(-\frac{\Delta_i}{2}\right) \tag{12.15}$$

Then the likelihood that model i is the best among the set considered is given by [17]:

$$w_i = \frac{l_i}{\sum_k l_k} \tag{12.16}$$

Then an approach called model averaging can be used, in which, given the sets of models, a weighted average of the ensemble predictions is computed, with each model i weighted by w_i. However, the use of model averaging is still undergoing theoretical development and may yet improve [19].

12.2 By-products

The application of inactivation processes, whether chemical or physical, may often result in the formation of chemical by-products during the treatment step. It is not the intention of this section to present an exhaustive discussion of such by-products, but to highlight some considerations from their formation that need to be taken into consideration in deciding which processes to use and under what circumstances.

12.2.1 Chlorine and Halogen Compounds

When free chlorine is in aqueous solution (either in bulk water or potentially in aqueous aerosols), it can react with many organic precursors to form chlorinated organics. Chlorine can also react with and oxidize iodide or bromide present to oxidized states, which in turn can undergo similar reactions. A broad spectrum of compounds may result depending on reaction conditions, including halo-methanes, halo-acetic acids, and halo-nitriles. A systematic literature review enumerated over 6300 reported compounds, of which less than 10% were confirmed by comparison with known standards [20]. This is likely an underestimate, since higher molecular products are known to be formed, but require specialized methods for their detection and characterization [21] chlorine by-products. Many of the known by-products have toxicological concerns, while the toxicity profiles of others are not yet known [22].

While chloramines tend to generate fewer by-products than free chlorine, they can still generate some, since in a chloramine solution there are equilibrium concentrations of HOCl and OCL^-. Furthermore, it is clear that chloramine decomposition and reaction may result in other by-products [23], including nitrogenous compounds, some of which may be of health concern [24].

The use of chlorine dioxide as a disinfectant, either in air or water, can generate by-products through several mechanisms. The generation of ClO_2 can result in a mixture with small amounts of free chlorine present. The free chlorine will react as previously described to produce by-products.

When applied, chlorine dioxide can decompose in part via disproportionation, forming chlorite and chlorate [25]. Chlorate in particular may be of health significance.

Chlorine dioxide can be applied as a gas for indoor air and surface decontamination, as was the case in response to the 2001 Amerithrax anthrax contamination event. Gaseous ClO_2 can react when used in such cases with various materials. In a study of 24 common indoor materials, the strongest reactions were observed with office partitions, ceiling tiles, medium-density fiberboard, and gypsum wallboard, although the reaction products were not identified [26]. The reaction rates decreased over time, suggesting multiple pathways and/or saturation.

12.2.2 Non Halogen Oxidants

Ozone is a highly reactive chemical, but relatively selective as an oxidant toward organic compounds with unsaturated bonds. As noted in Section 11.1.1, it can react to form hydroxyl ($OH^•$) radical. $OH^•$ in turn can react with organic materials present to form a spectrum of partially oxidized by-products such as aldehydes, ketones, and organic acids.

Another route to by-product formation is the oxidation of bromide that may be present to bromine, which, in turn, can form (brominated analogs of) some of the disinfection by-products observed with chlorine [27]. Ozone and hydroxyl radical can, by complex chains of further reactions, including organic materials, form inorganic bromate (BrO_3-), which has relatively high toxicity and is a regulated contaminant in drinking water [28].

Ozone also finds use in air disinfection. In this context, it can react with many indoor constituents, either directly or via the hydroxyl radical. The reactions may be with furnishings and finishings within the building, with volatile organics from emissions within the space, smoking, consumer products (fragrances, cosmetics), or volatile skin oils [29]. It has been found that the reaction of ozone with terpenols in the indoor environment may result in the formation of fine particulate aerosols [30], termed secondary particulates.

Hydrogen peroxide, being a weaker oxidant than ozone, has less by-product formation potential. Poppendieck et al. found that vaporized H_2O_2 used as an indoor air fumigant had a deposition velocity that was little affected by the types of furnishing and fixture materials present [31].

The use of peracetic acid, especially in vapor form, presents challenges related to its corrosive nature and chemical instability. It can degrade certain materials, including metals and polymers, necessitating careful consideration in material compatibility and the incorporation of corrosion inhibitors [32]. Since PAA is in equilibrium with acetic acid and hydrogen peroxide, if bromide is present, it is possible that the Br_- can be oxidized to bromine/hypobromous acid, and the latter can react with organics to form brominated by-products. Laboratory studies, however, suggest that in aqueous applications, the formation of such by-products may be well below any regulatory levels of concern [33].

12.2.3 UV and Other Radiation Systems

The radiation from far-UV lamps (222 nm), while useful for disinfection, has been shown to react as predicted by quantitative photochemistry, producing indoor air ozone concentrations in the ppb range [34]. The produced ozone can, in turn, react with volatile organics in a space to produce organic aerosol secondary contaminants [35]. The production rates can be predicted from known photochemical kinetics, and mitigation by ventilation can be effective [36].

12.3 Coincidental Inactivation by Other Processes

Microbial inactivation[15] occurs in other processes that may not have been designed for that primary objective. As was discussed in Section 8.4.2, the microbial decay rate may be influenced by various external conditions, including humidity, temperature, pH, water activity, and insolation. Therefore, processes that alter these parameters may change the degree of inactivation. In wastewater treatment, many biological processes, such as activated sludge and trickling filters, combine biotic and abiotic factors to inactivate pathogens. The basic literature on pathogen removal through these processes was systematically reviewed in an open-access online publication [37]. While that and other data compilations exist [38], there remains a need for a predictive microbiology of waste treatment to be able to design "fit for purpose" wastewater and excreta management for end-users beyond removal in deliberately designed disinfection processes [39]. In other words, the specific relationships between engineering design parameters and the estimate amount of pathogen inactivation need to be developed.

12.4 Applications

The aim of deploying inactivation technologies is to reduce exposure to potentially affected populations (Figure 3.4). Accordingly, there are different contexts for the application of agents. This will be divided into pathogen application in liquids, in air, in solids and semisolids, and on surfaces.

12.4.1 Liquids

There are numerous cases where liquids need to be disinfected. These include:

- Water to be treated for potable, agricultural, and industrial purposes.
- Wastewaters prior to discharge to the environment, or further treatment for reuse.
- Cooling waters, particularly when used in cooling towers, when the potential for dissemination of aerosols containing microorganisms exists.
- Pools, spas, whirlpools, water parks, and other recreational venues.
- Although not further covered here, waters used for food processing, where contamination of the product is of concern.

12.4.1.1 Chlorine

The most common agent used for disinfection of liquid streams is chlorine. For detailed design and analysis of systems using this technology, a key compendium is White's Handbook [40]. Key features to be considered in the application of chlorine will be summarized.

Chlorine can be applied to a liquid from chlorine gas, from a hypochlorite solution, or from solid (usually calcium) hypochlorite tablets or granules. It is also possible to continuously generate hypochlorite via on-site electrolysis of brines, although this is often a more expensive alternative, except for remote sites where chemical shipment is difficult.

Gaseous chlorine is supplied in small cylinders, in ton containers, and in rail cars (for very large users). The storage and handling of this pressurized mixture, containing chlorine in both liquid

---

15 Recall that removal may also occur by physical-chemical processes instead of or in addition to inactivation, which were discussed in Chapter 10.

and gas phases, present safety hazards and security issues and, as a result, have diminished in use in the water and wastewater sector [41].

To reduce the risk of hazardous gas leakage when chlorine is used as a disinfectant, a vacuum-dispensing chlorinator is often employed, with a schematic shown in Figure 12.6. In this system, gas is withdrawn from the headspace of the supply container into a pressure-reducing valve, and then the low-pressure gas stream is drawn by a vacuum created by the high-velocity flow of liquid through a venturi nozzle. The gas flow rate may be controlled either to be proportional to the liquid flow rate or using a feedback controller based on measured chlorine residual. Often, the flow through the venturi is only a small fraction of the total flow to be treated, and then the concentrated solution (of dissolved chlorine/$HOCl/OCl^-$) is mixed with the bulk liquid stream to be treated.
Sodium hypochlorite solutions are commercially available in strengths from 5 to 30%. Greater strengths are more corrosive and pose greater health and safety risks than lower strengths; however, higher strengths permit the use of smaller-volume feed tanks. The feed solution can be added directly to the bulk flow for disinfection.

In the case of both gas feed and sodium hypochlorite solution feed, the first step in the disinfection process is the rapid mixing of the disinfectant into the bulk flow.

There are several ways in which the disinfectant solution can be dispersed with the bulk flow. These are:

- Introduction into a pipe at turbulent flow
- Injection into a mechanically agitated rapid mixing chamber
- Introduction into a hydraulic structure inducing turbulence, such as a weir or a hydraulic jump
- Introduction into a pipe containing a static mixer

In all cases, the effectiveness of mixing can usefully be characterized by a "coefficient of variation" (C_v) of concentration. At any location in a system, the value of C_v is given by the (time to time) standard deviation of concentration divided by the mean concentration [42]. As mixing effectiveness increases, the coefficient of variation diminishes from its initial value.

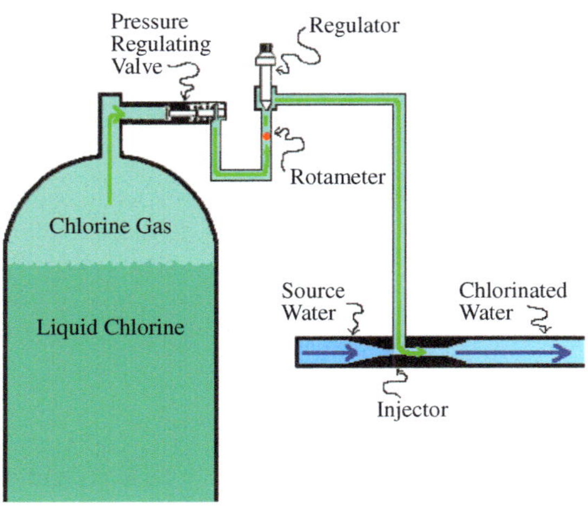

Figure 12.6 Schematic of a vacuum chlorinator. *Source:* IntechOpen / CC BY 4.0. https://www.intechopen.com/chapters/63788 (last accessed February 11, 2026).

In the first of these systems, the disinfectant solution is injected into a larger pipe with the bulk flow of liquid to be disinfected. The injection may occur at the sidewall, through a pipe inserted within the larger pipe, perhaps with a jet, or immediately upstream of a large venturi section. Each of these will result in a different pipe length needed to achieve a desired reduction in the C_v value relative to its initial value (at the point of injection), $C_{v,0}$.

Holley and students studied mixing in these systems. For simple in-pipe injection [42], the dimensionless downstream mixing length, Z, to achieve a coefficient of variance of 0.1 and 0.01 was 0.5–0.7 and 1–1.5, respectively. This, in turn, could be related to the physical length required by the following relationship:

$$z = \frac{D \cdot Z \cdot Sc_t}{0.048\sqrt{f}} \tag{12.17}$$

where D is larger pipe diameter, Sc_t is the turbulent Schmidt number (commonly assumed to be 1.0), and f is the pipe friction factor. The friction factor is, in turn, a function of the Reynolds number and the roughness of the interior pipe surface. Further work [43] showed that when the internal injector was a high velocity jet, the value for Z to achieve mixing could be reduced by a factor of 2–3.

Disinfectant solution can be introduced into a continuously mixed tank agitated by a turbine or propeller mixer. The mean residence time in such systems can be under a minute. A standard design basis for such systems is the use of the RMS velocity gradient, G (Equation 8.47).

For such rapid mix systems, common recommendations [44] have been to use a dimensionless product of the G value and the mean residence time of 20,000–35,000, with G values between 700 and 3500 s^{-1}. The G value achieved by a particular turbine or propeller mixture is a function of multiple design variables, including tank and turbine/propeller size and geometry, rotational velocity, and temperature (as it affects viscosity and, to a lesser degree, density). Detailed design calculations are presented in standard texts [45].

The third approach to achieving mixing, which is useful particularly in low-resource settings, is the introduction immediately downstream of a weir or the introduction before a hydraulic jump.

A schematic of a disinfectant diffuser downstream of a weir is shown in Figure 12.7. By introducing the disinfectant immediately in the tailwater, there is a highly turbulent flow that induces a degree of mixing.

A hydraulic jump is a dramatic shift in open-channel flow where fast, shallow water abruptly slows down and piles up into a deeper, slower-moving stream. This occurs when the flow transitions from *supercritical* (Froude number $Fr > 1$) to *subcritical* ($Fr < 1$), often triggered by a change in channel slope, width, or an obstruction. The Froude number, defined as

$$Fr = \frac{v}{\sqrt{gh}} \tag{12.18}$$

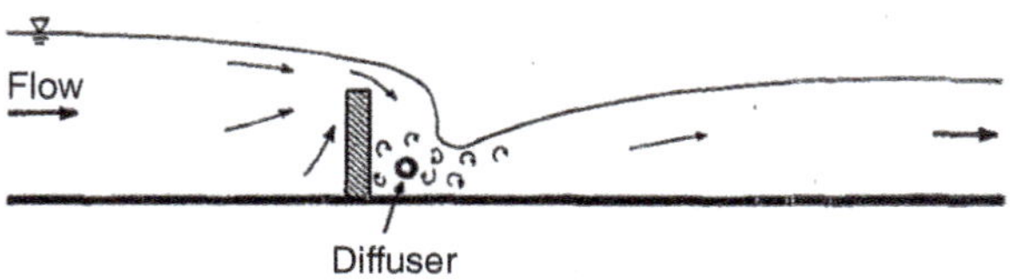

Figure 12.7 Schematic of a submerged diffuser downstream of a weir. *Source:* [46]/U.S. Environmental Protection Agency/Public domain.

which relates the flow velocity v to the gravitational wave speed $\sqrt{gh}$, where g is the acceleration due to gravity and h is the flow depth. The result of this transition is a churning, frothy zone characterized by violent turbulence and chaotic mixing that dissipates kinetic energy. By placing the solution diffuser at the point where this energy is being dissipated, effective mixing can occur. This can be realized by allowing a high velocity flow down a slanted chute, and at the bottom of the chute, a high energy dissipation rate will occur in the transition from supercritical to subcritical flow, at which the chemical diffuser should be placed. Design details have been presented in [46].

Static mixers, also called motionless mixers, are inserts installed within a pipe or closed channel to provide tortuous paths that result in more efficient mixing over a shorter pipe length (at the expense of an increased head loss). An example of a static mixer module is shown in Figure 12.8. The disinfectant solution is introduced to the bulk flow at the immediate upstream end of the mixer. There are numerous proprietary geometries for such mixers. The field was reviewed by Thakur et al. [47]. These authors suggest, in part informed by the work of Sir and Lecjaks [48], that the following relationship is useful for predicting the number of elements (modules), $\mathcal{E}$, required for homogenization (to a coefficient of variation of 0.05):

$$\mathcal{E} = a' Re^{a_1} Sc^{a_2} \left(\frac{Q_1}{Q_2}\right)^{a_3} \left(\frac{\mu_1}{\mu_2}\right)^{a_4} \tag{12.19}$$

In this equation, the Q's and μ's are the flow rates and viscosities of the two streams being mixed, Re is the Reynolds Number, and Sc is the Schmidt number.

For one class of mixer, Sir and Lecjaks [48] report a set of parameters in the turbulent region given in Table 12.8.

Figure 12.8 One type of static mixer insert. *Source:* Luigi Chiesa/CC BY 3.0. https://commons.wikimedia.org/wiki/File:StaticMixer.jpg (last accessed February 11, 2026).

Table 12.8 Correlation parameters for number of mixing modules to get to 5% coefficient of variation under turbulent conditions.

a'	a_1	a_2	a_3	a_4
270.5	−0.516	0.01688	−0.06352	0.04972

Source: Data from [48].

Based on the value of the exponents, the strongest dependency for the number of modules required is on the Reynolds number, with only a weak dependency on the flow and viscosity ratios and Schmidt number. As suggested by Thakur et al. [49], the functional form of Equation 12.19 is expected to hold for static mixers in general, with different coefficients for particular geometries.

Application of sodium hypochlorite solution for disinfection requires only a direct solution feed pump, injecting the chemical upstream of whichever mixing device described earlier is used.

Calcium hypochlorite, $Ca(OCl)_2$, called "bleaching powder" may be another source of chlorine for disinfection. This is supplied as a powder or granules and has a high weight percentage (60–70%) of chlorine; however, it is a strong oxidizer and therefore poses handling risks. In use, it must first be mixed with water to form a strong solution, which is then applied through an injector prior to rapid mixing.

In certain situations, a design with low maintenance and operational expenses is sought. This can be for small rural water systems, systems in developing countries, or the disinfection of small pools or spas. Either calcium hypochlorite granules or tablets can be used. Particularly with tablets, the flow system can be configured so that the dose dissolving from the tablets remains constant across varying flow rates, without requiring frequent operational adjustments. These have been reviewed in the context of use in less-developed countries by Lindmark et al. [50]. Subsequent to dosing, the treated water is passed to a contactor prior to dispersal.

One disadvantage of calcium hypochlorite compounds relative to sodium hypochlorite or dissolved chlorine gas is that the former adds hardness to the water and therefore increases the potential for undesirable precipitates to form. Hence, an understanding of the chemistry of the water to be treated is needed to avoid such unwanted side effects.

The multiple routes to production of chlorine dioxide described previously will generate a disinfectant solution. This solution can be applied in an identical manner to a chlorine or sodium hypochlorite solution.

Following dosing and rapid mixing in most situations (except perhaps for recreational water disinfection, where the treated water may be dispersed directly into the pool or spa), the water is held in a disinfection contactor for the desired level of inactivation to occur. The design of these systems proceeds via the concepts outlined in Section 11.2.2. Physically, there are different geometries for contactors that are often used, some of which are shown in Figure 12.9. In some situations, such as a drinking water system with a long transmission line before the first point of use, or in wastewater with a long outfall pipe prior to discharge, the pipe itself, which can be approximated as a plug-flow reactor, can serve as the disinfection contactor.

12.4.1.2 Ozone

While ozone gas can form a solution in cold water $\left(\text{typically around 5–20 } \frac{mg}{L}\right)$ [51] and can be used in this way in laboratory experiments, its reactivity generally means that it must be applied directly as a gas to the liquid flow being treated. Therefore, most frequently, the dissolution of gaseous ozone occurs in the same physical system in which disinfection contact time occurs.

While bubbles of oxygen or air enriched with ozone are introduced into a contactor, it has been established that the inactivation of microorganisms requires the dissolution of ozone into the liquid, rather than via the direct physical contact between the bubbles and the microorganisms [52]. However, inactivation can still occur even when the residual is below the detection limit [53]. Hence, the physical system must be designed to achieve both effective mass transfer and good hydraulics to ensure short reaction times.

A very common configuration to achieve both of these objectives is a multichamber contactor, an example of which is shown in Figure 12.10. In this example, there are three chambers connected

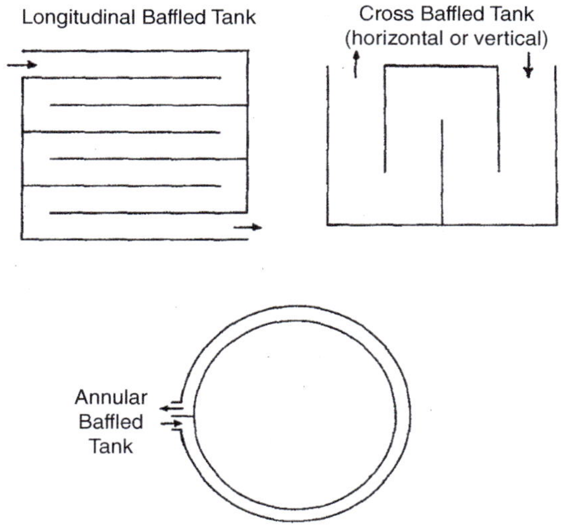

Figure 12.9 Common geometric configurations for contact tanks (top view). *Source:* [46]/U.S. Environmental Protection Agency/Public domain.

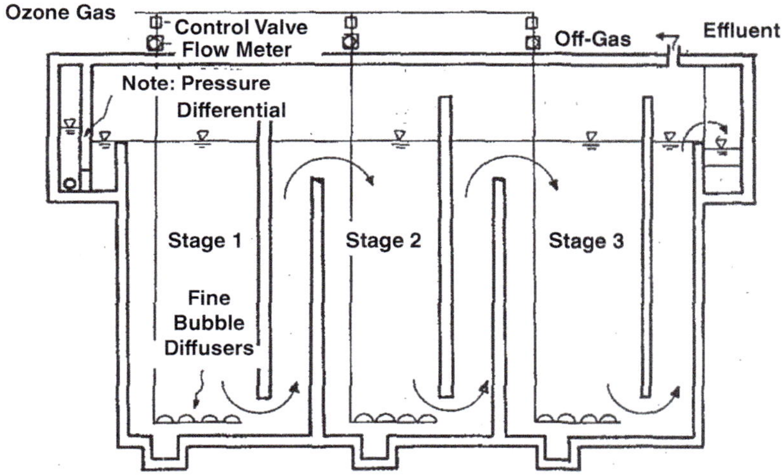

Figure 12.10 Schematic of three-chamber ozone contactor. *Source:* [46]/U.S. Environmental Protection Agency/Public domain.

in series with ozone added to all three. Due to the agitation caused by gas bubbles, each chamber will exhibit approximate CSTR behavior. In some designs, there may be more chambers, and only the first one or several may receive the ozone flow, thereby allowing the latter chambers to behave more closely as plug-flow systems with some dispersion. The choice of the number of chambers and the number into which ozone is introduced is an engineering optimization that balances increased ozone dose and residual with more chambers receiving gas against better residence time distribution for inactivation performance.

Other types of ozone contactors are also used. These have been reviewed by Bin and Roustan, along with the achievable mass-transfer coefficients for various designs [54].

Morrison et al. [55] have reviewed the inactivation kinetics of waterborne pathogens by ozone.

Given the tight coupling between mass transfer, ozone decay, and inactivation, the analysis of such systems by computational fluid dynamics has become very useful. An example is the work of Niazi et al. [56].

12.4.1.3 Peracetic Acid

Since peracetic acid is supplied as a concentrated solution, it can be mixed with the liquid to be treated in a manner similar to that used for sodium hypochlorite solutions. The same types of contactor designs as used for chlorination systems are appropriate for PAA. Consideration of materials compatibility with an acid oxidizer for feed tanks, pipes, and pumps must occur.

12.4.1.4 UV and Light

As noted previously, solar radiation (visible and principally near UV) can be used as a disinfectant, especially in low-resource settings and in emergency applications, and design guidelines [57, 58] and experiences have been reviewed [59]. However, such systems have not been demonstrated to be practicable in other circumstances and will not be considered further.

With UVC, several other design factors are important. Low-pressure vs. medium-pressure and monochromatic versus polychromatic sources have been discussed previously.

Some of the other design variables that influence the performance include:

- Enclosed pressurized contactor versus open channel flow through
- Flow axial versus transverse to lamps
- Electrical supply management
- Jacketing of lamps to protect them from water
- Cleaning of jackets on a periodic basis to reduce loss due to buildup of fouling
- Inclusion of UV sensors to control light emission as a function of flow and water quality

Detailed consideration of these factors can be found in dedicated monographs on UV system design [60, 61].

12.4.2 Air

There are two principal foci for the use of disinfection of air and associated surfaces.[16]

One application is the remediation and decontamination of the indoor environment that may have become contaminated with pathogens. This application has a long historical precedent, such as in the cleanup of patient rooms in infectious disease wards between uses, and in the decontamination of laboratories where experimental investigations of pathogens were being conducted. A notable decontamination application occurred in the 2001 release of spores of *Bacillus anthracis*, which has been called the "Amerithrax" episode. In response to that incident, five entire buildings and parts of two others underwent decontamination [62]. There were also almost 30 other buildings where the extent and nature of post-incident decontamination could not be well documented.

A second application is the continual disinfection of an indoor environment that may serve as a transmission venue, such as from infected individuals within the space.

These two applications clearly have different constraints. In the first use, decontamination, assuming individuals, pets, etc., can be evacuated during treatment, and that the residual chemicals

16 In this subsection, the focus is on inactivating organisms in *indoor* air. There have been few occasions where the need to decontaminate bulk air in the outdoors has been reported.

can be removed prior to re-occupancy; toxicity to humans and animals is only a minor issue.[17] In the second application, where disinfection occurs concurrently with occupancy, the lack of toxicity or protection of occupants from exposure is essential.

In both cases, the potential for the process to interact with, and possibly damage, paints, fabrics, furniture, electronics, etc., in the indoor environment is of concern.

12.4.2.1 Chemical Agents

Ethylene oxide is a cyclic ether available as a commercial gas under pressure and has been widely used for sterilizing medical devices and equipment and disinfecting certain heat-sensitive foods. In these applications, it can be employed in a contained and controlled setting with the collection and destruction of any off-gases.

Ethylene oxide (EtO), while an effective sterilant and fumigant, poses significant hazards in indoor environments. Acute inhalation can cause respiratory irritation, neurological symptoms, and in severe cases, pulmonary edema [63]. Chronic exposure markedly increases cancer risk, particularly lymphoid and breast cancers, and EtO is classified as carcinogenic to humans [63]. It is also mutagenic and teratogenic, with documented reproductive and developmental toxicity [64]. Its volatility and flammability present fire and explosion hazards, while its high odor threshold means dangerous concentrations may be undetectable without monitoring [65]. As a result, its use as a decontaminant is quite limited.

Methyl bromide is also commercially available as a liquefied gas. Historically, it has been used as a pesticide and has been shown to be sporocidal, including for whole indoor building decontamination [66]. However, due to its composition, it is an ozone-depleting gas, and its commercial use has been phased out under amendments to the Montreal Protocol [67].

The principal gaseous agents used are hydrogen peroxide, chlorine dioxide, paraformaldehyde, and ozone. Aerosolized peracetic acid has also found utility in smaller, more contained spaces.

Gas-phase hydrogen peroxide has been widely used for sterilizing heat-sensitive medical devices [68]. However, it has only recently been employed for large-scale in-room disinfection of both indoor air and adjacent surfaces. Vaporized hydrogen peroxide (VHP) generators use either flash vaporization or heated vaporization to convert liquid hydrogen peroxide (usually 30–35% solution) into vapor [69]:

- *Flash vaporization*: Liquid H_2O_2 is injected into a heated chamber, instantly converting it into supersaturated vapor.
- *Heated vaporization*: Liquid is spread over a heated plate, vaporizing as air is passed over, carrying the vapor into the target space.

Application of vaporized hydrogen peroxide to a room requires several steps [69]:

1. **Pre-conditioning and dehumidification:** The environment is dehumidified to below 30% RH to optimize the disinfection process.
2. **Conditioning:** Vaporized H_2O_2 is introduced, achieving target concentrations (typically 140–1,400 ppm).
3. **Holding:** The vapor concentration is held for 15–180 minutes to achieve microbial reduction.
4. **Aeration:** Residual H_2O_2 is removed using ventilation or catalytic converters, resulting in exhaust gas primarily of water and oxygen.

---

17 This also presumes that the chemicals used in the decontamination can be perfectly contained without any fugitive emissions — which is often not possible.

An alternative system uses aerosolized H_2O_2 generated as fog by passing a 5–12% hydrogen peroxide solution through fine nozzles, so that most of the generated material remains suspended [69]; additives such as silver ions can be added to the solution. Since aerosolized hydrogen peroxide fumigation occurs with the fog close to room temperature, there may be less of a problem with condensation and material damage.

For large spaces, such as a room or a floor, it is necessary to integrate multiple injection points, circulation fans, and sensors to ensure even vapor distribution and concentration control. Control of vapor concentration and temperature is required to minimize condensation. For complex environments, modeling the transport of H_2O_2 using computational fluid dynamics to optimize delivery and air flow has been found to be useful [70].

Vaporized hydrogen peroxide can act as a disinfectant by either the direct action of H_2O_2 or by the activity of free radicals generated upon reaction with constituents in a room and/or room light [31]. Being a reactive material, it can also decompose via reaction with room furnishings and finishes. The effective deposition velocity for a variety of room materials has been measured and found to have a median of 250 with a range of <10 to >1250 $\frac{\text{cm}}{\text{h}}$.

The efficacy of hydrogen peroxide has been frequently assessed by applying a known initial inoculum to surfaces, and measuring survival after contact in a gaseous environment at a known concentration and time. An example of this is the survival of spores of *Geobacillus stearothermophilus*, a spore-forming gram-positive bacterium frequently used as a sterilization indicator. Based on proprietary data, Chen and Chen [70] report the following equation for decimal reduction time (D_{10}) in minutes as a function of the air concentration of H_2O_2 in $\frac{\text{mg}}{\text{L}}$ at 30 °C:

$$D_{10}(C) = 45.1\ \exp(-1.78\,C) \tag{12.20}$$

Example 12.1 Calculating Hydrogen Peroxide Dose.

It is desired to achieve 6 logs kill of *Geobacillus stearothermophilus* spores in a room by use of H_2O_2. If this is to be done in 1 hour, at what concentration should the peroxide be maintained?

Solution:

1. If 4 logs inactivation needs to be accomplished, then the value of D_{10} needed is[18]:

$$D_{10} = \frac{1\,\text{h}}{6\,\text{logs}} = 0.167\,\text{h} = 10\ \text{min}$$

2. Substituting into Equation 12.20,

$$10 = 45.1\ \exp(-1.78\,C)$$

3. Solving, $C = 0.846$ mg/L or 619 ppm (v/v)

It should be noted that this peroxide concentration is substantially greater than the 8-hour time-weighted average permissible exposure limit of 1 ppm set by the Occupational Safety and Health Administration [71]. Therefore, during application and until thorough ventilation, occupants should be excluded, and the exhaust should undergo thorough peroxide destruction.

18 This calculation assumes that Chick's law holds.

Given the desired peroxide concentration in the air, the required flow rate of the chemical can be calculated. In the case of a simple room with a well-mixed volume, a mass balance can be constructed based on the input concentrations (if any), ventilation rate, and deposition velocity (reaction with surfaces); from this, the needed chemical flow can be calculated.

For more complex situations, a multi-zone or computational fluid dynamics model will need to be developed to ensure adequate chemical dispersion.

A starting point for assessing the application of a gaseous agent in a mixed room is Equation 8.5. In general, the inlet gas concentration (C_{in}) is zero, and $\frac{E}{V}$ is the internal gas emission rate per unit volume of the room. During the period of fumigation, ventilation is kept minor to minimize off-site emission of the agent (i.e., θ is small). The reaction rate (r) is given in terms of the deposition velocity (υ_{dep}), the room concentration, the total surface area, including internal surfaces, A_{surf}[19] and the room volume (V).

With these transformations, the mass flow rate of chemical (E) is given by Equation 12.21.

$$\frac{E}{V} = C \cdot \left(\frac{\upsilon_{dep} A_{surf}}{V} + \frac{1}{\theta} \right) \tag{12.21}$$

In general, during the fumigation operation, the inlet and outlet ventilation will be kept low (i.e., θ will be high).

Use of this relationship is illustrated by a follow-up to Example 12.1.

Example 12.2 Calculation of H_2O_2 Mass Flow to Acheive Disinfection in a Well-mixed Room.

For the problem in Example 12.1, to achieve the steady state concentration of $0.846 \frac{mg}{L}$ in the gas, what is the mass flow that needs to be evaporated or aerosolized? Assume the room has dimensions of 10 m x 15 m x3 m height, and that during the fumigation process, the residence time (θ) is 2 hours, and that no peroxide enters; the peroxide in the exhaust goes to a destruct system. A deposition velocity of the median from Poppendieck et al. [31] of 150 $\frac{cm}{h}$ can be assumed.

Solution:

1. A_{surf}

$$V = 10\,\text{m} \cdot 15\,\text{m} \cdot 3\,\text{m} = 450\,\text{m}^3$$

$$A_{surf} = 2(10 \cdot 15 + 15 \cdot 3 + 10 \cdot 3) = 225\,\text{m}^2$$

2. The right-hand side of Equation 12.21 is evaluated:

$$\left(0.846\,\frac{\text{mg}}{\text{L}}\right) \cdot \left(\frac{150\,\frac{\text{cm}}{\text{h}}}{450\,\text{m}^3} \times 225\,\text{m}^2 + \frac{1}{2\,\text{h}} \right)$$

(Continued)

19 If there are surfaces of different reactivities in the room, then this term can be divided into a summation of the specific surface deposition velocities each multiplied by the area of that type of surface.

Example 12.2 (Continued)

Adding conversion factors to express the calculation in consistent units and evaluating:

$$\left[\left(0.846\,\frac{\text{mg}}{\text{L}}\right)\cdot\frac{1000\,\text{L}}{\text{m}^3}\cdot\frac{0.001\,\text{g}}{\text{mg}}\right]\times$$

$$\left(\frac{150\,\frac{\text{cm}}{\text{h}}}{450\,\text{m}^3}\cdot\frac{0.01\,\text{m}}{\text{cm}}\frac{2.778\cdot10^{-4}\,\text{h}}{\text{s}}\times225\,\text{m}^2+\frac{1}{2\,\text{h}}\cdot\frac{\text{h}}{3600\,\text{s}}\right)$$

This can be evaluated as:

$$0.846\frac{g}{m^3}\times\left(2.08\cdot10^{-4}s^{-1}+1.39\cdot10^{-4}s^{-1}\right)=0.0002936\frac{g}{m^3 s}$$

3. Hence, given the room volume, the peroxide generation rate needs to be $450\times 0.0002936 = 0.132\,\text{g/s}$.

In practice, in critical situations such as the response to the Amerithrax incident, internal coupons were used in multiple locations in the space being treated to confirm effective sterilization. With any indoor fumigant, enclosed spaces that are inaccessible to the free movement of the chemical, such as drawers, closed cabinets, and file drawers, will not be well treated.

As noted, H_2O_2 has a relatively high deposition velocity with many materials. It is therefore not surprising that it has been found to have adverse effects on some materials. Chou et al. [72] found a spectrum of impacts on diverse textiles used in aviation interiors (Table 12.9).

Some degradation of wood, corrosion of metal surfaces, and damage to rubber and polymeric sealants have also been observed [73]. This was exacerbated under conditions favorable to hydrogen peroxide condensation.

Table 12.9 Comparison of textile damage from vaporized hydrogen peroxide (VHP) Exposure.

Textile	Mechanical damage	Chemical / structural changes	Flammability effects	Overall assessment
Wool	Moderate tensile loss (20–30%), increased elongation, surface cuticle damage	Oxidation of disulfide bonds → sulfonic acids (S=O), discoloration	Increased burn times, sometimes exceeding FAA §25.853 limits	Highly vulnerable to VHP damage
Leather	Severe tensile loss (~50%), reduced elongation (brittle)	Degradation from peroxide exposure (weight loss)	Increased burn length and flame time after VHP cycles	Highly vulnerable to VHP damage
Nylon	Minimal tensile loss (<10%), inconsistent ductility changes	No major structural changes	Generally stable in vapor; some increased flammability after liquid exposure	Moderately resistant; minor effects in vapor, more in liquid peroxide
Polyester	Mild tensile/elongation loss (~10–20%)	No major structural changes	Slight increases in burn length/time in some tests	Moderately resistant; minor effects
Nomex®	No significant changes; slight increase in elongation	Stable; no significant chemical changes	Unchanged; always within FAA §25.853 limits	Highly resistant; no significant impact

Source: Data from [72].

Chlorine dioxide gas can also be used as a fumigant. Relatively pure chlorine dioxide (with a Cl_2 concentration below 4%) can be generated by passing a chlorine/nitrogen gas mixture through a column of solid sodium chlorite [74]. This can be disbursed through a room volume in a manner similar to that of hydrogen peroxide. This was a common fumigant during response to the Amerithrax incident [62].

Rastogi et al. [75] tested coupons of six common building interior materials (carpet, acoustic ceiling tile, unpainted cinder block, painted I-beam steel, painted wallboard, and unpainted pinewood) for their behavior and the inactivation of organisms inoculated on them. They were exposed to chlorine dioxide generated by a dry generation system or a wet system. Chlorine dioxide gas concentrations ranged between 500 and 3000 ppm. It was found that times to reach 6 log inactivation of inoculated spores of *Bacillus anthracis* were longer on wood and cinder block than on other materials, with exposure times for some treatments exceeding 12 hours. Fumigation with ClO_2 was also found to be highly effective against aerosolized bacteria as well as surface bacteria, yeasts, and molds at lower concentrations [76].

Chlorine dioxide can react with constituents in a room, although at a somewhat lower rate than peroxide. Poppendieck et al. [31] have measured the deposition velocity of ClO_2 gas applied to a room and found a median value of 25 with a range of 0–135 cm/h.

The reaction of chlorine dioxide with interior materials and furnishings is somewhat less than with hydrogen peroxide. Some bleaching of carpet fibers was noted at high doses (12,000 ppm-hr), but at more moderate doses, no damage to paper, wood, epoxy, plastic, furniture, or wood occurred, while rugs, cloths, and some computing and office equipment may have damage [77].

The application process of chlorine dioxide follows a similar staging to that for peroxide application. An off-gassing destruction process is required, as is complete venting prior to reoccupation, since the OSHA PEL is 0.1 ppm citep NIOSH:2019aa.

Ozone gas generated onsite can be used as a fumigant/disinfectant, although it has been less widely deployed [77]. It was not used in response to the 2001 Amerithrax incident [62].

In laboratory testing, spores of *Bacillus subtilis* were inoculated on test coupons of various materials and exposed to ozone [78]. Tailing off inactivation kinetics were found, but for all materials, 3-log inactivation was obtained within 4 hour of exposure to 1500 ppm. The performance on glass, vinyl floor tile, and office paper was nearly the same. However, cut pile carpet and hardwood flooring surfaces were more difficult to decontaminate. Inactivation increased with relative humidity over the range of 70–90%.

The inactivation of spores of *B. subtilis* and *B. anthracis* by ozone on coupons of material were found to be similar [79]. At 12,000 ppm and 85% relative humidity, 9–12 hours of exposure were required to attain six logs inactivation, with laminate and galvanized metal surfaces being more difficult to decontaminate.

The ability of gaseous ozone to inactivate SARS-CoV-2 applied to surfaces has been reviewed. The required ozone time products for inactivation are substantially less than spores, and the inactivation rate has also been found to increase with RH [80].

As a potent oxidant, ozone can react with a variety of interior materials. The deposition velocity for 25 materials at air concentrations of 1000–1200 ppm was measured by Poppendieck et al. [81]. During the first hour of contact, the deposition velocity was a median for all materials of about 50 cm/h, with a range from 10 to 340. However, after five hours, this declined to a median of 25, with a range from 5 to 210, indicating some level of saturation of demand. This was accompanied by a production of by-products from 5 to 70 mg/m^2 during the course of a 24-hour exposure period.

Due to the formation of these by-products as well as the toxicity of O_3 itself, exhaust ventilation and gas destruction are also necessary for the implementation of such processes.

Use of gaseous formaldehyde ($H_2C = O$) (formalin) has well over 100 years of history as a disinfectant for medical spaces [82]. In early uses, aerosolization or simple evaporation from liquid or solution.[20] It can also be generated by heating paraformaldehyde, a polymeric solid form [84]. Due to its high toxicity and mobility, it has not been used outside controlled medical and medical device settings.

Vapor-phase PAA disinfection has emerged as a promising method for environmental and surface decontamination. Vapor-phase application of PAA provides uniform distribution and rapid microbial inactivation, making it effective in disinfecting large spaces or complex surfaces where liquid application would be impractical or insufficient. Studies have shown its efficacy against a variety of microorganisms, including bacterial spores, viruses, and molds, particularly in controlled environmental conditions and enclosed areas [85].

PAA fogging systems are available that produce aerosols with diameters less than 10 μm, which can remain suspended for long periods of time [86]. These have been found to be effective against viruses, fungi, and vegetative bacteria, achieving six logs of inactivation in a few hours, with longer times required for spores.

Ultraviolet

Ultraviolet radiation using 254 nm mercury vapor sources has been used for indoor air disinfection since the 1930s [87]. There are two principal modes of action for routine use of UV germicidal irradiation (UVGI):

- In duct use, in which a bank of lamps is mounted within a ventilation duct for the treatment of flowing air servicing a space. In this application, there is no possibility of the UV radiation impinging upon the eyes of occupants. Eye (particularly) and skin exposure represent hazards of 254 nm UV, and exposure standards of 6 mJ/cm^2 are recommended [88].
- Upper air UVGI in which the lamps are mounted on the support portion of a wall, or on the ceiling, with shrouding shielding occupants from direct illumination (Figure 12.11). The internal circulating flow patterns in the room will allow passage of recirculating air through the UV illumination zone and thus provide a reduction of the concentration of microorganisms.

More recently, efficient sources of 222 nm UV light have been developed. This far-UVC wavelength has effective germicidal properties against many organisms [89]. Furthermore, this wavelength appears to have less potential for skin and eye damage [90], with a suggested exposure limit of 160 mJ/cm^2 [88]. Therefore, such lights may be suitable for continuing use in occupied spaces without shrouding.

Due to the higher photon energy of 222 nm light (compared to 254 nm), there is greater potential for the formation of oxidized volatile organic by-products and for their secondary reactions with other organics present in indoor air. Such products can be reduced by other air purification devices [91].

12.4.3 Solids and Semisolids

Inactivation processes can be applied to a variety of solid and semisolid materials to reduce pathogenic load. Such materials can include manures, sewage sludges, or possibly fractions of solid wastes.

---

20 At room temperature, pure formaldehyde has a vapor pressure of 5 atm, so it is highly volatile from the pure liquid state or from solution [83].

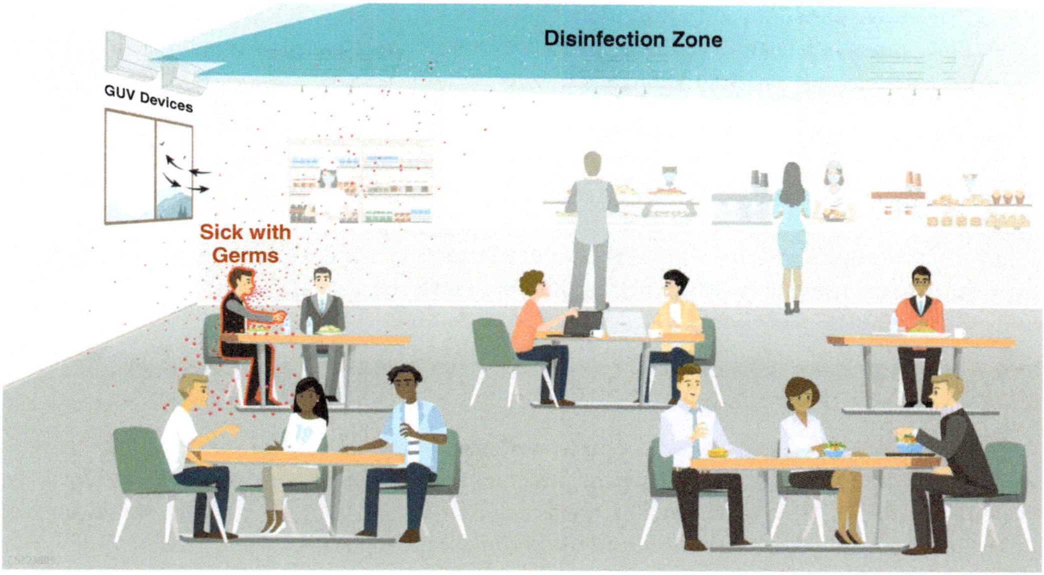

Figure 12.11 Upper air UV wall-mounted fixture. *Source:* Centers for Disease Control and Prevention/Public domain. https://www.cdc.gov/niosh/ventilation/germicidal-ultraviolet/index.html?CDC_AA_refVal=https%3A%2F%2Fwww.cdc.gov%2Fcoronavirus%2F2019-ncov%2Fcommunity%2Fventilation%2Fuvgi.html (last accessed February 10, 2026).

The processing of these materials depends on their consistency, which is in part related to water content. Sludge consistency is commonly classified into three categories [92]: liquid, characterized by its ability to flow under gravity or low pressure; paste-like, which can sustain continuous flow only when subjected to pressures above a certain threshold and exhibits limited shear resistance; and solid, defined by its high shear resistance.

The consistency of these materials influences the disposal options and the necessary treatments. In many cases, land application of such materials is a disposal option, and pre-processing to reduce pathogens is needed to mitigate exposure-related risks.

In the United States, sewage sludge (biosolids) are of two types—Class A and Class B. Class A biosolids may be directly applied to land, with no restriction as to public contact or further use of products derived from production on the land. Land application of Class B biosolids is subject to significant restrictions due to the presence of detectable levels of pathogenic organisms. Under the US Environmental Protection Agency's 40 CFR Part 503 regulations, Class B biosolids may only be applied to agricultural or reclamation sites under conditions that limit human and animal exposure. These conditions include mandatory site access restrictions (up to 1 year) to prevent public contact, as well as prescribed waiting periods before grazing livestock or harvesting crops to allow for natural pathogen die-off. In addition, occupational safeguards are required for workers handling or applying Class B material, including the use of protective equipment and hygiene practices [93].

To achieve Class B status, a process for significant reduction of pathogens (PSRP) must be used. To achieve Class A status, a process for further reduction of pathogens (PFRP) is needed. In the regulations (Code of Federal Regulations, Title 503), these are listed (along with specific design and operational conditions [95]:

PSRP: Aerobic digestion, air drying, anaerobic digestion, composting, and lime stabilization.

PFRP: Composting, heat drying, heat treatment, thermophilic aerobic digestion, beta ray irradiation, gamma ray irradiation, and pasteurization.

Additional processes have been, and can be granted either PFRP or PSRP status by a specific approval process. On an individual location basis, a treated sludge can be designated as Class A or Class B based on meeting the numeric and qualitative standards outlined in Table 12.10, rather than by including specifically designated processes.

Many of the processes designated as PSRP or PFRP are primarily designed to achieve other objectives, but also coincidentally reduce pathogen levels. However, several that are designed explicitly for pathogen reduction are worth describing in more detail.

12.4.4 Surface Disinfection

Pathogens can be deposited on surfaces from airborne particles, dust, or direct contact by individuals carrying them. These then have the potential for subsequent transmission as fomites. The decontamination of these surfaces can serve as a barrier to infection, although the significance of this route is dependent on the surface and the particular pathogen [96, 97].

One approach that has received attention is the development of surfaces or surface coatings with antimicrobial properties, especially in the healthcare setting. A systematic review of this approach in healthcare settings reported only mild reductions, with mixed-quality evidence [98]. However, there is active development of new materials and technologies, and progress in this approach may occur [99].

Table 12.10 Comparison of pathogen reduction processes: PSRP vs. PFRP.

Characteristic	PSRP (Class B)	PFRP (Class A)
Produce biosolids that meet disinfection requirements for...	Class B	Class A
Expected pathogen/indicator levels: fecal coliform	<2 million CFU[a] or MPN[b] per g total solids (dry wt)	<1000 MPN per g total solids (dry wt)
Salmonella spp.	Reduced by a factor of 10	<3 MPN per 4 g total solids (dry wt)
Enteric viruses	Reduced by a factor of 10	<1 PFU[c] per 4 g total solids (dry wt)
Viable helminth ova	Not applicable	<1 viable ova per 4 g total solids (dry wt)
Additional requirements for land application	Vector attraction reduction plus site restrictions after application	Vector attraction reduction
Microbiological monitoring requirements	None	Fecal coliform or *Salmonella* spp. — average of ≥7 samples over 2 weeks, frequency from monthly up to annually depending on biosolids production (or as directed by permitting authority)

[a]Colony forming units (CFU): from membrane filter method; only approved for Class B.

[b]Most probable number (MPN): from multiple tube fermentation; approved for all biosolids and untreated sludge.

[c]Plaque forming units (PFU): from plaque assay method of virus detection.

Source: Data from [94].

One approach to surface disinfection is surface cleaning, using wipes, cloths or similar materials, as covered in Section 10.4. With the addition of anti-microbial compounds during the cleaning process, inactivation can occur.

Disinfectant solutions can be used as an initial treatment prior to wiping or washing. For example, solutions of hypochlorite or peracetic acid have been found to be effective surface disinfectants [100]. Disinfectants can be added directly to single-use towelettes or wipes to improve the overall surface removal; however, for this to be efficacious, the kinetics of inactivation must be rapid [101]. Pretreating a surface with a detergent, such as a surface-active agent, prior to wiping can also increase overall removal [102].

When vaporized chemical agents are used to disinfect a space, they may condense on or come in contact with contaminated surfaces and thus also function as surface disinfectants. Many of the results cited in Section 12.4.2 for air disinfection were based on the use of coupons of material inoculated with test organisms and exposed to the disinfectant vapor or aerosol. For example, Goyal et al. [103] tested the virucidal capability of vaporized hydrogen peroxide by preparing disks with dried suspensions of feline calicivirus, human adenovirus type 1, transmissible gastroenteritis coronavirus of pigs, avian influenza virus, and swine influenza virus. These were dried on stainless steel disks and placed in an environmental chamber. They were exposed to VHP from a commercial generator, and after 3–2 hours of contact and re-aeration of the chamber, the chamber was titered for viability; no viable virus was detected, indicating greater than 3.8 log inactivation.

UV radiation can also be used to inactivate organisms on surfaces. However, in this application, correction for light attenuation in the materials in which microorganisms may be embedded on the surface, such as saliva, is necessary [104].

The frequency and intensity of surface cleaning and inactivation processes need to be determined by the extent and magnitude of re-contamination during use. Since many pathogens have multiple portals of entry, the application of this technique should be considered in concert with other interventions [105].

12.5 Questions and Problems

1 A municipal water treatment plant is evaluating the effectiveness of chlorine disinfection in a batch contact basin. The inactivation of a test organism follows the Hom kinetic model.

In a series of batch reactor tests with the organism:

- The initial concentration in each test was $N_0 = 10^5$ CFU/mL.
- Contact time was held constant at $t = 20$ min.
- The chlorine concentration and final measured microbial concentrations were:

Chlorine concentration, C (mg/L)	Final concentration, N (CFU/mL)
0.8	1.0×10^4
1.2	2.0×10^3
2.0	3.0×10^2

a Using the experimental data, determine the Hom model parameters.

b With these parameters, predict the microbial concentration after $t = 30$ min at $C = 1.2$ mg/L.

c Estimate the time required to achieve a 5-log reduction of organisms at $C = 1.2$ mg/L.

2 In laboratory batch tests with *E. coli*, the following inactivation data were observed:

Chlorine C (mg/L)	Time t (min)	Inactivation $\ln(N_0/N)$
0.5	10	0.85
0.8	10	1.42
1.2	10	2.10
2.0	10	3.25
0.5	30	1.65
0.8	30	2.70
1.2	30	3.95
2.0	30	6.10
0.5	50	2.25
0.8	50	3.65
1.2	50	5.45
2.0	50	8.15

Fit the Hom model parameters k, n, and m using nonlinear regression. How well does the Hom model fit this data?

3 Consider the data in the previous problem. Fit a Hom–power law model to this data set. Compare the fit of the HPL model to the Hom model and discuss which one of these two models should be used.

4 Consider the system defined in Example 11.3, but assume that the inactivation law is power law, with constants $k' = 0.1$ and $x = 0.8$ (disinfectant concentration units in mg/L and time units in s), and compute the required value of E_C.

5 Use the code for the IDDF example in Section 11.2.2 and compute the survival ratio at a range of initial concentrations from 0.5–5 mg/L. Is it possible to get 99% inactivation (i.e., 2 logs) in this range, and if so, what is the estimated dose required?

6 Consider the fit of inactivation models to the data from the Hibler *Giardia* experiments (Table 12.7). If a system of five identical CSTRs in series with a total residence time is used, and disinfectant demand is neglected, what should be the residual in each of the systems be (assumed identical) to achieve 3 logs inactivation using the parameters from the power law model?

7 For the previous problem, assume a disinfectant dose and use the maximum micromixed model with a residence time distribution equal to that of five CSTRs in series to compute the survival ratio.

8 For the example of analyzing count data in Section 12.1.1, revise the code to analyze the data using the power law model. Discuss the fit of the Hom model compared to the power law model.

9 The censored regression fit of the Anotai data using the Hom model appeared to exhibit systematic discrepancies (Figure 12.4). Modify the code for fitting to use the Hom–power law model and discuss whether this provides a superior fit.

10 Consider the inactivation data for *Giardia muris* that were analyzed by several models with results presented in Table 12.7. If a 3 $\frac{\text{mg}}{\text{L}}$ dose is used with a contact time of 20 minutes, compare the inactivation for the different fitted models. Using a model averaging approach, what would be the prediction?

11 A hydrogen peroxide generator is available that produces 50 g per hour, in a room of 1000 m^3 with total surface area of 800 m^2, and a deposition velocity of 100 cm/h. If the mean residence time is maintained at 30 minutes, what is the steady-state H_2O_2 that can be achieved?

References

1 Jakub Stoklosa, Wen-Han Hwang, and David I Warton. A general algorithm for error-in-variables regression modelling using Monte Carlo expectation maximization. *PLOS ONE*, 18(4):1–21, 04 2023.
2 Ralph B D'Agostino. *Goodness-of-Fit-Techniques*. CRC Press, Boca Raton, FL, 2017.
3 Hadley Wickham. Tidy data. *Journal of Statistical Software*, 59(10):1–23, 2014.
4 Charles N Haas. Maximum likelihood analysis of disinfection kinetics. *Water Research*, 22(6):669–77, 1988.
5 Jin Anotai. Effect of calcium ion on chemistry and disinfection efficiency of free chlorine at pH 10. *Ph.D. Thesis*, Drexel University, 1996.
6 C P Hibler. *Inactivation of Giardia Cysts with Chlorine at 0.5 to 5°C*. AWWA Research Foundation, Denver, CO, 1987.
7 Charles N Haas and B Heller. Kinetics of inactivation of *Giardia lamblia* by free chlorine. *Water Research*, 24:233–238, 1990.
8 S D Hodges and P G Moore. Data uncertainties and least squares regression. *Journal of the Royal Statistical Society Series C: Applied Statistics*, 21(2): 185–195, 1972.
9 Tao Wang, Jinhui Wu, Jiancheng Qi, Limei Hao, Ying Yi, and Zongxing Zhang. Kinetics of inactivation of *Bacillus subtilis* subsp. *niger* spores and *Staphylococcus albus* on paper by chlorine dioxide gas in an enclosed space. *Applied and Environmental Microbiology*, 82(10):3061–3069, May 2016.
10 Charles N Haas and Joseph Jacangelo. Development of regression models with below detection limit data. *Journal of Environmental Engineering*, 119(2):214–230, 1993.
11 Henry S Lynn. Maximum likelihood inference for left-censored HIV RNA data. *Statistics in Medicine*, 20(1):33–45, 2001.
12 Lijie Li. Effects of initial microbial density on disinfection efficiency in a continuous flow system and validation of disinfection batch kinetics in a continuous flow system. *PhD Thesis*, Drexel University, 2004.
13 N Facile. Evaluating bacterial aerobic spores as a surrogate for Giardia and Cryptosporidium inactivation by ozone. *Water Research*, 34(12):3238–3246, August 2000.
14 Eric R Ziegel and John W Gorman. Kinetic modelling with multiresponse data. *Technometrics*, 22(2):139–151, May 1980.
15 Sadanori Konishi and Genshiro Kitagawa. *Information Criteria and Statistical Modeling*. Springer, New York, 2008.

16 Yuhong Yang. Can the strengths of AIC and BIC be shared? A conflict between model indentification and regression estimation. *Biometrika*, 92(4):937–950, 2005.

17 Kenneth P Burnham, David R Anderson, and Kathryn P Huyvaert. AIC model selection and multimodel inference in behavioral ecology: Some background, observations, and comparisons. *Behavioral Ecology and Sociobiology*, 65:23–35, 2011.

18 Po-Hsien Huang. Asymptotics of AIC, BIC, and RMSEA for model selection in structural equation modeling. *Psychometrika*, 82(2):407–426, 2017.

19 Michael Schomaker and Christian Heumann. When and when not to use optimal model averaging. *Statistical Papers*, 61(5):2221–2240, October 2020.

20 Hechao Chen, Jidao Xie, Chengxiang Huang, Yining Liang, Yulin Zhang, Xiaoyan Zhao, Yuhua Ling, Lei Wang, Qi Zheng, and Xiaoqiu Yang. Database and review of disinfection by-products since 1974: Constituent elements, molecular weights, and structures. *Journal of Hazardous Materials*, 462:132792, January 2024.

21 William A Mitch, Susan D Richardson, Xiangru Zhang, and Michael Gonsior. High-molecular-weight by-products of chlorine disinfection. *Nature Water*, 1(4):336–347, 2023.

22 Indrajit Kalita, Andreas Kamilaris, Paul Havinga, and Igor Reva. Assessing the health impact of disinfection byproducts in drinking water. *ACS ES&T Water*, 4(4):1564–1578, 2024.

23 Julian L Fairey, Juliana R Laszakovits, Huong T Pham, Thien D Do, Samuel D Hodges, Kristopher McNeill, and David G Wahman. Chloronitramide anion is a decomposition product of inorganic chloramines. *Science*, 386(6724):882–887, 2024.

24 Alexandria L B Forster, Sheryl L Wiskur, and Susan D Richardson. Formation of eight classes of dbps from chlorine, chloramine, and ozone: Mechanisms and formation pathways. *Environmental Science & Technology*, 59(30):15594–15611, August 2025.

25 Fatima Al-Otoum, Mohammad A Al-Ghouti, Talaat A Ahmed, Mohammed Abu-Dieyeh, and Mohammed Ali. Disinfection by-products of chlorine dioxide (chlorite, chlorate, and trihalomethanes): Occurrence in drinking water in Qatar. *Chemosphere*, 164:649–656, 2016.

26 Heidi Hubbard, Dustin Poppendieck, and Richard L Corsi. Chlorine dioxide reactions with indoor materials during building disinfection: Surface uptake. *Environmental Science & Technology*, 43(5):1329–1335, 2009.

27 Xin-Yang Zhang, Ye Du, Yao Lu, Wen-Long Wang, and Qian-Yuan Wu. Characteristics of the formation and toxicity index of nine newly identified brominated disinfection byproducts during wastewater ozonation. *Science of The Total Environment*, 824:153924, 2022.

28 Christina M Morrison, Samantha Hogard, Robert Pearce, Aarthi Mohan, Aleksey N Pisarenko, Eric R V Dickenson, Urs von Gunten, and Eric C Wert. Critical review on bromate formation during ozonation and control options for its minimization. *Environmental Science & Technology*, 57(47):18393–18409, 2023.

29 Charles J Weschler and William W Nazaroff. Human skin oil: A major ozone reactant indoors. *Environmental Science: Atmospheres*, 3:640–661, 2023.

30 Y Yang and M S Waring. Secondary organic aerosol formation initiated by α -terpineol ozonolysis in indoor air. *Indoor Air*, 26(6):939–952, December 2016.

31 Dustin Poppendieck, Heidi Hubbard, and Richard L Corsi. Hydrogen peroxide vapor as an indoor disinfectant: Removal to indoor materials and associated emissions of organic compounds. *Environmental Science & Technology Letters*, 8(4):320–325, April 2021.

32 Helena Horn and Bernd Niemeyer. Corrosion inhibition of peracetic acid-based disinfectants. *Chemical Engineering & Technology*, 45(1):129–134, 2022.

33 Liang Meng, Jing Chen, Deyang Kong, Yuefei Ji, Junhe Lu, Xiaoming Yin, and Quansuo Zhou. Transformation of bromide and formation of brominated disinfection byproducts in peracetic acid oxidation of phenol. *Chemosphere*, 291:132698, 2022.

34 Zhe Peng, Douglas A Day, Guy A Symonds, Olivia J Jenks, Harald Stark, Anne V Handschy, Joost A de Gouw, and Jose L Jimenez. Significant production of ozone from germicidal UV lights at 222 nm. *Environmental Science & Technology Letters*, 10(8):668–674, 08 2023.

35 Sara Bjerre Sørensen, Frederik Rask Dalby, Søren Kristian Olsen, and Kasper Kristensen. Influence of germicidal UV (222 nm) lamps on ozone, ultrafine particles, and volatile organic compounds in indoor office spaces. *Environmental Science & Technology*, 58(45):20073–20080, November 2024.

36 Victoria P Barber, Matthew B Goss, Lesly J Franco Deloya, Lexy N LeMar, Yaowei Li, Erik Helstrom, Manjula Canagaratna, Frank N Keutsch, and Jesse H Kroll. Indoor air quality implications of germicidal 222 nm light. *Environmental Science & Technology*, 57(42):15990–15998, October 2023.

37 Global Water Pathogen Project. *Water and Sanitation for the 21st Century: Health and Microbiological Aspects of Excreta and Wastewater Management (Global Water Pathogen Project)*. UNESCO, 2017.

38 Laura Álvarez-Fraga, Gabriel Capson-Tojo, Malo Sanglier, Jérôme Hamelin, Renaud Escudié, Nathalie Wéry, Diana García-Bernet, Audrey Battimelli, and Felipe Guilayn. A meta-analysis of pathogen reduction data in anaerobic digestion. *Renewable and Sustainable Energy Reviews*, 207: 114982, 2025.

39 Wakana Oishi, Björn Vinnerås, Joan B Rose, and Daisuke Sano. Predictive environmental microbiology for safe use of sanitation products in agriculture: Challenges and perspectives. *Environmental Science & Technology Letters*, 8(11):924–931, November 2021.

40 Black & Veatch Corporation. *White's Handbook of Chlorination and Alternative Disinfectants*. Wiley, Hoboken, NJ, 5th edition, 2010.

41 AWWA Disinfection Systems Committee. Committee report: Disinfection survey, part 2—alternatives, experiences, and future plans. *Journal-American Water Works Association*, 100(11):110–124, 2008.

42 Edward R Holley and A Metin Ger. Circumferential diffusion in pipe mixing. *Journal of the Hydraulics Division*, 104(4):471–485, April 1978.

43 Steven D Fitzgerald and Edward R Holley. Jet injections for optimum mixing in pipe flow. *Journal of the Hydraulics Division*, 107(10):1179–1195, October 1981.

44 Mackenzie Leo Davis and David A Cornwell. *Introduction to Environmental Engineering*. McGraw-Hill, New York, 1991.

45 Kerry J Howe, David W Hand, John C Crittenden, R Rhodes Trussell, and George Tchobanoglous. *Principles of Water Treatment*. John Wiley & Sons, 2012.

46 Enos L Stover, Charles N Haas, Kerwin L Rakness, and O Karl Scheible. *Design Manual: Municipal Wastewater Disinfection*. Number EPA/625/1-86/021. US Environmental Protection Agency, Cincinnati, OH, 1986.

47 R K Thakur, C Vial, K D P Nigam, E Bruce Nauman, and G Djelveh. Static mixers in the process industries-A review. *Transactions of the Institution of Chemical Engineers*, 81(A):787–826, 2003.

48 Jiri Sir and Zdeněk Lecjaks. Pressure drop and homogenization efficiency of a motionless mixer. *Chemical Engineering Communications*, 16(1-6):325–334, August 1982.

49 Kristen K Coleman, Douglas Jie Wen Tay, Kai Sen Tan, Sean Wei Xiang Ong, The Son Than, Ming Hui Koh, Yi Qing Chin, Haziq Nasir, Tze Minn Mak, Justin Jang Hann Chu, et al. Viral load of severe acute respiratory syndrome coronavirus 2 (SARS-CoV-2) in respiratory aerosols emitted by patients with coronavirus disease 2019 (COVID-19) while breathing, talking, and singing. *Clinical Infectious Diseases*, 74(10):1722–1728, May 2022.

50 Megan Lindmark, Katya Cherukumilli, Yoshika S Crider, Perrine Marcenac, Matthew Lozier, Lee Voth-Gaeddert, Daniele S Lantagne, James R Mihelcic, Qianjin Marina Zhang, Craig Just, et al. Passive in-line chlorination for drinking water disinfection: A critical review. *Environmental Science & Technology*, 22(56):9164–9181, June 2022.

51 Gordon R Finch, Wang C Yuen, and Beverly J Uibel. Inactivation of *Escherichia coli* using ozone and ozone - hydrogen peroxide. *Environmental Technology*, 13(6):571–578, 1992.

52 S Farooq, E S K Chian, and R S Engelbrecht. Basic concepts in disinfection with ozone. *Journal (Water Pollution Control Federation)*, 49(8):1818–1831, 1977.

53 Charles N Haas and Joseph Jacangelo. Inactivation of *Giardia* cysts by ozone after residual disappearance. *Journal of Environmental Engineering*, 148(2):06021007, 2022.

54 A K Bin and M Roustan. Mass transfer in ozone reactors. In *Fundamental and Engineering Concepts for Ozone Reactor Engineering*, pages 99–131, Toulouse, France, International Ozone Association, March 2000.

55 Christina M Morrison, Samantha Hogard, Robert Pearce, Daniel Gerrity, Urs von Gunten, and Eric C Wert. Ozone disinfection of waterborne pathogens and their surrogates: A critical review. *Water Research*, 214:118206, 2022.

56 Shahram Niazi, Mansour Kalbasi, and Alireza Bahramian. Hydraulic and disinfection efficiency of an ozonation contactor for a municipal water treatment plant using computational fluid dynamics. *The Canadian Journal of Chemical Engineering*, 95(11):2063–2072, 2017.

57 Ángela García-Gil, Rafael A García-Muñoz, Kevin G McGuigan, and Javier Marugán. Solar water disinfection to produce safe drinking water: A review of parameters, enhancements, and modelling approaches to make SODIS faster and safer. *Molecules*, 26(11):3431, June 2021.

58 Yiding Wang, Greyson Xinghan He, Fernando Sanchez-Quete, and Stephanie K Loeb. Systematic review and meta-analysis on the inactivation rate of viruses and bacteriophage by solar wavelength radiation. *Environmental Science & Technology*, 59(15):7421–7439, April 2025.

59 Gouthami Rao, Emma Wells, Catherine Reynolds, Rebecca Yoo, Erin Kowalsky, Jennifer DeFrance, Karl Linden, and Joe Brown. Systematic review of the microbiological performance of household water treatment technologies. *Environmental Science & Technology*, 59(41): 21776–21789, 2025.

60 US Environmental Protection Agency. *Ultraviolet Disinfection Guidance Manual For The Final Long Term 2 Enhanced Surface Water Treatment Rule*. Number EPA 815-R-06-007. November 2006.

61 James R Bolton and Christine Anne Cotton. *The Ultraviolet Disinfection Handbook*. American Water Works Association, Denver, CO, 2008.

62 Ketra Schmitt and Nicholas A Zacchia. Total decontamination cost of the anthrax letter attacks. *Biosecurity and Bioterrorism: Biodefense Strategy, Practice, and Science*, 10(1):98–107, March 2012.

63 U.S. Environmental Protection Agency. Evaluation of the inhalation carcinogenicity of ethylene oxide. *Technical Report* EPA/635/R-16/350Fa, U.S. Environmental Protection Agency, 2016.

64 New Jersey Department of Health. Hazardous substance fact sheet: Ethylene oxide, 2016. Fact Sheet No. 0882.

65 Occupational Safety and Health Administration. Ethylene oxide. *29 CFR 1910.1047*, 2020.

66 S Serre, L Mickelsen, M W Calfee, J P Wood, M S Jr Gray, R H Scheffrahn, R Perez, W H Jr Kern, and N Daniell. Whole-building decontamination of *Bacillus anthracis* sterne spores by methyl bromide fumigation. *Journal of Applied Microbiology*, 120(1):80–89, January 2016.

67 E S Saltzman, M R Nicewonger, S A Montzka, and S A Yvon-Lewis. A post-phaseout retrospective reassessment of the global methyl bromide budget. *Journal of Geophysical Research: Atmospheres*, 127(3):e2021JD035567, 2022.

68 M R Karimi Estahbanati. Advances in vaporized hydrogen peroxide reusable medical device sterilization cycle development: Technology review and patent trends. *Microorganisms*, 11(10):2566, October 2023.

69 Daniel Kümin, Monika Gsell Albert, Benjamin Weber, and Kathrin Summermatter. The hitchhiker's guide to hydrogen peroxide fumigation, part 1: Introduction to hydrogen peroxide fumigation. *Applied Biosafety*, 25(4):214–224, 2020.

70 Xi Chen and Qingyan Chen. Comparison of different decontaminant delivery methods for sterilizing unoccupied commercial airliner cabins. *Building and Environment*, 45(9):2027–2034, September 2010.

71 Occupational Safety and Health Administration. OSHA occupational chemical data base: Hydrogen peroxide. *Technical Report*, Occupational Safety and Health Administration, December 2020.

72 S F Chou, R A Overfelt, W F Gale, H S Gale, C G Shannon, G Buschle-Diller, and J Watson. Effects of hydrogen peroxide on common aviation textiles. *Tech Report* DOT/FAA/AM-09/16, Federal Aviation Administration. Office of Aviation. Civil Aerospace Medical Institute, 2009.

73 Tohru Kimura, Hiroyuki Yahata, and Yoshimichi Uchiyama. Examination of material compatibilities with ionized and vaporized hydrogen peroxide decontamination. *Journal of the American Association for Laboratory Animal Science*, 59(6):703–711, 2020.

74 Joseph P Wood and G Blair Martin. Development and field testing of a mobile chlorine dioxide generation system for the decontamination of buildings contaminated with bacillus anthracis. *Journal of Hazardous Materials*, 164(2):1460–1467, 2009.

75 Vipin K Rastogi, Shawn P Ryan, Lalena Wallace, Lisa S Smith, Saumil S Shah, and G Blair Martin. Systematic evaluation of the efficacy of chlorine dioxide in decontamination of building interior surfaces contaminated with anthrax spores. *Applied and Environmental Microbiology*, 76(10):3343–3351, 2010.

76 Jeffrey Driver, Julie Lichterman, George Lukasik, Shawn Jones, Marie Bourgeois, and Raymond Harbison. Bactericidal and fungicidal efficacy of chlorine dioxide in various workspaces. *Occupational Diseases and Environmental Medicine*, 10(02):49–59, 2022.

77 James V Rogers, Young W Choi, William R Richter, Harry J Stone, and Michael L Taylor. *Bacillus Anthracis* spore inactivation by fumigant decontamination. *Applied Biosafety*, 13(2):89–98, June 2008.

78 Ahmet Aydogan and Mirat D Gurol. Application of gaseous ozone for inactivation of *Bacillus subtilis* spores. *Journal of the Air & Waste Management Association*, 56(2):179–185, February 2006.

79 Joseph P Wood, Morgan Wendling, William Richter, and James Rogers. The use of ozone gas for the inactivation of *Bacillus anthracis* and *Bacillus subtilis* spores on building materials. *PLOS ONE*, 15(5):e0233291, May 2020.

80 Saima Farooq and Chedly Tizaoui. A critical review on the inactivation of surface and airborne SARS-CoV-2 virus by ozone gas. *Critical Reviews in Environmental Science and Technology*, 53(1):87–109, January 2023.

81 D Poppendieck, H Hubbard, M Ward, C Weschler, and R L Corsi. Ozone reactions with indoor materials during building disinfection. *Atmospheric Environment*, 41(15):3166–3176, May 2007.

82 W Leslie Mackenzie. Methods of disinfection. *Public Health*, 12(6):438–448, 1900.

83 US Environmental Protection Agency. Chemistry, fate, and transport assessment for formaldehyde. *Technical Report*, US Environmental Protection Agency, 2024.

84 Larry A Taylor, Manuel S Barbeito, and Gardner G Gremillion. Paraformaldehyde for surface sterilization and detoxification. *Applied Microbiology*, 17(4):614–618, 1969.

85 Joseph P Wood, Michael Worth Calfee, Matthew Clayton, Nicole Griffin-Gatchalian, Abderrahmane Touati, and Kim Egler. Evaluation of peracetic acid fog for the inactivation of bacillus anthracis spore surrogates in a large decontamination chamber. *Journal of Hazardous Materials*, 250-251:61–67, 2013.

86 Jay Krishnan, Greg Fey, Carol Stansfield, Laura Landry, Hung Nguy, Stan Klassen, and Catherine Robertson. Evaluation of a dry fogging system for laboratory decontamination. *Applied Biosafety*, 17(3):132–141, September 2012.

87 Nicholas G Reed. The history of ultraviolet germicidal irradiation for air disinfection. *Public Health Reports®*, 125(1):15–27, January 2010.

88 David H Sliney and Bruce E Stuck. A need to revise human exposure limits for ultraviolet UV-C radiation. *Photochemistry and Photobiology*, 97(3):485–492, 2021.

89 Kimia Ghasemi Arzanani, Mohammad Reza Rashidian Vaziri, Samaneh Sharif, and Masoud Mollaee. A review on applications and safety of 222 nm far UVC light for surface and air disinfection. *Biophysical Reviews*, February 2025. https://doi.org/10.1007/s12551-025-01269-y.

90 Manuela Buonanno, Camryn Petersen, David Welch, Raabia Hashmi, and David J Brenner. 222 nm far- UVC light and skin health: Assessment of DNA damage across different skin types. *Photochemistry and Photobiology*, July 2025. https://doi.org/10.1111/php.70010.

91 Nadia Tahsini, Selena Zhang, Matthew B Goss, Seamus C Frey, Yaowei Li, Jessica B Smith, Norton T Allen, M Pang, Richard D Williamson, Frank N Keutsch, et al. Mitigation of indoor ozone and secondary products from 222 nm germicidal ultraviolet light using commercial air cleaners. *ACS ES&T Air*, 8:1750–1757, July 2025.

92 L Spinosa and K Wichmann. Sludge characterization: The rôle of physical consistency. *Water Science and Technology*, 49(10):59–65, 2004.

93 US Environmental Protection Agency. A plain English guide to the EPA part 503 biosolids rule. *Technical Report* EPA/832/R-93/003, US Environmental Protection Agency, 1994.

94 U.S. Environmental Protection Agency. Basic information: Pathogen equivalency committee. https://www.epa.gov/biosolids/basic-information-pathogen-equivalency-committee#whatare, 2025. Accessed: 21 08 2025.

95 Laura Boczek, Ronald Herrmann, Elizabeth Resek, and Tess Richman. Pathogens and vector attraction in sewage sludge. *Technical Report* EPA/600/R-22/194, US Environmental Protection Agency, Washington, D.C., 2023.

96 Alicia N M Kraay, Michael A L Hayashi, Nancy Hernandez-Ceron, Ian H Spicknall, Marisa C Eisenberg, Rafael Meza, and Joseph N S Eisenberg. Fomite-mediated transmission as a sufficient pathway: A comparative analysis across three viral pathogens. *BMC Infectious Diseases*, 18(1):540, December 2018.

97 Winnie Zambrana and Alexandria B Boehm. Occurrence of human viruses on fomites in the environment: A systematic review and meta-analysis. *ACS Environmental Au*, 3(5): 277–294, July 2023.

98 M P Muller, C MacDougall, M Lim, I Armstrong, A Bialachowski, S Callery, W Ciccotelli, M Cividino, J Dennis, S Hota, et al. Antimicrobial surfaces to prevent healthcare-associated infections: A systematic review. *Journal of Hospital Infection*, 92(1):7–13, 2016.

99 Ling Xin Yong and John Kaiser Calautit. A comprehensive review on the integration of antimicrobial technologies onto various surfaces of the built environment. *Sustainability*, 15(4):3394, 2023.

100 J N Morris and M A Esseili. Efficacy of peracetic acid and sodium hypochlorite against SARS-CoV-2 on contaminated surfaces. *Applied and Environmental Microbiology*, 89(7):e00622-23, July 2023.

101 Syed A Sattar and Jean-Yves Maillard. The crucial role of wiping in decontamination of high-touch environmental surfaces: Review of current status and directions for the future. *American Journal of Infection Control*, 41(5):S97–S104, May 2013.

102 Rachael L Hardison, Sarah W Nelson, Daniela Barriga, Jessica M Ghere, Gabrielle A Fenton, Ryan R James, Michael J Stewart, Sang Don Lee, M Worth Calfee, Shawn P Ryan, et al. Efficacy of detergent-based cleaning methods against coronavirus MHV-A59 on porous and non-porous surfaces. *Journal of Occupational and Environmental Hygiene*, 19(2):91–101, February 2022.

103 S M Goyal, Y Chander, S Yezli, and J A Otter. Evaluating the virucidal efficacy of hydrogen peroxide vapour. *Journal of Hospital Infection*, 86(4):255–259, April 2014.

104 John Gibson, Ramin Farnood, and Benoit Barbeau. Kinetics and Monte Carlo simulation of UV disinfection of *B. subtilis* spores and SARS-CoV-2 in dried saliva droplets. *Environmental Science and Pollution Research*, 28(43):61853–61859, November 2021.

105 Nan Zhang, Linan Zhuang, Marco-Felipe King, Hua Qian, and Min Zhu. Public surface disinfection every 2 hours can reduce the infection risk of norovirus in airports up to 83%. *PLOS Computational Biology*, 20(12):e1012561, 2024.

13

Exposure Assessment

The process of exposure assessment is a key subtask of risk assessment (Figure 3.1). The elements necessary to perform an exposure assessment have previously been covered in depth:

- Sources of pathogens (Chapter 7)
- Transport, growth, and attenuation in the environment from source to receptor (Chapter 8)
- Performance of processes that might be used to reduce the exposure (Chapters 9–11)

Once, via hazard assessment, the venue of exposure, the portal of entry, and the pathogen(s) of interest are defined (Figure 3.4), the task of exposure assessment is to quantify the dose, i.e., organisms/time (and the distribution) from the pathways and potential interventions.

The objective of exposure assessment is to determine the distribution of microorganism doses to which a population is exposed and the uncertainty and variability in that distribution. The dose distribution may, in part, be a function of environmental properties, resulting in attenuation (or, for vegetative organisms, amplification) between the source and the receptor.

Accurate exposure assessment requires not only the concentration or distribution of pathogens in an environmental compartment (e.g., air, water, surfaces) but also the quantity and potential distribution of that compartment to which individuals are exposed.

If there is a single homogeneous population where each individual is exposed to an amount (mass or volume) of medium per unit time, $\dot{m}$, containing an amount of pathogens (e.g., CFU, PFU, gene copies, etc.) per unit amount of medium, C, then the dose rate $\dot{d}$ to the population is exactly $\dot{d} = C\dot{m}$. However, rarely is there a situation where homogeneity of medium exposure and concentration in the medium can be assumed; therefore, such a simple product formulation, perhaps while useful pedagogically, is inappropriate. This convolution of products of random variables was discussed previously in the context of combining LRVs, each characterized by a probability distribution (Section 9.5). In general, as noted by Equation 9.7, the mean of the product of two quantities *is not* equal to the product of the mean of two quantities [1].

In the current context, the problem may be more complex than just taking the product of two random variables. While it is possible to perform the analytical process of constructing the probability distribution of functions of two or more random variables [2, 3], this rapidly becomes intractable. Therefore, Monte Carlo methods are more widely used. These methods will be discussed in Section 13.3, after consideration of sources of the underlying distributions is discussed.

There may also be systematic differences in either the nature of exposure or the exposed population that would justify considering them as separate subpopulations rather than a single population. For example:

- Consider emission of pathogens from a confined animal feeding operation into the air. If there is a community north of the site and a second community south of the site, it may be preferable to compute the exposure distributions for the north site and south site separately.
- There may be demographic groups with different propensities to exposure to a particular medium. For example, a survey in Salt Lake City (UT) found significant ethnic and economic factors that differentiate tap water consumption [4].
- Consider individuals who recreate at a beach site or in a community pool. There may be subpopulations that engage in primary contact by swimming or other intensive water activities, and others that are only exposed via spray from the shoreline or poolside. Therefore, the exposure distributions for these two groups could be different.[1]

The advantage of subdividing populations, where this is appropriate, is the reduction of the heterogeneity in the exposure assessment overall, replacing it with two or more exposure assessments, conditional on each subpopulation, each with intrinsically smaller heterogeneity.

At this point, sources of the distribution of microorganisms and of media contact will be considered.

13.1 Microorganism Distributions

There are two fundamental approaches to estimating the microorganism concentration in an environmental compartment at the point of exposure: direct sampling and measurement, and modeling the path from the source(s) to the point of exposure.

13.1.1 Direct Measurement

The first approach uses the methods discussed in Section 6.2 to fit the results from sampling to probability distributions that describe variability, for example, for concentration, log-normal, inverse Gaussian, Weibull, generalized inverse Gaussian, etc. An example of this approach is a dataset on the occurrence of *Campylobacter jejuni* in a recreational water, determined by qPCR in Table 13.1. Such data could be used directly to determine the best-fitting distribution to the exposure concentration.

Note, as discussed in Chapter 6, gene copies may overestimate (by some unknown amount) the concentrations of viable and infectious microorganisms.

An alternative to fitting a data set to a distribution is to actually use the experimental data points themselves as an empirical distribution function (EDF) to describe exposure. However, this has significant limitations. For example, the data set in Table 13.1 consists of only 13 observations. A greater exposure, and hence a greater risk, could occur at larger concentrations that are more rarely seen — for example, what would be the concentration at the upper 95th, 99th, or even 99.9th

1 A somewhat related question is whether the behavior of swimming versus nonswimming activity at a recreational site could be related to potential differences in susceptibility manifested by a difference in dose response. Or possibly that some who do not swim may have some prior health impairment.

Table 13.1 qPCR abundance of *Campylobacter jejuni* in water column at a state park recreational area. Units: $\log_{10}$ gene copies/100 mL.

5.04	4.35
5.86	4.97
5.51	4.44
5.68	5.12
5.37	4.56
4.68	4.01
5.29	

Source: Data from [5].

percentiles? This cannot be determined directly by the EDF, but only by fitting a distributional model to the data and using the results of that fit to compute the upper percentiles.[2]

For this particular data set, using direct calculations, the mean and standard deviation of a normal distribution are estimated as 4.99 and 0.556 (log likelihood = −9.98), respectively. It must be understood that the choice of models fitting a small data set may have different behaviors at upper (or lower) tails, and so this represents a form of epistemic uncertainty (Figure 3.5). Furthermore, there is uncertainty in the distribution parameters, which must be considered and will be discussed subsequently. Using the normal distribution parameters, upper tails can be estimated — for example, the upper 99th percentile is 2.33 standard deviations above the mean and therefore is calculated as:

$$4.99 + 2.33 \times 0.556 = 6.285$$

which is clearly above the observed range.

13.1.2 From Source to Receptor Models

In the second approach, the distribution(s) for the source(s) are ascertained, and considering intervening mitigation, fate, and transport processes between the source(s) and receptors, the distribution of microorganism concentrations at the point of exposure is estimated. This requires data to inform the construction of distributions for each step.

To illustrate, if it is desired to examine exposure to water processed for reuse from wastewater with respect to the pathogen *Cryptosporidium* that is to be treated by activated sludge, ozonation, and microfiltration, distributions are needed for the raw wastewater concentration, and the removal (LRVs) by activated sludge, ozonation, and microfiltration. For these individual steps, various distributions, informed by experimental measurements, have been reported:

- Darby et al. [6] reviewed high-quality data on *Cryptosporidium* oocyst concentrations in raw wastewaters and, based on aggregation of the results, found that the $\log_{10}$ oocysts per liter were described by a normal distribution with a mean of 1.9 and a standard deviation of 0.6.

  It might be expected that the distribution could be influenced by site-specific factors. These could include per capita water use, the presence of nonpoint inputs or industrial wastes from

2 The fitting method will depend on the type of data — count, quantal, or continuous, and whether there are censored data points. Alternative distributional fits can be compared using the AICc values to determine the most suitable model. If the data are amenable to this, the lack of fit can also be assessed.

pathogen-containing operations such as abattoirs or processing, and the nature of the sewer network.

- Based on intensive sampling at one facility in Australia, King et al. [7] found that the LRV in activated sludge could be described by a normal distribution for $\log_{10}$ reduction with a mean of 1.76 and a standard deviation of 0.46.
- Pecson et al. [8] found that the LRV of *Cryptosporidium* removal by first stage ozone treatment in a reuse pilot plant was described as an inverse Gaussian,[3] with pdf defined by:

$$f(x) = \sqrt{\frac{\lambda}{2\pi x^3}} \exp\left[-\frac{\lambda(x-\mu)^2}{2\mu^2 x}\right] \tag{13.1}$$

with best-fit parameters of $\mu = 3.38$ (mean) and $\lambda = 29.4$.[4]

- The LRV for oocyst removal by microfiltration was also found to be inverse Gaussian distributed with parameters $\mu = 4.68$ and $\lambda = 12.286$ [8].

A point estimate for the concentration at the end of the treatment system can be constructed by using the means of each of the steps, recognizing that when the variability and uncertainty are considered, this may not be equal to the mean of the derived quantity itself. The calculation by combining the means is given in Example 13.1.

Example 13.1 Using Means of Individual Distributions.

The mean of each input distribution is used to construct a point estimate. Based on the raw wastewater concentration distribution, the point estimate is:

$$10^{1.9} = 79.4 \quad \frac{\text{oocysts}}{\text{L}}$$

The fraction remaining after activated sludge treatment is:

$$10^{-1.76} = 0.017$$

The fraction remaining after the first-stage ozonation is:

$$10^{-3.38} = 4.17 \times 10^{-4}$$

The fraction remaining after microfiltration is:

$$10^{-4.68} = 2.09 \times 10^{-5}$$

Now all the previous numbers get multiplied together to get the final concentration:

$$79.4\frac{\text{oocysts}}{\text{L}} \times 0.017 \times 4.17 \times 10^{-4} \times 2.09 \times 10^{-5} = 1.18 \times 10^{-8}\frac{\text{oocysts}}{\text{L}}$$

Note that the aforementioned calculation assumes complete independence of all of the steps. The issue of correlations will be discussed in the context of a full uncertainty analysis of exposure in Section 13.3.

3 The Gaussian distribution is symmetric about the mean and defined over the entire real line. The inverse Gaussian distribution is skewed right and is defined over the positive real line. Recall this was first encountered in the context of residence time distributions (Equation 8.22).

4 Note that this is where ozone is used as a preliminary treatment step to achieve multiple objectives, including organic reduction and removal. It is expected, according to the principles in Section 11.2.2, that the performance distribution of ozone, if designed primarily for pathogen reduction, would be a function of factors such as residence time, ozone dose, demand, temperature, etc.

13.1.3 Uncertainty of Distributions

Using the direct measurement approach or the approach of combining individual steps in the pathway to get a final concentration via the steps earlier gives a point estimate. For a fuller characterization of exposure, the complete distributions need to be combined. Since this often occurs in combination with the distributions characterizing medium contact, the approaches for this fuller characterization will be presented in Section 13.3 following discussion of medium contact distributions.

13.2 Medium Contact

The amount of medium contact that results in exposure is typically not a function of whether the contaminant of concern is microbiological or non-microbiological. Hence, the principal sources for this information are much more general in nature. A foundational reference for such data is the US EPA Exposure Factors Handbook [9], which was compiled as a complete document in 2011; it is updated chapter by chapter electronically on a rolling basis at https://www.epa.gov/expobox/about-exposure-factors-handbook. There are numerous other international exposure compilations that have been reviewed by Phillips and Moya [10] and may be useful in settings outside the United States.[5] A listing of the key chapters of the EPA Exposure Factors Handbook is given in Table 13.2. A key feature of the EPA handbook is its critical analysis and synopsis of the primary literature on which its reference information is based. However, in most cases, only central tendency values, e.g., means, and an upper percentile are reported, rather than full distributions.

Table 13.2 Key chapters of the US EPA Exposure Factors Handbook.

Ingestion of Water and Other Select Liquids
Non-Dietary Ingestion Factors
Soil and Dust Ingestion
Inhalation Rates
Dermal Exposure Factors
Body Weight Studies
Intake of Fruits and Vegetables
Intake of Fish and Shellfish
Intake of Meats, Diary Products, and Fats
Intake of Grain Products
Intake of Home-Produced Foods
Total Food Intake
Human Milk Intake
Activity Factors
Consumer Products
Lifetime
Building Characteristics

Source: Data from [9].

5 For example, physiological measurements may differ internationally, as may ingestion rates, breathing rates, life expectancies, etc.

The discussion of medium exposure in the following subsections refers to information in the EPA handbook, unless otherwise noted.

13.2.1 Ingestion

The ingestion of tap water is a function of age, with mean values ranging from 145 to 956 mL/d. Higher ingestion occurs in pregnant and lactating women. There can also be indirect exposure via water used in food preparation, particularly if it is not boiled.

Recreational water exposure can result in inadvertent ingestion of water from the body of water. For swimming, defined as the immersion of the head in water, a mean of 28–38 mL/h is used. For other water-based recreational activities, inadvertent ingestion is lower, although the data to inform such estimates are much sparser [11].

13.2.2 Inhalation

Inhalation exposure is also a function of age and activity level. For individuals aged 6 and above, the mean long-term breathing rate ranges between 12 and 18 m^3/d, with the upper 95th percentiles between 15.7 and 24. Short-term breathing rates are a function of activity level as shown in Table 13.3. As activity level increases, inhalation increases to keep pace with metabolic rate.

Closely related to inhalation is a route of exposure via aspiration. This is the inadvertent entry of material originally ingested (food or water) into the respiratory tract, possibly exacerbated by other physiological conditions. This may be more significant for respiratory microorganisms that occur in water (or food).

Quon et al. [12] reviewed the literature on aspiration and concluded that there is a wide uncertainty in the frequency and volume of inadvertent aspiration. However, for overall exposure to *Legionella* in drinking water, it could be a significant route. More work is needed to develop precise quantitative estimates.

13.2.3 Fomites

Exposure via fomites occurs via a multistep process. While the Exposure Factors Handbook provides data on the surface area of hands, face, etc., and also on the adherence of soil and dust to hands during touching, this is less relevant than measurements explicitly made in a microbial context.

Table 13.3 Short-term inhalation rates versus activity level, individuals 6 years and older.

	L/m^3	
Activity level	**Mean**	**Upper 95th percentile**
Sleep or nap	4.5–5.3	6.3–7.5
Sedentary	4.8–5.4	6.4–7.5
Light activity	11–13	15–17
Moderate activity	22–29	29–40
High activity	42–53	59–76

Source: Data from [9].

Unlike ingestion and inhalation, exposure via fomites is quite venue and situational-dependent. There is also a multistep process by which microorganisms on fomites are transported to a portal of entry. In a very typical generic sense:

1. An individual physically contacts a surface containing pathogens.
2. A portion of the pathogens on the surface in physical contact with the individual, such as on fingertips or hands, is transferred to the hands.
3. The transferred organisms may colonize on the hands and cause dermal effects, or enter through damaged skin.
4. The individual with transferred organisms on the hand or fingers may touch their eyes, nose or mouth, and some pathogens may then transfer to these surfaces and enter the host by an ocular, respiratory, or ingestion route.

Therefore, at least two partitioning phenomena are important: transference from surface to hands (step 1) and transference from hands to the portal of entry (step 2). Note also that the use of gloves or handwashing may pose a barrier to transmission via this route.

Numerous factors may influence the fomite to hand transference, including the type of microorganism, relative humidity, type of surface, and pressure between finger/hand and surface [13], with ranges of transference between 5% and 80%. Since fomite exposure is infrequently the sole route of exposure, and due to the situational specificity of considering fomites, it is best to consider this route in the context of specific cases. Such examples will be presented in Section 14.4.

13.2.4 Miscellaneous Other Routes

There are other routes of potential exposure that, as with fomites, are often application-specific.

Inadvertent ingestion of soils contaminated with animal wastes (without adequate handwashing) can present a route of exposure. This has been found to be important in the transmission of *Toxoplasma gondii* from soils contaminated with feline excreta [14]. It is also a significant path of exposure in communities with improper sanitation and waste disposal practices [15], especially in lower-income or resource-challenged settings.

If skin is damaged or breached, e.g., by an injury or an accidental needle stick or puncture wound, exposure to infectious agents could result in a blood-borne infection. However, in such a situation, it would also frequently require the breach of personal protective equipment, such as gloves, that are used.

13.3 Formal Computation of Uncertainty and Variability

The concepts and sources of uncertainty were discussed in Section 3.3. Both contribute to the lack of precision in the determination of exposure and risk. Given a set of data on microbial occurrence, how can the best-fitting distribution be determined, what are the uncertainties in estimates for its parameters, and what is the impact of the choice of distribution form on the resulting exposure estimates (particularly at the upper tails)? Estimation of the upper tails is essential if the exposures and risks of concern are influenced by the observations that can occur at a probability greater than $\frac{a}{N_{obs}}$, where N_{obs} is the number of observations in a data set.

13.3.1 Determining Best Distributional Forms

Distributions can be fit to data using a maximum likelihood approach. This was seen in Section 6.2 in fitting microbial data, whether count, quantal, or concentration data. The same approach will be used here. To select among alternative distributions, the AICc statistic described in Section 12.1.6 will be used. The number of potential distributions is infinite; however, a broad spectrum of forms is included within R or Python. In R, the library `fitdistplus` can be used for fitting, and in Python, the package `scipy.stats.fit` can be used. Among the common two-parameter continuous distributions are the normal, lognormal, Weibull, gamma, inverse Gaussian, and Cauchy.[6] Generalizations and extensions of these to three or more parameters and other three or more parameter distributions are available. Comprehensive discussions of broad categories of the more complex distributions may be found in Johnson and Kotz [16, 17].

Fitting of different distributions and comparing fits will be discussed in the context of an example. Sylvestre et al. [18] reported data on *Cryptosporidium* oocyst concentrations in raw water of several Canadian drinking water utilities. Data for utility "C1" were abstracted and converted to concentrations shown in Table 13.4. To simplify the discussion, two observations with non-detects have been deleted—if they were retained, this analysis would need to be done using a censored likelihood approach.

These data were fitted to several common two-parameter distributions using maximum likelihood. The best-fit parameters, log likelihoods, and AICc values are shown in Table 13.5. The best-fit distribution to the data among those considered is the inverse Gaussian, with the second-best fit

Table 13.4 Oocyst concentration in raw water of utility C1.

Oocyst concentration (1/L)	
0.8095	0.0333
0.7010	0.0154
0.6190	0.1356
0.3770	0.0167
0.0938	0.1728
0.0794	0.0142
0.0164	0.1280
0.0348	0.1633

Source: Data from [18].

Table 13.5 Fit of two parameter distributions to data of Sylvestre et al.

Distribution	Parameters	Best values		log L	AICc
Normal	mean, s	0.2131	0.2568	−0.9495	6.8220
Uniform	min, max	0.0142	0.8095	3.6646	−2.4061
Gamma	shape, rate	0.7285	3.4178	9.3334	−13.7437
Weibull	shape, scale	0.7947	0.1856	9.5235	−14.1240
Lognormal	mean log, s log	−2.3709	1.3700	10.1942	−15.4652
Inverse Gaussian	mean, dispersion	0.2131	18.5761	10.8239	−16.7248

6 The Cauchy and the log-Cauchy distributions are two-parameter, symmetric, "fat tail" distributions that are considered "pathological" in that the mean, variance, and higher order moments are not defined, and furthermore, sums of random observations from these distributions do not obey the central limit theorem.

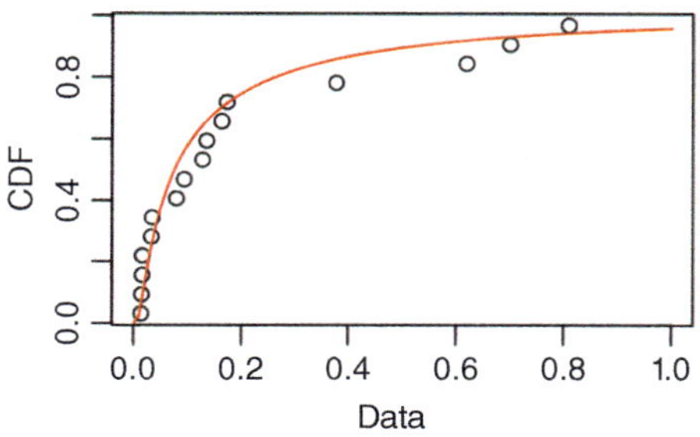

Figure 13.1 Fit of Sylvestre *Cryptosporidium* data to inverse Gaussian distribution.

provided by the lognormal. A comparison of the data to the fitted inverse Gaussian cumulative distribution functions is shown in Figure 13.1. Formal goodness-of-fit testing shows that this cannot be rejected.

13.3.2 Parametric Uncertainty of Distributional Parameters

Having ascertained that, among the distributions considered, the inverse Gaussian is preferred, the next question is what the uncertainty is in the distributional parameters that have been estimated. There are several approaches to this.

Asymptotically, the confidence limits to the parameters estimated via maximum likelihood are normally distributed with a variance-covariance matrix equal to the inverse of the Hessian matrix of the log-likelihood function [19]. If LL is the log-likelihood, and the maximum likelihood estimators of parameters are given by the vector $\bar{\theta}$, then the Hessian matrix has elements (i,j) given by:

$$H_{i,j} = \frac{\partial^2 LL}{\partial \theta_i \, \partial \theta_j} \tag{13.2}$$

The advantage of this approach is that once the ML parameter estimates have been determined, the Hessian matrix can be evaluated directly, either algebraically or numerically. However, for small samples, particularly in the presence of nonlinearity, this approach is inappropriate and may yield less accurate confidence ranges.

A second approach is likelihood-based [20, 21]. If $\widehat{\Theta}$ is the ML parameter estimate(s) (possibly a vector, with dimension k), then all other parameter values Θ are in the upper p percentile confidence region if satisfied by Equation 13.3.

$$2\left[LL(\widehat{\Theta}) - LL(\Theta)\right] < \chi^2_{(1-p),k} \tag{13.3}$$

This method was long disfavored because the direct application of this equation required root-finding to determine the locus of points. However, with current computational methods, the left-hand side of Equation 13.3 can be evaluated on a (multidimensional) grid of points, and all those satisfying the inequality are retained in order to bound the region satisfying the relation.

This approach is implemented for the Sylvestre data in Listing 13.1. A grid is constructed over 10,000 data points, logarithmically spaced in two dimensions, and at each point (of mean and dispersion), the likelihood is calculated. Then all points outside the desired confidence region are dropped, and the remaining points are plotted.

```
library(stats)
library(actuar)
CryptoConcentration<-c(0.8095,0.7010,0.6190,0.3770,0.0938,0.0794,
        0.0164,0.0333,0.0154,0.1356,0.0167,0.1728,
        0.0142,0.1280,0.1633,0.0348)
bestmean<-0.2131298; Lm=log(bestmean)
bestdispersion<-18.5760673; Ld=log(bestdispersion)
bestLL<-10.82394
Npoints<-10000
meanvalues<-exp(Lm-5+runif(Npoints)*(Lm+15))
dispersionvalues<-exp(Ld-5+runif(Npoints)*(Ld+10))
LogLR<-0*meanvalues
points<-data.frame(meanvalues,dispersionvalues,LogLR)
LLdiff<-function(theta){
  LL<-dinvgauss(CryptoConcentration,mean=theta[1],dispersion=theta[2],log=TRUE)
  LLdiff<-bestLL-sum(LL)
}
for (i in 1:Npoints) {
  points$LogLR[i]<-LLdiff(c(points$meanvalues[i],points$dispersionvalues[i]))
}
Chi2Crit<-qchisq(0.90,2)
points_in<- subset(points, LogLR <= 0.5*Chi2Crit)
plot(points_in$meanvalues,points_in$dispersionvalues,log='xy',xlim=c(0.1,500),
     ylim=c(5,50),xlab='Mean',ylab='Dispersion')
```

Listing 13.1 Likelihood ratio confidence region for Sylvestre fit to inverse Gaussian.

The results are plotted in Figure 13.2. Several things of importance are noteworthy. First, the dispersion of the inverse Gaussian is well defined (within a factor of two or so from the best value). However, the confidence limit to the mean is virtually open to the right, indicating that, assuming the inverse Gaussian is accepted, very high values well beyond the observed range are possible. Finally, while the application of the Hessian approximation would yield an elliptical confidence limit consistent with the assumption of asymptotic normality, the likelihood-based confidence region is highly asymmetric.

A third approach to estimating uncertainty is the bootstrap. This concept, developed by Efron [22], is shown schematically in Figure 13.3. It is desired to estimate some statistic[7] of the

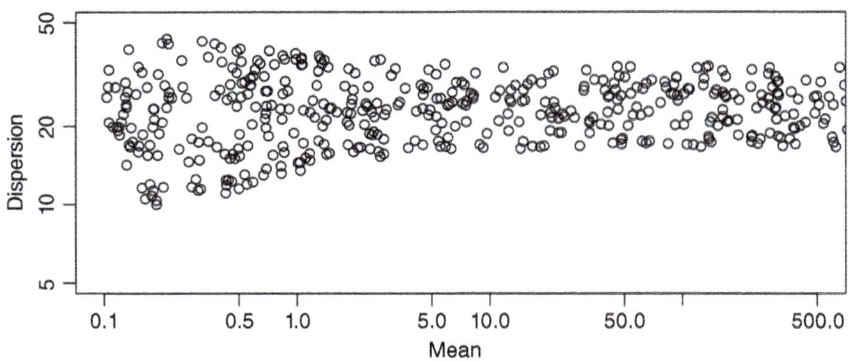

Figure 13.2 Parameter pairs for the inverse Gaussian that are in the 90 percentile confidence region of the fit to Sylvestre data.

7 In this context, a statistic is some quantity or set of quantities computed from a set of data.

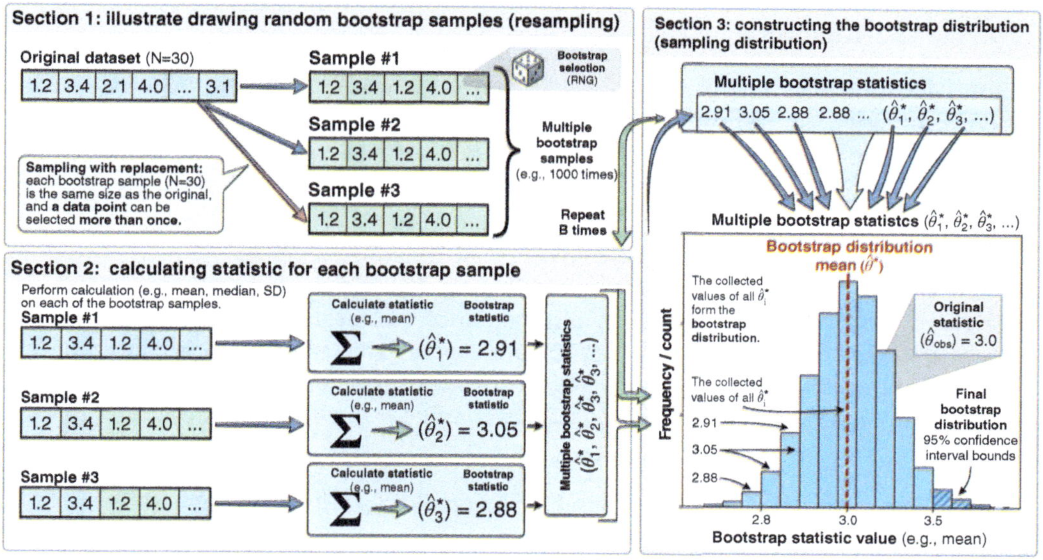

Figure 13.3 Schematic of the simple bootstrap method.

population. Rather than the full population, only a (representative) sample is available, from which the statistic can be calculated. But it is also desired to compute the uncertainty of this sample statistic. The simple bootstrap principle involves deriving multiple sets (termed bootstrap replicates) of data from the original sample *with replacement*, and then computing the statistic for each replicate. The distribution of the sampling statistic then asymptotically describes the confidence region for that statistic *in the population.*

To illustrate the concept of a bootstrap sample, consider again the data on *Campylobacter* presented in Table 13.1, from which the mean and standard deviation were computed. In Table 13.6, four bootstrap replicates are constructed, and the mean and standard deviation of each replicate are calculated. If this were repeated a large number of times, the set of pairs of <mean, standard deviation> would form the estimated confidence distribution for these statistics in the population from which the original sample was drawn.

A key aspect of the bootstrap approach is the assumption that the original observations were representative of the population itself. With small sample sizes, this might not hold. A simple illustration is shown in Example 13.2.

Example 13.2 Bootstrapping from Small Sample.

If the original sample is itself small, it may not be sufficiently representative of the underlying population, and therefore, bootstrapping from this small sample could produce misleading results. The key issue is representativeness: the bootstrap reproduces the quirks of the observed sample, not the unobserved population variability. When n is small, a sample can easily land in one tail of a distribution; the resampling distribution will then be centered at a biased value, producing confidence intervals that are too narrow and potentially exclude the truth.

(Continued)

Example 13.2 (Continued)

Consider estimation of the mean for data drawn from a well-behaved population: $X \sim \mathcal{N}(0, 1)$ with true mean 0 with a sample of 5.

Suppose the sample has values that all happen to be negative:

$$x = \{-1.0, -0.6, -0.4, -0.2, -0.1\}.$$

There is a 6.25% chance that all positives or all negatives could have been drawn in a sample of five from the underlying true distribution. The sample mean is $\bar{x} = -0.46$. If we compute the nonparametric bootstrap distribution of the mean by resampling these five values with replacement, the resamples contain only negative numbers. The bootstrap distribution is therefore centered near -0.46, and a nominal 95% percentile interval may be $(\ell, u) = (-0.80, \ -0.20)$, which entirely excludes the true mean 0. This figure illustrates this behavior.

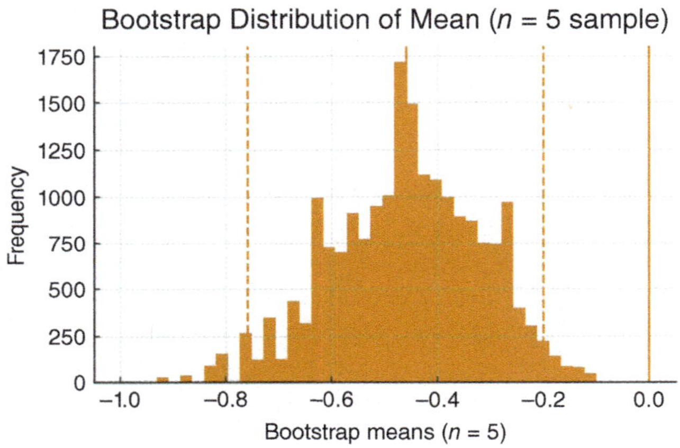

Bootstrap distribution of the sample mean for the $n = 5$ dataset. Solid vertical lines mark the sample mean (center) and the true mean 0; dashed lines mark the 95% bootstrap percentile CI. The interval excludes 0.

Many other applications and embellishments of the simple bootstrap are available. The foundational book of Efron and Tibshirani should be consulted for details [23]. Bootstrapping of residuals from a regression model can be used to assess uncertainty in model parameters, which will be discussed later in this section. Bootstrapping of dose–response models can also be used to assess uncertainty in these models; a discussion of this application will be in Section 14.1.

In implementing a resampling method, such as the simple bootstrap, one question is how many bootstrap samples to take. As computational speed has increased, it has become more tractable to handle greater numbers of samples; for example, Hesterberg suggests 10,000 [24]. The impact of the number of bootstrap samples on the resulting estimate of the statistic of interest can also be addressed experimentally. One can draw repeated bootstrap samples of size n and then compute the variance of the statistic across those samples; by repeating this with larger values of n, the bootstrap sample size required to achieve the desired variance can be determined.

Table 13.6 Bootstrap replicates for *Campylobacter* data in Table 13.1.

	Original	Bootstrap Replicates			
		A	B	C	D
	5.86	5.86	4.01	5.29	5.86
	5.68	4.35	5.51	5.68	4.01
	5.51	4.44	4.68	4.35	4.44
	5.37	5.29	5.86	4.56	4.97
	5.29	4.56	4.35	4.35	5.51
	5.12	4.35	5.86	5.12	4.68
	4.97	5.68	5.12	5.37	4.68
	4.68	5.86	4.97	5.12	5.29
	4.56	4.35	4.01	4.97	5.51
	4.44	5.12	5.68	4.68	4.56
	4.35	4.68	5.86	4.56	4.01
	4.01	5.29	5.37	4.35	4.44
Mean	4.987	4.986	5.107	4.867	4.830
Standard deviation	0.581	0.603	0.703	0.453	0.602

If the original sample being bootstrapped is small (subject to the caveats discussed in Example 13.2), the number of distinct bootstrap samples from a dataset with n distinct values is small and is given by the quantity in Equation 13.4 [24]:

$$\frac{(2n-1)!}{n! \cdot (n-1)!} \tag{13.4}$$

For example, for $n = 8$, there are 6435 distinct samples. So, in principle, for a small data set, exhaustive enumeration of all alternatives is tractable rather than relying on random sampling with replacement.

The bootstrap estimation of distribution parameters can be performed in R using the `boot` package, or in Python using `scipy.stats.bootstrap`. Listing 13.2 presents the code for applying this to the inverse gaussian fit to the Sylvestre data.

```
library(fitdistrplus)
library(stats)
library(actuar)
library(boot)
CryptoConcentration<-c(0.8095,0.7010,0.6190,0.3770,0.0938,0.0794,0.0164,
                       0.0333,0.0154,0.1356,0.0167,0.1728,0.0142,0.1280,
                       0.1633,0.0348)
fitted_parameters<-function(concentrations,boot_indices){
  A<-fitdist(concentrations[boot_indices],'invgauss')
  return(A$estimate)}
b_results<-boot(CryptoConcentration,fitted_parameters,1000)
b_params<-b_results$t
plot(b_params[,1],b_params[,2],log='xy',xlab = 'mean',ylab = 'dispersion')
```

Listing 13.2 Code for bootstrapping Sylvestre fit to inverse Gaussian.

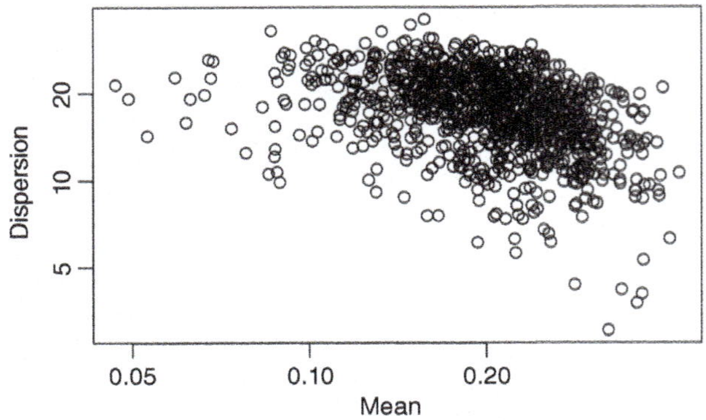

Figure 13.4 Bootstrapped parameters (1000 replicates) for inverse Gaussian fit to Sylvestre data.

The resulting scatter plot of the parameters from the bootstrapping is shown in Figure 13.4. Note that since this is a random resampling, repeated runs of 1000 samples will exhibit slightly different shapes. However, compared with the likelihood-based confidence limit, the bootstrap distribution is more tightly defined. Furthermore, each point in this figure can be taken as a realization of the exposure distribution parameters, which can be incorporated into a full risk characterization. Or the entire set of bootstrap points can be used to construct the confidence limits to the probability distribution of the original data by the following process:

- Define the range of values over which the PDF or CDF is sought.
- For each value within that range, use all of the bootstrapped parameters of the probability distribution to calculate the value of the PDF or CDF at that value.
- Find the upper and lower percentile of the values for the PDF or CDF, i.e., is it desired to calculate the 95 percentile symmetric confidence region, find the PDF/CDF value that is exceeded 2.5% of the time, and the PDF/CDF value that exceeds the minimum 2.5% of the values.
- Repeat for the entire range of desired values, and connect the loci of the upper and lower confidence ranges.

Often, the exposure and ultimately the risk can be strongly influenced by the upper tails of a distribution. Since, in general, inference from (relatively) small samples is required to ascertain the distribution, the question is how many samples, under what conditions, may be needed to reliably distinguish among alternative distributions. This is a function of the relative standard deviation (RSD), defined as the ratio of standard deviation to mean. Based on a simulation study, the following general conclusions have been reported [25]:

- The proper identification of a distribution among the potential alternatives requires a larger sample size as the RSD decreases. In particular, for RSD values of 0.1 or less, more than 200 samples are needed to achieve a high level of reliability in distributional attribution. However, in this circumstance, different alternative distributions show similar tail behavior.
- At high RSD levels, the sample size required for correct identification decreases to modest values. At an RSD of 100, a sample size of 50 may be adequate.

Therefore, for small samples, good practice is to examine fits of alternative distributions for tail behavior. The distributional form is then considered as a component of overall model uncertainty.

A number of the process and transport models relate the outcome (e.g., concentration or removal fraction) to some descriptive and design variables. For example, disinfection processes as a function of residual and contact time. Bootstrapping can be used to estimate uncertainty in these models.

Bootstrapping can be used to estimate the uncertainty of statistical estimators and construct confidence intervals for regression models, e.g., of microbial removal in processes, or changes in fate and transport. Two primary approaches are commonly employed: **bootstrapping observations** and **bootstrapping residuals**. Each method has distinct assumptions and implications for inference [24, 26].

This bootstrapping observations (or cases) method involves resampling the original data pairs (x_i, y_i) with replacement to generate new datasets. The nonlinear model is refitted to each bootstrap sample, and the variability of parameter estimates is assessed across replicates.

Advantages:

- Minimal assumptions about the error structure or model correctness
- Captures both model and data variability
- Robust to model misspecification

Limitations:

- May distort the predictor–response relationship, especially in nonlinear models
- Can lead to wider confidence intervals due to increased variability, especially when a few cases are influential or disparate

The bootstrapping residual approach first fits the nonlinear model to the original data, then resamples the residuals $e_i = y_i - \hat{y}_i$ with replacement. These resampled residuals are added to the fitted values to generate new response vectors, keeping the original predictors fixed.

Advantages:

- Preserves the structure of the predictor variables
- Efficient when the model is correctly specified
- Can yield tighter confidence intervals and more stable estimates

Limitations:

- Assumes residuals are independent and identically distributed
- May underestimate variability in finite samples, especially in prediction tasks

In nonlinear regression, the choice between these methods depends on the confidence in the model's structure and the nature of the data. Residual bootstrapping is often preferred when the model accurately captures the underlying nonlinear relationship, as it maintains the design matrix and avoids resampling predictor values that may have nonlinear effects.

However, in models with heteroscedasticity or complex error structures, case bootstrapping may be more robust because the residuals from all cases cannot be assumed to be identically and independently distributed.

To some degree, the choice between the methods is based on the underlying nature of the system being analyzed. In designed experiments, the investigator has chosen to fix the values of the predictor variables *a priori*, for example, in a disinfection bench test by setting times and doses, and then measuring the resulting survival. In other situations, frequently with observational data, the investigator takes samples from a system and seeks to relate the levels of predictor variables to an outcome variable — for example, consider monitoring a disinfection system in the field over time and collecting data on residual, time, etc., to develop a model. The first system would be more amenable to a bootstrapping residuals approach, whereas the second, in which there is some randomness in both the predictor and the outcome variables, would be more amenable to a bootstrapping case approach. Asymptotically, the two methods are regarded as equivalent [24, 26].

As an example of bootstrapping residuals, recall the experiment on inactivating *Bacillus subtilis* spores by gas phase chlorine dioxide presented in Table 12.4. These data were fit to the Hom model. It is desired to determine the confidence distribution of the kinetic parameters, k, m, and n, and since this was a designed experiment, bootstrapping residuals is appropriate. The R code to perform this is presented in Listing 13.3.

```
library(stats)
library(boot)
experiment<-read.csv('fig2_digitized.csv')
Hom<-as.formula(Log10S~log10(exp(-k*(C^n)*(t^m))))
results<-nls(Hom,experiment,start=list(k=0.1,n=1,m=1),trace=TRUE)
fitted_values<-fitted(results)
residual_values<-residuals(results)
HomSS<-function(dummylog10S,bindex){
  bootstrap_log10S<-fitted_values+residual_values[bindex]
  bexperiment<-experiment
  bexperiment$log10S<-bootstrap_log10S
  Hom<-as.formula(bootstrap_log10S~log10(exp(-k*(C^n)*(t^m))))
  results<-nls(Hom,bexperiment,start=list(k=0.1,n=1,m=1),trace=FALSE)
  return(coef(results))
  }
boot_results<-boot(experiment,HomSS,R=1000)
A<-data.frame(boot_results$t); colnames(A)<-c('k','n','m');
pairs(A)
```

Listing 13.3 Code for bootstrapping residuals for Hom model fit to data of Table 12.4.

Since there are three parameters being estimated, the results can be presented graphically in a pairs plot given in Figure 13.5.

13.3.3 Combining Multiple Distributions

If multiple distributions characterize the individual steps from source to receptor, as in the cascade shown in Figure 9.8, they must be combined to obtain an overall exposure estimate. This can be done using a Monte Carlo sampling technique as illustrated in Figure 9.9. The distributions may include uncertainties as estimated by bootstrapping.

There are well-regarded principles for Monte Carlo analysis of exposure, and ultimately risk, that were outlined by Burmaster and Anderson [27]. These are excerpted and modified in Table 13.7, recognizing that since the original principles were delineated in 1994, advances in computational resources have allowed modification of some of the concepts.[8]

When exposure assessment is combined with dose–response assessment to develop a risk estimate, a full Monte Carlo analysis incorporating all source distribution terms, fate and transport processes, media exposure distributions, and the dose–response model is necessary. This ultimately yields a complete risk characterization, which will be the focus of Chapter 14.

8 For example, in 1994, running 100 iterations of a complex Monte Carlo containing simultaneous ordinary differential equations could require hours of computational time, whereas now it might only require minutes, and many more iterations have become feasible.

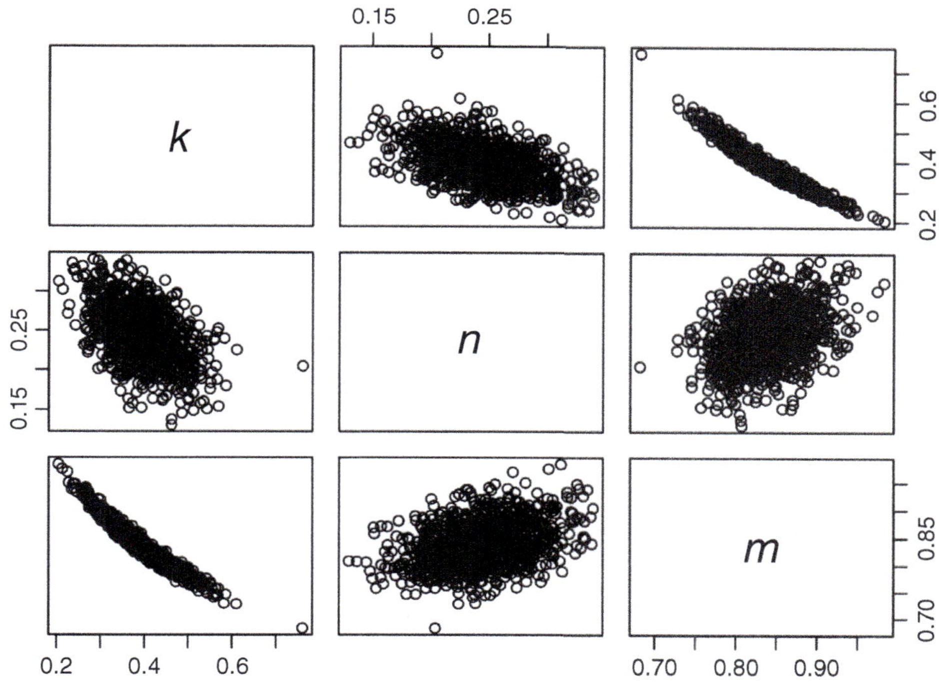

Figure 13.5 Pairs plot for bootstrapping regression residuals of Hom model fit to data of *B. subtilis* inactivation. Note that *k* is $\log_{10}$ basis.

13.3.4 Advanced Concepts

13.3.4.1 Correlated Variables

In general, the distributions of two quantities can be correlated. The concept of being correlated in general terms is defined as, "The condition of being correlated; mutual relation of two or more things" [28]. In the absence of such a relationship or condition, two quantities are independent, meaning that the value of one bears no relationship to the bearing of the other.

If two quantities are related, each of them, considered separately, will have a probability distribution, which is termed a marginal distribution, i.e., $f(x)$ is the pdf for a variable x. If x and y are related, their marginal distributions might be $f(x)$ and $g(y)$, where f and g each contain parameters (e.g., mean, standard deviation, etc.) that describe their marginal distribution.

The relationship can be depicted by a scatter plot of the pairs $< x_i, y_i >$. This is shown by example in Figure 13.6 for a five-year data set of monitoring fecal coliform and total coliform in the Illinois River at Peoria, IL [29]. The marginal distributions are shown by the curves at the top and right-hand sides of the graph; not shown is that each marginal can be well described by a log-normal distribution. The scatter shows a relationship between the two quantities. It is clearly shifted from a line of equality (as would be expected since fecal coliforms are principally a subset of total coliforms), and there are also some apparent outliers.

A previous example of correlations between distributions was observed in time-series data for the removal of *Giardia* (Figure 9.6). In that case, correlations could be (mostly) removed by looking at the distribution of pairwise log reduction values.

Table 13.7 Best principles for Monte Carlo methods in risk assessment.

Principle	
1	Show all formulae used for the computation, preferably in mathematical form rather than code.
2	Do an initial point estimate of the output.
3	Do a preliminary univariate sensitivity analysis to limit the Monte Carlo analysis to the variables that have the dominant effect on the range.
4	Restrict the analysis to the most important paths for the decision to be made.
5	Provide detailed information on each of the input distributions and the method for their selection and the evaluation of their parameters.
6	Capture both elements of variability and uncertainty.
7	To the degree possible, use actual measured quantities to select and specify the distributions used.
8	Report goodness-of-fit statistics for the chosen input distributions.
9	Discuss the presence or absence of correlations among the input distributions and their potential to affect the output quantities.
10	Provide graphs of the input distributions used.
11	Present sensitivity analyses of the output to the variability and uncertainty in the inputs.
12	Determine whether enough Monte Carlo trials have been run to assure stability, within the desired precision, of the output quantities.
13	Document the random number generator used, and verify that it has been determined to be of adequate quality.
14	Discuss additional sources of uncertainty and variability, including qualitative, which were not considered in the analysis.

Source: Adapted from [27].

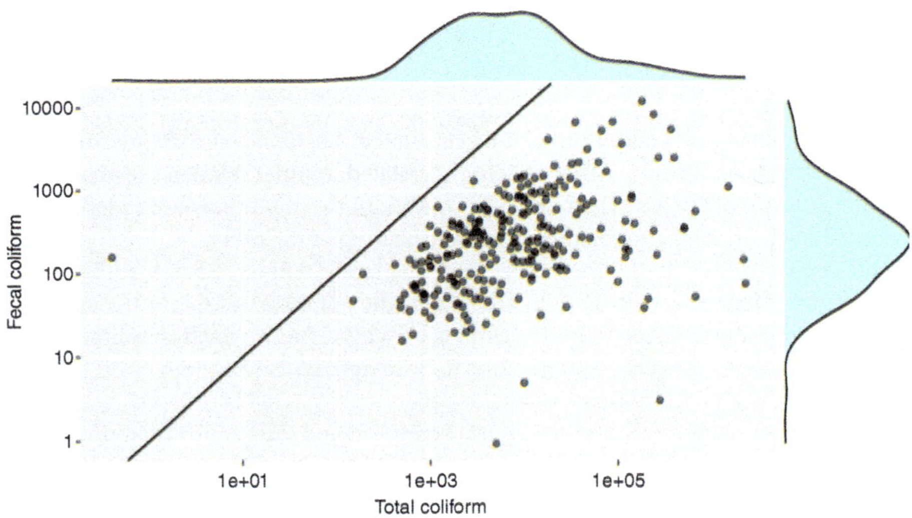

Figure 13.6 Scatter plot of five years of fecal and coliform organisms measured at Peoria, IL. The upper and right-hand edges show marginal density distributions. The diagonal line is the equality line between the variables. *Source:* Data from [29].

A conventional way of describing relationships between variables is the use of correlation, which is closely related to the concept of regression. The concept of correlation goes back at least to the early 19th century work of Gauss on the properties of bivariate normal distributions [30]. There are now several different types of correlation coefficients or alternative forms of describing relationships between two variables, only one of which arises from the bivariate normal distribution of Gauss. In the case of the data in Figure 13.6, there appears to be a positive correlation between the two bacterial measurements—a larger value in one *tends to* be accompanied by a larger value in the other. Negative correlations between variables are possible, as is no association.

There are multiple mechanisms that can lead to correlation between variables. A first caveat is that correlation between variables does not *per se* imply causality. There could be one or more unobserved (also called latent or lurking) variables that act in a related manner on both of the observed variables. Here are some plausible examples in environmental data:

- Imagine a monitoring program has found a positive association between pathogen levels at a water intake and total dissolved solids (TDS). Upon further investigation, the water is fed by two upstream sources, one with high concentrations of both TDS and pathogens. Therefore, the association is really an index of the fraction of water at the intake coming from the more heavily contaminated upstream source.
- Water treatment frequently involves multiple treatment processes in series, for example, the coupled coagulation–flocculation–sedimentation (CFS) process followed by filtration. However, the performance of filtration is related to the efficient destabilization of particles by CFS. Therefore, poor performance of CFS could result in poor filtration performance, leading to positive associations in LRVs between the two processes.
- Microbial contamination of a surface may co-occur with contamination by visible dust or debris. Hence, individuals might have a lower propensity to be exposed to a fomite because of withdrawing from frequent or extensive contact.

Correlation can be computed in multiple ways. Each approach has different underlying assumptions and can lead to different numerical results.

Galton and Pearson, following the work of Gauss, developed the concept of a product–moment correlation coefficient, which has come to be known as the Pearson correlation coefficient. For two variables (vectors) x and y, with respective means $\bar{x}$ and $\bar{y}$ and standard deviations s_x and s_y, containing n paired observations, the covariance, s_{xy} is defined by:

$$s_{xy} = \frac{1}{n-1}\sum(x_i - \bar{x})(y_i - \bar{y}) \tag{13.5}$$

The Pearson correlation coefficient ρ_{xy} is defined as Equation 13.6.

$$\rho_{xy} = \frac{s_{xy}}{\sqrt{s_x \cdot s_y}} \tag{13.6}$$

ρ_{xy} can range between -1, indicating perfect inverse correlation, and $+1$, indicating perfect positive correlation.[9] A value of $\rho_{xy} = 0$ indicates the absence of a relationship between the variables.

There are multiple ways to interpret the Pearson correlation coefficient, including as a parameter of the bivariate normal distribution. This is a distribution of two random variables, each with a marginal normal, characterized by the single interaction parameter, ρ_{xy}. However, there are several features of the Pearson correlation that need to be understood, which have led to the development of other measures of association between variables:

9 As will be shown momentarily, this actually refers to a linear relationship between the quantities, and if there are nonlinear relationships, then there can still be strong relationships while the Pearson coefficient will be small.

- It is a measure of the *linear* association between the variables, and nonlinear associations may be inadequately characterized. This will be illustrated by example.
- Being a measure of linear association, the numerical value of the Pearson correlation will be affected by variable transformations, e.g., log, power, reciprocal, etc.
- The Pearson correlation may be unduly influenced if there are one or a small number of influential outlying points.
- While there are tests of the significance of correlation coefficients, these are not robust to the presence of non-normal errors, and alternative significance tests may need to be used [31].

Two alternatives for measuring association between variables are the Spearman correlation and Kendall's Tau. Most software, including R and Python, has libraries or packages that can calculate all three of these measures of association.

The Spearman correlation is obtained by replacing the raw values of each variable with its ranking among the set of values of that variable. In the case of ties, the average of the tied ranks is used for each of the ties — for example, if two values are tied for ranks 9 and 10 among the set of observations, then each would be given a rank of 9.5. Having ranked each variable separately, the Spearman correlation is computed by applying the Pearson formula to the ranked values.

Kendall's τ is a measure of the discordance between pairs of the two variables. It is computed by the double summation in Equation 13.7.

$$\tau_{ij} = \frac{1}{n(n-1)} \sum_{i=1}^{n} \sum_{j=1}^{n} \text{sgn}(x_j - x_i)\text{sgn}(y_j - y_i) \tag{13.7}$$

In this equation, sgn(z) is the mathematical sign function, assuming a value of -1 if $z < 0$ and a value of $+1$ if $z > 0$.

Computation of these three correlations will be illustrated with the data set in Table 13.8, which are paired measurements of *Enterococcus* in wastewaters and in receiving waters in Southern California using culture and qPCR methods [32].

For the data in Table 13.8, the values of the Pearson, Spearman, and Kendall's coefficients are 0.7203, 0.7776, and 0.5954, respectively.

Table 13.8 Measurements of *Enterococcus* in wastewater and environmental waters by culture and qPCR.

Sample and date/location	qPCR (gc/100 mL)	Culture (#/100mL)
9/15/03 Primary effluent — plant I	350000	810000
9/15/03 Secondary effluent — plant I	140000	600000
9/15/03 Secondary effluent — plant II	170000	32000
9/30/03 Primary effluent — plant I	2600000	1800000
9/30/03 Secondary effluent — plant I	350000	140000
9/30/03 Primary effluent — plant II	160000	170000
10/1/03 Primary effluent — plant I	640000	2000000
10/1/03 Secondary effluent — plant I	210000	23000
10/1/03 Primary effluent — plant II	810000	420000
Coastal waters San Diego Creek	4500	880
Coastal waters Middle Newport Bay	2400	7.2
Coastal waters Lower Newport Bay	730	59

Source: Data from [32].

If there are more than two variables, then any of these correlations can be used to construct a correlation matrix between all pairs. If there are k variables, then there will be $\frac{k(k-1)}{2}$ pairwise correlations.

The numerical values of correlation coefficients can portray an incomplete story of association, or lack thereof. This is illustrated by several examples in Figure 13.7. The top left panel in this figure shows two variables with a distinctive periodic relationship and small values of both Pearson and Spearman coefficients. The upper-right panel shows a data set dominated by two outliers with large values of both variables, while other observations are clustered with somewhat lower association, yet the values of both the Pearson and Spearman correlations are high. The panel at the bottom right shows two groups of observations, each with low correlation, whereas the pooled data shows much higher values for both Pearson and Spearman correlations. The panel on the lower left shows two variables with a strong parabolic relationship; however, the values of both the Pearson and Spearman coefficients are low, indicating a breakdown when associations are neither linear nor monotonic.

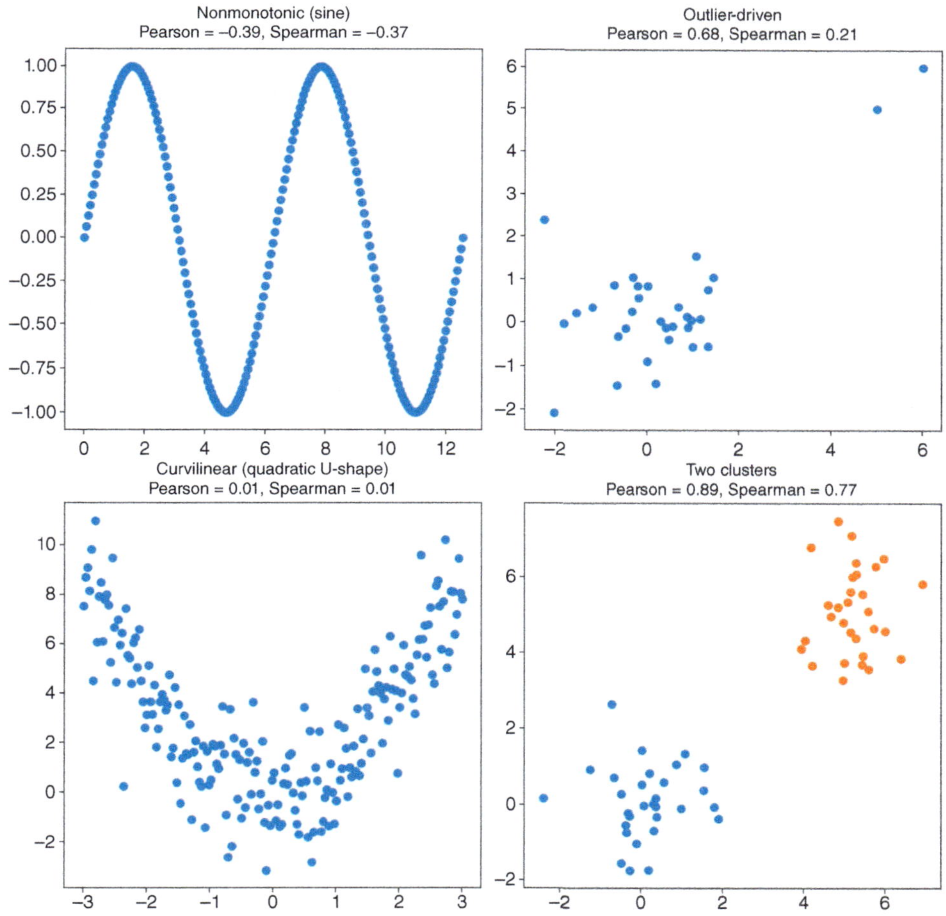

Figure 13.7 Examples of associations between random variables with misleading correlations.

Associations can emerge intrinsically from fitting a model with multiple parameters. This was shown in Figure 13.5 for the bootstrap parameter distributions fitting a data set to the Hom model. Notable associations are apparent between the values of k, m, and n.

For the Pearson, Spearman, and Kendall measures, there are statistical tests for their significance, i.e., against the null hypothesis of no association. However, particularly for the Pearson coefficient, the significance levels are not highly robust to violations of the underlying assumptions, such as the normality of errors [33].

When two variables are correlated, this association must be considered in the estimation of exposure and risk derived from those variables. Burmaster and Anderson [27], based on the work of Smith et al. [34], propose that if the absolute value of correlation (Pearson or Spearman) is greater than 0.6, then it should be incorporated into a Monte Carlo analysis, and when tail probabilities are of concern, this cutoff should be reduced.

13.3.4.2 Copulas

Another approach to assess relationships between variables is via a mathematical construct called a copula. Lai and Balakrishnan [35] provide a concise verbal definition:

> "What are copulas? Briefly speaking, copulas are functions that join or 'couple' multivariate distributions to their one-dimensional marginal distribution functions. Equivalently, copulas are multivariate distributions whose marginals are uniform on the interval (0, 1)."

The formal mathematical exposition of copulas is presented concisely by Genest and Favre [36] and more extensively by Lai and Balakrishnan [35]. Let x and y be the random variables of interest, and they are characterized by marginal cumulative distribution functions F and according to:

$$u = F(x) \tag{13.8}$$

$$v = G(y) \tag{13.9}$$

Note that, as CDFs, the values of u and v lie in the interval [0,1]. F and G can be any univariate cumulative distribution function, and the forms of F and G can be different for each variable.

Then a copula function, C, can be defined, which combines the marginal distributions into a bivariate cumulative distribution, H, via Equation 13.10.

$$H(x, y) = C[F(x), G(y)] \tag{13.10}$$

H is bounded by [0,1]. The function C may have some parameters that determine the nature of the relationship. A key property of copulas is that they capture a relationship between the variables that is invariant to monotonic transformation of the variables [36], and that this relationship is separable from the underlying marginal distributions of each of the variables.

Since H is a bivariate CDF, the corresponding density function, h, can be obtained by differentiating with respect to both variables, i.e.,

$$h(x, y) = \frac{\partial^2 H(x, y)}{\partial x\, \partial y} \tag{13.11}$$

Since h is a density function, it must be non-negative throughout its range of support.

Just as there are a large (infinite) number of univariate distribution functions, there are a large (infinite) number of potential copula functions — the key characteristics are that they are monotonic functions of two variables bounded by zero and one, returning a value between zero and one, and symmetric in arguments. A key subset of potential functions is those that are termed Archimedean copulas.

An *Archimedean copula* is specified by a generator and has the form

$$C(u_1, \dots, u_d) = \varphi^{-1}\left(\varphi(u_1) + \cdots + \varphi(u_d)\right), \qquad (u_1, \dots, u_d) \in [0, 1]^d \tag{13.12}$$

where the generator $\varphi : [0, 1] \to [0, \infty]$ is continuous, strictly decreasing, convex, with $\varphi(1) = 0$, and φ^{-1} denotes the (generalized) inverse on $[0, \infty]$. Equivalently, one may work with $\psi = \varphi^{-1}$: $[0, \infty) \to [0, 1]$, decreasing and convex with $\psi(0) = 1$ and $\lim_{t\to\infty} \psi(t) = 0$, and write

$$C(u_1, \dots, u_d) = \psi\left(\psi^{-1}(u_1) + \cdots + \psi^{-1}(u_d)\right).$$

Archimedean copulas offer (i) flexible shapes of dependence, including asymmetric tails, (ii) closed-form C and often tractable densities for likelihood inference, and (iii) simple simulation via the generator's latent-variable representation (e.g., when ψ is completely monotone). Furthermore, the additive nature based on generation functions permits potential extension to three or more variables. Examples of some copulas follow.

13.3.4.2.1 Frank's Copula

Frank's copula is Archimedean [37] and is the only known one-parameter Archimedean copula spanning (rank) correlations from −1 to +1. It is described by the CDF and density functions (with respect to its marginal uniforms, u and v):

$$H(u, v) = -\frac{1}{\theta} \ln\left(1 + \frac{(e^{-\theta u} - 1)(e^{-\theta v} - 1)}{e^{-\theta} - 1}\right) \tag{13.13}$$

$$h(u, v) = \frac{\theta\,(1 - e^{-\theta})\, e^{-\theta(u+v)}}{[(e^{-\theta u} - 1)(e^{-\theta v} - 1) + (1 - e^{-\theta})]^2} \tag{13.14}$$

The parameter θ can assume any real value.

13.3.4.2.2 Joe's Copula

Joe developed a two-parameter extension of Frank's copula, which is also Archimedean and spans the entire range of correlations [38]. Its CDF and density functions are defined by:

$$H(u, v) = \frac{1}{\delta}\left\{1 - \left[1 - \frac{(1 - (1 - \delta u)^{\theta})\,(1 - (1 - \delta v)^{\theta})}{1 - (1 - \delta)^{\theta}}\right]^{1/\theta}\right\} \tag{13.15}$$

$$h(u, v) = \frac{\delta\, q'(\delta u)\, q'(\delta v)}{\theta\, q(\delta)}\left[1 - \frac{q(\delta u)\, q(\delta v)}{q(\delta)}\right]^{\frac{1}{\theta}-2}\left[1 - \frac{1}{\theta}\,\frac{q(\delta u)\, q(\delta v)}{q(\delta)}\right] \tag{13.16}$$

$$\text{Where} \quad q(z) = 1 - (1 - z)^{\theta} \tag{13.17}$$

The restrictions on the parameters are $\theta \geq 1,\ 0 < \delta \leq 1$.

13.3.4.2.3 Plackett's Copula

Plackett [39] developed a one-parameter non-Archimedean copula that also can span the entire range of negative and positive correlations. Its CDF and density functions are:

$$H(u, v) = \frac{1}{2(\theta - 1)}\left[1 + (\theta - 1)(u + v) - \sqrt{(1 + (\theta - 1)(u + v))^2 - 4uv\,\theta(\theta - 1)}\right] \tag{13.18}$$

$$h(u, v) = \frac{\theta\,(1 + (\theta - 1)(u + v - 2uv))}{\left((1 + (\theta - 1)(u + v))^2 - 4uv\,\theta(\theta - 1)\right)^{3/2}} \tag{13.19}$$

The parameter θ can be any positive value. For $\theta < 1$, the correlation is negative, while for $\theta > 1$, the correlation is positive.

What is apparent is that there are a variety of choices using copulas to describe the association between two variables. It is noted in passing that the bivariate normal distribution with a Pearson correlation coefficient of ρ can also be considered as a particular copula form.

For univariate data, it is known that computing the first and second moments (or even the first through fourth moments) does not completely specify the distribution that best fits the data. Analogously, for bivariate data, information on the first bivariate moment (i.e., the correlation) is insufficient to specify the full nature of the association between variables. Also, as with univariate data, fits to alternatives can be examined and used as a basis for judging. Genest and Favre [36] provide an introduction to fitting copula models. R has a library for the construction and generation of random observations from copulas, `copula`. Support in Python is limited.

13.3.4.3 Sampling Methods

The concept of Monte Carlo methods was introduced in Section 9.5. There are several broad strategies for sampling, i.e., generating random points in d-dimensional space (where d is the number of dimensions to be considered). If the inputs are independent, then random points on a line are generated for each input. If the dimensions are coupled, then points in a multidimensional space must be generated. The concept of quasi-random sampling was shown in a basic two-dimensional scatter plot in Figure 9.10. There are multiple algorithms for producing of these points, including Halton, Sobol, and Hammersley. Furthermore, performance can be improved by skipping and/or shuffling. Theoretically, for estimating the mean of a distribution, quasi-random methods are more efficient than pseudorandom methods, since the rate of convergence is approximately proportional to the reciprocal of the number of points, whereas for pseudorandom numbers, convergence is more typically approximately proportional to the reciprocal of the square root of the number of points [40]. Python, R, and Matlab all have routines to generate such sequences.

The relative performance, in a simple example, of pseudorandom sampling and quasi-random sampling was shown in the *Giardia* example in Listing 9.1 and its output.

In the pseudorandom approach, random samples are drawn from a uniform distribution using a vetted algorithm. In both R, Python and Matlab, a high-quality algorithm called the Mersenne Twister [41] is used, which has a periodicity of $2^{19937} - 1$, i.e., $\sim 10^{6001}$. While no algorithmic generation of a pseudorandom sequence can be truly random, this generator has passed many tests and so has been widely implemented. These three platforms, and many others, allow the random number stream to be initialized by specifying a "seed," which is particularly useful for testing a simulation to ensure reproducible results. These methods are called pseudorandom, since the algorithms themselves are purely deterministic, but they are designed to produce streams of uncorrelated outputs covering the range.

The stream of uniformly distributed numbers from either a pseudo-random or quasi-random approach must be transformed to deviate from the corresponding probability distribution to obtain the actual variate. All software packages have efficient algorithms for doing this for standard distributions. Here, broad approaches are outlined for implementing other distributions that may not be covered.

Hörmann et al. extensively treat the problem of generating random deviates from distributions and should be referred to for theoretical bases and pseudocode for implementation [42]. There are two broad classes of methods — inversion and acceptance–rejection, as well as special methods for particular distributions. The inverse method is suitable for any distribution that has a cumulative distribution, $F(x)$, that is easily analytically invertible and computable, i.e., $F^{-1}(u)$. This is illustrated by Example 13.3.

Example 13.3 Example of Random Deviates from a Weibull Distribution Using the Inversion Method.

There are data described by a Weibull CDF:

$$F(x) = 1 - \exp\left[-\left(\frac{x}{x}\right)^{b}\right]$$

The values for the data to be simulated are $a = 1,\ b = 2$. Construct random deviates from this distribution.

Solution:

The CDF can be inverted to yield the following if *F* is a uniform random number between zero and one:

$$x = a[-\ln(1-u)]^{\frac{1}{b}}$$

With a few lines of code, uniform deviates can be produced from a random number generator and then inverted to produce the Weibull deviates:

```
weibull_inverse<-function(u,a,b){
return(a*(-log(1-u))^(1/b))}
u<-runif(1000)
x<-weibull_inverse(u,1,2)
hist(x)
```

A resulting histogram of 1000 random deviates is shown in the following figure.

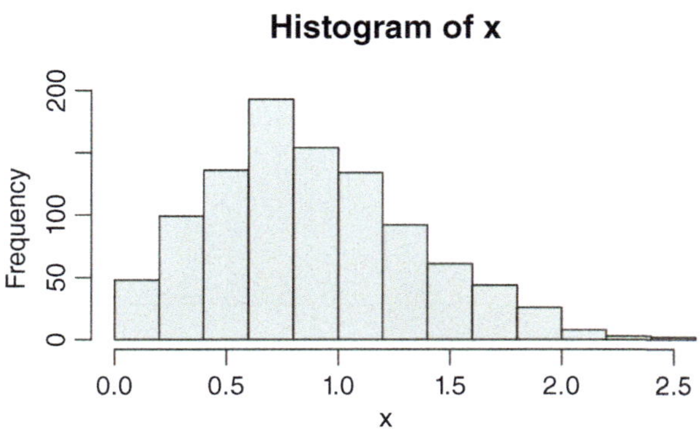

(Continued)

Example 13.3 (Continued)

Note that R (function `rweibull`), as well as Python and Matlab, have built-in functions for directly providing Weibull random variates.

Also, rather than using a pseudorandom number generator (`runif`), a quasi-random number generator (e.g., by the R toolbox `randtoolbox`) could have been used.

A second approach to generating random deviates is the acceptance–rejection function. This can be used when the *density* function $f(x)$ can be evaluated analytically, and it is particularly useful when the distribution is defined only over a finite region.[10] The method can be defined by a series of steps:

1. Let $f_{\max} = \max\ f(x)$ over its range of support $< L, U >$.
2. Draw a random variate (x) over the range $< L, U >$. Draw a second random variate, z, over the range $< 0, f_{\max}$.
3. If $z < f(x)$, accept x as a random variate; otherwise reject it.
4. If the desired number of random variates has been accepted, stop. Else return to step 2.

The downside of the simple acceptance–rejection method outlined earlier is that there are regions in the interval, $< L, U >$, particularly where the distribution extends over a broad range, for which few random draws will result in an acceptance. There are modifications of this approach, using what Hormann [42] terms a "hat" function whose density is everywhere greater than the target distribution, and whose integral (CDF) is invertible.

For the normal distribution, a special procedure, developed by Box and Muller [43], is available. Since the normal CDF cannot be inverted analytically, and since its support is over the entire real line, this algorithm has been frequently implemented in software, although there are alternative methods using a precise numerical representation of the inverse CDF function that have also been implemented [44].

Muller Method for Producing Random Deviates from the Standard Normal Distribution.

1. Draw two random variates from a uniform random distribution, y_1 and y_2.
2. Then x_1 and x_2, computed from the following, are normally distributed random numbers.

$$x_1 = \sqrt{-2\ln(y_1)} \qquad \cos(2\pi y_2)$$

$$x_2 = \sqrt{-2\ln(y_1)} \qquad \sin(2\pi y_2)$$

For bi- and multivariate distributions, there are also multiple approaches for generating pairs or tuples of random variates. Iman and Conover developed algorithms, involving matrix decomposition, that can produce random tuples having a specified Spearman correlation (matrix) using pseudorandom generators. Since the Spearman correlation is invariant to monotone transformation,

10 Even if distributions are defined over the positive real line, they can be truncated at sufficiently large distances from the mean, e.g., 5 standard deviations.

such as from a uniform CDF to an alternate CDF, this approach is useful for any set of marginal distributions. This approach is implemented in many software platforms:

- In R, the `MultiRNG` package can draw multivariate uniform tuples, as well as variates from a variety of other distributions.
- In Python, the `numpy.random.multivariate_normal` function can generate tuples of correlated random deviates. These can be converted to other distributions by first inverting the normal variates to the uniform distribution and then applying the transformation to an alternative distribution by using its CDF.
- Matlab has the function `mvnrnd` that can similarly return tuples of multivariate normal deviates with defined correlations.

An example of generating paired deviates is shown in the R code in Listing 13.4. This is for two variables with a Spearman correlation of 0.7, one with a marginal beta distribution with parameters 0.8 and 0.3 and the other with a gamma distribution with parameters 0.75 and 0.5. Using the `MultiRNG` package, paired uniform random deviates with the given correlation are generated, and then these are used to calculate the values of the two variables according to the beta and gamma distributions, respectively.

```
 library(MultiRNG)
spearman_correl<-0.7
correlation_matrix<-matrix(c(1,spearman_correl,spearman_correl,1),
          nrow = 2,ncol = 2)
print(correlation_matrix)
biuniform_deviates<-draw.d.variate.uniform(1000,2,correlation_matrix)
beta_variable<-pbeta(biuniform_deviates[,1],0.8,0.3)
gamma_variable<-pgamma(biuniform_deviates[,2],shape=0.75,scale=0.5)
plot(beta_variable,gamma_variable)
```

Listing 13.4 Example of generating correlated random variates in R.

The scatter plot generated from this calculation is shown in Figure 13.8.

Copulas provide more flexibility in describing bivariate relationships. An acceptance-rejection method can be used to generate random deviates [45]. The R package `copula` can efficiently generate such deviates as well, as can analogous resources in Python and Matlab.

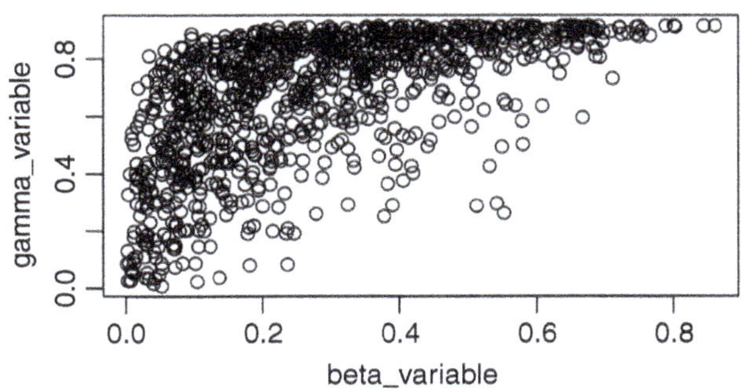

Figure 13.8 Beta and gamma correlated random deviates with a Spearman correlation of 0.7.

13.4 Discussion Questions and Problems

1 From the data in Table 13.1, determine whether this is best described (the log transform gene copy values) by the normal, inverse Gaussian, Weibull, or gamma distributions. For the best and the second best-fitting distribution, determine the upper 95th percentile and compare.

2 The Burr Type III distribution was defined in Table 8.7. Derive the density function, and use this to fit the data in Table 13.4 and determine how the fit using AICc compares to that of the other distributions considered.

3 Replicate the analysis leading to Figure 13.2 with respect to the likelihood-based confidence region for the parameters of the best-fitting lognormal distribution to the Sylvestre data.

4 For the bootstrapped fit of the inverse Gaussian to the Sylvestre data, construct the 95th percentile confidence limits to the cumulated distribution function of oocyst concentrations.

5 Develop the bootstrap distribution for fitting a lognormal distribution to the Sylvestre *Cryptosporidium* data.

6 The numerical value of some correlation coefficients is not invariant to the transformation of variables. Consider the data in Table 13.8. Perform a log transform of both variables, and calculate the Pearson, Spearman, and Kendall τ values. Discuss reasons for any difference from the correlations of the raw concentrations reported in the text.

7 Use the inverse method to generate 1000 random variates from the Burr Type III distribution (Table 8.7). Choose a set of distribution parameters and plot a histogram. Consider a beta distribution with parameters $\alpha = 2$ and $\beta = 3.5$. This is defined over the range $< 0, 1 >$. Use the acceptance–rejection method to compute 1000 random variates and plot a histogram.

8 Using software of your choice, generate 1000 random deviates from Frank's copula with a correlation of 0.8 (which corresponds to a θ of 6.15). Do the same thing with Joe's copula with $\delta = 0.05$ and $\delta = 0.95$. Compare the results of the scatter plots.

References

1 H H Ku. Notes on the use of propagation of error formulas. *Journal of Research of the National Bureau of Standards - C: Engineering and Instrumentation*, 70C(4):263–273, 1966.

2 Milan Holický. *Functions of Random Variables*, pages 79–93. Springer, Berlin, Heidelberg, 2013.

3 Andrew G Glen, Lawrence M Leemis, and John H Drew. Computing the distribution of the product of two continuous random variables. *Computational Statistics & Data Analysis*, 44(3):451–464, January 2004.

4 Wendy L Hobson, Miguel L Knochel, Carrie L Byington, Paul C Young, Charles J Hoff, and Karen F Buchi. Bottled, filtered, and tap water use in latino and non-latino children. *Archives of Pediatrics & Adolescent Medicine*, 161(5):457–461, May 2007.

5 Ryan J Oster, Rasanthi U Wijesinghe, Sheridan K Haack, Lisa R Fogarty, Taaja R Tucker, and Stephen C Riley. Bacterial pathogen gene abundance and relation to recreational water quality at seven great lakes beaches. *Environmental Science & Technology*, 48(24):14148–14157, December 2014.

6 Emily Darby, Adam Olivieri, Charles N Haas, George Di Giovanni, Walter Jakubowski, Menu Leddy, Kara L Nelson, Channah Rock, Theresa Slifko, and Brian M Pecson. Identifying and aggregating high-quality pathogen data: A new approach for potable reuse regulatory development. *Environmental Science: Water Research & Technology*, 9:1646–1653, 2023.

7 Brendon King, Stella Fanok, Renae Phillips, Melody Lau, Ben Van Den Akker, and Paul Monis. Cryptosporidium attenuation across the wastewater treatment train: Recycled water fit for purpose. *Applied and Environmental Microbiology*, 83(5):e03068-16, March 2017.

8 B M Pecson, S C Triolo, S Olivieri, E C Chen, A N Pisarenko, C C Yang, A Olivieri, C N Haas, R S Trussell, and R R Trussell. Reliability of pathogen control in direct potable reuse: Performance evaluation and QMRA of a full-scale 1 MGD advanced treatment train. *Water Research*, 122:258–268, June 2017.

9 United States Environmental Protection Agency. Exposure factors handbook. Technical Report EPA/600/R-09/052, United States Environmental Protection Agency, Washington, D.C., 2011.

10 Linda J Phillips and Jacqueline Moya. Exposure factors resources: Contrasting EPA's exposure factors handbook with international sources. *Journal of Exposure Science & Environmental Epidemiology*, 24(3):233–243, 2014.

11 Neha Sunger and Charles N Haas. Quantitative microbial risk assessment for recreational exposure to water bodies in Philadelphia. *Water Environment Research*, 87(3):211–222, March 2015.

12 Hunter Quon, Divya Ram, Émile Sylvestre, and Kerry A Hamilton. The role of aspiration in Legionnaires' disease: A quantitative microbial risk assessment (QMRA)-based comparison with inhalation exposures. *Environmental Science & Technology Letters*, 12(7):785–791, June 2025.

13 G U Lopez, C P Gerba, A H Tamimi, M Kitajima, S L Maxwell, and J B Rose. Transfer efficiency of bacteria and viruses from porous and nonporous fomites to fingers under different relative humidity conditions. *Applied and Environmental Microbiology*, 79(18):5728–5734, 2013.

14 Huifang Deng, Kitty E Exel, Arno Swart, Axel A Bonačić Marinović, Cecile Dam-Deisz, Johanna W B van der Giessen, and Marieke Opsteegh. Digging into Toxoplasma gondii infections via soil: A quantitative microbial risk assessment approach. *Science of The Total Environment*, 755:143232, 2021.

15 Drew Capone, Aaron Bivins, Jackie Knee, Oliver Cumming, Rassul Nalá, and Joe Brown. Quantitative microbial risk assessment of pediatric infections attributable to ingestion of fecally contaminated domestic soils in low-income urban Maputo, Mozambique. *Environmental Science & Technology*, 55(3):1941–1952, February 2021.

16 Norman L Johnson, Samuel Kotz, and N Balakrishnan. *Continuous Univariate Distributions*, volume 1. Wiley Interscience, New York, 2nd edition, 1994.

17 Norman L Johnson, Samuel Kotz, and N Balakrishnan. *Discrete Univariate Distributions*, volume 2. Wiley Interscience, New York, 2nd edition, 1994.

18 Émile Sylvestre, Michèle Prévost, Patrick Smeets, Gertjan Medema, Jean-Baptiste Burnet, Philippe Cantin, Manuela Villion, Caroline Robert, and Sarah Dorner. Importance of distributional forms for the assessment of protozoan pathogens concentrations in drinking-water sources. *Risk Analysis*, 41(8):1396–1412, August 2021.

19 Larry Wasserman. *All of Statistics: A Concise Course in Statistical Inference*. Springer Science & Business Media, 2003.

20 Art Owen. Empirical likelihood ratio confidence regions. *The Annals of Statistics*, 18(1):90–120, March 1990.

21 William Q Meeker and Luis A Escobar. Teaching about approximate confidence regions based on maximum likelihood estimation. *The American Statistician*, 49(1):48–53, February 1995.

22 Bradley Efron and Gail Gong. A leisurely look at the bootstrap, the jackknife and cross-validation. *The American Statistician*, 37(1):36–48, 1983.

23 Bradley Efron and Robert J Tibshirani. *An Introduction to the Bootstrap*. Chapman and Hall/CRC, 1994.

24 Tim C Hesterberg. What teachers should know about the bootstrap: Resampling in the undergraduate statistics curriculum. *The American Statistician*, 69(4):371–386, October 2015.

25 Charles N Haas. Importance of distributional form in characterizing inputs to Monte Carlo risk assessments. *Risk Analysis*, 17(1):107–113, February 1997.

26 Philip M Dixon. Bootstrap resampling. In Abdel H El-Shaarawi and Walter W Piegorsch, editors, *Encyclopedia of Environmetrics*. Wiley, 1st edition, October 2001.

27 David E Burmaster and Paul D Anderson. Principles of good practice for the use of Monte Carlo techniques in human health and ecological risk assessments. *Risk Analysis*, 14(4):477–481, August 1994.

28 Oxford English Dictionary. correlation, n., March 2025.

29 Shun Dar Lin and Ralph L Evans. Coliforms and fecal streptococcus in the Illinois River at Peoria, 1971-1976. *Report of Investigation (Illinois State Water Survey) RI-093*, 1980.

30 Joseph Lee Rodgers and W Alan Nicewander. Thirteen ways to look at the correlation coefficient. *American Statistician*, 42(1):59–66, 1988.

31 Anthony J Bishara and James B Hittner. Testing the significance of a correlation with nonnormal data: Comparison of Pearson, Spearman, transformation, and resampling approaches. *Psychological Methods*, 17(3):399–417, 2012.

32 Jian-Wen He and Sunny Jiang. Quantification of enterococci and human adenoviruses in environmental samples by real-time PCR. *Applied and Environmental Microbiology*, 71(5):2250–2255, May 2005.

33 Edwin Van Den Heuvel and Zhuozhao Zhan. Myths about linear and monotonic associations: Pearson's r , Spearman's ρ , and Kendall's τ. *The American Statistician*, 76(1):44–52, January 2022.

34 Andrew E Smith, P Barry Ryan, and John S Evans. The effect of neglecting correlations when propagating uncertainty and estimating the population distribution of risk. *Risk Analysis*, 12:467–474, 1992.

35 Chin Diew Lai and Narayanaswamy Balakrishnan. *Continuous Bivariate Distributions*. Springer, 2009.

36 Christian Genest and Anne-Catherine Favre. Everything you always wanted to know about Copula modeling but were afraid to ask. *Journal of Hydrologic Engineering*, 12(4):347–368, July 2007.

37 Maurice J Frank. On the simultaneous associativity of f (x, y) and x+ y- f (x, y). *Aequationes Mathematicae*, 19(1):194–226, 1979.

38 Harry Joe. *Multivariate Models and Multivariate Dependence Concepts*. CRC Press, London, 1997.

39 Robin L Plackett. A class of bivariate distributions. *Journal of the American Statistical Association*, 60(310):516–522, 1965.

40 Ying-Chao Hung. A review of Monte Carlo and quasi-Monte Carlo sampling techniques. *WIREs Computational Statistics*, 16(1):e1637, January 2024.

41 Makoto Matsumoto and Takuji Nishimura. Mersenne twister: A 623-dimensionally equidistributed uniform pseudo-random number generator. *ACM Transactions on Modeling and Computer Simulation (TOMACS)*, 8(1):3–30, 1998.

42 Wolfgang Hörmann, Josef Leydold, and Gerhard Derflinger. *Automatic Nonuniform Random Variate Generation*. Springer Science & Business Media, 2004.

43 George E P Box and Mervin E Muller. A note on the generation of random normal deviates. *The Annals of Mathematical Statistics*, 29(2):610–611, 1958.

44 Christian P Robert and George Casella. *Introducing Monte Carlo methods with R*, volume 18 of *Use R!* Springer, 2010.

45 Charles N Haas. On modeling correlated random variables in risk assessment. *Risk Analysis*, 19:1205–1214, 1999.

Part IV

Balancing and Deciding

14

From Dose–Response to Risk Characterization

In previous chapters, sources of pathogens, their fate and transport from source to receptor, and the effectiveness of physical removal and inactivation processes were discussed. Application of these concepts results in an exposure assessment (distribution). The last step in risk assessment (Figure 3.1) is to combine this exposure assessment with dose–response information to produce an estimate of risk, along with its uncertainty and variability. These steps are the focus of this chapter.

First, the types of dose–response models, their fitting to data, and the evaluation of parametric uncertainty will be discussed. Then the combination of this information with exposure will be addressed.

14.1 Dose–Response

A dose–response relationship connects an individual's exposure to a pathogen with the likelihood that the individual will experience an effect. Alternatively, it may be viewed as connecting the average exposure to a population with the expected proportion of the population that would experience the effect.

It is now essential to specify the different ways in which both the dose (exposure) and the response (effect) may be characterized.

Dose has multiple attributes. One is the scale of interest (individual vs. population). The second is where and how the dose is measured.

To assess individual risk, dose could refer to the average exposure for that individual, i.e., based on the average environmental concentration and average contact with it. Or alternatively, and rarely, it might refer to the actual integral number of organisms constituting an exposure.

At the population level, every member of the population may (though unlikely) be exposed to *exactly* the same number of organisms. Alternatively, the distribution of environmental concentrations and the extent of contact with the environment are known, and the average exposure is known.

The second attribute of dose is where it is measured. Most commonly, dose refers to the immediate amount that an individual or population intakes from the environment. This can be referred to as the *experienced dose*. However, several other methods of dose are conceivable, although rarely elucidated:

- For inhalation exposure, not all organisms that are inhaled are retained—some may be discharged in exhaled breath. So the *retained dose* may be used as a metric.

- Once in (or on) the body, the organism(s) may multiply and possibly disseminate to multiple sites where they can do so, and the total body burden at any time, or the maximum body burden attained, can be used as a dose metric.
- For organisms that cause pathogenicity via excretion of toxins *in vivo*, a time-integrated body burden may be a proxy for the toxin production.

In most cases, the experienced dose is used as a metric because it is the most convenient to measure or control (in well-defined experiments).

Two attributes of outcomes describe response:

- Who are the subjects in which the outcome is being measured?
- What is the effect on the subjects that are being assessed, and how is this being done?

The subjects under study can be either animal or human. In animal studies, generally, these are experiments in which subgroups of animals are exposed to known doses of pathogens, and the effects are noted. A good protocol requires randomizing individual animals to dose groups and providing appropriate housing and ventilation to reduce the impact of contagious agents on cross-transmission between animals.

For contagious agents, if the primary risk is of interest, each individual animal should be housed separately. If the interest is assessing the potential for secondary transmission, several designs are available, such as infecting "donor" animals and then co-housing them with "recipient" animals to evaluate response in the latter group [1]. Another design for agents known or suspected to be transmitted by air is to physically separate the donor and recipient animals, allowing air to flow from the former group to the latter, and then measure outcomes in the latter group [2].

Human subjects can also be used. There are two fundamental types of such studies—experimental and observational.

Experimental studies are similar to those outlined for animals. Human subjects are randomized into dose groups, and each group receives a defined dose, after which the effects are measured. Currently, there are limited types of pathogens that can be used due to ethical considerations: there should be known therapies, and the effects should be mild to moderate. In addition, subjects should be pre-screened for factors that may exacerbate effects or for prior immunity to the pathogen under study. Participation should be voluntary.

There are historical data sources that may still be useful, for example, in prisoners or other confined populations [3, 4]. While these data can be informative, it is crucial for users of such data to be cognizant of the different ethical standards under which they were obtained.

The interactions between pathogens and their hosts can follow a complex trajectory. Casadevall and Pirofski [5], on the basis of a review of the literature, outline the following framework:

- Initial contact between a host and pathogens may result in successful acquisition, which is termed **infection**. For most microorganisms (an exception being helminths), this is accompanied by *in vivo* multiplication of the pathogen.
- Immune processes may result in the **elimination** of the infection.
- If not eliminated, **colonization** can occur. This may result in a spectrum of effects, ranging from none to severe. If there is no effect but continued persistence, this results in a short- or long-term commensal relationship with the host.

The state of infection can be monitored by measuring excretion or respiratory emission of pathogens or, for animals, by necropsy and assay at specific sites. Colonization may result in physiological changes such as biochemical or immune markers, presence of fever, or, if the effects include disease (morbidity), the particular signs and symptoms characteristic of that illness. In the extreme case, the host may die.

Some of these responses may be able to be monitored in members of a population well after the exposure occurred. For example, Frost et al. [6] measured antibodies to *Cryptosporidium* in blood banked samples and related that to the type of drinking water serving the individual blood donor to develop epidemiology correlations.

For most responses, the interval between exposure and response, if available, provides additional data useful for the analysis.

For the actual ascertainment of morbidity, a formal case definition is required as a prerequisite to conducting a study. For example, in the assessment of gastrointestinal effects from recreational water exposure, definitions of "highly credible gastrointestinal illness" have incorporated some combination of the following: (1) vomiting, (2) watery diarrhea, (3) soft diarrhea and abdominal cramps, and (4) nausea and abdominal cramps" [7].

Effects can be either quantal, i.e., all-or-none, or continuous. Examples of the former are ill (by some criterion)/not ill, dead/live, excreting pathogens/not excreting, etc. Examples of continuous data may be levels of a biochemical marker, core body temperature, weight (change), etc. To date, the primary thrust of dose–response modeling has been with quantal data, and this will be assumed from this point. More experience and testing are needed with microbial dose–response data for continuous effect measures.

For quantal dose–response, whether human or animal, whether designed experiment or opportunistic observation, the structure of the data has a common form. If time to effect is not recorded (this will be further discussed later), then the data set consists of one row per group, with the following elements (where "i" is the group index):

$\boldsymbol{d_i}$ is the average dose administered to the group,[1] which is a non-negative real number.
$\boldsymbol{T_i}$ is the number of subjects in the dose group. This is a non-negative integer.
$\boldsymbol{P_i}$ is the number of subjects in the dose group evidencing the effect. For example, this could be the number of subjects ill or dying. Obviously $P_i \leq T_i$.

These data should be accompanied by metadata describing, for example, the health and physiologic status of the hosts, the time at which observation of the effect was made, preparation of the pathogen dose, and details of how the effect was measured. Given such experiments, the question is how well various dose–response models describe the data, which model(s) are preferred, and what the parametric uncertainty in the dose–response model(s) is.

14.1.1 Dose–Response Models

The theoretical derivation of mechanistically based dose–response models for infection has been covered elsewhere [8, 9], and will not be repeated here. Instead, the high-level assumptions and the results will be presented.

There are several levels of dose–response models that have been conceptualized [9], designated as generations 0–3 and beyond generation 3.

1 Or, alternatively, if the exact *number* of organisms is known, $N_{d,i}$ is the number of pathogens administered to each member of the group.

14.1.1.1 Generation 0

This class of models dates back at least a century [10]. The concept is that there is a "minimum infectious dose," which is the most significant value of dose in an experiment such that all subjects administered doses less than that value have zero observed effect.

Unfortunately, this idea has several conceptual flaws, yet some continue to espouse it. For example, it is highly dependent on both the number of subjects tested at each dose and the dose spacing. Its early attractiveness likely stemmed from the thought that a single large aggregate bolus was necessary to overcome the resistance of the host; this has been overwhelmed by many subsequent investigations.

The nature of competing hypotheses was stated succinctly by Meynell and Stocker [11]:

> "The relationship between inoculated bacteria might take one of two extreme forms. Either they could be assumed to be acting co-operatively, death being a consequence of their joint action, or they could be regarded as acting independently, more than one bacterium usually being needed because the probability of a given bacterium being lethal is less than unity... The situation when the LD_{50} dose contains many organisms is analogous to that of a poor marksman firing at a bottle. Since his aim is poor, the bottle is unlikely to have been broken after a small number of shots has been fired but if he persists he will probably hit the bottle eventually. A local observer might be aware that the bottle was broken by the action of one bullet. On the other hand, a distant observer, informed only of the total number of shots fired before the bottle broke, would not be able to exclude the hypothesis that the breakage was due to the accumulated stresses produced by all the bullets fired."

The clear preponderance of evidence has favored the hypothesis of independent action and so forms the basis of more current models.

14.1.1.2 Generation 1 Dose–Response

Generation 1 dose–response models arise from mechanistic considerations, including the following:

- Individual pathogens are discrete.
- *In vivo*, each microorganism has a stochastic probability of surviving to successfully proliferate or die off (a birth–death model).
- The probability of such survival is identical among all pathogens, and the survival of one organism is independent of the survival of other organisms.

Given these assumptions, if the exact integral dose is known (N_d), then applying simple probability laws gives the probability of adverse effect[2] P_e, where r is the survival probability for a single administered organism:

$$P_e = 1 - (1 - r)^{N_d} \tag{14.1}$$

Dose–response equations where the exact integral dose is known have been termed "conditional dose–response" relationships [12].

This approach of Equation 14.1 uses an assumption that the survival of only one organism to initiate an infectious focus is sufficient for the adverse effect. In principle, this can be generalized if

---

2 The term adverse effect will be used to describe the outcome of a specific experiment, which may be infection, morbidity, or death.

more than one survivor is needed [8]; however, data have yet to be encountered that would appear to justify such an approach. If the integral dose is not known, but each host is administered a dose considered to be a Poisson sample having an average of d, then Equation 14.1 is convoluted with a Poisson distribution having that average for N_d. This yields a simple exponential, which is termed the exponential dose–response function (Equation 14.2)

$$P_e = 1 - \exp(-rd) \tag{14.2}$$

The r value in Equation 14.2 is a location parameter. Another way to view the significance of "r" is to relate it to a median effective dose, N_{50}, via

$$N_{50} = -\frac{\ln(0.5)}{r} \tag{14.3}$$

As an aside, in the field of indoor air quality, the Wells–Riley equation has long been used to assess risk in a well-mixed room containing one or more individuals exhaling infectious microorganisms [13]. This is equivalent to the exponential dose–response function as illustrated as follows.

Equivalence Between Wells–Riley Relationship and Exponential Dose–Response Relationship for a Well-mixed Room

Consider a well-mixed room at steady state with a ventilation flow (volume/time) Q, containing l individuals each emitting q infectious quanta per unit time.[3] A susceptible individual is exposed in the room, has a breathing rate Q_B (volume per time), and is exposed for time t in the room. With this formulation, Wells–Riley describes the risk via Equation 14.4.

$$p = 1 - \exp\left[-\left(\frac{l \cdot q}{Q}\right)Q_B \cdot t\right] \tag{14.4}$$

Using the concepts from Section 8.3.3, for a well-mixed volume with a continuous source (the infectors) defined as earlier, the steady state concentration of infectious agents would be given by Equation 14.5, where ζ is the number of viable infectious pathogens per infectious quanta.

$$C = \frac{\zeta l \cdot q}{Q} \tag{14.5}$$

Substituting this result into the exponential dose–response relationship produces Equation 14.6.

$$p = 1 - \exp\left[-r\left(\frac{l \cdot q \cdot \zeta}{Q}\right)Q_B \cdot t\right] \tag{14.6}$$

Hence, it is concluded that, under the underlying assumptions, the two approaches are identical in form.

In the author's opinion, the standard exponential is preferable, as the Wells–Riley approach conflates the physical nature of the room and its ventilation with the dose–response relationship and does not readily extend to more complex exposure situations or dose–response models.

3 By definition, an infectious quantum that is breathed by a susceptible individual will have a $1 - \exp(-1) = 0.632$ probability of causing illness.

The assumption of a constant survival probability, r, may not always be realistic. There may be variability due to heterogeneity in the pathogens or the hosts,[4] or both. Therefore, r itself may be given by a probability distribution. Furumoto and Mickey [14, 15] were apparently the first to suggest that the beta distribution, being one with support over the interval <0,1>, is the most appropriate.

Beta Distribution

The density function, which is defined for arguments between zero and one, of the beta distribution can be written as follows [16]:

$$P(r) = \frac{r^{\alpha-1}(1-r)^{\beta-1}}{B(\alpha, \beta)} \tag{14.7}$$

where α and β are parameters that are greater than zero and B is the mathematical beta function defined as:

$$B(\alpha, \beta) = \frac{\Gamma(\alpha)\Gamma(\beta)}{\Gamma(\alpha+\beta)} \tag{14.8}$$

The mean and variance of this distribution in terms of the parameters are:

$$\mu(r) = \frac{\alpha}{\alpha+\beta} \tag{14.9}$$

$$\sigma^2(r) = \frac{\alpha\beta}{(\alpha+\beta)^2(\alpha+\beta+1)} \tag{14.10}$$

If the constant "r" conditional dose–response relationship (Equation 14.1) is convoluted with a beta distribution for the parameter, then the beta conditional dose–response distribution results:

$$P_e = 1 - \frac{B(\alpha, \beta + N_d)}{B(\alpha, \beta)} \tag{14.11}$$

Parenthetically, there are generalizations [17] of the beta distribution with three or more parameters that permit more flexibility; however, it is unclear whether fitting to available dose–response experimental data would benefit from such extensions.

While in certain laboratory studies, the exact number of organisms administered to test subjects is known, this is rare. For assessing risks to populations, it is also more common to know the average concentration from which an exposure occurs. Therefore, the conditional dose–response distribution must be combined with another distribution that describes the frequency with which a discrete number of organisms can be sampled from a system at a given average dose.

This topic was discussed in Section 4.1 and employed in the analysis of microbial enumeration data in Section 6.2. A baseline model for the distribution of discrete organisms in a well-mixed system is the Poisson. In the context of discussion here, the distribution of discrete doses of microorganisms from a system with an average amount per dose[5] "d" is given by:

$$P(N_d) = \frac{d^{N_d}}{N_d!}\exp(-d) \tag{14.12}$$

---

4 It is impossible based solely on the characteristics of the dose–response relationship to differentiate among these sources of heterogeneity [12].

5 Note that, as discussed previously, if the average concentration of organisms in the medium is known, and if the amount of medium, e.g., volume, constituting the exposed is known, then d is given by the product of these two quantities.

Then the product of the Poisson distribution for microorganisms can be multiplied by the conditional dose–response function and summed over all values of N_d to obtain dose–response functions as a function of the mean dose. With the constant "r" conditional dose–response, the result is the exponential dose–response model of Equation 14.13.

$$P_e = 1 - \exp(-rd) \tag{14.13}$$

In a similar manner, if the beta conditional dose–response distribution is combined with a Poisson distribution of microorganisms, the exact beta-Poisson dose–response model results (Equation 14.14).

$$P_e = 1 - {}_1F_1(\alpha, \alpha + \beta, -d) \tag{14.14}$$

Evaluation of the Confluent Hypergeometric Function

Not all platforms have direct support to evaluate ${}_1F_1$. However, R and Python (among others) do:

R: The `gsl` library (Gnu Scientific Library) has the function `hyperg_1F1`.

Python: The function `scipy.special.hyp1f1` can be used.

The function ${}_1F_1$ is the confluent hypergeometric function [18]. When the exact beta- Poisson model was developed, its numerical evaluation was relatively intractable, and therefore an approximate form was developed by Furumoto and Mickey [15] given in Equation 14.15. This is termed the approximate beta-Poisson model.

$$P_e = 1 - \left(1 + \frac{d}{\beta}\right)^{-\alpha} \tag{14.15}$$

The approximation is reasonable for large values of β or risk. The adequacy of this approximation will be illustrated as follows and was discussed in detail by Teunis and Havelaar [19]. Particularly in the construction of confidence limits for the dose–response curve at low risk, the differences could become significant.

It is convenient to redefine the approximate and exact beta-Poisson models in terms of parameters N_{50} and α by making a substitution via Equation 14.16. The parameter N_{50} has the meaning of the median effective dose, and α can be regarded as a slope parameter.

$$\beta = \frac{N_{50}}{2^{\frac{1}{\alpha}} - 1} \tag{14.16}$$

With these substitutions, the exact beta-Poisson equation and the approximate beta-Poisson equation become, respectively:

$$P_e = 1 - {}_1F_1\left(\alpha, \alpha + \frac{N_{50}}{2^{\frac{1}{\alpha}} - 1}, -d\right) \tag{14.17}$$

$$P_e = 1 - \left[1 + \frac{d}{N_{50}}\left(2^{\frac{1}{\alpha}} - 1\right)\right]^{-\alpha} \tag{14.18}$$

It should be noted, as was stated in prior chapters, that the distribution of microorganisms in an environment may not be random, as described by the Poisson, and may be either more variable or less variable. Other mixing distributions, besides the Poisson, may be combined with either the fixed "r" or beta-distributed conditional dose–response model. Among these distributions could be the negative binomial, Poisson-lognormal, or inverse Gaussian. Further discussion on these more complex relationships between effect and average dose has been presented elsewhere [12].

A comparison of the exponential and the beta-Poisson models is shown in Figure 14.1. Several aspects of the plots are noteworthy:

- On the semi-log plot (left-hand panel), it is clear that the effect of α is to produce a shallower dose–response curve as its value decreases. Hence, for the same N_{50}, the risk at low dose is greater for beta-Poisson than for exponential.
- Only looking at the left-hand plot, there is an appearance of a "threshold," i.e., an average dose below which the risk is zero. However, this is clearly not the case when the same curves are plotted on a log-log scale (right). Hence, both the exponential and beta-Poisson models are no-threshold models. This appears to be true for all models built from the foundational mechanistic assumptions. This results from the concept that there is a finite, nonzero probability (however small) that one organism persists and initiates an infection.
- For the exponential and beta-Poisson models, at sufficiently low dose, there is a linear relationship between dose and response, as shown by the 1:1 slope on a log-log plot (right-hand panel). This arises from the model basis that only one persisting organism is required to initiate an effect. If multi-hit dose–response models were used, which have yet to be justified by any data, then low-dose linearity would not hold [8].

The development of the exponential and the beta-Poisson models can be viewed as special cases of a generalized framework defined by Equation 14.19.

$$P_e(d) = \sum_{k_{\min}=1}^{\infty} \sum_{k=k_{\min}}^{\infty} \sum_{j=k}^{\infty} P_1(j|d)p_2(k|j)P(k_{\min}) \tag{14.19}$$

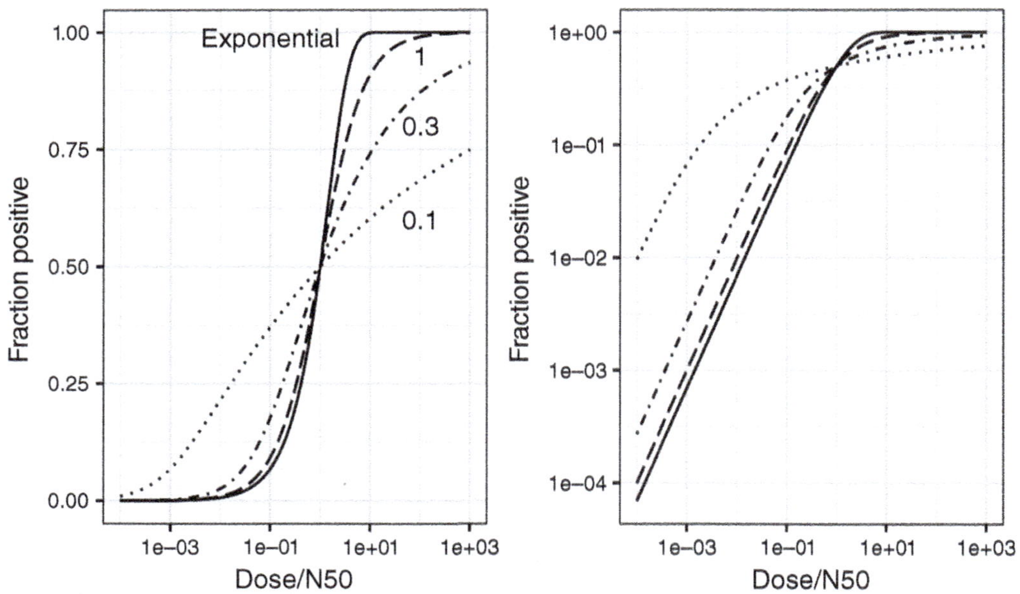

Figure 14.1 Exponential vs. beta-Poisson dose–response on semilog (left) and log-log (right) scales. Legend on left indicates value of α or exponential.

In this equation, the various terms are as follows:

$P(k_{\min})$ is the fraction of subjects that require $k_{\min}$ organisms to persist to initiate infections. In the case of the exponential and beta-Poisson, these are point distributions at $k_{\min} = 1$; alternatives could be point distributions at other values, Poisson, truncated Poisson, etc.

$P_1(j|d)$ is the fraction of subjects exposed to an average dose d who actually have an administered dose of j organisms. In exponential and beta-Poisson distributions, this is assumed to be Poisson-distributed; alternatives could be negative binomial, Poisson log-normal, Poisson inverse Gaussian, or others.

$P_2(k|j)$ is the fraction of subjects with an administered dose of j organisms, of which k organisms persist to initiate infection. This was assumed to be an independent and identically distributed probability, and a generalization of this assumption could require a more complex framework.

A dose–response model is fundamentally a mathematical function that takes as an argument a non-negative real number (the dose) and returns a value between 0 and 1 (the response). If a further restriction is made that the function be monotonically non-decreasing with its argument, then in principle, any cumulative distribution with support over the positive real line can be considered as a potential dose–response model.[6] However, such models are not necessarily mechanistically derivable, nor do they always exhibit low-dose linearity and non-threshold behavior. Additionally, experience to date does not show that such models provide (statistically) superior descriptions of microbial data compared to the others discussed here. Nevertheless, three can be cited that have been used for comparison.

Log-probit. This model is based on the cumulative standard normal distribution function $\Phi(x)$ and has two positive constants q_1 and q_2 as follows:

$$P_e = \Phi\left[\frac{1}{q_2}\ln\left(\frac{d}{q_1}\right)\right] \tag{14.20}$$

Log-logistic. This model is based on the logistic distribution and contains two constants: q_1 can be any real number, and q_2 must be non-negative:

$$P_e = \frac{1}{1 + \exp[q_1 - q_2\ \ln(d)]} \tag{14.21}$$

Weibull. The Weibull distribution has been discussed in multiple contexts previously. As a dose–response model with non-negative parameters, q_1 and q_2, it can be written as:

$$P_e = 1 - \exp(-q_1\, d^{q_2}) \tag{14.22}$$

The process of selecting a dose–response model and determining the best-fitting parameters and their uncertainties will be covered in Section 14.1.2. Once a model is selected and its best-fit parameters are estimated, these can be used to generate a point estimate of risk given a known exposure. Or alternatively, if the desired risk to be attained is specified, the corresponding exposure can be determined. This is illustrated by Example 14.1.

6 Parenthetically, it should be noted that this is not only true for a microbial dose–response model but also for a chemical dose–response model.

Example 14.1 Point Estimate of Exposure Given Fitted Dose–Response Model and Specified Risk

The dose–response model for inhalation of *Legionella pneumophila* was found to be exponential with $r = 0.06$ based on animal data for administered dose [20]. If the risk from an exposure is to be kept at or below 10^{-5}, what is the maximum acceptable exposure?

Solution

Equation 14.13 can be rearranged to solve for average exposure to yield:

$$d = -\frac{1}{r}\log(1 - P_e)$$

Since P_e is specified, and r is known, the exposure dose can be calculated:

$$d = -\frac{1}{0.06}\log(1 - 10^{-5}) = 7.24 \cdot 10^{-5}$$

14.1.1.3 Generation 2 Dose–Response with Modulating Factors

These models are built from Generation 1 models by expressing the values of the dose–response parameters, e.g., r for the exponential and α and N_{50} for the beta-Poisson, as functions of other characteristics of either the pathogen or the host. The functions may be qualitative, in other words by a characteristic descriptor of the interaction, or based on quantitative characteristics.

An example of an innate qualitative factor is the effect of the FUT2 (fucosyltransferase 2) gene on susceptibility to norovirus infection [21]. Individuals who express this gene are susceptible to infection. Non-expression, and this non-susceptibility, is estimated to occur in 20% of North American/European Caucasian and 2% of Meso-American/Hispanic populations [21].

It is possible that susceptibility to other pathogens can be modulated by (as yet unknown) genetic factors. In such cases, the overall population dose–response can be regarded as a mixture of multiple individual dose–response relationships for each of the genetically distinct subpopulations, or alternatively, the conditional risk to each subpopulation can be computed separately.

Host physiological factors can qualitatively influence susceptibility to infection. Not all of these may be known. But some examples of these influences are:

- It has long been known that infection with *Cryptosporidium* in individuals with AIDS and not on antiretroviral therapy is frequent [22], although whether this is due to intrinsic shifts in the dose–response relationship, or the greater severity once infected, is not known.
- With many ingested bacterial pathogens, stomach pH can serve as a barrier to further penetration into the gastrointestinal tract. Dietary habits, such as excessive dairy consumption or the use of antacid medications that increase stomach pH, may facilitate further penetration into the gastrointestinal tract and promote infection by these agents [23].
- Individuals with Type 2 diabetes are more susceptible to infection with *Burkholderia pseudomallei* , the agent causing a potentially severe tropical disease, melioidosis [24]. This has also been confirmed by differential dose–response curves in animal models [25].

Concomitant exposures to infectious agents with chemical or physical contaminants can increase the ability of those microorganisms to persist and cause adverse effects. This has been demonstrated for various chemical and physical air pollutants and a diverse spectrum of respiratory pathogens [26]. Whether these interactions occur with more frequency, including for gastrointestinal pathogens, is unclear, and the impact of the intrinsic human host microbiome on this interaction needs to be better understood [27].

In a few cases, quantitative models for the interaction with modulating factors have been developed. In the case of inhaled pathogens, it is well known that different-sized aerosols have different deposition fractions and also may have deposition in different areas of the respiratory tract (Figure 14.2). Depending upon the pathogen, greater biological effects may occur based on where the organisms are deposited, for example, alveoli, upper respiratory tract, etc. These effects have been documented in several studies for pathogens. Note also that the site of deposition and infection can affect the potential for exaltation of organisms grown in the respiratory tract back into the environment and potentially infecting additional individuals (Section 7.3). While the exact shapes of these curves differ with breathing rate, health status, nose vs. mouth breathing, etc., the broad features are generally applicable, i.e.:

- There is typically a size at which the overall deposition is a minimum between 0.1 and 1μm.
- Smaller particles below this minimum tend to deposit deeper in the respiratory tract, while particles at sizes above this minimum tend to deposit in the upper respiratory tract.

There are at least two approaches to consider the effect of particle size on the administered dose in the host response.

One approach that can be used when underlying experimental data are available on administered (inhaled) doses for different particle sizes is to modify either the exponential or the beta-Poisson model with a particle size-dependent parameter. Teske et al. [29] used this approach for

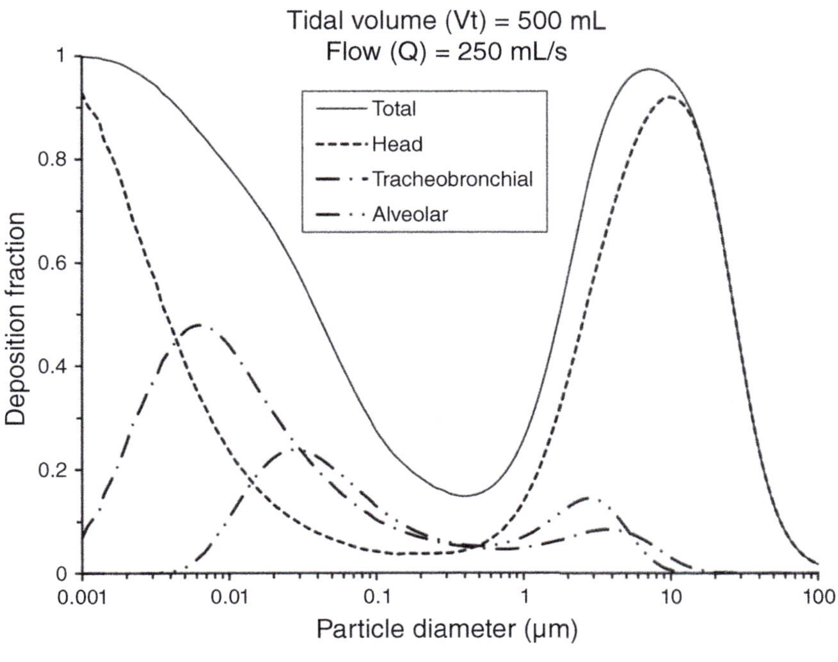

Figure 14.2 Deposition fraction of particles in different portions of the human respiratory tract from nasal inhaled exposures. *Source:* [28]/U.S. Environmental Protection Agency/Public domain.

fitting inhaled doses of aerosolized *Francisella tularensis* to animals using the following relationship between r in the exponential model and particle diameter (d_p):

$$r = (A \cdot d_p)^{-B} \tag{14.23}$$

An advantage of this approach over the second approach to be discussed is that if there are actual experiments with administered particles of different sizes, not only does the size dependency correct for the physics of deposition, but it also implicitly accounts for the differential impacts of immune and clearance responses, depending on location within the respiratory tract.

A second approach is to correct the administered dose to the delivered dose at the *in vivo* location where infection will occur. Curves such as from Figure 14.2 make this correction; for example, if the inhaled particles containing microorganisms are 1 μm in diameter, and the site of infection is the alveoli, then less than 10% of the inhaled organisms (administered dose) will be deposited. More detailed modeling can be done of particle deposition *in vivo* (for either human or animal approaches) in several ways to get such a ratio:

- The respiratory tract can be subdivided into multiple compartments, and a compartment (box) transport model between them can be formulated.[7] In one example, three compartments, nares to larynx, larynx to main bronchi, and main bronchi to alveoli were used to model deposition of inhaled spores of *Bacillus anthracis*[30].
- More detailed compartment models, or CFD models of the respiratory tract can be used to predict particle deposition [31].

An advantage of using delivered dose, rather than administered dose, for inhaled pathogens is the greater potential for fidelity in extrapolating from animal to human studies to assess exposure and thereby risk.

The age of the host can be an important modulator of the response to pathogens. For the exponential dose–response function, the parameter r can be written as a function of host age, in a manner similar to that in which inhaled particle size was treated as a modulating factor. With the beta-Poisson dose–response, either or both of the α and N_{50} parameters can be written as a function of host age. This latter approach was found to be useful in describing the age dependency of sensitivity to intraperitoneally injected variola major virus (smallpox virus) in suckling mice (neonate to 168-hour age). The following modification was found to be useful, where τ is the host age [32]:

$$\alpha = \alpha_0 + \alpha_1 \cdot \tau \tag{14.24}$$

$$N_{50} = N_{50,0} \cdot \exp(b \cdot \tau) \tag{14.25}$$

What was noteworthy was that the median infectivity decreased (N_{50} increased) with host age, and that α increased with host age, reflecting reduced variability as the suckling mice aged. Application of host age modulated dose–response models was more mixed when tested against data sets on equine encephalitis virus [33].

14.1.1.4 Generation 3 Dose–Response Models with Dynamics

Generation 3 dose–response models predict not only the proportion of subjects who will experience an adverse effect but also the distribution of times following dose administration at which such an effect will manifest. The temporal distribution of effect can be important in assessing the impact on medical response activities and, for contagious agents, will also influence disease spread

7 This fundamentally uses the same approaches as the box models discussed in Section 8.3.1.

in a population (Section 14.3). This incubation time has long been known to be a skewed distribution. Sartwell [34] found the lognormal distribution to be a useful descriptor. However, he did not test alternative distributions, and more recent workers noted that other skewed distributions could provide a superior fit to empirical outbreak data [35, 36].

The effect of incubation time and its variability can affect the consequences of exposure to a population. This is illustrated by a hypothetical example in Figure 14.3. Both curves have a total case burden of 1000 cases. However, one case is delayed and less peaked. The consequence of population exposure may be more than a linear function of total cases; it may also depend on the daily case rate. This was a motivation behind the strategy of "flattening the curve" during the COVID-19 pandemic [37]. Furthermore, in the case of contagious diseases, the potential for secondary transmission may be a function of the instantaneous case prevalence.

For a simple growth model of microorganisms in the host, recalling Figure 8.19, there will be a distribution of *in vivo* growth trajectories. If infection requires that at least a certain critical number of organisms be reached, then there will be a distribution of times at which this occurs, corresponding to a phenomenological incubation time. In random processes, this has been termed a "first passage" problem [38], and in the context of microbial infection, one of the first analyses was provided by Williams [39, 40]. Although this approach yielded an incubation time distribution as a function of initial dose, testing of this model against the Sartwell data found statistical identifiability problems [41].

The first passage approach depicts adverse responses resulting from the internal body burden of microorganisms crossing some particular level and has been successful to date in describing the temporal distribution of effects. But other approaches may be useful. Where the adverse effect occurs from the accumulation of damage, perhaps by release of toxic products *in vivo*, the time integral of body burden (i.e., area under the curve) may be applicable in describing time to effect. Further data and models are needed for this approach.

An alternative approach is to modify Generation 1 models to incorporate time after dose as a modulating parameter. Another approach, to be discussed under the beyond Generation 3 models, is to describe, in finer detail, the physiological balances that occur *in vivo* between microbial growth and host responses against that growth.

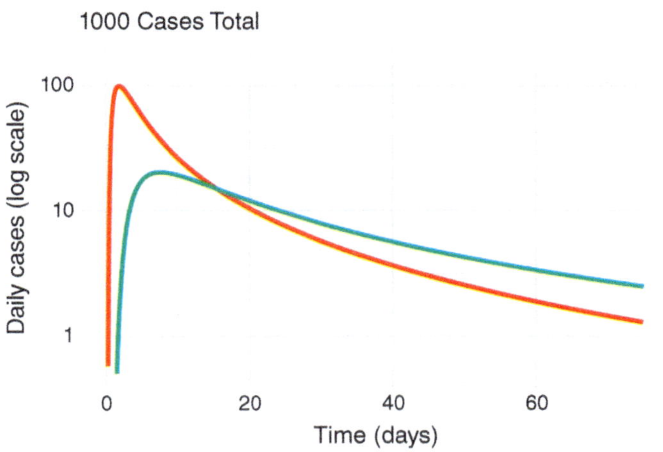

Figure 14.3 Effect of incubation time distribution on the case distribution.

Incorporation of a time-dependent modulating function for either r (for the exponential dose–response) or N_{50} (for the beta-Poisson) by a time to response parameter was elaborated upon by Huang and Haas [42]. Increasing the dose–results in a *decreasing* time to observe an adverse effect. Several one- and two-parameter extensions of the exponential and beta-Poisson models were developed and found to fit a large number of datasets in which the time to adverse effect was measured. These functions are shown in Table 14.1.

It is noteworthy that when the time dependence was coupled to the beta-Poisson model, it affected only N_{50}, not α. Although there are very few datasets measuring time to effect *and also* body burden during the incubation phase, testing these models against such data shows correspondence between the initial dose/time-to-effect relationship and the internal body burden resulting from exposure [43].

As an example, data on *Francisella tularensis* 2.1 μm aerosols in monkeys were fit by an exponential dose–response with exponential power time dependency and best-fit parameters of $a_0 = -2.934$, $a_1 = -375.659$, and $a_2 = -3.572$ with time in days (from 1 to 21) [42]. Plots of the model predictions are shown in Figure 14.4, with the left plot showing probability of response versus time for several doses, and the right plot showing the density function (versus time) for the effect. As dose increases, as expected, response increases (left). Also, the incubation time distribution (right) is skewed, and as the dose increases, the mode shifts to the left and the distribution becomes more peaked (less variable).

14.1.1.5 Beyond Generation 3

The second approach to incorporating time to effect is to quantitatively describe internal host fate and transport processes that modulate growth and/or serve as barriers to microorganisms' movement to sites within a host where they can infect. When microbial growth overcomes inhibitory modulating forces, adverse effects can occur.

Physical transport from the initial portal of entry until a site of potential infection can form one such barrier. This has been shown clearly for inhaled pathogens, where regions of the respiratory tract can each be modeled as a completely mixed compartment, with flows between compartments and potentially growth and/or loss in each compartment can be included.

In one application, the inhalation of spores of *Bacillus anthracis* spores was modeled using three compartments: nasopharynx, tracheobronchial, and pulmonary (including bronchioles and alveoli), and this was modeled using a dynamic Markov Chain[8] approach [30]. On a steady-state basis, this allows determination of the ratio between the administered dose and the deposited dose (in this case, in the alveolar regions), which are the primary sites of infection for this pathogen. As importantly, having such a dynamic model provides a platform for determining the time-integrated delivered dose in the presence of a variable administered dose — for example, a susceptible host in

Table 14.1 Modifications of dose–response models for time to effect (τ).

Model	Equation for r or N_{50} vs. time to effect (τ)
Exponential	$\exp(a_0 + a_1 \cdot \tau)$
Exponential-reciprocal	$\exp\left(a_0 + \frac{a_1}{\tau}\right)$
Exponential-power	$\exp(a_0 + a_1 \cdot \tau^{a_2})$

Source: Modified from [42].

8 Recall the use of a Markov Chain approach in modeling microbial growth and decay from Section 8.4.4.

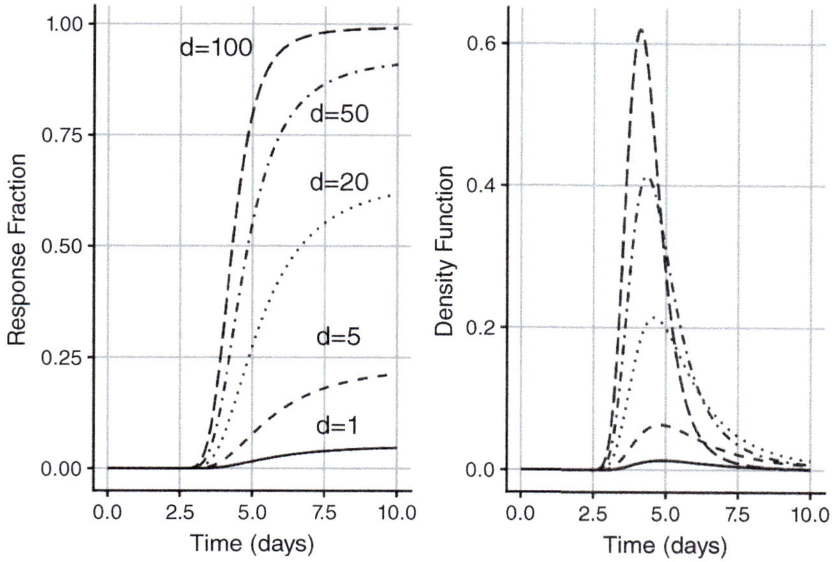

Figure 14.4 Plot of fitted dose–response time model to 2.1 μ *Francisella tularensis* model. The left panel is the cumulative response, and the right-hand panel is the density function for the response. *Source:* [42]/with permission of John Wiley & Sons.

the vicinity of an infected individual who may be releasing pathogen-laden particles via episodic coughing or sneezing. Basing dosimetry on the delivered dose also results in closer correspondence of dose–response relationships across animal species.

Akpeimeh et al. [44] developed an extended Markov Chain model by also incorporating surfaces in different respiratory regions with the mucociliary clearance process, resulting in movement of pathogens from the respiratory to the gastrointestinal tract. They applied this to estimating exposure (and ultimately risk) among workers at municipal solid waste disposal facilities to the fungus *Aspergillus fumigatus* and *E. coli*.

Transport between *in vivo* locations is only one type of process that is relevant to detailed understanding of fate and transport of pathogens within a host. Julien et al. [45] introduced the "key events dose-response framework" (KEDRF), in which the spectrum of within-host processes from the administered dose to successful infection is outlined. In addition to physical transport between sites, the intervention and interaction with various physiological defense mechanisms are also included. This framework was initially developed with a particular focus on ingested foodborne contaminants — both chemical and biological. Several examples illustrating aspects of this concept for pathogens via other exposure media have been developed and are summarized below.

Rahman et al. [46] developed a two-region model for the infection by ingested *Listeria monocytogenes*. Ingestion of the organisms was followed by inactivation in the stomach by acidity. Subsequently, the persistence in the intestine was modeled using logistic growth kinetics (Equation 8.69) coupled with a pathogen concentration-dependent inactivation rate of Michaelis–Menten form. The resultant model gave reasonable fits to the data.

For infection by pathogenic *Salmonella*, the process was conceptualized as ingestion and passage through the esophagus, entry into the stomach with inactivation as a function of gastric pH, and entry into the small intestine with potential infection of the intestinal epithelium, inactivation by bile fluid, interaction with other microflora, and potential multiplication [47].

The dynamics of ingested *Campylobacter jejuni* are similar to those for *Salmonella*. Abe et al. [48] developed a model considering transit through the stomach, through the intestine, and transfer and invasion to intestinal cells. Included in this framework was consideration of the vehicle in which the pathogen was contained (food versus liquid), and the effect of that vehicle on the gastric pH and the retention time in each compartment. They also considered differentials in these characteristics between adult and elderly populations and found reasonable correspondence with infectivity observed in experimental trials.

Leptospira is a pathogen that can be transmitted to humans from contact with urine from infected animals (often rodents). Gostic et al. [49] developed a dose–response model incorporating the breaching of the upper layers of the skin, allowing for colonization and ultimate systemic infection. Using this model, they were also able to harmonize dose–response data from other routes and provide a model that was in concordance with epidemiological data.

A key aspect of infection and ultimate disease is the successful competition between the pathogen and the host immune system. This is being modeled with increasing detail.

Pujol et al. [50] developed a generic model in which the *in vivo* pathogens interact with "immune particles." The presence of pathogens induces the formation of these particles, while their interaction with the particles results in the destruction of both. They then developed a dose–response model that could be used for either a bolus (single) dose or a time-varying set of multiple doses. However, further work on calibrating these types of models using complementary *in vivo* measurements, rather than only the overall outcome (e.g., infection), would be desirable.

Coxiella burnetii is a gram-negative bacterium and the causative agent of Q fever, which can be persistent in the environment and transmitted via multiple routes. Heppell et al. [51] developed a model for risk from inhalation of these organisms, including deposition into the lung, uptake by lung macrophages, multiplication, and inactivation by immune processes. They included host variability in each of these steps, described the process by a series of differential equations, developed a time-dependent variant of the exact beta-Poisson model, and showed concordance with experimental studies on dose–response and incubation time.

It has become apparent that the human body harbors a rich set of mostly non-pathogenic microorganisms, termed the human microbiome, which can modulate the impact of pathogens [52]. Alterations in the normal microbiome by other factors, including nutrition, medication, chemical toxins, or age, can alter pathogen responses. However, approaches for quantitatively considering these are still under development. Akritidou et al. [53] reported that diet impacted the persistence and multiplication of *Salmonella* and *Listeria* in an *in vitro* simulated intestinal model, presumably by, in part, altering the dynamics of the microbiome components.

14.1.2 Fitting Dose–Response Models

Experimental dose–response studies can be used to fit model parameters, assess model fit, and compare potential models. In addition, the uncertainty of the parameters given the data can be determined. These tasks are analogous to those discussed for fitting quantal enumeration data (Section 6.2.2), fitting growth and decay models (Section 8.4), and fitting inactivation data (Section 12.1.2).

For single bolus dose–response experiments in which response is measured only by presence or absence, such as infection, illness, or death, the data elements are typically:

- Average dose of microorganisms administered to group "i," d_i. The actual number each member of a group receives is assumed to be Poisson-distributed.
- The total number of subjects in group "i," T_i
- The number of subjects in group "i" who have the selected response, P_i

Given this, a data table can be constructed in tidy data format Section 12.1.

As an example, Ward et al. investigated responses in human volunteers to calibrated doses of ingested rotavirus and determined response by infection (determined by shedding of virus, sero-conversion, or both) [54]. The data obtained are shown in Table 14.2 in tidy format.

These data are quantal and can be fitted using the likelihood approach, as discussed several times previously. Assuming the responses of subjects in a dose group are identical and independent from each other, then a negative likelihood function can be written as Equation 14.26.

$$\ln(L) = -\sum_i \left[p_i \ln(\pi_i) + (T_i - P_i)\ln(1-\pi_i) \right] \tag{14.26}$$

In this equation, π_i is the predicted fraction positives at dose i given from the dose–response model to be fit with a set of parameters, i.e., r for the exponential or α and N_{50} for the approximate beta Poisson model. The best-fit parameters are those that minimize the value of Equation 14.26. The best possible model would be one that passes exactly through the observed points; hence, if $\pi_i^o := \frac{P_i}{T_i}$, then this saturated model negative log likelihood would be:

$$\ln(L^o) = -\sum_i \left[p_i \ln(\pi_i^o) + (T_i - P_i)\ln(1-\pi_i^o) \right] \tag{14.27}$$

Using these relationships, fitting and assessment proceed via the following steps, using concepts defined for previous quantal data analysis (Section 12.1.2):

- The optimum value(s) of the dose–response parameters for the select model(s) are determined by minimizing Equation 14.26.
- For each model fit, using the optimized likelihood, the value of AICc can be computed using Equation 6.10, in which N is interpreted as the total number of subjects summed over all doses, and k is interpreted as the number of fitted dose–response parameters (1 for the exponential, 2 for the beta-Poisson, etc.). The AICc for the saturated model can be computed using k as the number of dose groups.
- Among the models considered (including the saturated model), the one that yields the lowest value of AICc should be regarded as the best. If the saturated model *does not* have the lowest AICc, it indicates the superiority of one of the actual models.
- A formal goodness of fit for the selected best model can be performed by computing $\Delta = 2[\ln(L) - \ln(L^o)]$. A null hypothesis of adequate fit is *rejected* if $Delta$ exceeds the critical χ^2 statistic for degrees of freedom equal to the number of dose groups minus the number of parameters.

Table 14.2 Human response to rotavirus.

Dose (PFU)	Total subjects	Positive subjects
90000	3	3
9000	7	5
900	8	7
90	9	8
9	11	8
0.9	7	1
0.09	7	0
0.009	7	0

Source: Data from [54].

Table 14.3 Model fit and dose–response parameter estimates for ward rotavirus data (N = 59).

Model	k	$-\log L_{max}$	$\hat{\alpha}$ or k	$\hat{N}_{50}$	AICc
Saturated	8	19.66	—	—	58.20
Exponential	1	86.95	0.001088 (k)	637 (ED_{50})	175.94
Approx. β-Poisson (α, N_{50})	2	23.0655	0.265	5.597	50.35
Exact β-Poisson (α, N_{50})	2	22.181	0.173	10.091	50.06

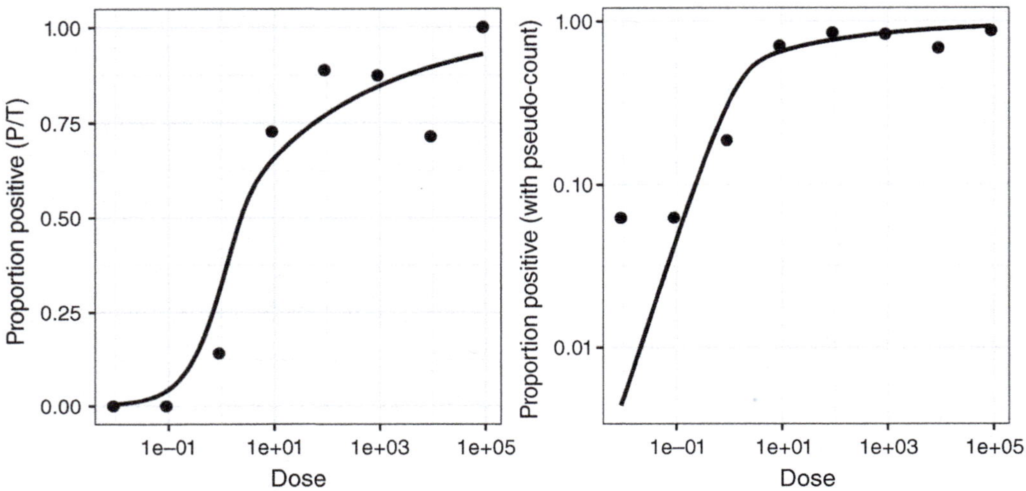

Figure 14.5 Observed proportion of positives compared to exact beta-Poisson best fit. Left panel on semilog scale. Right panel on log-log scale (with zero responses assumed to be one divided by the total subjects at that dose).

An example of this approach is shown for the rotavirus data in Table 14.2. This is fit to the exponential, the approximate beta-Poisson, and the exact beta-Poisson model, with results shown in Table 14.3. Python code for executing the fitting is shown in Listing 14.1. To assure convergence, a small tolerance is used along with exact large-digit floating-point precision.

Examination of the fitting shows that the beta-Poisson models are much better (lower AICc) than the exponential. Furthermore, the exact beta-Poisson model is slightly superior to the approximate (due to the small value of α). The exact beta-Poisson model has a lower AICc than the saturated model, so it is accepted. A formal likelihood goodness-of-fit test gives $\Delta = 2[22.181 - 19.66] = 5.042$, which is less than the upper 95th percentile of the χ^2 distribution (12.592), so the null hypothesis of adequacy of fit cannot be rejected.

Plots of the fitted exact beta-Poisson dose–response function compared to observed fraction positives are shown in Figure 14.5. There is good qualitative agreement between the observed data and the fitted dose–response model.

```
import numpy as np
import pandas as pd
from math import log, exp
from scipy.optimize import minimize
```

```
from mpmath import hyp1f1, mp

mp.dps = 80

dose = np.array([90000, 9000, 900, 90, 9, 0.9, 0.09, 0.009], dtype=float)
T = np.array([3, 7, 8, 9, 11, 7, 7, 7], dtype=int)
P = np.array([3, 5, 7, 8, 8, 1, 0, 0], dtype=int)
n = int(T.sum())
LN2 = log(2.0)

def binom_negloglik(pi):
    pi = np.clip(pi, 1e-12, 1-1e-12)
    return -np.sum(P*np.log(pi) + (T-P)*np.log(1-pi))

def saturated_loglik(T, P):
    s = 0.0
    for t,p in zip(T,P):
        if 0 < p < t:
            s += p*log(p/t) + (t-p)*log((t-p)/t)
    return s

def AICc(negloglik, k, n):
    AIC = 2*k + 2*negloglik
    return AIC + (2*k*(k+1))/(n - k - 1)

LL_sat = saturated_loglik(T,P)

def fit_with_restarts(fun, starts, bounds):
    best = None
    for x0 in starts:
        res = minimize(fun, x0=np.atleast_1d(x0), method="L-BFGS-B",
                       bounds=bounds,
                       options={"maxiter": 5000, "ftol": 1e-14})
        if (best is None) or (res.fun < best.fun):
            best = res
    return best

def safe_c(alpha):
    if alpha <= 0:
        return np.inf
    val = LN2/alpha
    if val > 40:
        return np.inf
    return exp(val) - 1.0

# Exponential
def nll_exp(theta):
    k = theta[0]
    if k <= 0 or not np.isfinite(k):
        return np.inf
    pi = 1 - np.exp(-k * dose)
    return binom_negloglik(pi)

# Approximate beta-Poisson (alpha, N50)
def nll_abp_alphaN50(theta):
    alpha, N50 = theta
    if alpha <= 0 or N50 <= 0 or not np.isfinite(alpha) or not np.isfinite(N50):
        return np.inf
    c = safe_c(alpha)
    if not np.isfinite(c) or c <= 0:
        return np.inf
    x = (c/N50) * dose
```

```
    logterm = np.log1p(x)
    pi = 1.0 - np.exp(-alpha * logterm)
    return binom_negloglik(pi)

# Exact beta-Poisson (1F1)
def nll_exact_1F1_alphaN50(theta):
    alpha, N50 = theta
    if alpha <= 0 or N50 <= 0 or not np.isfinite(alpha) or not np.isfinite(N50):
        return np.inf
    c = safe_c(alpha)
    if not np.isfinite(c) or c <= 0:
        return np.inf
    beta = N50 / c
    if not np.isfinite(beta) or beta <= 0:
        return np.inf
    pis = []
    for d in dose:
        val = 1.0 - float(hyp1f1(alpha, alpha + beta, -d))
        val = min(max(val, 1e-12), 1-1e-12)
        pis.append(val)
    pi = np.array(pis)
    return binom_negloglik(pi)

# Run fits
exp_starts = [1e-6, 1e-5, 1e-4, 5e-4, 1e-3, 2e-3, 5e-3]
res_exp = fit_with_restarts(nll_exp, exp_starts, bounds=[(1e-12, None)])

abp_starts = [(a, N) for a in [0.03, 0.05, 0.07, 0.1, 0.2]
              for N in [100, 300, 600, 1000, 1500, 3000]]
res_abp = fit_with_restarts(nll_abp_alphaN50,
              abp_starts, bounds=[(1e-9, None), (1e-9, None)])
exact_starts = [(a, N) for a in [0.03, 0.05, 0.07, 0.1, 0.2]
              for N in [300, 600, 800, 1000, 1500, 3000]]
res_exact_1f1 = fit_with_restarts(nll_exact_1F1_alphaN50, exact_starts, bounds=[(1e-9, None), (1e-9, None)])

# Compile results
rows = []
rows.append({"Model":"Saturated","k":8,"-logLmax":-LL_sat,"alpha_or_k":np.nan,
               "N50":np.nan,"AICc":AICc(-LL_sat,8,int(T.sum()))})
khat = float(res_exp.x[0])
rows.append({"Model":"Exponential","k":1,"-logLmax":float(res_exp.fun),
               "alpha_or_k":khat,"N50":float(LN2/khat),"AICc":float(AICc(res_exp.fun,1,int(T.sum())))} )
alpha_abp, N50_abp = map(float, res_abp.x)
rows.append({"Model":"Approx. beta-Poisson (alpha, N50)","k":2, "-logLmax":float(res_abp.fun),"alpha_or_k":alpha_abp, "N50":N50_abp,"AICc":float(AICc(res_abp.fun,2,int(T.sum())))} )

alpha_ex, N50_ex = map(float, res_exact_1f1.x)
rows.append({"Model":"Exact beta-Poisson 1F1 (alpha, N50)","k":2,"-logLmax":float(res_exact_1f1.fun),"alpha_or_k":alpha_ex,"N50":N50_ex,"AICc":float(AICc(res_exact_1f1.fun,2,int(T.sum())))} )
results_df = pd.DataFrame(rows, columns=["Model","k","-logLmax","alpha_or_k","N50","AICc"])
results_df.to_csv("final_results_1F1_correct.csv", index=False)
print(results_df)
```

Listing 14.1 Python code for fitting Ward rotavirus data.

Table 14.4 Test of pooling multiple strains.

Strain	Subjects	Deviance	Parameters	AICc
Alaska 6/77	8	0.01	2	
Washington 897/80	27	5.89	2	
Korea 1/82	14	1.89	2	
Bethesda 85	36	4.70	2	
Total	85	12.49	20	65.62
All pooled	85	31.79	2	35.94

Source: Data from [55]. Adult human responses to attenuated influenza H3N2 strains.

If there are experiments with multiple strains or under multiple conditions, each experiment can be fit separately to a dose–response model. Then it can be determined whether the ensemble of experiments is suitable for a single dose–response relationship. This is analogous to the pooling test for inactivation models discussed in Section 12.1.6.

To perform this test, each experiment is fit separately to a dose–response model, and its $-\ln(L)$ is determined. The sum of these individual likelihoods is then taken and termed $(-\ln(L_{\text{individual}}))$. This can be used to determine an $AICc_{\text{individual}}$.[9] Then the entire set of experiments is fit to a common model, optimum, with the optimum negative log likelihood determined, $-\ln(L_{\text{pooled}})$. The pooled AICc is then computed with k equal to the number of parameters in the common dose–response model. If the AICc for the pooled fit is lower than the AICc of the individual fits, pooling into a common model is deemed appropriate.

This procedure is illustrated with data from Watanabe et al. who reviewed available human dose–response experiments using influenza virus [55]. Results of computations are shown in Table 14.4. Adult human testing was performed on responses to doses of influenza virus H3N2 of different strains. Each strain separately could be fit to the approximate beta-Poisson model, and the resulting deviance computed.[10] For the fit to the models separately, the total deviance of 12.49, with 20 parameters (five strains, each with two parameters for the beta-Poisson), yields an AICc of 65.62. When all the data are pooled and fit to a single approximate beta-Poisson model, the resulting deviance and AICc are 31.79 and 35.94, respectively. Since the AICc for the pooled fit is (much) less than the AICc for the individual fits, it is concluded that the pooled model is superior and should be adopted.

Generation 2 data differs from Generation 1 by the incorporation of an additional characteristic observation, such as particle size, host age, etc., that modulates a dose–response relationship, such as the expression of the exponential "r" in terms of particle size as in Equation 14.23. In this case, the data contains an additional descriptor, and the dose–response model to be fit incorporates that modulating descriptor with additional parameters to be fit to the data. Analysis of model comparisons and goodness of fit (relative to a saturated model) proceeds as in the case of a Generation 1 model.

Table 14.5 shows an example data set from Teske et al. [29] on rhesus monkey response to aerosol exposure to *Francisella tularensis* of different diameters. The dose metric used is the deposited dose.

---

9 With N equal to the total number of subjects in the set of experiments, and k equal to the sum of the number of parameters for each experiment.

10 Deviance is defined as $-2[\ln(L) - \ln(L_o)]$, where L_o is the likelihood of the saturated model. Since the values of L_o are not dependent on fit, the AICc can be calculated by using the deviance directly in the AICc formula.

Table 14.5 Dose–response data for rhesus monkey exposure to aerosolized *Francisella tularensis* of different particle sizes.

Aerosol particle diameter (μm)	Deposited dose	Subjects	Positives
2.1	4	6	1
2.1	9	6	3
2.1	25	6	4
2.1	52	6	6
7.5	157	6	3
7.5	471	6	4
7.5	1446	6	6
7.5	2892	6	6
12.5	306	6	3
12.5	628	6	5
12.5	1510	7	6
12.5	16424	3	3
24.0	74	6	0
24.0	445	6	1
24.0	1180	6	2
24.0	5918	5	4

Source: Data from [56] as summarized in [29].

Table 14.6 Response of mice to interperitoneal injection of *Yersinia pestis*.

		Deaths observed on day									
Animals	**Dose (CFU)**	**0**	**1**	**2**	**3**	**4**	**5**	**6**	**7**	**8**	**9**
20	2	0	0	0	0	0	0	0	0	0	0
20	8	0	0	0	1	2	1	0	0	0	1
20	26	0	0	0	1	6	3	2	0	1	0
10	74	0	0	0	4	5	0	0	0	0	0
10	257	0	0	0	2	8					

Source: Data summarized by Huang et al. [57] from [58].

These were fit to the exponential model using the power law dependence of the dose parameter on size (Equation 14.23).

Generation 3 models incorporate the time to response as an observed variable. A summary of a common layout is shown in Table 14.6. Note that this is not in "tidy" form, for ease of presentation. The typical experiment will not observe the precise time at which an effect occurs, but only an interval. For example, at the dose of 8 CFU, one death was observed on day 3 that happened in the interval between the end of day 2 and the start of day 3. Due to the complexity of these experiments, the detailed likelihood formulation for fitting will not be presented here, but can be found in the original papers [42, 57].

14.1.3 Multiple Exposures

The presentation to this point has centered on estimating the response when individuals are challenged with a single dose of a pathogen. In real exposures, pathogen doses may occur multiple times. For example, there may be a small, possibly sporadic, concentration of a pathogen in drinking water to be ingested or in the air that is breathed. Therefore, the question arises as to how to describe the risk resulting from such a time course of exposure. There are very few actual controlled experimental trials that shed light on these situations.

The most parsimonious modeling approach is to assume that each exposure results in a risk that is independent of prior and subsequent exposures [59]. In this case, if there are "j" separate exposures, each resulting in a risk of P_j (from a dose d_j), then the combined risk can be written (using probability laws of independent events) as Equation 14.28.

$$P_t = 1 - \prod_j (1 - P_j) \tag{14.28}$$

If each of the risks, P_j (and each corresponding dose, d_j is small, then the product can be expanded and all but the first linear terms discarded, to yield, under these conditions, an additive relationship in risk:

$$P_t = \sum_j P_j \tag{14.29}$$

Under these same conditions, with all d_j small, both the exponential and beta-Poisson models are in the linear range, and the total dose can be written as a simple sum of the individual doses, i.e.:

$$d_t = \sum_j d_j \tag{14.30}$$

If the dose–response relationship is exponential, then, with independence, the total risk can be obtained by substituting the total dose into the exponential, as in Equation 14.31.

$$P_t = 1 - \exp(-r \cdot d_t) \tag{14.31}$$

If neither the individual doses and risks are small, nor the exponential dose–response relationship holds, then to evaluate each value of P_j in Equation 14.28, the dose–response function must be evaluated for each value of d_j. If the individual doses are the same for every exposure and equal to d', then for the approximate beta-Poisson, there is a straightforward relationship given by Equation 14.32.

$$P_t = 1 - \left[1 + \frac{d_t}{j \cdot N_{50}}(2^{1/\alpha} - 1)\right]^{-j \cdot \alpha} \tag{14.32}$$

The impact of splitting a single bolus into multiple doses if the approximate beta-Poisson model is applicable and if each exposure is regarded as equal and independent is shown in Figure 14.6. Note that at low (total) dose, there is minimal effect of dose splitting, as would be expected from the additivity property at low doses (Equation 14.30). At higher doses, dose splitting increases the risk compared to a single bolus, with an upper limit set by the exponential dose–response.

There are few published data sets on controlled multiple exposures to test the assumption of independence. However, there are several reasons why this assumption might not (always) be correct:

- Infection, and ultimately any adverse effects, result from the *in vivo* multiplication of pathogens. Depending on the spacing of multiple doses, there may or may not be a substantial impact on the time course of the "body burden" from exposure.

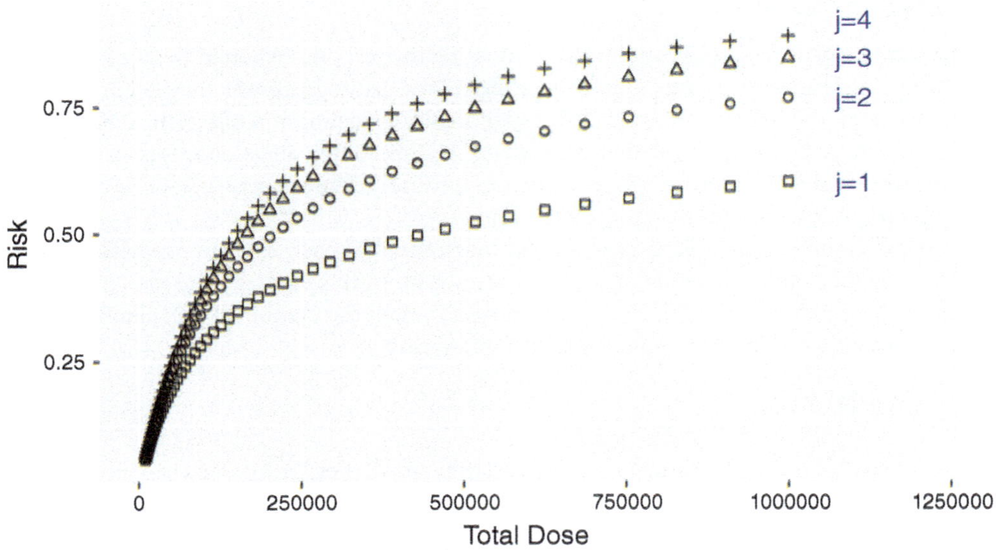

Figure 14.6 Effect of dose splitting on approximate beta Poisson risk. Number of individual doses given as "*j*." Dose-response parameters from Watanabe et al. [55] for adult response to influenza virus H3N2.

- If the time for host response to a dose via immune system upregulation is short, then successive doses beyond the initial (one or few) may be less effective. An approach such as in Pujol et al. [50] might be used to describe such processes.
- For some pathogens, a prior exposure may result in enhanced sensitivity to subsequent exposures. The classic example of this is dengue fever, resulting from a flavivirus [60], which can also be presensitized by an exposure to the related Zika virus.

14.1.4 Uncertainty in Dose-Response Models

Fitting a dose-response model to a data set yields the best-fit parameters. However, as with any parametric model, these parameters represent the best estimates of the true parameters, and it is desirable to estimate their uncertainty. Since, when combining dose-response and exposure to obtain an overall characterization, it is desirable to obtain the actual sampling distribution for these parameters, the most applicable method is bootstrapping, as discussed previously in Section 13.3.2.

To conduct a bootstrap analysis, a large number of bootstrap pseudosamples are constructed, and at each sample, the best-fitting dose-response parameter(s) is/are determined. Then the distribution of the parameter(s) across the pseudosamples provides an estimate of the overall uncertainty in these quantities. Construction of a pseudosample for dose-response data is illustrated in Figure 14.7.

A pseudosample is constructed by keeping the total number of subjects at each dose the same as in the original sample, but randomly sampling from the observed affected and unaffected at each dose.[11]

Using the exact beta-Poisson model, the rotavirus data (Table 14.2) bootstrap distribution of parameters was determined. A scatter plot from 500 trials[12] is shown in Figure 14.8. Note that there

11 This is equivalent to drawing a random integer from a binomial distribution with total members equal to the total subjects at that dose, and a proportion positive from the observed proportion positive.

12 In actual applications, it would not be uncommon to use 10,000 bootstrap trials.

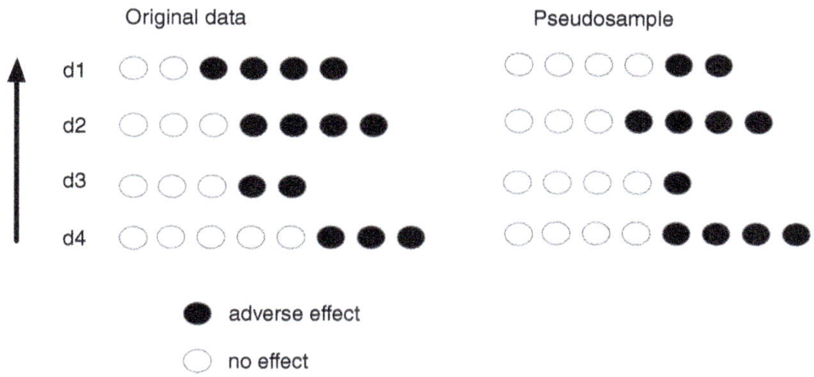

Figure 14.7 Schematic of construction of bootstrap pseudosamples from a dose–response experiment. Doses increasing from d4 to d1.

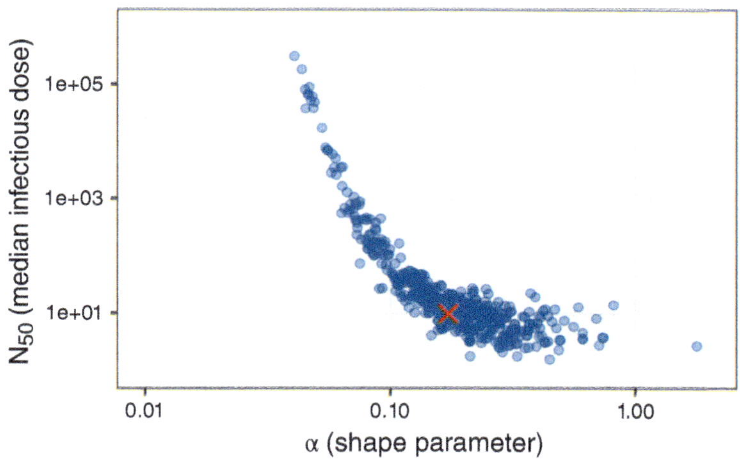

Figure 14.8 Bootstrap parameters for exact beta-Poisson fit to rotavirus data. 500 Trials (44 trials outside the range of the plot). Best fit to original data shown with "X."

is a small proportion of bootstrap trials that have an extreme combination of very small values of α and very large values of N_{50}, which would produce a very shallow dose–response relationship. Also note that this bivariate distribution has a complex form, not likely to be well described by a functional relationship and therefore the set of bootstrapped points itself can be used as points for the empirical distribution. The shape of a bootstrapped sampling distribution will depend in significant degree on the nature of the underlying experiment, and how well it serves to precisely define model parameters.

Using the bootstrapped distribution of parameters, it is possible to get confidence limits to the dose–response curve by the following approach:

- Define a vector of (average) doses for construction of the confidence region.
- At each dose in the vector, use the set of bootstrapped parameters to calculate the associated probability of response.

- If the upper and lower "z" percentile confidence limits are desired, where 1-z is the proportion to include, then trim the upper and lower z/2 fraction from the probabilities and accept the included range as the confidence range for that dose.
- Repeat at all doses, and construct the locus of points defining the upper and lower range of the dose–response relationship.

A significant use of the bootstrapped dose–response parameters is that their empirical distributions serve as one of the inputs for integrating with exposure assessment to construct a risk characterization. For a single-parameter dose–response curve, such as the exponential, the distribution of "r" values from the bootstrap is used. For two (or more) parameter dose–response relationships, the set of parameters, e.g., $\langle \alpha, N_{50} \rangle$, form samples from a bi(multi)variate distribution.

14.2 Combining with Exposure

Risk characterization results from combining exposure assessment, with its variability and uncertainty (Chapter 13), and dose–response assessment, with its variability and uncertainty (Section 14.1). At a very basic level, point values can be used to generate a point estimate for the characterized risk; however, modern best practices focus on using full input probability distributions to obtain the output distribution of risk, incorporating the uncertainty and variability of the inputs.

A complete risk characterization requires identifying which inputs are to be characterized using point values versus distributions. For each of the latter, using either available information or expert judgment, the distribution (with its parameters and any uncertainty therein) is determined. Then a Monte Carlo approach (Figure 9.9) can be used to construct the estimated risk distribution. The process will be shown by example.

14.2.1 Example Risk Characterization

A scenario is considered in which a wastewater treatment plant discharges into a river, downstream of which is a recreational beach. The treatment plant has primary and secondary treatment, but no disinfection. It is desired to estimate the risk to swimmers at the beach to determine the estimated risk and to ascertain whether additional disinfection is needed, such as effluent disinfection. The individual inputs to this characterization are described sequentially from source to receptor.

14.2.1.1 Virus Concentration in Raw Wastewater

Li et al. [61] measured viruses in raw wastewater at locations in Switzerland and analyzed other data for locations in the United States and Japan. Either qPCR or dPCR was used. The distributions were found to fit the lognormal. In the case of rotavirus (measured at only one facility), they reported an arithmetic mean concentration of 9700 gene copies per mL, with a lognormal σ of 2.3 (based on natural log units). This would correspond to a (natural) log mean of 6.535.

14.2.1.2 Removal of Virus in Treatment

The removal of culturable enteric virus by activated sludge treatment of wastewater was studied in two full-scale treatment plants [62]. The median and lower 5th percentile LRVs were found to be 2.2 and 0.85 $\log_{10}$ units, respectively. Assuming a lognormal distribution, this corresponds to a ln mean of 5.07 and a lognormal σ of 1.89. It will be assumed that this is applicable to the removal of rotavirus.

14.2.1.3 Dilution with River

Upon discharge, the effluent will be diluted with the flow of the river,[13] which is presumed not to contain rotavirus. If the dilution factor is defined as the ratio of effluent flow to the sum of effluent flow plus river flow, this factor multiplies the effluent concentration to get the concentration after mixing. This will be a function both of the volumetric flow of the wastewater effluent (and its variability) and the river flow and its variability. For this example, assume the dilution factor (d_f) is given by a beta distribution[14] with parameters $\alpha = 1.659$ and $\beta = 2.537$.

14.2.1.4 Dieoff Between Discharge and Swimming Location

It is presumed that the contaminants in the river are conveyed in a plug flow manner; otherwise, an estimate of the dispersion coefficient would be needed. The virus will decay via first-order kinetics (k_d), and the travel time between the discharge is given by θ. These will be treated as independent distributions.

The first-order die-off rate in hr^{-1} will be described by a gamma distribution with a shape parameter of 6.25 and a scale parameter of 0.08 hr. The transit time (θ) will be described by a Weibull distribution with shape parameter of 3.304 and scale parameter of 26.754 hr. These parameters will depend on the biological, chemical, and morphological characteristics of the river, as well as climatic conditions. With the assumption of plug flow, the concentration of pathogens in the river at the point where swimming occurs is given by Equation 14.33, considering the effect of both dieoff and dilution, where C_{eff} is the pathogen concentration at the end of wastewater treatment.

$$C = C_{\text{eff}} \cdot d_f \cdot \exp(-k_d \cdot \theta) \tag{14.33}$$

14.2.1.5 Ingestion of Water by Swimmers

The amount of water ingested by swimmers (defined as full immersion of the head in water) depends on the water body, duration of swimming, gender, and age. In this example, informed by the work of Schets et al. [63], this will be treated as a constant value of 28 mL.

14.2.1.6 Dose–response Relationship

For the dose–response relationship, the empirical bivariate bootstrap distribution for the parameters of the exact beta-Poisson dose–response model depicted in Figure 14.8 will be used.

The computation can be conducted using a variety of tools. R has a library `mc2D` for conducting Monte Carlo risk assessment, although it has not been recently updated. It is possible to do the calculations in standard R or Python. For this example, Python will be used.

In Listing 14.2, the input distributions are combined to conduct the Monte Carlo calculations. One of the inputs is the set of dose–response parameters from the bootstrap analysis — the 500 bootstrapped pairs of $\langle \alpha, N_{50} \rangle$, which were saved to a file, and are now read in to the Monte Carlo program; a random integer is used to select the pair of parameters (thereby preserving the correlation) that are used as input.

---

13 It is assumed that this occurs instantaneously. In a real situation, depending on hydraulics, this mixing may occur over time and distance downstream.

14 Since the beta distribution is defined for a variable over the range $\langle 0, 1 \rangle$, it is convenient for this application. For more complex distributions, a generalized beta or a mixture of beta distributions might be used.

```
import numpy as np
import scipy as sp
import pandas as pd
import matplotlib.pyplot as plt

MCtrials=10000
dose_response_parameters = pd.read_csv("BootstrapData.csv")
# ---------
raw_wastewater_virus= np.exp(np.random.normal(6.535, 2.3, MCtrials)) # per mL
LRV=np.random.normal(5.07,1.89,MCtrials) # Natural log basis
dilution=np.random.beta(1.659,2.537, MCtrials)
dieoff_rate=np.random.gamma(6.25, 0.08, MCtrials) #units of 1/hr
transit_time=sp.stats.weibull_min.rvs(3.304, 26.37, size=MCtrials)  #units of hr

concentration_at_swimmers=(raw_wastewater_virus*np.exp(-LRV)*dilution
        *np.exp(-dieoff_rate*transit_time))
# ---------
dose=concentration_at_swimmers*28
DRindex = np.random.randint(1, 500+1,MCtrials)
alpha=dose_response_parameters.iloc[DRindex-1,0]
N50=dose_response_parameters.iloc[DRindex-1,1]
beta=N50/(2**(1/alpha)-1)
risk=1-sp.special.hyp1f1(alpha, alpha+beta, -dose)
log10_risk=np.log10(risk)
log10_risk = log10_risk[np.isfinite(log10_risk)]
# ---------
plt.rcParams.update({
    "font.size": 14,
    "font.family": "serif"})
plt.hist(log10_risk,bins=50)
plt.xlabel("log10(Risk)")
plt.ylabel("Number of 10,000 Simulations")
print(np.mean(risk))
```

Listing 14.2 Risk characterization for recreational water with wastewater effluent.

The histogram of $\log_{10}$ values of risk (per an individual exposure) is shown in Figure 14.9. While the mean of this distribution is 0.028, it is clear that there is a large probability of greater risks, approaching certitude. This suggests that an intervention above and beyond secondary treatment would be needed to protect public health. This could include disinfection or the inclusion of polishing ponds.

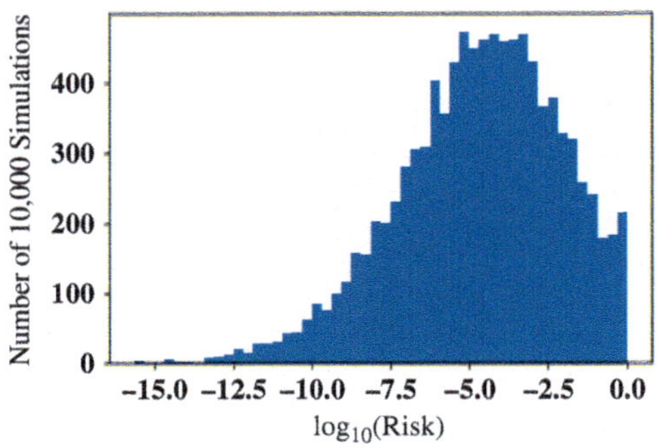

Figure 14.9 Distribution of $\log_{10}$ risk from 10,000 simulations.

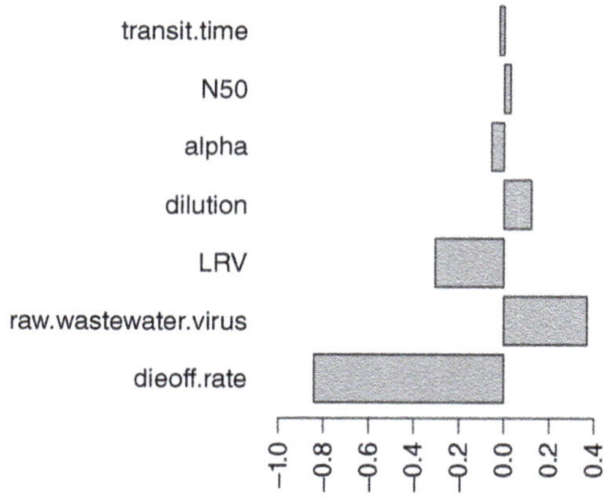

Figure 14.10 Tornado plot of Spearman rank correlation coefficients for inputs to rotavirus risk characterization from 10,000 Monte Carlo simulations.

During the simulation, the results of the 10,000 simulations are preserved, along with the random draws from each of the primary input variables. The Spearman rank correlation between the output (risk) and each of the inputs can be calculated, and the results displayed in a tornado plot shown in Figure 14.10. Examination of this plot shows that the largest magnitude correlations are for the die-off rate, the raw wastewater virus concentration, and the LRV. To reduce the imprecision of the risk distribution, it would be prudent to focus on increasing the certainty of the estimates of these three inputs, i.e., reducing the variance of the distributions used for their input. The impact of the heterogeneity in the dose–response parameters, α and N_{50}, is substantially less.

This example treated each of the inputs, except for the dose–response parameters, as independent. Should there be correlated inputs, then the random draws in the Monte Carlo simulation need to account for such effect, possibly using a copula approach as discussed in Section 13.3.4. There are several mechanisms in the aforementioned example that could result in such correlations, including:

- Increasing in dilution could result from larger upstream flow, which could in turn change the die-off rate and/or reduce the travel time downstream.
- If the concern was with the total number of cases, then the number of recreators could be affected by the stream flow rate (perhaps increased due to recent increased precipitation).

14.3 Integrating to Populations

For the assessment of consequences, a key outcome is not merely the risk *per individual*, but the total number of people experiencing adverse effects, possibly the severity of the adverse effects and the duration of those effects. In addition, the temporal distribution of those effects unique to microbes is important (Figure 14.3).

It is therefore essential to know how many people may be exposed when, and also for contagious diseases, the dynamics of spread among people not necessarily exposed to the initial source of infection. This requires the use of disease transmission models.

While what are regarded as infectious diseases were known to the ancients, the earliest systematic consideration of disease dynamics appears to be Bernoulli's 1766 paper on smallpox [64]. However, the modern theories of disease transmission dynamics commenced with the recognition of the germ theory of disease; Serfling has provided a history of these developments [65]. Foundational to these approaches were works by Lotka [66] and Kermack and McKendrick [67]. These approaches treated a population as existing in several states depending on its properties with respect to an infectious disease and characterized the population in each state and the flows between states as continuous quantities governed by a set of differential equations.

As an aside, this approach can also be used stochastically, given the fluctuations intrinsic to small populations; the foundation of this work was presented by Bailey [68]. The mathematical framework uses coupled differential-difference equations.

The presentation here will rely upon the deterministic approach. Frameworks for this approach have been set forth in a review by Dietz and Schenzle [69] and a monograph by Anderson and May [70].

The simplest example is illustrated by the flow diagram in Figure 14.11. A population (which is assumed to be constant,[15] with no births, deaths, immigration, or emigration) is comprised of individuals in one of three states: susceptible, infectious, and recovered. It is also assumed that all infectious individuals are symptomatic. This is termed the SIR model. The arrows indicate the rate processes, which will be quantified more formally:

- Susceptible individuals can become infected by contact with infectious individuals.
- Infectious individuals can recover and no longer be infected or infectious.
- Depending on the disease, hence the dotted arrow, recovered individuals may have only transient immunity and may return to becoming susceptible.

There are several limitations of the basic SIR model, particularly for pathogens relevant to this book. Firstly, the SIR model in Figure 14.11 assumes that any successful exposure renders an individual immediately infectious, and secondly, it assumes preexisting infectious case(s) in the population and does not consider new cases resulting from environmental contamination.

With respect to the first issue, Cooke [71] was apparently the first to introduce a delay in the conversion to an infectious state following exposure [72]. In current terminology, this introduces what has been termed an exposed state according to the state diagram of Figure 14.12.

Burke [72] has discussed the use of the term "exposed" in the SEIR model, which is potentially misleading. Of individuals who potentially come into contact with an infectious agent or infectious agents released by an individual in proximity, not all will ultimately progress to symptomatic or asymptomatic infection. Therefore, what is termed the SEIR model, in Burke's analysis, should rather be termed an SLIR model (where L stands for "latent"). In subsequent discussion, the SEIR nomenclature will be used, with the understanding that the "E" compartment represents a latent state.

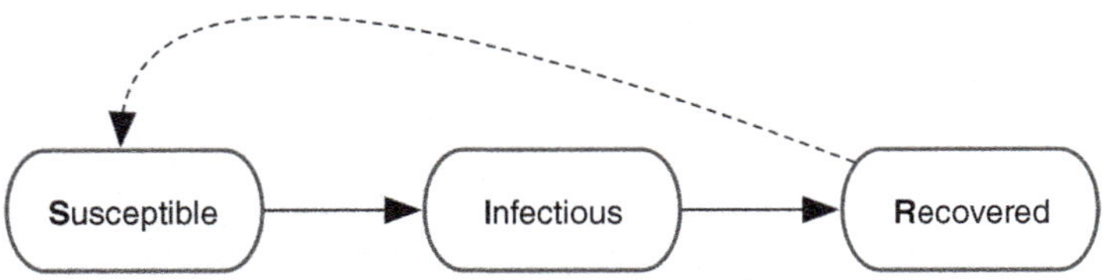

Figure 14.11 Basic SIR model with possibility of incomplete immunity.

15 This is reasonable for relatively short times; however, this assumption can be relaxed by considering immigration, emigration, births, and mortality both from the target disease and from other causes.

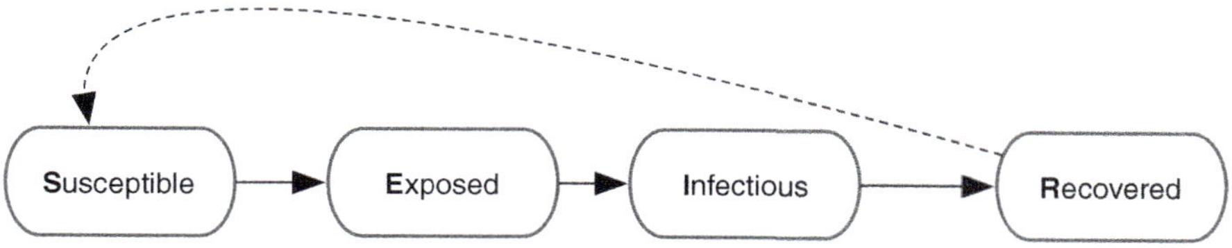

Figure 14.12 Basic SEIR model with possibility of incomplete immunity.

The term latent has a precise definition that differs from other temporal characteristics of disease progression, as defined by Anderson and May [70]. Following the time at which an exposure occurs:

- The **incubation time** is the period before symptoms occur.
- The **latent time** is the period before an individual becomes infectious.

Furthermore, the incubation time and the latent time may be different; the time when symptoms are evident and the time during which an individual is infectious may not coincide, and the incubation+symptomatic period may differ from the latent+infectious period.

The second issue concerns how the rates of formation of the exposed (latent) state are quantified. First, the overall set of differential equations describing the SEIR model in Figure 14.12 will be written, and then the rates of exchange between states will be defined.

Considering each of the four states, and defining rates between compartments, assuming each state is homogeneous, the following four differential equations can be written:

$$\frac{dS}{dt} = -r_{SE} + r_{RS} \tag{14.34}$$

$$\frac{dE}{dt} = r_{SE} - r_{EI} \tag{14.35}$$

$$\frac{dI}{dt} = r_{EI} - r_{IR} \tag{14.36}$$

$$\frac{dR}{dt} = r_{IR} - r_{RS} \tag{14.37}$$

Note that $\frac{dS}{dt} + \frac{dE}{dt} + \frac{dI}{dt} + \frac{dR}{dt} = 0$, indicating a constant total population (over longer time periods, additional terms reflecting births, deaths, immigration, and emigration would need to be added). To completely specify this model, values or functions of each of the five r values are required.

The classical SIR and SEIR models, as defined, assume the presence of one or more exposed or infectious individuals in a population, who then transmit the disease to susceptible individuals. The transmission was presumed to occur by direct or near-direct contact between the infectious cases and the susceptible individuals.

In the context of the discussion in this book, this framework omits several essential features that are our primary concern and represent important points of intervention. Except in situations such as sexually transmissible diseases, which are out of scope here, inevitably transmission occurs when an infectious case emits pathogens into the environment, such as by excretion, exhalation, etc., as discussed in Chapter 7. A susceptible individual then acquires a dose of pathogens via inhalation, ingestion, or contact. Subsequently, that individual has a probability (defined by the dose–response relationship) of acquiring an infection, with the propensity of being infectious. To describe this process, and to connect the environment, dose–response models, and the SEIR approach, an extended model termed the environmentally mediated infectious disease (EMID)

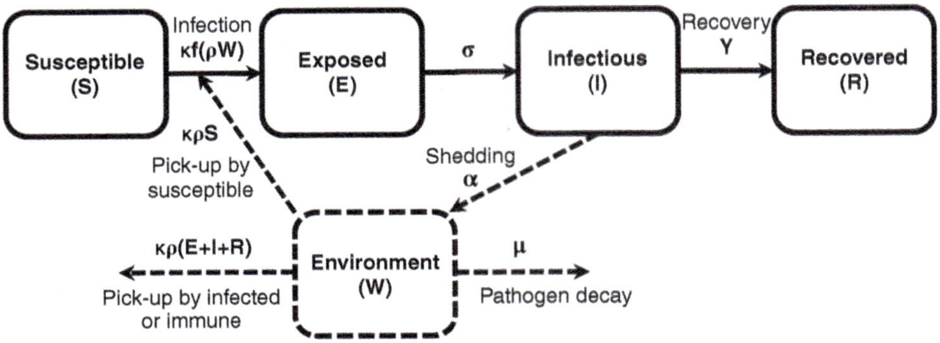

Figure 14.13 Environmentally mediated infectious disease model. *Source:* [73]/Public Library of Science (PLOS)/CC BY 4.0.

model has been developed [73]. This is shown in Figure 14.13. The individual processes are labeled and described as follows:

- The compartments for susceptible, exposed (latent), infectious, and recovered are defined as in the SEIR state.
- The assumption of a constant population is maintained, and since the focus is on an early outbreak wave, the reversion of some in the recovered to susceptible is neglected.
- There is an "environmental" compartment, which refers to the concentration of pathogens (or perhaps for fomite transmitted infection, the pathogen density per unit area) that receives shed pathogens from infectious individuals, and from which susceptible (as well as other persons) receive exposure. There is also an intrinsic decay rate of pathogens in this compartment. Although not considered in the initial conceptualization, it is possible that vegetative pathogens, such as bacteria or fungi, could grow. There is depletion, which may or may not be a significant factor, due to the pathogens' uptake by the population.

The formulation of this model is given by Equations 14.38–14.42. In formulating these equations, the dose–response is parameterized as a single quantity, π, which is the low-dose slope of the dose–response relationship — a characteristic of both the beta-Poisson and exponential models. For other, non-mechanistic, models, the low-dose relationship may be nonlinear, which can have dramatic effects on the outbreak curve [73]. Perhaps a more significant assumption is that all exposures are regarded as statistically independent; further work is needed to incorporate potential changes in host sensitivity into this transmission approach.

$$\frac{dS}{dt} = -\pi\kappa\rho\, WS \tag{14.38}$$

$$\frac{dE}{dt} = \pi\kappa\rho\, WS - \sigma E \tag{14.39}$$

$$\frac{dI}{dt} = \sigma E - \gamma I \tag{14.40}$$

$$\frac{dR}{dt} = \gamma I \tag{14.41}$$

$$\frac{dW}{dt} = \alpha I - \xi W \tag{14.42}$$

Table 14.7 summarizes the individual parameters and variables in the EMID model. In particular, note that the parameter ξ embodies the overall losses from pathogen decay and efflux in the environment and uptake by the population (whether susceptible, exposed, infectious, or recovered).

Other loss terms can be added to Equation 14.42 to account for engineering interventions. For example, in fomite exposures, periodic cleaning is used to reduce microbial amounts on a surface. For indoor air exposures, ventilation or air purification systems will add an additional loss term. For water systems, water treatment as an intervening barrier may be considered.

The system of equations in the EMID model is a set of first-order ordinary differential equations with defined initial conditions. As such, they can be integrated numerically to produce the curves of susceptible, exposed (latent), infectious, and recovered, as well as the concentration of infectious agents in the environmental compartment.

A limitation of the EMID model as defined by Equations 14.38–14.42 is that the exposed, infectious, and recovered stages are each treated as a single homogeneous compartment. This effectively yields an exponential distribution of the times a host spends in each stage. However, as noted earlier in this chapter, the incubation times, and presumably the time spent in the exposed state, are not exponential but are more appropriately described by distributions with stronger central tendency (e.g., gamma, Weibull, lognormal, inverse Gaussian).

One approach to achieving a more realistic distribution of the exposed, infectious, and recovered states is to treat each of them as a series of substages [74]. For each substage in a state, there is an identical rate into and out of the state. With this, the effective distribution of transit time in a state becomes a gamma, analogous to the treatment of continuous flow systems (Figure 8.6). This will result in a set of differential equations for each substage, and an additional parameter, being the integer number of substages in a state. As noted by Lloyd [74], this will affect the overall dynamics of an outbreak wave. This approach needs greater consideration in modeling EMIDs.

Table 14.7 Model variables and parameters for an EMID transmission model with dose–response and a latency period.

Variables	
$S(t)$	Number of susceptible people
$E(t)$	Number of exposed people
$I(t)$	Number of infectious people
$R(t)$	Number of recovered people
$W(t)$	Concentration of pathogens in the environment
Parameters	
γ	Recovery rate (per day)
κ	Rate at which individuals contact the environment (per day)
π	Low-dose linear dose–response slope
ρ	Volume of environment consumed (per contact)
α	Deposition rate of pathogens per unit volume of environment (per day)
μ	Pathogen decay rate (per day)
ξ	Removal rate of pathogens; sum decay and pick-up rates (per day)
σ	Rate of moving from a latent to infectious state (per day)

Source: Adapted from [73].

An alternative approach is to explicitly incorporate a delay function based on an incubation time distribution (or a distribution of time in the infectious stage). Depending on how this is parameterized, this leads to a delay differential equation [75] or an integrodifferential equation [76, 77]. Although this provides increased flexibility, the numerical solution is more computationally intensive.

If data on the occurrence of cases versus time following an exposure is available, it is possible to take a disease transmission model with a functional form for the exposure versus time (containing unknown parameters) and determine the best-fitting model parameters. Execution of this process shows, however, that similar qualities of fit can be obtained for very disparate functional forms for exposure (force of infectivity) and incubation time, leading to a lack of identifiability [76, 77]. This has also shown to be the case in situations where an outbreak is driven principally by direct person-to-person contact [69].

The transmission models presented earlier assume each population state is homogeneous. For large-scale and complex systems, there may be considerable heterogeneity, for example:

- Demographic variables (age, gender, etc.) may influence exposure intensity and duration in particular venues.
- Disease characteristics (dose–response, distributions for the latent and infectious periods) may differ in subgroups.
- In cases where the direct contact between susceptible and infectious individuals provides a significant route of disease spread, the contact rates between and among different subgroups may differ.

The importance of such heterogeneity was highlighted in the modeling of the early stages of the COVID-19 pandemic [78, 79]. In the extreme, every individual in a population can be modeled, describing their pathway through an environment and their exposures to different extents and times of contamination to develop probabilities of conversion between epidemic states using an agent-based model framework [80]. While very computationally intensive, such an approach can produce predictions (and their uncertainties) at fine spatial and temporal resolutions.

14.4 Risk Characterization Applications

Risk characterization of microbial exposures has been conducted across a broad range of contexts and applied in regulatory settings. Rather than giving a systematic review of those applications here, a brief note of several reviews and studies is presented to provide an entrée to the literature.

Owens et al. performed a systematic review of all drinking water-related QMRA studies reported in English and published from 2003 to 2019 [81]. Thirty-nine usable studies were identified and summarized. These studies included systems in North and South America, Asia, Europe, and Africa. They included viruses, bacteria, and protozoa, as well as a diversity of source water types and treatment trains. Outcome reference levels included both probability of infection and disability-adjusted life years (DALYs).[16] The authors offered a qualitative checklist of factors that future risk assessors of drinking water systems should consider.

The assessment of risk from recreational waters has been dominated by studies that use direct epidemiological methods and generally characterize water quality using indicator organisms (such as coliform or enterococci) to assess exposure. A systematic review summarized 35 such studies

16 The concept of DALY will be discussed in Section 15.3.2.

Table 14.8 Selected recreational water QMRA examples.

Author	Venue	Organisms	Outcomes	Comments
Viau et al. [83]	Hawaii beaches, swimming	Adenovirus, enterovirus, norovirus	Gastroenteritits	Waters impacted by septic tanks and animal polluted streams
Kundu et al. [84]	Southern California, swimming and secondary contact	Adenovirus	Gastroenteritis	Sewage effluent and agriculture impacted. Measurement by qPCR
Sunger and Haas [85]	Urban rivers	Indicator-based	Gastroenteritis	Primarily secondary contact; dry and wet weather
Suppes et al. [86]	Swimming pool	*Cryptosporidium*	Cryptosporidiosis	Risk estimated for children and adults
Burch et al. [87]	Iowa, impacted by human and animal sources	Pathogen concentrations imputed from indicators	Infections	Source tracking used to estimate pathogens
Ngo et al. [88]	Freshwater recreation in Ontario	Virus, protozoa, bacteria	Gastroenteritits	Province-wide estimates
Brown et al. [89]	Southern California	Indicator organisms modified by source tracking	Gastroenteritis	qPCR

[82] with venues in the Americas, Europe, and Asia, and outcomes including gastrointestinal, respiratory, skin, ear, and eye infections.

There has not been a systematic review of recreational water QMRA. However, a large number of such studies have been done.[17] Some example studies are summarized in Table 14.8.
Even prior to the COVID-19 pandemic, QMRA was applied to many respiratory pathogens, including bioterrorist agents, influenza, measles, and *Legionella*. Respiratory pathogen QMRA work and its influence on policy were reviewed by Tang et al. [90]. Ninety-three such studies were identified, along with needs for improvement in data and methodologies to improve the utility of QMRA in these applications.

Risk assessment of fomite exposure is much more venue-specific, and therefore, the number of systematic reviews is limited, other than with respect to particular elements of QMRA and/or specific venues.

Kramer et al. conducted a systematic review of the retention of microbial viability on surfaces in the healthcare environment [91]. They summarized results from 171 studies on bacteria, viruses, and fungi across different surfaces and conditions.

Zambrana and Boehm conducted a systematic review of the occurrence of viruses on fomites [92]. Although 134 papers were identified covering multiple virus families and venues, many lacked sufficient data to quantify per-unit-area rates. In general, concentrations were higher in the healthcare environment than in general environmental areas, and the presence of nonenveloped viruses exceeded that of enveloped viruses.

During the initial phase of the SARS-CoV-2 pandemic, there was great concern about the role of fomites in transmission (a concern that has diminished regarding general environmental surfaces relative to other routes). Mohamadi et al. did a scoping review, identifying 25 studies (as of

17 As of November 2025, there were over 1000 "hits" in Google Scholar using the search phrase: "`QMRA and (swim* or recreat*) and illness.`"

February 2021) where some assessment of fomite transmission was discussed [93]. Many of the studies were qualitative or semiquantitative in nature, and conclusions regarding risk were highly variable.

A CDC-sponsored workshop on fomite transmission of viruses was held in 2024 and led to 15 broad recommendations for future work [94]. These covered many areas discussed previously, including methods for sampling, the efficacy of intervention strategies, the effect of occupant behavior on deposition and transmission, and improvements to methods for characterizing risk. Many of these recommendations could readily be applied to other classes of microorganisms beyond the viruses.

14.5 Research Needs for Risk Assessment

Since risk characterization is the culminating step in risk assessment (Section 3.1.2), it is important to ask what improvements in knowledge and data would facilitate more precise characterizations. Hamilton et al. [95] reported on results of an expert workshop on needs and opportunities for improvement, and the following were identified:

- Use of sequencing methods for scoping and prioritizing hazards and for evaluation of pathogenicity
- Advanced monitoring methods for persistence and viability
- Rapid methods for evaluating and updating models
- Additional data for assessing dose–response and severity of outcomes
- Rapid methods for assessing exposure

It had been proposed to move beyond a single-pathogen risk characterization paradigm toward a new paradigm for "bigger, better, faster" QMRA [96]. This will leverage advantages in omics, environmental analytics, modeling, and artificial intelligence/machine learning. This will clearly require long-term dedicated effort to develop new methods, tools, and workflows.

14.6 Discussion Questions and Problems

1 A pathogen is characterized by an "r" that is beta distributed with a mean of 0.2 and a variance of 0.4. Calculate α and β. Plot the distribution function for r. Then use the beta conditional dose–response equation (Equation 14.11) and plot P_e versus N_d.

2 Consider the approximate beta-Poisson dose–response model. If the acceptable effect level is 10^{-4}, compute the corresponding value of $\frac{\text{dose}}{N_{50}}$ for values of α from 0.1 to 1.0, and plot your data.

3 Repeat the calculation earlier for the exact beta-Poisson model, and compare the results.

4 Consider a pathogen whose dose–response relationship has been found to be an approximate beta-Poisson, with $\alpha = 0.3$ and $N_{50} = 1200$. A population is exposed to an average dose of 200 organisms. What is the expected proportion of affected individuals?

5 Exposure of rhesus monkeys to aerosols of *Francisella tularensis* yielded particle-dependent dose–response parameters for the modified exponential relationship (Equation 14.23) of A = $0.184\,\mu m^{-1}$ and B = 3.556. If exposure to $3\,\mu m$ particles occurs, what would be the inhaled (administered) dose producing a risk of 0.1?

6 Using R, Python, or another available platform, write the code to fit the data in Table 14.2 to the approximate beta-Poisson distribution. Hint: The approach in Listing 12.2 for analysis of quantal inactivation data may provide some useful code approaches that can be used.

7 Consider the data in Table 14.5, and fit the data *for each* particle diameter by an exponential dose–response function. Using the AICc approach, test the ability to use a common exponential model for the entire pooled data.

8 Fit the data in Table 14.5 to the exponential dose–response with power law dependency on aerosol diameter (Equation 14.23). Test the goodness-of-fit relative to a saturated model.

9 Consider a room with dimensions 15 m × 15 m × 2.5 m height, which is well mixed with an ACH of 2 hr^{-1}. An infectious emitter exhales 100 viable pathogens per hour, and its dose–response is approximately beta-Poisson, with $\alpha = 0.3$ and $N_{50} = 1500$. What is the daily risk to a susceptible adult doing office work who occupies the same office for 8 hours? If this is in excess of 0.1%, how much would the ACH need to be increased to achieve that? Note that it would also be possible to increase removal by one of the interventions discussed earlier, such as air purifiers or UV irradiation.

References

1 Andrey Gudymo, Galina Onkhonova, Alexey Danilenko, Ivan Susloparov, Natalia Danilchenko, Maxim Kosenko, Anastasia Moiseeva, Natalia Kolosova, Svetlana Svyatchenko, Vasily Marchenko, et al. Quantitative assessment of airborne transmission of human and animal influenza viruses in the Ferret model. *Atmosphere*, 14(3):471, February 2023.

2 Mathilde Richard, Adinda Kok, Dennis de Meulder, Theo M Bestebroer, Mart M Lamers, Nisreen M A Okba, Martje Fentener van Vlissingen, Barry Rockx, Bart L Haagmans, Marion P G Koopmans, et al. SARS-CoV-2 is transmitted via contact and via the air between ferrets. *Nature Communications*, 11(1):3496, December 2020.

3 Allen M Hornblum. *Acres of Skin: Human Experiments at Holmesburg Prison*. Routledge, 2013.

4 John W Powell. Japan's germ warfare: The U.S. cover-up of a war crime. *Bulletin of Concerned Asian Scholars*, 12(4):2–17, 1980.

5 Arturo Casadevall and Liise-anne Pirofski. Host-pathogen interactions: Basic concepts of microbial commensalism, colonization, infection, and disease. *Infection and Immunity*, 68(12):6511–6518, December 2000.

6 F J Frost, T R Kunde, T B Muller, G F Craun, L M Katz, A J Hibbard, and R L Calderon. Serological responses to *Cryptosporidium* antigens among users of surface- *vs* . ground-water sources. *Epidemiology and Infection*, 131(3):1131–1138, December 2003.

7 John Bylund, Jonas Toljander, Maria Lysén, Niloofar Rasti, Jannes Engqvist, and Magnus Simonsson. Measuring sporadic gastrointestinal illness associated with drinking water–An overview of methodologies. *Journal of Water and Health*, 15(3):321–340, 2017.

8 Charles N Haas, Joan B Rose, and Charles P Gerba. *Quantitative Microbial Risk Assessment*. John Wiley & Sons, Hoboken, NJ, 2nd edition, 2014.

9 Charles N Haas. Microbial dose response modeling: Past, present, and future. *Environmental Science & Technology*, 49(3):1245–1259, February 2015.

10 E Weil and F Breinl. Experimental studies on infection and immunity in typhus. *Journal of Infectious Diseases*, 33:60–74, 1923.

11 G G Meynell and B A D Stocker. Some hypotheses on the aetiology of fatal infections in partially resistant hosts and their application to mice challenged with *Salmonella paratyphi-B* or *Salmonella typhimurium* by intraperitoneal injection. *Journal of General Microbiology*, 16:38–58, 1957.

12 Charles N Haas. Conditional dose-response relationships for microorganisms: Development and application. *Risk Analysis: An International Journal*, 22(3):455–463, 2002.

13 G N Sze To and C Y Chao. Review and comparison between the Wells-Riley and dose-response approaches to risk assessment of infectious respiratory diseases. *Indoor Air*, 20:2–16, February 2010.

14 W A Furumoto and R Mickey. A mathematical model for the infectivity-dilution curve of tobacco mosaic virus: Experimental tests. *Virology*, 32:224, 1967.

15 W A Furumoto and R Mickey. A mathematical model for the infectivity-dilution curve of tobacco mosaic virus: Theoretical considerations. *Virology*, 32(2):216, 1967.

16 Norman L Johnson, Samuel Kotz, and N Balakrishnan. *Continuous Univariate Distributions*, volume 1. Wiley Interscience, New York, 2nd edition, 1994.

17 James B McDonald and Yexiao J Xu. A generalization of the beta distribution with applications. *Journal of Econometrics*, 66(1-2):133–152, 1995.

18 Adri B Olde Daalhuis. Confluent hypergeometric functions. *NIST Handbook of Mathematical Functions*, 321:349, 2010.

19 P F M Teunis and A H Havelaar. The Beta Poisson dose-response model is not a single-hit model. *Risk Analysis*, 20(4):513–520, 2000.

20 T W Armstrong and C N Haas. A quantitative microbial risk assessment model for legionnaires' disease: Animal model selection and dose-response modeling. *Risk Analysis*, 27(6):1581–1596, 2007.

21 Nicole Van Abel, Mary E Schoen, John C Kissel, and J Scott Meschke. Comparison of risk predicted by multiple norovirus dose-response models and implications for quantitative microbial risk assessment: Comparison of risk predicted by multiple norovirus dose-response models. *Risk Analysis*, 37(2):245–264, February 2017.

22 Roberta M O'Connor, Rehaan Shaffie, Gagandeep Kang, and Honorine D Ward. Cryptosporidiosis in patients with HIV/AIDS. *AIDS*, 25(5):549–560, 2011.

23 Mahendra H Kothary and Uma S Babu. Infective dose of foodborne pathogens in volunteers: A review. *Journal of Food Safety*, 21(1):49–68, 2001.

24 Morris Jodie, Williams Natasha, Rush Catherine, Govan Brenda, Sangla Kunwarjit, Norton Robert, and Ketheesan Natkunam. Burkholderia pseudomallei triggers altered inflammatory profiles in a whole-blood model of type 2 diabetes-melioidosis comorbidity. *Infection and Immunity*, 80(6):2089–2099, 2012.

25 S B Tamrakar and Charles N Haas. Dose–response model for *Burkholderia pseudomallei* (melioidosis). *Journal of Applied Microbiology*, 105(5):1361–1371, November 2008.

26 Alexys Monoson, Evangeline Schott, Kerry Ard, Brita Kilburg-Basnyat, Robert M Tighe, Sonal Pannu, and Kymberly M Gowdy. Air pollution and respiratory infections: The past, present, and future. *Toxicological Sciences*, 192(1):3–14, March 2023.

27 Beth J Feingold, Vegosen Leora, Davis Meghan, Leibler Jessica, Peterson Amy, and K Silbergeld Ellen A niche for infectious disease in environmental health: Rethinking the toxicological paradigm. *Environmental Health Perspectives*, 118(8):1165–1172, 2010.

28 US Environmental Protection Agency. Particle pollution exposure, April 2025.

29 Sondra S Teske, Mark H Weir, Timothy A Bartrand, Yin Huang, Sushil B Tamrakar, and Charles N Haas. Dose-response models incorporating aerosol size dependency for *Francisella tularensis*: Dose-response models incorporating aerosol size dependency for *F. tularensis*. *Risk Analysis*, 34(5):911–928, May 2014.

30 Mark H Weir and Charles N Haas. A model for in-vivo delivered dose estimation for inhaled *Bacillus anthracis* spores in humans with interspecies extrapolation. *Environmental Science & Technology*, 45(13):5828–5833, 2011.

31 Mihalis Lazaridis. Modelling approaches to particle deposition and clearance in the human respiratory tract. *Air Quality, Atmosphere & Health*, 16(10):1989–2002, October 2023.

32 Mark H Weir and Charles N Haas. Quantification of the effects of age on the dose response of *Variola major* in suckling mice. *Human and Ecological Risk Assessment: An International Journal*, 15(6):1245–1256, 2009.

33 Mark H Weir, Alexis L Mraz, Sharon Nappier, and Charles N Haas. Dose-response models for eastern US, western US and Venezuelan equine encephalitis viruses in mice–Part II: Quantification of the effects of host age on the dose response. *Microbial Risk Analysis*, 9:38–54, March 2018.

34 P E Sartwell. The distribution of incubation periods of infectious Disease. *American Journal of Hygiene*, 51:310–318, 1950.

35 H Nishiura. Early efforts in modeling the incubation period of infectious diseases with an acute course of illness. *Emerging Themes in Epidemiology*, 4:2, 2007.

36 Rachel M Lee, Justin Lessler, Rose A Lee, Kara E Rudolph, Nicholas G Reich, Trish M Perl, and Derek AT Cummings. Incubation periods of viral gastroenteritis: A systematic review. *BMC Infectious Diseases*, 13(446):1–11, 2013.

37 Auni Aslah Mat Daud. Five common misconceptions regarding flattening-the-curve of covid-19. *History and Philosophy of the Life Sciences*, 44(3):41, 2022.

38 Arnold J F Siegert. On the first passage time probability problem. *Physical Review*, 81:617–623, February 1951.

39 Trevor Williams. The basic birth-death model for microbial infections. *Journal of the Royal Statistical Society Part B*, 27:338–360, 1965.

40 Trevor Williams. The distribution of response times in a birth-death process. *Biometrika*, 52(3/4):581, December 1965.

41 Byron J T Morgan and Simon A Watts. On modelling microbial infections. *Biometrics*, 36(2):317–321, June 1980.

42 Yin Huang and Charles N Haas. Time-dose-response models for microbial risk assessment. *Risk Analysis*, 29:648–661, 2009.

43 Yin Huang and Charles N Haas. Quantification of the relationship between bacterial kinetics and host response for monkeys exposed to aerosolized *Francisella tularensis*. *Applied and Environmental Microbiology*, 77:485–490, 2011.

44 G F Akpeimeh, L A Fletcher, B E Evans, and I E Ibanga. Quantitative microbial risk assessment (QMRA) of workers exposure to bioaerosols at MSW open dumpsites. *Risk Analysis*, 41(10):1911–1924, October 2021.

45 E. Julien, A. R. Boobis, S. S. Olin, and Group ILSI Research Foundation Threshold Working The key events dose-response framework: A cross-disciplinary mode-of-action based approach to examining dose-response and thresholds. *Critical Reviews in Food Science and Nutrition*, 49:682–689, September 2009.

46 S M Ashrafur Rahman, Daniel Munther, Aamir Fazil, Ben Smith, and Jianhong Wu. Unraveling the dose-response puzzle of L. monocytogenes: A mechanistic approach. *Infectious Disease Modelling*, 1(1):101–114, 2016.

47 Lucas M Wijnands, Peter F M Teunis, Angelina F A Kuijpers, Ellen H M Delfgou-Van Asch, and Annemarie Pielaat. Quantification of *Salmonella* survival and infection in an *In vitro* model of the human intestinal tract as proxy for foodborne pathogens. *Frontiers in Microbiology*, 8(1139):1–11, June 2017.

48 Hiroki Abe, Kohei Takeoka, Yuto Fuchisawa, Kento Koyama, and Shigenobu Koseki. A new dose-response model for estimating the infection probability of Campylobacter jejuni based on the key events dose-response framework. *Applied and Environmental Microbiology*, 87(20):e01299-21, September 2021.

49 Katelyn M Gostic, Elsio A Wunder, Vimla Bisht, Camila Hamond, Timothy R Julian, Albert I Ko, and James O Lloyd-Smith. Mechanistic dose–response modelling of animal challenge data shows that intact skin is a crucial barrier to leptospiral infection. *Philosophical Transactions of the Royal Society B: Biological Sciences*, 374(1782):20190367, September 2019.

50 Josep M Pujol, Joseph E Eisenberg, Charles N Haas, and James S Koopman. The effect of ongoing exposure dynamics in dose response relationships. *PLoS Computational Biology*, 5(6):e1000399, 2009.

51 Charles W Heppell, Joseph R Egan, and Ian Hall. A human time dose response model for Q fever. *Epidemics*, 21:30–38, December 2017.

52 Louis-Patrick Haraoui and Martin J Blaser. The microbiome and infectious diseases. *Clinical Infectious Diseases*, 77(Supplement_6):S441–S446, December 2023.

53 Theodora Akritidou, Simen Akkermans, Cindy Smet, Sotiria Gaspari, Chahat Sharma, Eimear Matthews, and Jan F M Van Impe. Gut microbiota of the small intestine as an antimicrobial barrier against foodborne pathogens: Impact of diet on the survival of *S. typhimurium* and *L. monocytogenes* during *in vitro* digestion. *Food Research International*, 173(11):113292, November 2023.

54 R L Ward, D L Bernstein, E C Young, J R Sherwood, D R Knowlton, and G M Schiff. Human rotavirus studies in volunteers: Determination of infectious dose and serological response to infection. *Journal of Infectious Diseases*, 154(5):871–880, 1986.

55 Toru Watanabe, Timothy A Bartrand, Tatsuo Omura, and Charles N Haas. Dose-response assessment for influenza A virus based on data sets of infection with its live attenuated reassortants. *Risk Analysis*, 32(3):555–565, 2012.

56 William C Day and Richard F Berendt. Experimental tularemia in macaca mulatta: Relationship of aerosol particle size to the infectivity of airborne pasteurella tularensis. *Infection and immunity*, 5(1):77–82, 1972.

57 Yin Huang, T A Bartrand, Charles N Haas, and M H Weir. Incorporating time postinoculation into a dose-response model of Yersinia pestis in mice. *Journal of Applied Microbiology*, 107(9):727–735, September 2009.

58 James V Rogers, Young W Choi, Lenore F Giannunzio, Patrick J Sabourin, Daniel M Bornman, Emily G Blosser, and Carol L K Sabourin. Transcriptional responses in spleens from mice exposed to *Yersinia pestis* CO92. *Microbial Pathogenesis*, 43(2-3): 67–77, 2007.

59 Charles N Haas. How to average microbial densities to characterize risk. *Water Research*, 30(4):1036–1038, April 1996.

60 Braulio M Valencia, Ponsuge C Sigera, Praveen Weeratunga, Nicodemus Tedla, Deepika Fernando, Senaka Rajapakse, Andrew R Lloyd, and Chaturaka Rodrigo. Effect of prior Zika and dengue virus exposure on the severity of a subsequent dengue infection in adults. *Scientific Reports*, 12(1):17225, October 2022.

61 Chaojie Li, Tamar Kohn, Shotaro Torii, Htet Kyi Wynn, Alexander J Devaux, Charles Gan, Timothy R Julian, and Émile Sylvestre. Lognormal distributions capture site-specific variability in enteric virus concentrations in wastewater. *Environmental Science: Water Research & Technology*, 11:2973–2985, 2025.

62 Julio A Polanco, Jana Safarik, Jason S Dadakis, Claire Johnson, and Megan H Plumlee. Enteric virus removal by municipal wastewater treatment to achieve requirements for potable reuse. *PLOS Water*, 2(9):e0000052, September 2023.

63 Franciska M Schets, Jack F Schijven, and Ana Maria de Roda Husman. Exposure assessment for swimmers in bathing waters and swimming pools. *Water Research*, 45:2392–2400, 2011.

64 Daniel Bernoulli and Sally Blower. An attempt at a new analysis of the mortality caused by smallpox and of the advantages of inoculation to prevent it. *Reviews in Medical Virology*, 14(5):275–288, September 2004.

65 Robert E Serfling. Historical Review of epidemic theory. *Human Biology*, 24(3):145–166, September 1952.

66 Alfred J Lotka. A contribution to quantitative epidemiology. *Journal of the Washington Academy of Sciences*, 9(3):73–77, 1919.

67 W O Kermack and A G McKendrick. A contribution to the mathematical theory of epidemics. *Proceedings of the Royal Society A*, 115(772):700–721, 1927.

68 Norman T J Bailey. A simple stochastic epidemic. *Biometrika*, 37(3/4):193–202, 1950.

69 K Dietz and D Schenzle. Mathematical models for infectious disease statistics. In Anthony C. Atkinson and Stephen E. Fienberg, editors, *A Celebration of Statistics*, pages 167–204. Springer, New York, NY, 1985.

70 R M Anderson and Robert M May. *Infectious Diseases in Humans: Dynamics and Control*. Oxford University Press, Oxford, 1991.

71 Kenneth L Cooke. Functional-differential equations: some models and perturbation problems, 1967.

72 Donald S Burke. Origins of the problematic E in SEIR epidemic models. *Infectious Disease Modelling*, 9(3):673–679, September 2024.

73 A F Brouwer, M H Weir, M C Eisenberg, R Meza, and J N S Eisenberg. Dose-response relationships for environmentally mediated infectious disease transmission models. *PLOS Computational Biology*, 13:e1005481, 2017.

74 Alun L Lloyd. Realistic distributions of infectious periods in epidemic models: Changing patterns of persistence and dynamics. *Theoretical Population Biology*, 60(1):59–71, August 2001.

75 Frank Hoppensteadt and Paul Waltman. A problem in the theory of epidemics. *Mathematical Biosciences*, 9:71–91, January 1970.

76 Mukul Gupta and Charles N Haas. The Milwaukee Cryptosporidium outbreak: Assessment of incubation time and daily attack rate. *Journal of Water and Health*, 2(2):59–69, 2004.

77 Bidya Prasad and Charles N Haas. Incorporating time-dose-response into *Shigella flexneri* and *Shigella sonnei* outbreak models. *Journal-American Water Works Association*, 109(12):E548–E562, 2017.

78 Adam J Kucharski, Timothy W Russell, Charlie Diamond, Yang Liu, John Edmunds, Sebastian Funk, Rosalind M Eggo, Fiona Sun, Mark Jit, James D Munday, et al. Early dynamics of transmission and control of COVID-19: A mathematical modelling study. *The Lancet Infectious Diseases*, 20(5):553–558, March 2020.

79 Gerrit Großmann, Michael Backenköhler, and Verena Wolf. Heterogeneity matters: Contact structure and individual variation shape epidemic dynamics. *PLOS ONE*, 16(7):e0250050, July 2021.

80 Liliana Perez and Suzana Dragicevic. An agent-based approach for modeling dynamics of contagious disease spread. *International Journal of Health Geographics*, 8(1):50, 2009.

81 Christopher E L Owens, Mark L Angles, Peter T Cox, Paul M Byleveld, Nicholas J Osborne, and Md Bayzid Rahman. Implementation of quantitative microbial risk assessment (QMRA) for public drinking water supplies: Systematic review. *Water Research*, 174(115614):1–18, May 2020.

82 Ripon Kumar Adhikary, Md Sarower-E Mahfuj, Danswell Starrs, Barry Croke, Kathryn Glass, and Aparna Lal. Risk of human illness from recreational exposure to microbial pathogens in freshwater bodies: A systematic review. *Exposure and Health*, 14(2):325–343, June 2022.

83 Emily J Viau, Debbie Lee, and Alexandria B Boehm. Swimmer risk of gastrointestinal illness from exposure to tropical coastal waters impacted by terrestrial dry-weather runoff. *Environmental Science & Technology*, 45(17): 7158–7165, September 2011.

84 A Kundu, G McBride, and S Wuertz. Adenovirus-associated health risks for recreational activities in a multi-use coastal watershed based on site-specific quantitative microbial risk assessment. *Water Research*, 47:6309–6325, October 2013.

85 Neha Sunger and Charles N Haas. Quantitative microbial risk assessment for recreational exposure to water bodies in Philadelphia. *Water Environment Research*, 87(3):211–222, March 2015.

86 L M Suppes, R A Canales, C P Gerba, and K A Reynolds. *Cryptosporidium* risk from swimming pool exposures. *International Journal of Hygiene and Environmental Health*, 219:915–919, July 2016.

87 Tucker R Burch, Joel P Stokdyk, Aaron D Firnstahl, Sarah A Opelt, Rachel M Cook, Joseph A Heffron, Amanda Brown, Claire Hruby, and Mark A Borchardt. Quantitative microbial risk assessment with microbial source tracking for mixed fecal sources contaminating recreational river waters, Iowa, USA. *ACS ES&T Water*, 4(7): 2789–2802, June 2024.

88 Henry Ngo, E Jane Parmley, Nicole Ricker, Charlotte Winder, and Heather M Murphy. Quantitative microbial risk assessment of acute gastrointestinal illness attributable to freshwater recreation in Ontario. *Canadian Journal of Public Health*, 116(4):582–597, December 2024.

89 Kendra I. Brown, Katherine E. Graham, Jeffrey A. Soller, and, Alexandria B. Boehm. Estimating the probability of illness due to swimming in recreational water with a mixture of human-and gull-associated microbial source tracking markers. *Environmental Science: Processes & Impacts*, 19(12):1528—1541, 2017.

90 Lizhan Tang, William J Rhoads, Antonia Eichelberg, Kerry A Hamilton, and Timothy R Julian. Applications of quantitative microbial risk assessment to respiratory pathogens and implications for uptake in policy: A state-of-the-science review. *Environmental Health Perspectives*, 132(5):056001, May 2024.

91 Axel Kramer, Franziska Lexow, Anna Bludau, Antonia Milena Köster, Martin Misailovski, Ulrike Seifert, Maren Eggers, William Rutala, Stephanie J Dancer, and Simone Scheithauer. How long do bacteria, fungi, protozoa, and viruses retain their replication capacity on inanimate surfaces? A systematic review examining environmental resilience versus healthcare-associated infection risk by "fomite-borne risk assessment". *Clinical Microbiology Reviews*, 37(4):e00186-23, October 2024.

92 Winnie Zambrana and Alexandria B Boehm. Occurrence of human viruses on fomites in the environment: A systematic review and meta-analysis. *ACS Environmental Au*, 3(5):277–294, September 2023.

93 Marjan Mohamadi, Awa Babington-Ashaye, Agnès Lefort, and Antoine Flahault. Risks of infection with SARS-CoV-2 due to contaminated surfaces: A scoping review. *International Journal of Environmental Research and Public Health*, 18(21):11019, October 2021.

94 Sarah E Abney, Sally F Bloomfield, Stephanie A Boone, Todd Cutts, Charles P Gerba, M Khalid Ijaz, Amanda K Lyons, Jean-Yves Maillard, Kristina D Mena, Julie McKinney, et al. Fomite workshop recommendations addressing the role of surfaces in virus transmission in the built environment. *mSphere*, 10(19):e00927-24, August 2025.

95 Kerry A Hamilton, Joanna Ciol Harrison, Jade Mitchell, Mark Weir, Marc Verhougstraete, Charles N Haas, A Pouyan Nejadhashemi, Julie Libarkin, Tiong Gim Aw, Kyle Bibby, et al. Research gaps and priorities for quantitative microbial risk assessment (QMRA). *Risk Analysis*, 44(11):2521–2536, November 2024.

96 Kerry A Hamilton, Hunter Quon, Muhammad Atif Nisar, Michael Jahne, Jay Garland, Benjamin Davis, Clinton Williams, Nicholas J Ashbolt, Joseph N S Eisenberg, Qian Zhang, et al. Next generation quantitative microbial risk assessment (QMRA): Bigger, better, faster. *Environmental Science & Technology Letters*, 12(11):1471–1480, November 2025.

15

Balancing Interventions and Risk

15.1 Introduction

Having considered the risks from (potential) exposure to pathogens in the environment, and potential interventions to reduce such risk, it is now appropriate to consider *how and whether* such interventions to reduce the risks should be implemented. This decision goes beyond purely technical considerations and falls within the realm of risk management (Section 3.1.3). As such, the decision is not purely a technical one, but incorporates a variety of social, economic, legal, and political factors. The focus of this book is engineering, and therefore, this chapter will cover a variety of tools and approaches that ultimate decision-makers may use as inputs to their risk management decisions.

There are three broad, though not mutually exclusive, approaches to risk management [1]:

- Risk-informed (via risk assessment) decision-making
- Applying cautionary and precautionary principles
- Discursive

Any real decision will likely have characteristics of all three approaches. *How* risk management decisions are made should be recognized to have an ethical dimension [2]. Risk-informed decisions, operationalized via a cost-benefit or analogous approach, implicitly adopt a utilitarian ethic, whereas the other approaches also incorporate deontological ethical concepts.

The concept of risk-informed decision-making is generally regarded as a process in which the reduction in risk from taking a proposed action is balanced against the costs of that action. The "devil is in the details" in that, for this to be a fair evaluation, all the benefits from the risk reduction, both the target of the action and any co-benefits, are considered along with all the costs, including the unintended consequences of such an action.

The cautionary principle, as stated by Ersdal and Aven [2], is that

> "In face of uncertainties, caution should be the guiding principle.... Following the cautionary principle, one should seek solutions that are robust in the sense that [if hazardous situations or accidents occur] such events are avoided and the consequences reduced."

The precautionary principle is regarded as a subset of this. It has been stated as, "when an activity raises threats of harm to human health or the environment, precautionary measures should be taken even if some cause and effect relationships are not fully established scientifically" [3].

However, neither the cautionary nor precautionary principles consider the possibility of unintended adverse consequences of an intervention. Therefore, they are limited in complex situations.

In the discursive, or what has been termed the analytic deliberative process [4, 5], there is intense two-way interaction among the various stakeholders and the decision-makers, who are all fully informed regarding the characterization of risks (and benefits) and their uncertainties before weighing the options. Ideally, communication with these same parties occurs during the risk assessment and characterization steps to incorporate their knowledge and concerns into the assessment process.

The risk-informed and, to some degree, the cautionary approaches (operationalized as bright lines of acceptability) will be the focus of subsequent discussions. Furthermore, it is assumed that potential direct costs of interventions, such as treatment processes, can be quantified; hence, the focus will be on the benefits that may be attained from risk reduction.

15.2 Implementation of Strategies

The earliest strategy for reducing people's exposure to risky materials may have occurred in the food sector, with the passage of the Delaney Clause in 1958. This built upon earlier food laws (from the early 20th century) prohibiting food adulteration, which were as much driven by concern for quantity as for human health. This was a blanket prohibition against food additives that induced cancer in animals, humans, or other suitable assay systems. However, with improvements in both analytical methods and bioassay systems, it was soon recognized that this precautionary approach was unrealistic [6]. Even shortly after, in 1961, it was argued that a better procedure would be to use the results of animal dose–response experiments to define a "virtually safe dose" (which the authors suggested could be one in 100,000,000) as a point of regulation [7]. No justification for the 100,000,000 value was provided, but at that time the US population was 184 million, so this would correspond to one to two cases per year from exposure.

As a generalization of this concept, if an "acceptable" risk is defined, then this approach of a bright line would directly define the decision to be taken. This will be elaborated upon in the following section.

However, Kasperson early on recognized the limitations of a purely scientific approach to this problem [8]. The acceptable risk depends on contextual factors such as background risks, comparisons with similar risks, cost-effectiveness and cost-benefit, equity, and the assumption that risks accepted in the past, even for similar hazards, are in fact those that would be consciously tolerated in the future.

Chauncey Starr was perhaps the earliest to undertake comparisons between risks incurred by activities and social benefits [9]. This can be regarded as a forerunner of cost-benefit analysis (CBA) applied to risk management. More than 50 years later, some of his findings remain cogent:

- Among hazardous occupations, there is a documented trade-off between risk and salary.
- For more general risks, there are different trade-off levels of acceptance between those perceived as voluntary and those perceived as involuntary.
- For a particular hazardous technology, over time, the degree of participation increases, and the risk per unit participation decreases. Although this may not necessarily be causal, it is phenomenological.

It should also be recognized that legal constraints exist on the process used for regulatory decisions.[1] For example (broadly speaking):

- Clean Air Act (CAA) NAAQS: pure health-based bright-line standards that exclude consideration of cost.
- Clean Water Act (CWA) effluent limits (BAT/BPT/BCT): technology-based bright-line standards where costs inform feasibility but not benefit justification.
- CWA also requires the application of the best available technology economically achievable (a cautionary approach, with elements of cost-effectiveness) when setting effluent limits for industrial pollutants.
- The Safe Drinking Water Act requires guidelines (MCLGs) that are purely health-based (precautionary). But regulatory MCLs are set by balancing cost and benefit information. Also, when an enforceable MCL is not feasible, such as due to analytical limitations, an economically feasible one must follow the same cost-benefit considerations.
- Resource Conservation and Recovery Act Subtitle C regulations for hazardous waste treatment requirements incorporate cost-effectiveness considerations.
- FIFRA pesticide registration requires balancing risks against benefits to determine whether use is acceptable.

The various approaches will now be discussed in the context of pathogens.

15.3 Concept of Acceptable Risk

The first significant numerical standard, albeit not risk-based, for environmental exposure to pathogens appears to be the use (in the United States) of coliforms for assessing the quality of drinking water supplied on passenger railroads. This mandated that in five tubes of 10mL samples, i.e., an MPN assay, not more than one tube be positive [10]. This can be regarded as a first application of a cautionary/precautionary standard.

15.3.1 Historical

In the case of swimming, a series of epidemiological studies, beginning with the work of Stevenson [11], have been conducted, both in the United States and internationally. These studies, using varying methods, conducted onsite surveys of individuals at recreational sites who were engaged in swimming or other primary-contact activities. They were followed up to assess subsequent illnesses, and in many studies, these illness rates were associated with concentrations of indicator organisms at the time of exposure.

These studies, and others, were used to develop initial criteria for recreational water quality using coliforms[2][13]. Subsequently, and based on further epidemiological studies, the then-extant

---

1 However, except for FIFRA when dealing with biocides, none of these programs has yet developed pathogen-specific standards.

2 It should be noted that the bulk of these studies were conducted where the dominant source of pollution was from wastewater effluents receiving some degree of treatment. In these circumstances, there has been an implicit supposition that the ratio of indicators to individual pathogens is *relatively* consistent between venues, and hence the risk based on indicators is not necessarily location-specific. However, in circumstances where the dominant source(s) of pollution are not from such effluents, and instead from animals or nonpoint runoff, the ratios and risks may be different. This will not be discussed further; however, the use of microbial source tracking approaches may be helpful in interpreting or modifying such criteria [12].

coliform criteria, which were regarded as acceptable, were found to be associated with a risk of 32–36 illnesses per 1000 primary-contact recreation days [14]. This represents the bright-line criteria for primary-contact water quality acceptability.

Having established such an acceptance threshold, it is possible to develop pathogen-specific criteria when dose–response information is available, along with medium exposure per event. The process, using reverse QMRA, is illustrated by Denpetkul et al. [15], although these authors did not consider situations where multiple pathogens might be present.

An early derivation of acceptable risk for treated drinking water relied upon the per capita illness rate in reported waterborne outbreaks, which at the time was approximately 10^{-4} cases per person per year. It was known that the total number of illnesses was substantially under-reported, and in a desire to protect sensitive subpopulations, the same rate applied *to infections* was regarded as acceptable [16]. At that point, the primary risks of concern were enteric virus and *Giardia* cysts.

It soon became apparent that the total disease burden was much greater (by at least an order of magnitude) than the 10^{-4} level, so arguably this benchmark may have been overly stringent. A 2021 study estimated the annual total waterborne disease burden (from all causes, including potable water and recreational exposure) to be 7.15–20 million cases per year, or a yearly incidence rate of 0.02–0.06. Nonetheless, the 10^{-4} value has been widely used in the United States and internationally as a regulatory action level.

From a purely technical view, movement away from such a bright line for drinking water and related exposures, such as wastewater treated for reuse, is strongly justified [17].

In other media, such as air (outdoor or indoor), direct contact, or exposure via fomites, there has been no movement to define absolute acceptable risk levels in terms of either incidence or other numerical measures of consequence.

15.3.2 HALY, QALY, DALY Concepts

Using only infection, or even disease, as a metric has limitations in several respects:

- Infection with a pathogen can result in a spectrum of effects from mild to moderate to death.
- The extent, progression, and severity of these effects will differ between pathogens (and among individuals).
- The spectrum of severity may itself differ between subpopulations being exposed, due, for example, to other medical conditions (such as immunosuppression), nutritional status, or age.

Due to these limitations, other, more integrative metrics have been developed. In a broad sense, these are termed "health-adjusted life years" (HALYs) [18].

The most common HALY metrics are the "quality-adjusted life year" (QALY) and the "disability-adjusted life year" (DALY). These metrics share some broad features in their computation. Conceptually, they are computed as the sum of two terms:

- In the case of mortality, the years of life lost due to the exposure. This depends on the baseline used for the calculation. It could be based on the average life expectancy for the full population, or on the life expectancy at death for the affected individual; for the World Health Organization (WHO) Global Burden of Disease estimates, a universal "idealized" life expectancy is used [19]. These will lead to different numbers.
- For non-fatal effects, the severity is characterized by a weight, with zero being unaffected and one being tantamount to fatality, and the contribution is integrated over the duration of the effect. Depending on the disease, as shown in Figure 15.1, the effect may be prolonged then recede (upper left), may be sudden and fatal (upper right), or there may be a short acute phase

with delayed sequelae (bottom). Combinations of these effects are possible depending on the illness and the population. This is termed years lived with disability.

The significant difference between DALYs, which are more widely used for characterizing population impacts, and QALYs, which are more widely used for individual impacts, is that the weights for DALYs (termed disability weights) are disease-specific and intended to be objective in nature, whereas for QALYs the weights (termed quality-adjusted) are based on perceptual impact by individuals [18].

For an individual disease, it is possible to compute the DALY impact per case (or per infection). This is shown in Table 15.1. The difference in effects among diseases is apparent. For example, giardiasis and cryptosporidiosis, both important waterborne pathogens, have low (0.002 and 0.003, respectively) ratios of DALYs per infection. At the same time, hepatitis A and legionellosis, also

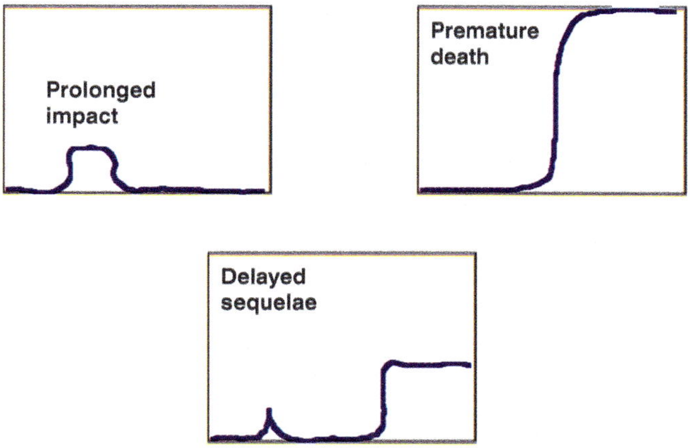

Figure 15.1 Hypothetical time course of disability weights for three circumstances.

Table 15.1 Ratio between DALYs and infections for selected pathogens transmissible by ingestion or inhalation.

	DALY/Infection
Campylobacteriosis	0.035
Cryptosporidiosis	0.003
Giardiasis	0.002
Hepatitis A	0.17
Norovirus	0.003
Shigellosis	0.026
Salmonellosis	0.035
Toxoplasmosis	4.52
Influenza	0.026
Legionellosis	0.97
Q fever	0.19

Note. Based on data from The Netherlands.
Source: Adapted from [20].

transmissible by water (in the case of *Legionella*, mainly by inhalation of aerosolized water), have higher ratios.

A simple example of the calculation of DALYs is given in Example 15.1

Example 15.1 Example Calculation of DALYs

A population of 100,000 is exposed to an infectious agent. Hundred persons experience a two-day mild illness (disability weight 0.05) and an expected 0.001 death (with an expected lifetime at death at 30 years). What is the consequence in DALYs per capita?

Solution:

To compute the years of life with disability, the incidence is multiplied by the duration and the disability weight:

$$\text{YLD} = 100 \cdot \left(\frac{2}{365}\right) \cdot 0.05 = 0.027$$

To compute the years of life lost, the number of deaths is multiplied by the expectancy lost:

$$\text{YLL} = 0.001 \cdot 30 = 0.03$$

The total DALYs lost is the sum of these two, i.e., 0.057 years. Therefore, the annual rate is taken by dividing by the number exposed (10^5), and hence the risk can be expressed as: $\frac{0.057}{10^5} = 5.7 \times 10^{-7} \frac{DALY}{person - year}$ or $0.57 \frac{\mu DALY}{person - year}$.

The DALY approach has been adopted by the WHO. As a target, a risk of 10^{-6} DALY per person per year, i.e., 1 μDALY, is regarded as an acceptable risk target for drinking water [21].

DALYs have not been used in the United States for enforceable risk targets in any environmental application. However, a 2025 analysis by the US EPA developed log reduction value (LRV) targets for wastewater reuse using a DALY approach, and the WHO acceptability recommendation, for several target pathogens [22].

Similarly, Hamilton et al. [23] developed risk-based acceptable water concentrations for exposure to aerosolized *Legionella* from domestic uses, including showers, sinks, and toilets, derived from the WHO target. The DALY-based criteria were substantially lower (more stringent) than those based on a 0.0001 annual infection risk line, driven by the greater severity weight of legionellosis (as shown in Table 15.1).

A significant aspect of the DALY approach is that it can be used to assess the consequences of other exposures, such as chemical or radiological exposures. This makes it a very useful tool in deciding on interventions using a pure risk–risk trade-off approach without necessarily considering economics.

In finished drinking water, the presence of oocysts of *Cryptosporidium parvum* poses a risk in the general population, predominantly gastroenteritis of a spectrum of severity, but much more severe in persons with AIDS. One control is the disinfection of water with ozone; however, this can result in the formation of disinfection by-products, including carcinogens such as bromate. Havelaar et al. set forth an explicit risk balancing with application to the Netherlands using DALYs to account for the infection risk and the illness from carcinogenic by-products [24]. They concluded that, when overall DALYs from both risks were computed, applying ozone yielded a net benefit despite the formation of carcinogenic by-products; however, they noted considerable uncertainties regarding both risks.

15.3.2.1 Critiques of HALY

The use of DALYs (or QALYs) is more complex than calculating total cases and contains a number of details within the calculation that may introduce bias and/or be subject to contextual factors. One issue is that the disability weights are assumed to be universal. However, Reidpath et al. note that in a global context they would differ with the degree of development of a country [25]. Very likely, even within a developed country, localized disability weights might be appropriate to account for disparities in access to medical care and, as was done by Havelaar et al. [24], preexisting and concomitant medical conditions.

Chen et al. [19] have reviewed the use of DALYs and have noted a variety of other issues, including some of the following:

- Some calculations of DALYs incorporate a "discounting" factor that values life lost in the future, or disability occurring in the future, at a lesser current value. This, ethically, may imply that future adverse consequences are given lesser significance than current ones.
- If a fixed value for the total expected life for all members of the population is used, this results in a situation where adverse effects in individuals of ages beyond their expected life who incur mortality from the hazard are regarded as inconsequential to the DALY total. If the conditional life expectancy prior to exposure to the hazard is used, the calculation becomes more complex.
- Accounting for comorbidities, or coincident exposures in a DALY framework, is not well defined, particularly where interaction between the multiple stressors exists.
- Some applications of a DALY approach incorporate age weighting, in which a disability occurring during a "higher value" (e.g., adults during their working years) age is accorded greater numerical significance, which leads to further ethical issues.

15.4 Balancing Interventions with Benefits

There are two broad approaches to evaluating the costs of interventions against the benefits in terms of risk reduction that may be achieved. When the risk reduction (i.e., impact) is kept on a natural scale, such as mortality, morbidity, or DALYs, then a cost-effectiveness approach is used. When both risk reduction and costs can be expressed in a common scale, such as currency, then formal cost-benefit methods can be used [26].

While it is possible to translate health benefits, such as DALYs, into currency, as will be shown, this raises additional sources of uncertainty. Further, there are philosophical issues, such as those noted by Ashby (Section 3.1.3) [27], e.g., it is difficult, if not impossible, to assign economic costs to some attributes of risk management. Examples might include, for example, enhancement or diminution of an ecological resource.

Also, as noted by Cameron, "However, not everything that can be counted counts, and not everything that counts can be counted" [28]. Therefore, any attempt to value risk reduction, or to assign costs to an intervention, is likely to be incomplete and imperfect.

Nevertheless, economic valuation is often desired, or even required in some circumstances, in order to more fully inform risk management decisions.

15.4.1 Direct Economic Valuation

The aim of economic valuation of consequences is to assign a value as fully as possible to the adverse impacts (prevented by an intervention) from an exposure. There are multiple layers of such costs that could be considered:

- Direct costs to the individual associated with mortality and morbidity
- Healthcare costs
- Other direct, non-healthcare costs
- Indirect societal economic impacts

These become progressively more difficult to evaluate and will be considered in succession.

15.4.1.1 Direct Adverse Effects

The methodology for valuation of direct effects is frequently expressed in terms of "value of a statistical life (VSL)." The VSL is typically obtained by surveying people on their willingness to forgo other benefits for a diminution/increase in their life expectancy [29]. This varies very much between countries and appears to correlate strongly with the gross national per capita income. Table 15.2 tabulates VSL used for policymaking among US government agencies.

There are significant methodological issues in arriving at such values. These include conveying complete information to individuals being surveyed and distinguishing between expressed preferences in surveys and revealed preferences, as, for example, in differential wages across occupations exposed to different levels of risk (as noted early by Starr [9]). Methodological methods used to derive such values have been reviewed by Kniesner and Viscusi [30].

In principle, the economic value of a DALY (or QALY) can be obtained by dividing the VSL by the average life expectancy. For example, using the midpoint EPA value from Table 15.2 of $7.4 million and a US life expectancy of 78 years, a per-DALY value of $96,000 is obtained in 2006 dollars. Using a consumer price index inflation factor of 1.6, this would result in a 2025 estimate of $154,000. Neumann and Kim reviewed economic estimates for life years used in cost-effectiveness analysis in medical contexts and noted a range of $100,000–$150,000 in 1990–2021 studies [31]. It is also important to note that this results in equal valuation regardless of the age at which an adverse effect occurs; it has been argued, based both on empirical evidence and equity arguments, that an age-specific value of a statistical life year is more appropriate [32].

15.4.1.2 Healthcare Costs

Healthcare costs have been estimated for diseases transmitted by water. Collier et al. [33] calculated the costs from emergency room visits and hospitalizations of 17 of the most frequently associated waterborne diseases. They estimate the total cost to be about $500 per case. This excluded the costs of mild cases that did not seek emergency room or hospital care.

Table 15.2 Value of statistical life by several US agencies.

Agency	Year	VSL range (million $)
US EPA	2006	2.7–12.1
US Dept. of Health and Human Services	2014	4.4–14.2
US Dept. of Transportation	2015	5.4–13.4

Source: Adapted from [29].

Table 15.3 Medical and productivity costs for 1993 Milwaukee *Cryptosporidium* outbreak per case (costs in 1993 dollars[3]).

Severity	Medical	Productivity
Mild	2	113
Moderate	62	413
Severe	6399	1409

Source: Adapted from [35].

Using a different methodology, focusing on recreational waterborne diseases, DeFlorio-Barker et al. [34] developed separate cost estimates for mild, moderate, and severe cases. Their estimates also included the economic impacts of lost productivity. For the three severities, the per-case costs were (excluding mortality for the severe cases) $9.50, $233, and $12,000, respectively.

15.4.1.3 Indirect Factors

There can be a number of indirect costs associated with diseases. These include lost wages (which were included in DeFlorio-Barker et al. [34]), caregiver costs (i.e., direct costs of individuals so employed, or lost wages from family and friends), and macroeconomic effects resulting from lost productivity. The few estimates available come from in-depth studies of specific outbreaks.

The 1993 drinking water outbreak of cryptosporidiosis in Milwaukee, Wisconsin, was the largest such outbreak in the United States and has been extensively studied. The per-case cost (other than direct morbidity and mortality) was estimated by Corso et al. [35]. The results are summarized in Table 15.3. Productivity costs include lost wages for affected individuals and caregivers. Note that, for mild-to-moderate cases, productivity losses substantially dominate medical costs.

The costs associated with a 2007 drinking water outbreak in Galway, Ireland, were studied, and costs accruing to both individuals and public institutions were estimated [36]. This study also considered factors such as the cost of alternative water during a boil-water notice, costs to businesses (hotels, nurseries, etc.) from lost business, and public sector costs to modify and enhance inspection of drinking water systems. The principal cost elements are shown in Table 15.4. It is noteworthy that the cost of a precautionary shift to alternative water supplies (such as bottled or boiled water) was the largest cost element (other than upgrading the water treatment system, not shown). The direct and indirect costs to actual cases (either lab-confirmed or presumed) were exceeded by the various indirect costs. This may not be true for other pathogens with greater degrees of severity. When the total costs from Table 15.4 are divided by the total cases (lab-confirmed and other), a per-case estimated cost of €8425 (in 2007 costs) is found.

Using US data from 2014, with an estimated 11,000 cases of legionellosis, of which 98% were estimated to require hospitalization and 9% resulted in death, Baker-Goering et al. [37] estimated the total economic burden. The medical costs were estimated at $36,500 per case. The direct productivity costs[4] were estimated at $1970 per case. The economic losses from premature deaths, considering the age distribution of legionellosis fatalities, were $37,400 per case.[5] Importantly, it

3 To convert costs to 2025 dollars, the Consumer Price Index ratio of 2.24 might be used.

4 Only based on forgone wages, rather than full macroeconomic impacts.

5 When expressed on a per fatal case basis, $413,700, which is considerably less than the VSL since fatalities tend to occur in older individuals.

Table 15.4 Estimated costs for *Cryptosporidium* outbreak in Galway, Ireland. Total population of 120,000 persons, with 242 laboratory-confirmed cases and 498 estimated other cases.

Cost element	Amount (2007, €)
Direct costs of illness	374,238
Medical costs	1,258,641
Lost wages and school days	180,308
Caregiver list wages	179,233
Productivity losses from cases	195,500
Productivity losses from caregivers	94,335
Community alternative water supply	3,952,461

is worth noting that the costs from non-fatal cases were of a similar magnitude to those associated with premature deaths.

Infectious diseases, especially when highly prevalent or in the context of an outbreak (or, especially, a pandemic), will have broader economic impacts. The use of econometric computable general equilibrium models to estimate these broader impacts has been illustrated in the case of influenza [38]. This study estimated that a pandemic influenza outbreak in the United States could result in a loss in gross domestic product (GDP) of $45 billion, and with vaccination,[6] the impact could be reduced to $34 billion.

A final cost element of relevance in many cases is that associated with institutional responses and modifications necessitated by the disease itself, particularly during significant outbreaks. This will be quite location-specific. However, there are a number of examples of such costs:

- After the 1993 *Cryptosporidium* outbreak from Milwaukee drinking water, it was estimated that upgrading the water treatment plants in that city would cost $90 million [39]. Arguably, since that outbreak triggered a broader national response, with concomitant additional costs incurred by many other water agencies, the economic consequences have been substantially greater.
- Similarly, the outbreak in Galway, Ireland, triggered expenses at the utility in excess of €6 million.
- The costs of investigating the sources of an outbreak, with respect to both environmental sampling and epidemiological assessment, can be high. This has been documented in the case of a legionellosis outbreak in London, UK [40].

15.4.1.4 Propagation of Uncertainties

As with other aspects of risk characterization, the valuation of benefits will introduce uncertainties. These might include the values used for VSL, disability weights for DALYs, disease severity, and economic valuation of morbidity and mortality. Therefore, in conducting the valuation, these additional sources of uncertainty should be propagated through the analysis, for example, by including input distributions in a Monte Carlo analysis.

6 It must be noted that this study was published before the COVID-19 pandemic, and it is likely that the post-COVID vaccine resistance may be greater and would reduce the differential in savings.

15.4.1.5 Secondary and Indirect Costs and Benefits

Intervention to reduce a microbial risk may introduce secondary benefits and/or adverse side effects. These should be accounted for in any true reckoning of costs and benefits. While it is impossible to foresee *all* possible outcomes of a potential decision, there should be a thorough effort to do so. Some examples of these ancillary costs and benefits in the context of interventions to reduce microbial risk are listed as follows:

- The disinfection of drinking water by chlorine can result in the formation of a spectrum of disinfection by-products, some of which can have adverse health effects [41]. While other disinfectants, such as ozone, can be used, they can also produce by-products of health concern [42].
- Arguably, the concern about disinfection by-products that developed in the mid- to late 1970s resulted in a reduction in the use of chlorination for water disinfection in South and Central America, which has been a factor attributed to the cholera epidemic in the early 1990s [43].
- During the COVID-19 pandemic, indoor air filters, including the do-it-yourself Corsi-Rosenthal Box (Section 10.2.1), were widely deployed. It has been found that these systems can be effective in reducing indoor particulates from wildfire smoke [44] and presumably other sources.
- The use of indoor air purifiers using ionizing technologies, or far UV light, deployed for microbial reductions, has been found to produce potentially carcinogenic organic by-products and/or ultrafine particulate matter [45].

15.4.1.6 Future Costs and Benefits

When valuing costs and benefits, it is important to take into account *when* the costs and benefits occur. In a general sense, costs and benefits in the future are "worth" less in the present than they would be if they happened in the present. Costs of an intervention may occur in the present, such as capital investment or the purchase of equipment, and may also recur in the future as operational and maintenance costs. Benefits in terms of risk reduction would accrue to a population over time.

For purely economic costs and benefits, it is standard practice to use a discount rate to evaluate future events. In other words, how much money would be needed today to fund the future stream of expenses, or how much money received today would be equivalent to the future stream of income. One common approach in engineering is to compute the "net present value" (NPV) based on a stream of income and/or expenses and a discount rate.[7] Given a stream of benefits (B_i) and costs (C_i) occurring over time, given a discount rate (interest rate) r, the NPV is given by Equation 15.1. The parameter "n" is the time horizon, which can be thought of as the duration over which the costs and benefits occur. The time horizon may be based on the intervention's design life or the policy's durability.

$$\mathrm{NPV} = \sum_{i=0}^{n} \frac{B_i - C_i}{(1+r)^i} \tag{15.1}$$

This formulation uses a common discount rate for costs and benefits, which is the most frequent practice used in regulation.

7 NPV is most commonly used in public-related investments. In purely private sector investments, another approach, the internal rate of return has been used, which will not be discussed further.

The effect of any discount rate of benefits philosophically is to regard benefits that occur at a future time as being of less value than they would be if they occurred at present. Some have argued that it is therefore appropriate to use a different, often lower, "social discount rate" when assessing the value of future benefits [46].

The value of the discount rate "*r*" may influence the intervention's favorability or unfavorability. For example, a capital-intensive intervention that yields future benefits will be relatively disfavored under a higher discount rate than under a lower one. Since the discount rate may be specified by regulators as a policy matter, this impacts the valuation of choices. For many years in the United States, a discount rate of 3% was used; however, as of 2025, the US Office of Management and Budget policy has reverted to a historical value of 7%.

The effect of the discount rate is illustrated by Example 15.2.

Example 15.2 How Discount Rate Influences Net Present Value: An Example

Consider an intervention that will cost $10,000 in initial capital costs and $150 in annual operating costs. It produces a benefit of $1000 annually. The intervention lasts 20 years (before needing replacement), and the annual inflation rate on operating costs is 4%. What is the NPV (benefit) to this intervention using a discount rate from 3% to 15%?

Solution:

Equation 15.1 is used for this problem. For each year from 1 to 20, the inflation rate is applied to the costs and benefits, and then the NPV is computed. The results for the range of discount rates are shown in the following figure.

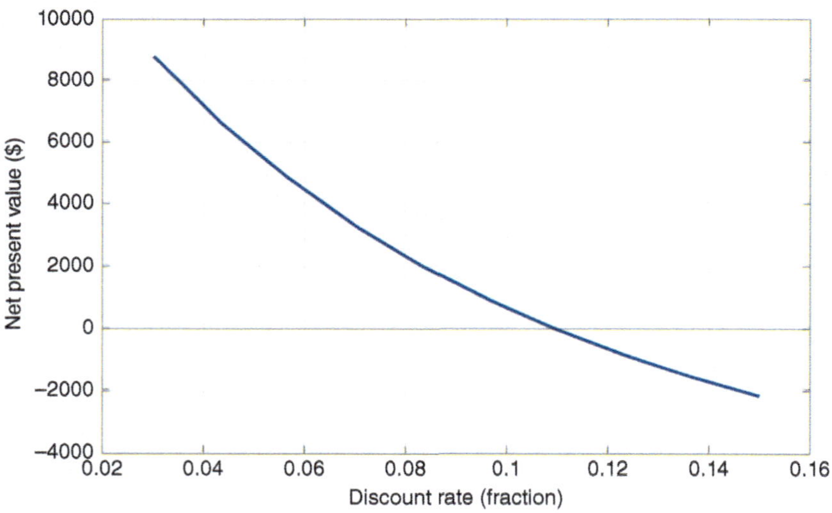

Note that the investment has positive NPV (i.e., net benefits) if the discount rate is less than 11%, whereas at higher discount rates, the investment is not favorable.

15.4.2 Cost-effectiveness Analysis

Cost-effectiveness is a useful approach when it is either not possible or not desirable to evaluate benefits and costs in commensurate units, such as currency. Perhaps its earliest uses were deciding between discrete choices in a medical context, and basic concepts were outlined by Shepard and Thompson [47]. These are restated in an environmental context as follows:

1. Define the program and alternatives. The alternatives might be different types of interventions. For example, consider the mitigation of indoor air transmission of infectious disease. Alternatives might include increasing ventilation by a certain amount, mandating masking, or introducing Corsi–Rosenthal boxes.
2. Compute the costs of each of the proposed alternatives, using NPV to properly account for the initial as well as future operational costs. If there are monetary benefits, they should be considered reductions in initial and/or future costs, as appropriate.
3. Compute the benefits in terms of risk reduction using natural units, such as reduction in illnesses, deaths, or DALYs. Any negative side effects should be considered as well. The stream of net benefits should also be discounted to the present.
4. The cost-effectiveness for each of the alternatives under consideration is computed by dividing the NPV of costs by the present value of the risk reduction.

Using a pure cost-effectiveness approach, the alternative yielding the lowest ratio would be preferred. However, within the framework of analytic deliberation, risk managers and/or stakeholders may prefer another option.

In situations in which policy alternatives do not consist of a set of discrete options but instead lie along a continuum—such as the selection of a specific regulatory, guidance, or operational numerical value—the manner in which cost-effectiveness analysis is applied becomes correspondingly more complex and must be adapted to account for this continuous decision space.

Conceptually, consider a decision to be made about the level of ventilation (air changes per hour) in a room or the extent of cleaning and disinfection of a surface for reduction in fomite exposure. Both of these may be expressed on a continuum. As the intervention's stringency increases and its cost rises, risk decreases. However, for many interventions, as shown in Figure 15.2, the marginal effectiveness (slope of the curve) diminishes with increasing stringency. Using a cost-effectiveness approach, a decision might be made somewhere at the "knee" of the curve, which is clearly a subjective measure. Alternatively, the reciprocal of the slope of the curve, $\Delta \text{Risk}/\Delta \text{Cost}$, is the marginal cost of additional risk reduction. This may be compared with marginal costs from other risk interventions and, after deliberation by risk managers and stakeholders, may be at a point where similar marginal costs are regarded as appropriate.[8]

As noted previously, action to reduce a primary (microbial) risk may result in the emergence of a secondary countervailing risk, which may increase with the stringency of the primary intervention. This is shown conceptually by Figure 15.3. If the primary and countervailing risks can both be expressed in the same units, such as DALYs, then one decision point may be to set policy at the point of total minimum risk. However, given the uncertainties, both risk curves might have broad

8 However, recall that there may be amplification of the perception of such risks as shown in Figure 3.3. Therefore, it may be that a particular risk accepted at a given marginal cost of reduction may be less than or greater than another risk.

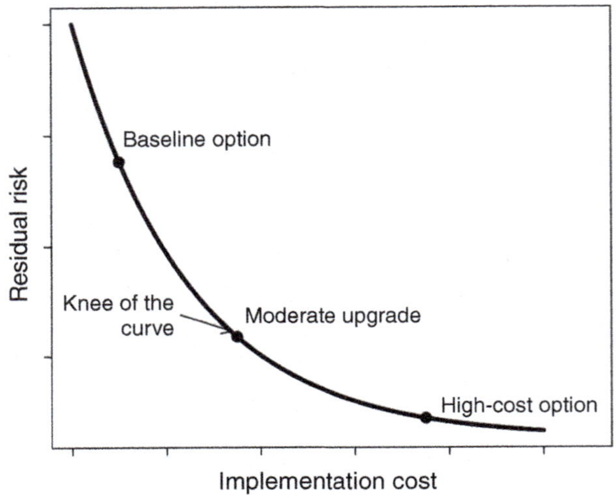

Figure 15.2 Schematic cost-effectiveness curve with a continuum of alternatives.

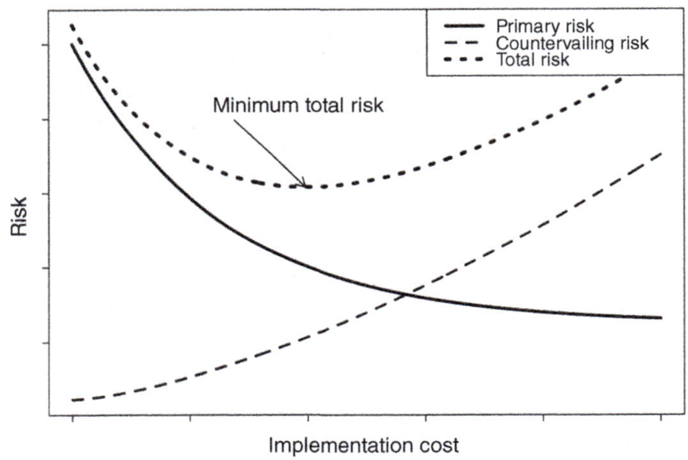

Figure 15.3 Cost-effectiveness curve with countervailing risk.

confidence bands, and thus there may be a large region in which the total risk is plausibly a minimum. In such cases, other factors might be used to inform a policy decision.

As an example of the application of cost-effectiveness analysis, Haller et al. [48] estimated the costs of various levels of improved water and sanitation services in underserved regions and also estimated the potential DALY reduction for individual interventions. These were compared with the regional per capita GDP. On this basis, point-of-use water disinfection was highly effective in all locations, as was the combination of improved water supply and sanitation plus disinfection. Piped water and sewer services were cost-effective only in some regions.

15.4.3 Formal Cost-benefit Analysis

15.4.3.1 History

Cost-benefit analysis (CBA) has its conceptual roots in early 19th century public works economics, but its earliest applications emerged first in public health sanitation and, slightly later, in environmental infrastructure and flood control. Chadwick's "Report on the Sanitary Condition of the Labouring Population of Great Britain" (1842) is widely regarded as the first systematic application of cost-benefit reasoning to public health in which he qualitatively compared the costs of sanitary reforms (drainage, sewerage, refuse removal) with the economic benefits of reduced mortality, fewer funerals, increased labor productivity, and reduced poor-law expenditures [49]. In the 1850s in the United Kingdom, Farr produced mortality reports linking population health to economic productivity. Farr developed life tables and estimated the monetary value of reduced death rates, laying quantitative foundations for later public-health CBA. In the 1910s, during the construction of the Panama Canal, the utility of vector control programs for tropical diseases such as yellow fever was justified by comparing costs and benefits [50]. By the 1930s, US municipalities were performing explicit benefit–cost calculations for vaccination campaigns and vector control programs. By the time of the 1936 enactment of The Flood Control Act, it was widely recognized that a comparison of costs and benefits, both direct and indirect, was an appropriate metric for deciding on large infrastructure projects [51]. This established the first legal benefit-cost test in environmental management. In the 1960s, cost-benefit testing for environmental decision-making became more widely adopted, with books being written [52] and the direct connection to bacteriological standards for water being expounded [53].

15.4.3.2 Methodology

Unlike cost-effectiveness analysis discussed earlier, and multicriteria decision analysis (MCDA) (to be addressed in Section 15.4.4, CBA requires placing the benefits of risk-reduction (and possibly increases in countervailing risks) on the same monetary scale as costs through approaches discussed in Section 15.4.1. Once that is done, the "optimal" decision, from a CBA perspective, is obtained by maximizing the NPV among a set of candidate decision options.

Given the previously discussed valuation of effects, conducting a CBA involves comparing the NPV of the change in risk (the reduction in the primary impact minus the increase in any countervailing effects) to the NPV of implementation costs. The specifics of implementation will be problem-specific and can be illustrated by summarizing several individual cases.

Analysis of Boil Water Orders When finished drinking water sampling finds the presence of pathogens or high levels of indicator organisms, a frequent response has been for the water utility to issue a boil-water advisory. This recommends that consumers heat their water to a vigorous boil for at least one minute [54] prior to consumption (after cooling) or food processing.

Ryan et al. [55] used CBA to determine at what concentration of *Cryptosporidium parvum* oocysts, and what corresponding level of risk, could justify the issuance of a boil-water advisory. They considered the health benefits of the risk reduction and the costs of preparing boiled water (or purchasing and hauling bottled water). This study did not consider indirect costs of time, nor the potential adverse effects (from boiling) of scalding or fires. Nevertheless, they determined that a positive benefit-cost existed when the projected daily risk from oocysts exceeded 9/10,000, corresponding to an oocyst concentration of 4.6 oocysts per 100 L in finished water.

Modifying Drinking Water Treatment Processes Bergion et al. considered drinking water treatment options for pathogen reduction in the case of a small city (330,000 connections) in Sweden [56]. The community water was supplied by a small lake, which contained about 2800 onsite wastewater treatment systems (septic tanks, etc.). The options considered in their analysis were the connection of 25%, 50%, or 75% of the onsite systems to the local wastewater treatment plant, or installing UV as an additional treatment barrier at the drinking water treatment plant.

These authors accounted for the removal of pathogens (their reference organisms were *Cryptosporidium*, *Campylobacter*, and Norovirus) in the source water lake, in the drinking water treatment plant, and in subsurface infiltration from the onsite wastewater systems. Risk was quantified using a QALY approach and the valuation of consequences. The NPV associated with each proposed change was also quantified. Uncertainties in all of the inputs were also considered.

The analyses showed that only the installation of UV could reliably meet a WHO target of 10^{-4} infections per year, and that all of the alternatives resulted in a negative NPV, i.e., they were economically disfavored. However, the authors suggest that favorability might have been achieved if other benefits had been monetized, including increased property values and improved animal health (by diverting wastewater treated only via onsite systems from surface water).

15.4.3.3 Critiques

CBA is not without its criticisms. Aside from methodological issues in the development of cost functions and the assumptions of discount rates, as noted earlier, there are substantive philosophical and ethical issues.

Possibly one of the earliest critiques of CBA for environmental and health standard setting was Kelman [57], who noted three principal objections, paraphrased as follows:

- Pure CBA requires putting all costs and benefits in commensurate units and deciding based on pure maximization of net benefits. This derives from the ethical framework of utilitarianism, frequently phrased as "the greatest good for the greatest number." However, there may be instances in which some decisions are clearly wrong, even though they lead to maximization. The many variants of the trolley problem [58] illustrate circumstances in which net maximization may lead to decisions that many may find discomfiting.[9] The idea of right and wrong actions stems from an alternative ethical framework of deontology, as expressed initially by Kant.
- For certain benefits and costs that are not marketed, there are strong ethical bases for not forcing them into a monetary frame.
- If some benefits and costs are not readily assigned value by a market, it may not even be justifiable to devote efforts to develop a (fictitious) value to them.

Given these critiques, several alternatives exist. It is possible to modify the analytical objective functions, as noted in the following section, to reflect deontological constraints. It is possible to use a comprehensive MCDA approach (Section 15.4.4) to incorporate incommensurate preferences and involve multiple stakeholders. However, as others have noted [2], this leads to an analytic-deliberative approach in which the risk assessors' roles are to educate all stakeholders in the process, rather than primarily to provide data for a calculation.

9 The simplest version of the trolley problem, paraphrased from Bruers and Braeckman [59], is: A trolley is moving toward five people on a main track. You (the decision-maker) are standing at a switch. If you turn the switch, the trolley will be diverted to a side track, but there is one person on this side track. Turning the switch will result in that person's death, and the five people on the main track will be saved. Pure utilitarianism would lead to a decision to divert the switch. However, the active intervention to *cause* a death may have a strong moral objection.

Lowry and Peterson [60] argue that constraining the space over which CBA optimization can occur allows the incorporation of absolute deontological principles.

One limitation of classical CBA is the lack of consideration of who pays and who benefits, and particularly of the wealth of those who pay and benefit. It can be argued that the "pleasure"[10] derived by giving $100 to an individual who has a wealth of one billion dollars is less than that derived by giving the same $100 to an individual who has a wealth of $10,000. It is possible to incorporate this type of differential in CBA by defining different utility functions for different subpopulations and then seeking to maximize total social utility rather than total social (monetary) benefits [61].

Kasperson [8] argued at an early stage that the entire framework of identifying a single acceptable risk, whether by CBA or cost-effectiveness, was fraught with difficulties for multiple reasons. Paraphrasing his arguments:

- The concept of finding the best solution requires reconciliation of competing social values, the scientific task of estimating and measuring risk, and developing information that needs to be separate for an acceptability decision. These competing values may justify alternative approaches, including what has come to be called MCDA, *infra*.
- Acceptability as a social construct incorporates other values such as equity, availability of alternatives, and public preferences.
- Risks cannot always be made entirely voluntary, and so a prime consideration should be to minimize unavoidable risks and fairly compensate those who may bear a residual risk.
- Fairness is best achieved by active participation of all in the decision process, which leads to an analytical deliberative framework.

Fischoff [62] notes the importance of full communication at all stages to enable an effective analytic deliberative framework:

> "To realize the potential of their science, analysts must create partnerships with the stakeholders who depend on their work, so that analyses are relevant and understood. To that end, analysts' professional standards should require full disclosure of the scientific and ethical judgments made in formulating and executing analyses. However, those disclosures may not satisfy stakeholders unless they are preceded by continued engagement with them. Together, analysts and stakeholders need to determine how best to invest analytical resources, resolve definitional issues, and interpret the resulting estimates."

15.4.4 Formal Multicriteria Decision Analysis

Due to difficulties in quantifying *all* costs and benefits in commensurate units, alternative approaches that preserve the multidimensionality of a decision have been developed over the last several decades. Some of these methods also lend themselves to active collaboration with multiple stakeholders. The key is that multidimensional methods use weighting techniques to avoid explicitly requiring all costs and benefits to be expressed in the same units.

The central concept is that the performance of individual metrics for a decision option is evaluated against defined criteria. Each criterion may be described by performance on several attributes.

---

10 Economists would use the term "hedonic value."

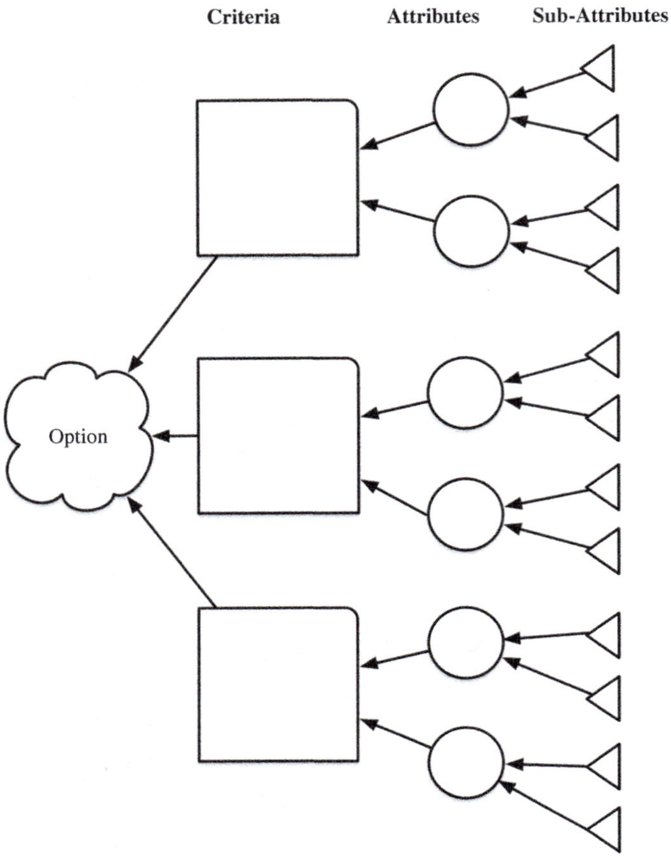

Figure 15.4 Hierarchy of criteria, attributes, and sub-attributes.

Each attribute may be described by performance on several sub-attributes. This forms a hierarchical tree, as depicted in Figure 15.4. By developing such a tree and rules for "rolling up" the performance metrics, the suitability of one option can be assessed. As an example, Julien et al. [63], in developing aggregations for health effects from both microorganisms and disinfection by-products (a criterion) associated with drinking water treatment, used a hierarchy with two attributes: toxicity (for the disinfection by-products) and infectivity (for the microorganisms). The sub-attributes for infectivity include dose–response potency, presence of virulence factors, and presence of antibiotic resistance genes. Formal costs might be a criterion, with attributes of capital cost, operating cost, etc. Operating costs might include sub-attributes such as labor, energy, etc. Other environmental impacts might be considered criteria with attributes such as land requirements, local noise, construction-related disturbance, air emissions, etc. The air emissions might have sub-attributes for particular constituents (including greenhouse gases).

Multiobjective optimization is commonly employed in environmental engineering to examine technical trade-offs, such as the balance between cost and risk or performance and reliability. However, environmental decision-making is typically an MCDA problem. MCDA selects from a set of management or policy options by evaluating them against diverse, often incommensurate criteria.

Such criteria may include public health protection, ecological impacts, equity, regulatory feasibility, and stakeholder acceptability. With MCDA, it is essential to explicitly represent preferences and trade-offs. A foundational reference in this area has been published [64], along with more recent reviews (some cited as follows), so only a high-level summary is provided here.

Typical steps in undertaking an MCDA, extending and elaborating the approach by Edwards [65], are listed as follows:

1. Identify all the organizational and individual stakeholders whose preferences should be maximized.
2. Identify the issue(s) or decision(s) to be resolved or made.
3. Identify the criteria/attributes/sub-attributes that contribute to the overall utility of the solution.
4. Identify the measures or characteristics that affect the utility of the criteria/attributes/sub-attributes.
5. Rank the importance of each of the criteria/attributes/sub-attributes within their tier. This may differ among the individual stakeholders.
6. Quantify how the measures or characteristics of each criterion/attribute/sub-attribute contribute to the *utility* of that factor. This may be a linear or nonlinear function. The purpose of these utility functions is to express each underlying input on a common metric.
7. Within each attribute, determine its overall utility by combining the utility of its sub-attributes, considering their weights. Within each criterion, similarly combine the utilities of the attributes, weighted by their weights. The contributions at each stage may be formulated using an additive or multiplicative approach (see the following).

Significant effort is required to ascertain the weights of the individual criteria, attributes, and sub-attributes, as well as the pertinent utility functions as they relate to the characteristics of the options (and their underlying design variables). This can be approached through combinations of expert judgment, polls, and surveys (the latter two options are particularly well suited to situations involving large numbers of stakeholders). Early work compared approaches to generating weights using three methods: ranking, rating, and paired comparisons [66].

As an example of weighting, Linkov et al. [67] used a multicriteria approach to assess risks of uses of particular nanomaterials. They developed criteria and attributes and formed weighting factors as summarized in Table 15.5. They did not incorporate economic factors, nor did they go down to the level of detail of sub-attributes. However, for this problem, the construction of sub-attributes for some of the attributes could have been devised—for example:

- Public health effects could have been decomposed into acute and chronic.
- Environmental effects could have been decomposed into those occurring during manufacturing and those from use, or perhaps via media (air, water, and land).
- Both public and political preferences could be divided into particular classes of those entities.

The process of generating utility functions for individual attributes or sub-attributes requires complex discussion with decision-makers and stakeholders, as explicated by Keeney [68]. The objective is to translate a range of an individual performance metric (e.g., chronic health effects, capital cost, etc.) into a value between zero and one.[11] This may be linear over the range; for example, in the case of chronic health effects, utility would be zero for any value deemed utterly

---

11 Note that the performance metric itself may be a function of the system design variables. For example, if the performance metric is the infection risk from a pathogen to occupants in a room with various potential ventilation rates and intensities of ultraviolet germicidal radiation, the performance would be a function of the particulars of ventilation rate, light intensity field, etc.

Table 15.5 Weighting factors for criteria and sub-attributes for nanomaterial risk assessment.

Criterion	Attribute	Weighting	Sub-weighting
Health and environmental effects		0.582	
	Public health effects		0.5
	Occupational effects		0.25
	Environmental effects		0.25
Societal importance		0.309	
	Manufacturing		0.474
	Consumer products		0.376
	Alternatives		0.149
Stakeholder preference		0.109	
	Political preferences		0.55
	Public preferences		0.21
	Scientific preferences		0.24

Source: Adapted from [67].

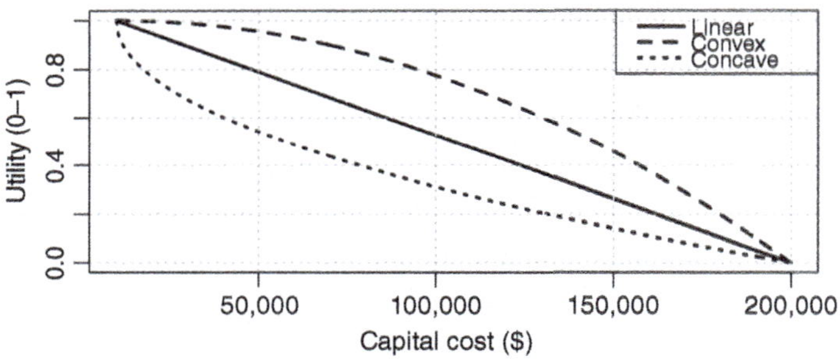

Figure 15.5 Example sub-attribute utility functions for cost.

unacceptable, and one for a value regarded as either *de minimum* or zero. Or the utility could be monotonic but nonlinear over the range. An example of these functions, for linear, concave, and convex forms, is shown in Figure 15.5. The nomenclature adopted is that $u_{ijk}(Z_{ijk})$ is the utility function of the kth sub-attribute of the jth attribute of the ith criterion, where Z_{ijk} is the corresponding performance function.

In the case of sub-attributes that are characterized not by continuous measurement, but perhaps only by discrete characteristics, rather than a continuous function, the utility will be a set of individual points at each of the discrete characteristics. Also note that the property of the utility function is identical to that of a cumulative distribution function (if utility *increases* with value of the argument) or, like cost, identical to a complementary cumulative distribution function (if utility *decreases* with the argument), so methods used for estimating such distributions can be used to estimate utility functions.

Measurement Scales

There are different types of measurement scales that might describe decision outcomes or the design characteristics of the alternatives. One purpose of the utility function is to transform the outcome variable to a consistent scale. Stevens [69] defined four types of measurements. These can be expressed as follows:

Nominal: These are only described by a label that conveys no numeric information. For example, granular filtration versus fabric filtration (Section 10.2), or the use of UV radiation versus electron beam radiation (Section 11.2.4).

Ordinal: Information conveyed only regarding ranking. For example, Class A versus Class B biosolids (Section 7.7.2), where there is a relative ranking with regard to the potential contaminant concentrations.

Interval: Where the differences in values between measurements convey information. Numerous examples exist, such as temperature measured in Fahrenheit or Celsius scales.

Ratio: Typically, these occur where there is a defined absolute range. Temperature in Kelvin is one such example. Mass, concentration, and flow rate are other examples.

Once the utilities are at the most detailed level of the tree (e.g., Figure 15.4), they are aggregated to the next level (e.g., sub-attributes to attributes) by the weights (derived, such as by one of the approaches described by Eckenrode [66], or more modern methods [70]). There are two main approaches to this.

As a preliminary, given the sub-attribute utilities, u_{ijk}, it is first desired to aggregate them to get the utilities of each attribute, designated as $u_{ij.}$, and then use the attribute utilities to get criterion utility, $u_{i..}$, and finally the get the utility of the decision option, $u_{...}$. The weights provide the linkages between levels in the hierarchy.

The principal approaches to "rolling up" utilities are additive and multiplicative. The additive approach produces a weighted sum of the component utilities. For example, for the overall option performance in terms of the criteria, Equation 15.2 can be written, where w_i is the criterion weight.

$$u_{...} = \sum_i w_i \cdot u_{i..} \tag{15.2}$$

Analogously, the utility of a particular criterion can be written in terms of the attribute utilities and their weights.

The multiplicative form of combining utilities is written as Equation 15.3.

$$u_{...} = \prod_i (u_{i..})^{w_i} \tag{15.3}$$

The use of additive and multiplicative forms may, but not always, result in similar rankings of decision alternatives [70].

Over the last 50+ years, MCDA has developed in numerous directions. The approach presented earlier represents one common approach. There have been recent comprehensive reviews of multiple other approaches for those who seek to delve further into the methodologies [71–73].

The multicriteria approach presented here treats the parameters in the utility function(s) and the weights as problem-specific constants. As with parameters used for exposure and dose–response assessment, it is entirely appropriate to regard the parameters used in decision-analytic models as potentially uncertain.

It is possible to analyze the performance of different alternatives incorporating such uncertainties in the utility and weight parameters in a variety of ways. They can be modeled as probability distributions and analyzed using a Monte Carlo approach. It is also possible to use interval analysis or fuzzy set theory to model these uncertainties [74]. This would lead to a set of options that, given the uncertainties, are indistinguishable from one another.

Additionally, in the case of multiple stakeholders, each stakeholder (group) may have a different set of utilities and weights. In principle, this could result in different optima for various groups. Hence, in such circumstances, the multicriteria approach could result in a subset of possibilities that are partially acceptable, excluding others. In this manner, a potential consensus can be facilitated.

15.5 Discussion Questions and Problems

1 Consider the Ward rotavirus data from Table 14.2 and its analysis. Assume each rotavirus infection results in 0.0025 DALY. If only one infection occurs per year from consumption of drinking water, and the acceptable risk is 1 μDALY per person-year, what should the rotavirus be in finished drinking water?

2 Consider the scenario presented in Example 15.1.
   a What is the estimated economic cost associated with the computed DALYs?
   b Using the data presented in this chapter, if 75% of cases were mild, 20% were moderate, and 5% severe, estimate the total healthcare costs based on the information from DeFlorio-Barker et al. [34].

3 Consider the scenario in Example 15.2. What would be the "breakeven" discount rate if the inflation rate applied to both costs and benefits were 2%?

4 In an MCDA of municipal water treatment upgrades, one attribute of interest is the annual increase in household cost ($/household·year). Stakeholder elicitation produced the preference information shown in the table.

Elicited utility values for annual household cost increases.

Annual cost increase ($)	Utility
0	1.00
50	0.85
100	0.65
200	0.35
400	0.10
600	0.00

The utility scale is normalized so that a value of 1 represents the most preferred outcome and zero the least preferred. The stakeholder further indicated that small increases in cost at low cost levels are less objectionable than equal increases at higher cost levels.

- **a** Explain why a linear utility function is unlikely to be appropriate for this attribute.
- **b** Propose a functional form for a utility function $u(c)$ that is consistent with the information provided. Clearly state any assumptions.
- **c** Estimate the parameters of the proposed utility function using this data.
- **d** Plot the resulting utility function over the range $0 \leq c \leq 600$.
- **e** Briefly discuss whether alternative functional forms could also be consistent with the elicited preferences, and the implications of this non-uniqueness for decision analysis.

5 In an MCDA of environmental pathogen control strategies, one attribute of interest is the *annual population health burden* expressed in DALYs per person-year attributable to exposure.
Stakeholder elicitation produced the preference information shown in the table.

Elicited utility values for annual DALY burden.

Annual μDALYs (per person-year)	Utility
0.0	1.00
0.5	0.90
1.0	0.75
2.0	0.50
4.0	0.20
6.0	0.00

The utility scale is normalized so that a value of 1 represents the most preferred outcome and 0 the least preferred. The stakeholder further indicated that each additional DALY is more objectionable than the previous one, reflecting increasing marginal disutility with increasing health burden.

- **a** Explain why a linear utility function is unlikely to be appropriate for DALYs in this context.
- **b** Propose a functional form for a utility function $u(d)$ that is consistent with the information provided, where d denotes annual DALYs.
- **c** Estimate the parameters of the proposed utility function using these data.
- **d** Plot the resulting utility function over the range $0 \leq d \leq 6$.
- **e** Briefly discuss whether the resulting utility function reflects societal preferences, individual risk preferences, or a hybrid of the two.

6 Using the results of the previous two problems, if the weighting is 0.3 for cost and 0.7 for health effects, compute the optimum cost and μDALY per person-year using both the arithmetic and the multiplicative combination approaches.

References

1 Terje Aven and Shital Thekdi. *Risk Science: An Introduction*. Routledge, 2024.
2 Gerhard Ersdal and Terje Aven. Risk informed decision-making and its ethical basis. *Reliability Engineering & System Safety*, 93(2):197–205, February 2008.

3 D Kriebel, J Tickner, P Epstein, J Lemons, R Levins, E L Loechler, M Quinn, R Rudel, T Schettler, and M Stoto. The precautionary principle in environmental science. *Environmental Health Perspectives*, 109(9):871–876, 2001.

4 Harvey V Fineberg and Paul C Stern. *Understanding Risk: Informing Decisions in a Democratic Society*. National Academies Press, 1996.

5 Ortwin Renn. A model for an analytic-deliberative process in risk management. *Environmental Science & Technology*, 33(18):3049–3055, 1999.

6 B L Oser. An assessment of the Delaney Clause after 15 years. *Food Drug and Cosmetic Law Journal*, 29(4):201–209, 1974.

7 Nathan Mantel and W Ray Bryan. "Safety" testing of carcinogenic agents. *JNCI: Journal of the National Cancer Institute*, 27(2):455–470, August 1961.

8 Roger E Kasperson. Acceptability of human risk. *Environmental Health Perspectives*, 52(Oct.):15–20, 1983.

9 Chauncey Starr. Social benefit versus technological risk: What is our society willing to pay for safety? *Science*, 165(3899):1232–1238, September 1969.

10 W F Monfort. A special water standard. *Journal AWWA*, 2(1):65–73, March 1915.

11 A H Stevenson. Studies of bathing water quality and health. *American Journal of Public Health and the Nations Health*, 43:529–538, 1953.

12 Qian Zhang, Javier Gallard, Baolei Wu, Valerie J Harwood, Michael J Sadowsky, Kerry A Hamilton, and Warish Ahmed. Synergy between quantitative microbial source tracking (qMST) and quantitative microbial risk assessment (QMRA): A review and prospectus. *Environment International*, 130:104703, September 2019.

13 John M Henderson. Enteric disease criteria for recreational waters. *Journal of the Sanitary Engineering Division*, 94(6):1253–1276, December 1968.

14 US Environmental Protection Agency. *Recreational Water Quality Criteria*. Number 20-F-12-058. US Environmental Protection Agency Washington, DC, 2012.

15 Thammanitchpol Denpetkul, Monchai Pumkaew, Oranoot Sittipunsakda, Thitima Srathongneam, Skorn Mongkolsuk, and Kwanrawee Sirikanchana. Risk-based critical concentrations of enteric pathogens for recreational water criteria and recommended minimum sample volumes for routine water monitoring. *Science of The Total Environment*, 950:175234, November 2024.

16 Stig Regli, Joan B Rose, Charles N Haas, and Charles P Gerba. Modeling the risk from Giardia and viruses in drinking water. *Journal (American Water Works Association)*, 83(11):76–84, 1991.

17 Kerry A Hamilton, Hunter Quon, Nicholas J Ashbolt, Patrick L Gurian, Eva Reynaert, Charles N Haas, Eberhard Morgenroth, and Amanda M Wilson. Making waves: Moving beyond the 1 in 10,000 benchmark in quantitative microbial risk assessment (QMRA) through evidence-informed risk approaches and systems decision-making. *Water Research*, 289:124903, 2026.

18 Marthe R Gold, David Stevenson, and Dennis G Fryback. HALYs and QALYs and DALYs, Oh My: Similarities and differences in summary measures of population health. *Annual Review of Public Health*, 23(1):115–134, May 2002.

19 Ariel Chen, Kathryn H Jacobsen, Ashish A Deshmukh, and Scott B Cantor. The evolution of the disability-adjusted life year (DALY). *Socio-Economic Planning Sciences*, 49:10–15, March 2015.

20 Alies van Lier, Scott A McDonald, Martijn Bouwknegt, EPI group, Mirjam E Kretzschmar, Arie H Havelaar, Marie-Josée J Mangen, Jacco Wallinga, and Hester E de Melker. Disease burden of 32 infectious diseases in the Netherlands, 2007-2011. *PLOS ONE*, 11(4):e0153106, April 2016.

21 World Health Organization, editor. *Guidelines for Drinking-Water Quality*. World Health Organization, Geneva, fourth edition incorporating the first and second addenda edition, 2022.

22 Michael Jahne, Sharon Nappier, Jay Garland, Mary Schoen, and Jeffrey A Soller. Risk-based framework for developing microbial treatment targets for water reuse. *Technical Report* EPA/600/R-25/009, U.S. Environmental Protection Agency, Office of Research and Development, January 2025.

23 Kerry A Hamilton, Mark T Hamilton, William Johnson, Patrick Jjemba, Zia Bukhari, Mark LeChevallier, Charles N Haas, and P L Gurian. Risk-based critical concentrations of *Legionella pneumophila* for indoor residential water uses. *Environmental Science & Technology*, 53(8):4528–4541, February 2019.

24 A H Havelaar, A E De Hollander, P F Teunis, E G Evers, H J Van Kranen, J F Versteegh, J E Van Koten, and W Slob. Balancing the risks and benefits of drinking water disinfection: Disability adjusted life-years on the scale. *Environmental Health Perspectives*, 108(4):315–321, 2000.

25 D D Reidpath. Measuring health in a vacuum: Examining the disability weight of the DALY. *Health Policy and Planning*, 18(4):351–356, 2003.

26 John Graham, P Johansson, and Junko Nakanishi. The role of efficiency in risk management. In Timothy McDaniels and Mitchell J. Small, editors, *Risk Analysis and Society*, chapter 8, pages 251–288. Cambridge University Press, 2004.

27 Eric Ashby. What price the Furbish lousewort? *Environmental Science & Technology*, 14(10): 1176–1181, October 1980.

28 William Bruce Cameron. *Informal Sociology: A Casual Introduction to Sociological Thinking*. Random House, New York, 1963.

29 Lisa A Robinson, James K Hammitt, and Lucy O'Keeffe. Valuing mortality risk reductions in global benefit-cost analysis. *Journal of Benefit-Cost Analysis*, 10(S1):15–50, 2019.

30 Thomas J Kniesner and W Kip Viscusi. The value of a statistical life. In *Oxford Research Encyclopedia of Economics and Finance*. Oxford Research Encyclopedias, 2019.

31 Peter J Neumann and David D Kim. Cost-effectiveness thresholds used by study authors, 1990-2021. *JAMA*, 329(15):1312, April 2023.

32 Thomas J Kniesner and W Kip Viscusi. A tale of the tails: The value of a statistical life at the tails of the age distribution. *Journal of Benefit-Cost Analysis*, 15(1):204–222, 2024.

33 Sarah A Collier, Li Deng, Elizabeth A Adam, Katharine M Benedict, Elizabeth M Beshearse, Anna J Blackstock, Beau B Bruce, Gordana Derado, Chris Edens, Kathleen E Fullerton, et al. Estimate of burden and direct healthcare cost of infectious waterborne disease in the United States. *Emerging Infectious Diseases*, 27(1):140–149, January 2021.

34 Stephanie DeFlorio-Barker, Coady Wing, Rachael M Jones, and Samuel Dorevitch. Estimate of incidence and cost of recreational waterborne illness on United States surface waters. *Environmental Health*, 17(3):1–10, December 2018.

35 Phaedra S Corso, Michael H Kramer, Kathleen A Blair, David G Addiss, Jeffrey P Davis, and Anne C Haddix. Cost of Illness in the 1993 waterborne *Cryptosporidium* outbreak, Milwaukee, Wisconsin. *Emerging Infectious Disease*, 9(4):426–431, 2003.

36 Aksana Chyzheuskaya, Martin Cormican, Raghavendra Srivinas, Diarmuid O'Donovan, Martina Prendergast, Cathal O'Donoghue, and Dearbháile Morris. Economic assessment of waterborne outbreak of cryptosporidiosis. *Emerging Infectious Diseases*, 23(10):1650–1656, October 2017.

37 Madeleine Baker-Goering, Kakoli Roy, Chris Edens, and Sarah Collier. Economic burden of legionnaires' disease, United States, 2014. *Emerging Infectious Diseases*, 27(1):255–257, January 2021.

38 Fynnwin Prager, Dan Wei, and Adam Rose. Total economic consequences of an influenza outbreak in the United States. *Risk Analysis*, 37(1):4–19, January 2017.

39 Kathleen Blair. *Cryptosporidium* and public health. *Journal of Environmental Health*, 58(2):34, September 1995.

40 K Lock, C Millett, R Heathcock, C A Joseph, T G Harrison, J V Lee, G Rao, S Surman-Lee, and on behalf of the Outbreak Control Team. Public health and economic costs of investigating a suspected outbreak of Legionnaires' disease. *Epidemiology and Infection*, 136(10):1306–1314, October 2008.

41 Xing-Fang Li and William A Mitch. Drinking water disinfection byproducts (DBPs) and human health effects: Multidisciplinary challenges and opportunities. *Environmental Science & Technology*, 52(4):1681–1689, February 2018.

42 Christina M Morrison, Samantha Hogard, Robert Pearce, Aarthi Mohan, Aleksey N Pisarenko, Eric R V Dickenson, Urs von Gunten, and Eric C Wert. Critical review on bromate formation during ozonation and control options for its minimization. *Environmental Science & Technology*, 57(47):18393–18409, 2023.

43 Horst Otterstetter and Gunther Craun. Disinfection in the Americas: A necessity. *Journal AWWA*, 89(9):8–10, 1997.

44 Brett W Stinson, Amity L Deters, and Elliott T Gall. Evaluating diy air cleaner variability and potential for post-construction emission of aromatic VOCs during wildfire events. *Indoor Environments*, 1(3):100023, 2024.

45 Michael F Link, Rileigh Robertson, Megan S Claflin, and Dustin Poppendieck. Quantification of byproduct formation from portable air cleaners using a proposed standard test method. *Environmental Science & Technology*, 58(18):7916–7923, May 2024.

46 Arthur E Attema, Werner B F Brouwer, and Karl Claxton. Discounting in economic evaluations. *PharmacoEconomics*, 36(7):745–758, July 2018.

47 Donald S Shepard and Mark S Thompson. First principles of cost-effectiveness analysis in health. *Public Health Reports*, 94(6):535–543, 1979.

48 Laurence Haller, Guy Hutton, and Jamie Bartram. Estimating the costs and health benefits of water and sanitation improvements at global level. *Journal of Water and Health*, 5(4):467–480, December 2007.

49 Edwin Chadwick and Great Britain. *Report on the Sanitary Conditions of the Labouring Population of Great Britain. A Supplementary Report on the Results of a Special Inquiry into the Practice of Interment in Towns. Made at the Request of Her Majesty's Principal Secretary of State for the Home Department*. Number vii, 279, [1] p. Printed by W. Clowes and sons for H. M. Stationery off., London, 1843.

50 William C Gorgas. Sanitation at Panama. *Journal of the American Medical Association*, 58(13):907–909, 1912.

51 Nathan B Jacobs. The economic aspects of flood control. *Transactions of the American Society of Civil Engineers*, 103(1):570–577, 1938.

52 Allen V Kneese. *Water Pollution: Economics Aspects and Research Needs*. Routledge, London, 1962.

53 Harold A Thomas. The animal farm: A mathematical model for the discussion of social standards for control of the environment. *Journal AWWA*, 56(9):1087–1091, September 1964.

54 Centers for Disease Control and Prevention. Drinking water advisories: An overview, March 2024.

55 Michael G Ryan, Patrick L Gurian, Charles N Haas, Joan B Rose, and Phillip J Duzinski. Acceptable microbial risk: Cost-benefit analysis of a boil-water order for *Cryptosporidium*. *Journal of the American Water Works Association*, 105(4):E189–E194, 2013.

56 Viktor Bergion, Andreas Lindhe, Ekaterina Sokolova, and Lars Rosén. Risk-based cost-benefit analysis for evaluating microbial risk mitigation in a drinking water system. *Water Research*, 132:111–123, April 2018.

57 Steven Kelman. Cost-benefit analysis: An ethical critique, February 1981. https://www.aei.org/articles/cost-benefit-analysis-an-ethical-critique/.

58 Judith Jarvis Thomson. The trolley problem. *Yale Law Journal*, 94:1395, 1984.

59 Stijn Bruers and Johan Braeckman. A review and systematization of the trolley problem. *Philosophia*, 42(2):251–269, 2014.

60 Rosemary Lowry and Martin Peterson. Cost-benefit analysis and non-utilitarian ethics. *Politics, Philosophy & Economics*, 11(3):258–279, 2012.

61 Matthew D Adler. Benefit–cost analysis and distributional weights: An overview. *Review of Environmental Economics and Policy*, 10(2):264–285, July 2016.

62 Baruch Fischhoff. The realities of risk-cost-benefit analysis. *Science*, 350(6260):aaa6516, October 2015.

63 Emily Julien, Kate Wernicke, John Norton, Dienye Tolofari, and Jade Mitchell. Exploring emerging contaminant exposure in a drinking water risk-prioritization framework. *Society for Risk Analysis Annual Conference*, December 2025.

64 Ralph L Keeney and Howard Raiffa. *Decisions with Multiple Objectives: Preferences and Value Trade-offs*. Cambridge University Press, 1993.

65 Ward Edwards. How to use multiattribute utility measurement for social decisionmaking. *IEEE Transactions on Systems, Man, and Cybernetics*, 7(5):326–340, 1977.

66 Robert T Eckenrode. Weighting multiple criteria. *Management Science*, 12(3):180–192, November 1965.

67 Igor Linkov, F Kyle Satterstrom, Jeffery Steevens, Elizabeth Ferguson, and Richard C Pleus. Multi-criteria decision analysis and environmental risk assessment for nanomaterials. *Journal of Nanoparticle Research*, 9(4):543–554, August 2007.

68 Ralph L Keeney. The art of assessing multiattribute utility functions. *Organizational Behavior and Human Performance*, 19(2):267–310, 1977.

69 S S Stevens. On the theory of scales of measurement. *Science*, 103(2684):677–680, 1946.

70 Evangelos Triantaphyllou. Multi-criteria decision making methods. In Evangelos Triantaphyllou, editor, *Multi-Criteria Decision Making Methods: A Comparative Study*, pages 5–21. Springer US, Boston, MA, 2000.

71 Nayli Adriana Azhar, Nurul Asyikin Mohamed Radzi, and Wan Siti Halimatul Munirah Wan Ahmad. Multi-criteria decision making: A systematic review. *Recent Advances in Electrical & Electronic Engineering*, 14(8):779–801, December 2021.

72 Ivy B Huang, Jeffrey Keisler, and Igor Linkov. Multi-criteria decision analysis in environmental sciences: Ten years of applications and trends. *Science of The Total Environment*, 409(19):3578–3594, September 2011.

73 Salvatore Greco, Roman Słowiński, and Jyrki Wallenius. Fifty years of multiple criteria decision analysis: From classical methods to robust ordinal regression. *European Journal of Operational Research*, 323(2):351–377, June 2025.

74 Ian N Durbach and Theodor J Stewart. Modeling uncertainty in multi-criteria decision analysis. *European Journal of Operational Research*, 223(1):1–14, November 2012.

Index

Wiley The manufacturer's authorized representative according to the EU General Product Safety Regulation is Wiley-VCH GmbH, Boschstr. 12, 69469 Weinheim, Germany, e-mail: Product_Safety@wiley.com.

Printed and bound by CPI Group (UK) Ltd, Croydon, CR0 4YY
30/07/2026
02181222-0001